Energy Efficiency and Management in Food Processing Facilities

Energy Efficiency and Management in Food Processing Facilities

Lijun Wang

CRC Press is an imprint of the
Taylor & Francis Group, an **informa** business

CRC Press
Taylor & Francis Group
6000 Broken Sound Parkway NW, Suite 300
Boca Raton, FL 33487-2742

First issued in paperback 2019

ISBN-13: 978-1-4200-6338-7 (hbk)
ISBN-13: 978-0-367-38625-2 (pbk)

Library of Congress Cataloging-in-Publication Data

Energy efficiency and management in food processing facilities / Lijun Wang.
p. cm.
Includes bibliographical references and index.
ISBN-13: 978-1-4200-6338-7
1. Food processing plants--Energy conservation--Congresses. I. Wang, Lijun. II. Title.

TJ163.5.F6 W35 2008
664.0068/2 22 2008023570

Visit the Taylor & Francis Web site at
http://www.taylorandfrancis.com

and the CRC Press Web site at
http://www.crcpress.com

Contents

PART II Energy Conservation Technologies Applied to Food Processing Facilities

PART III Energy Consumption and Saving Opportunities in Existing Food Processing Facilities

PART IV Energy Efficiency and Conservation in Emerging Food Processing Systems

Preface

The food industry is one of the energy-intensive industries. Energy efficiency, environmental protection, and food processing waste management have attracted increasing attention in the food industry. Effective energy utilization and energy source management in food processing facilities are desirable for reducing processing costs, conserving nonrenewable energy resources, and reducing environmental impact. The food processing industry, on the one hand, consumes large amounts of energy and, on the other hand, generates large amounts of processing waste. Energy conservation and energy recovery from processing wastes have become two important issues to reduce production costs, maintain economic growth, and improve sustainability in the food processing industry. The food processing industry does not have sufficient resources to obtain knowledge and skills on advanced energy conservation and conversion technologies. The goal of *Energy Efficiency and Management in Food Processing Facilities* is to provide comprehensive knowledge and skills on advanced energy conservation and conversion technologies to students, researchers, engineers, and managers in food-related areas for improving energy efficiency and energy recovery from processing waste in food processing facilities.

This book covers five key topics: (1) fundamentals of engineering principles, energy auditing, and project management for energy analysis and management; (2) energy conservation technologies applied to the food processing industry; (3) energy use and energy saving opportunities in existing food processing facilities; (4) energy efficiency and conservation in emerging food processes; and (5) conversion of food processing waste into energy.

Part I covers fundamentals of engineering principles, energy auditing, and project management. The development of energy conservation technologies starts from the comprehensive understanding of the principles of heat transfer, fluid mechanics, and thermodynamics underlying food processing systems. These principles are concisely reviewed in Chapter 1. Energy auditing is one of the first tasks to be performed in an energy conservation program. The energy audit is to examine how a facility uses energy, what the facility pays for that energy, and what changes can be made to effectively reduce energy costs. The fundamentals of energy auditing are covered in Chapter 2. An energy conservation and conversion project reduces energy consumption in a food processing facility but requires investment for implementation. In the implementation of an energy conservation and conversion project, it must be kept in mind that bottom-line decisions are based on economics as well as energy and environmental considerations. Management of energy conservation and conversion projects is addressed in Chapter 3.

Part II focuses on various energy conservation technologies that can be applied to food processing facilities. Steam, compressed air, and electricity are the three main direct energy sources used in a food processing facility. Energy conservation technologies in generation and distribution of steam, compressed air, and power in food processing facilities are covered in Chapters 4, 5, and 6, respectively. Many unit operations in food processing such as refrigeration, freezing, thermal sterilization, drying, and evaporation involve the transfer of heat between food products and the heating or cooling medium. Heating and cooling of foods are achieved in heat exchangers. Energy conservation technologies for heat exchangers are covered in Chapter 7. Food processing facilities usually generate large amounts of waste heat. Technologies for recovering waste heat and storing thermal energy in food processing facilities are addressed in Chapter 8. Various novel thermodynamic cycles such as low-grade heat powered refrigeration cycles, heat pumps, heat pipes, and heat and power cogeneration cycles have been developed to save energy in the food industry. These novel thermodynamic cycles are examined in Chapter 9.

Parts III and IV examine energy efficiency and conservation in existing and emerging food processes, respectively. Chapter 10 in Part III gives an overview of the energy consumption in existing food processing facilities. Chapters 11 through 16 then examine the energy use and saving opportunities in six main food processing sectors in terms of food products, which include grains and oilseeds milling, sugar and confectionary processing, fruits and vegetables processing, dairy processing, meat processing, and bakery processing. Specific energy consumption in each food processing sector is examined. Energy-saving opportunities are identified and energy conservation measures are discussed for each sector. Various emerging food processes such as membrane processing, food irradiation, pulsed electrical field, high-pressure processing, microwave heating, and supercritical fluid processing have been widely investigated. Chapters 17 through 21 in Part IV briefly address the working principles, applications, energy efficiency, and energy conservation of these emerging food processes.

Part V discusses energy conversion technologies for utilization of food processing waste. Current food processing facilities depend on large inputs of high-quality energy resources such as electricity and natural gas. Waste streams generated in food processing facilities are abundant renewable energy sources. Food waste can be converted to marketable liquid and gaseous fuels, and heat and power that can be used in food processing facilities. Chapter 23 gives an overview of food processing wastes and their utilization in various food processing sectors. Chapters 24 through 27 then discuss various energy conversion technologies used for energy recovery from food processing wastes.

This book can be used as a senior level elective or graduate course in food science, food and bioprocess engineering, and industrial engineering programs. It is an essential resource for energy efficiency improvement and management for plant engineers and managers in food processing facilities. It can also be used as a valuable reference book by researchers, energy engineers, and energy management professionals.

Part I

Fundamentals of Engineering Analysis and Management

1 Fundamentals of Heat Transfer, Fluid Mechanics, and Thermodynamics in Food Processing

1.1 INTRODUCTION

Energy can be defined as the potential for providing useful work or heat. There are six basic forms of energy: heat or thermal energy, work, kinetic energy, potential energy, chemical energy, and electromagnetic energy. Energy can neither be created nor destroyed. However, it can be changed from one form to another. Analyses of energy consumption and its efficiency in food processing facilities involve the application of scientific and engineering principles such as physics, chemistry, heat transfer, fluid mechanics, and thermodynamics.

In a food processing facility, high energy demand, underutilization of facility capacity, and underutilization of individual unit operations may cause unnecessary energy consumption. Heat transfer models can be used to quantify transient energy consumption during different food processes such as pasteurization/sterilization, chilling and freezing, and dehydrating/drying (Davey and Pham, 1997; Wang and Singh, 2004; Simpson et al., 2006). Transient energy consumption profiles should be an important factor for optimizing the unit operations and minimizing energy consumption (Simpson et al., 2006). Heat transfer calculations are also essential in selecting an insulation material and determining the thickness of the insulation layer. Pumps and fans are widely used to deliver liquid foods, such as milk and juices, and processing media, such as steam, water, and air, in food processing facilities. It is critical to understand how to choose the correct pump or fan, calculate the frictional energy loss, and design an efficient pumping system in food processing facilities. Power requirement and energy loss of a pumping system can be calculated by engineering principles of fluid mechanics (Singh and Heldman, 2001). Energy loss can be interpreted by the loss of energy content using the first law of thermodynamics and by the loss of energy quality using the second law of thermodynamics. The quality of energy is described by a parameter of exergy and is defined as its maximum potential to perform work (Dincer, 2000). This chapter gives an overview of heat transfer, fluid mechanics, and thermodynamics. Applications of these scientific and engineering principles for energy analyses are demonstrated through examples.

1.2 HEAT TRANSFER IN FOOD PROCESSING

1.2.1 Modes of Heat Transfer

Heat is a form of energy, which causes temperature change and phase change of a substance when heat is added or removed from the substance. Many unit operations of food processes involve the transfer of heat into or out of foods to improve eating quality and safety and to extend shelf life. Sterilization and pasteurization are heating processes to inactivate or destroy enzyme and microbiological activities in foods (Wang and Sun, 2006). Cooking (including baking, roasting, frying) is a heating process to alter the eating quality of foods and to destroy microorganisms and enzymes for food safety. Dehydration and drying are heating processes to remove the majority of water in foods by evaporation (or by sublimation for freeze-drying). The resulting reduction in water activity extends the shelf life of foods. Cooling and freezing are processes to remove heat from foods for the inactivation of microorganisms in foods (Gould, 1996; Singh and Heldman, 2001).

Heat transfer is energy movement due to temperature difference. Heat will move from a high-temperature region to a low-temperature region. Heat may be transferred by one or more of the three mechanisms of conduction, convection, and radiation. Most industrial heat transfer operations involve a combination of these but it is often the case that one mechanism is dominant (Singh and Heldman, 2001; Wang and Sun, 2006).

1.2.2 Conductive Heat Transfer

Conduction is the transfer of heat through solids or stationary fluids due to lattice vibration or particle collision. On a microscopic scale, the molecules at the hot region increase in vibrational energy as the temperature increases. The heated molecules then interact with their slower-moving neighboring molecules at a lower temperature and share some of their energy with them. The energy thus passes from the hot region to the cold region while each individual molecule remains in its original position (Wang and Sun, 2006).

The heat flux due to conduction in the x direction through a uniform homogeneous slab of a material, as shown in Figure 1.1a, is given by Fourier's first law of conduction (Wang and Sun, 2006):

$$q = -kA\frac{\mathrm{d}T}{\mathrm{d}x} \tag{1.1}$$

where

- q is the heat flux (W)
- k is the thermal conductivity of the material (W/m°C)
- A is the total cross-sectional area for conduction (m^2)
- $\mathrm{d}T/\mathrm{d}x$ is the temperature gradient in the x direction (°C/m)

Fourier's law of heat conduction may be solved for a rectangular, cylindrical, or spherical coordinate system, depending on the geometrical shape of the object studied. Thermal conductivity of a material is an important property used in calculating the

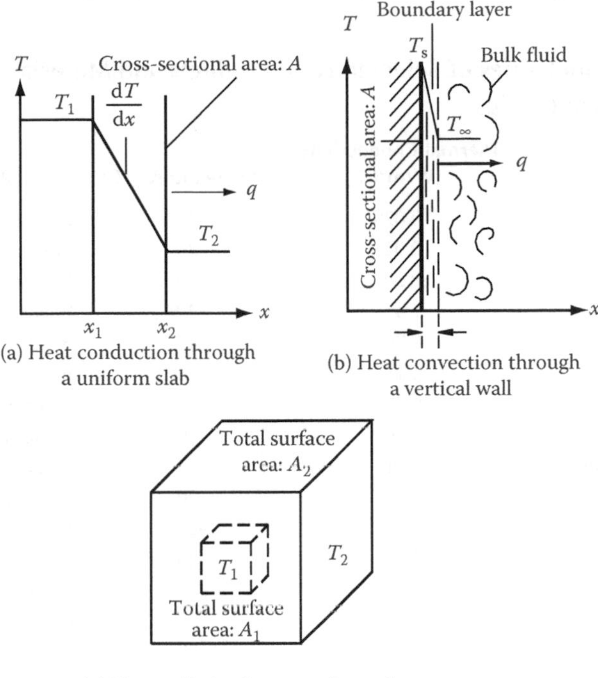

FIGURE 1.1 Schematic of three basic heat transfer modes. (Reprinted from Wang, L.J. and Sun, D.W., *Thermal Food Processing: New Technologies and Quality Issues*, CRC Press, Boca Raton, 2006. With permission.)

rate of heat transfer. Thermal conductivity is a measure of the ease with which heat travels through a material of a specified thickness. Measurement techniques for thermal conductivity can be grouped into steady state, transient, and quasi-steady state (Sweat, 1995; Wang and Weller, 2006). A guarded hot plate and a heated probe are two popular experimental instruments used for the measurement of thermal conductivity. As food materials normally have low conductivities, they take longer (e.g., 12 h) to reach the steady state, resulting in moisture migration and property changes due to long exposure at high temperature. Therefore, some well-established, standard techniques for measuring thermal conductivity, such as the guarded hot plate, work well for nonbiological materials but are not well suited for foods because of the long time required for temperature equilibration, moisture migration in the sample, and the need for a large sample size. Transient and quasi-steady state techniques are more popular for measuring the thermal conductivity of food materials because they require short measurement times and relatively small samples. The line heat source thermal conductivity probe, which is based on transient techniques, is recommended for most food applications (Sweat, 1995; Wang and Weller, 2006). The thermal conductivities of selected food products and materials used in the food industry are given in Table 1.1.

TABLE 1.1
Thermal Conductivities of Selected Food Products and Materials Used in Food Industries

Substance	Thermal Conductivity (W/m°C)	Temperature (°C)	References
Air	0.025	20	Fryer et al., 1997
Steam	0.023	20	Fryer et al., 1997
Water	0.60	20	Fryer et al., 1997
Ice	2.4	–20	Wang and Weller, 2006
Protein	0.20	20	Fryer et al., 1997
Fat	0.18	20	Fryer et al., 1997
Carbohydrate	0.58	20	Fryer et al., 1997
Salt	0.36	20	Wang and Weller, 2006
Apple (75% of water)	0.51	20	Fryer et al., 1997
Chicken meat	0.49	20	Fryer et al., 1997
Frozen salmon	1.43	–20	Wang and Weller, 2006
Stainless steel	42	20	Fryer et al., 1997
Aluminum	200	20	Fryer et al., 1997
Copper	386	20	Fryer et al., 1997
Glass	0.8	20	Fryer et al., 1997
Brick	0.7	20	Fryer et al., 1997
Wood	0.3	20	Fryer et al., 1997
Dry concrete	0.13	20	Fryer et al., 1997
Glass fiber blanket	0.035	24	Mont and Harrison, 2006
Urethane foam blocks	0.023	24	Mont and Harrison, 2006
Polyisocyanurate foam	0.020	24	Mont and Harrison, 2006
Phenolic foam	0.037	24	Mont and Harrison, 2006

As seen from Table 1.1, there is a wide variability in the magnitude of thermal conductivities for different materials:

- Fresh fruits and vegetables (e.g., apple with 75% water): 0.5 W/m°C
- Frozen foods: 1.5 W/m°C
- Metals (e.g., stainless steel, aluminum, and copper): 50–400 W/m°C
- Insulation materials (polyisocyanurate foam and concrete): 0.02–0.15 W/m°C

Some materials, such as metals and glass fiber blankets, conduct heat more readily than others. Therefore, metals are used in cookware while glass fiber blankets are used as insulators. For nonfrozen food materials, thermal conductivity varies between 0.02 W/m°C for air and 0.6 W/m°C for water. The thermal conductivity of ice, which is usually the main component of a frozen food, is about 2–2.5 W/m°C depending on the temperature. The thermal conductivity of frozen foods can be 1.5 W/m°C or higher (Wang and Weller, 2006).

The thermal conductivity of a food material depends on its chemical composition, the physical arrangement or structure of each chemical component, and the temperature of the material. The structure of a food material has a significant effect on its thermal conductivity. Foods that contain fibers exhibit different thermal conductivities parallel to fibers compared to conductivities perpendicular to fibers. Porosity has a major influence on the thermal conductivity of foods. Since food systems vary widely in composition and structure, it is difficult to find an accurate model for predicting the thermal conductivity of a broad range of foods. Existing models can generally be divided into four groups: (1) models assuming structural arrangement, (2) models using a structural parameter, (3) empirical correlations for a specific food, and (4) generalized correlations (Wang and Weller, 2006).

Most models are based on volume rather than mass fractions. Volume fractions of each component are calculated by

$$\varepsilon_j = \frac{\rho X_j}{\rho_j} \tag{1.2}$$

where

ε_j, X_j, and ρ_j are the volume fraction, mass fraction, and density of the jth component in a food product, respectively
ρ is the density of a food product

Since the porosity of a food material can strongly influence its density, the density of a food product can be determined by (Mannaperuma and Singh, 1990):

$$\frac{1}{\rho} = \frac{1}{1-\phi}\sum_{j=1}^{n}\frac{X_j}{\rho_j} \tag{1.3}$$

where

ϕ is the porosity
j denotes the jth component
n is the total number of components in the food

The presence of ice in a frozen material has been found to greatly influence the thermal conductivity of the material. Since the thermal conductivity of ice (2–2.5 W/m°C) is about four times that of liquid water (0.6 W/m°C), the accuracy of a thermal conductivity model largely depends on the accuracy of ice content prediction. Murakami and Okos (1989) considered various models for different nonporous foods. Above the freezing point, the simple parallel model was recommended as the best, which can be expressed as

$$k = \sum_{j=1}^{n} k_j \varepsilon_j \tag{1.4}$$

where

ε_j and k_j are the volume fraction and thermal conductivity of the jth component in the foods, respectively
n is the total number of components in the food

Below the freezing point, a combined parallel–series model was suggested (Murakami and Okos, 1989). The nonliquid water components are arranged in parallel. The parallel model for calculating the thermal conductivity of the nonliquid water part in food products is given by

$$k_s = \sum_{j=1}^{n_s} k_j \varepsilon_j \tag{1.5}$$

where

ε_j and k_j are the volume fraction and thermal conductivity of the jth nonliquid water component in the foods, respectively

k_s is the thermal conductivity of the nonliquid water part

n_s is the total number of nonliquid water components in the foods

Then the nonliquid water and liquid water components are perpendicular to each other in series. The parallel–series model is given by

$$\frac{1}{k} = \frac{1-\varepsilon_w}{k_s} + \frac{\varepsilon_w}{k_w} \tag{1.6}$$

where

ε_w and k_s are the volume fractions of liquid water and nonwater components in the foods, respectively

k_w and k_s are the thermal conductivities of water and nonwater components, respectively

Equations 1.2 through 1.6 require the values of the density and thermal conductivity of food components. Equations for predictions of temperature-dependent density and thermal conductivity of major food components are given in Table 1.2 (Choi and Okos, 1986).

Example 1.1

A slab of cooked meat was placed into a refrigerator at a chamber temperature of 1°C. Both sides of the slab are 10 cm and its thickness is 1 cm. The thermal conductivity of the cooked meat is 0.45 W/m°C. The central and surface temperatures of the slab were 70°C and 55°C at the beginning of cooling, 65°C and 5°C after 30 min cooling, and 5°C and 2°C at the end of cooling. Calculate the heat release rates at the three selected times. Assume that the heat transfers only from the center to the surface of the slab.

Solution 1.1

The total area of the two surfaces of the slab for heat transfer is

$$A = 2 \times 0.1\,\text{cm} \times 0.1\,\text{cm} = 0.02\,\text{cm}^2$$

Using Equation 1.1, the heat release rates at the beginning of cooling, after 30 min cooling, and at the end of cooling are

TABLE 1.2
Equations for Estimating Thermal Properties of Major Food Components as a Function of Temperature over the Range of −40°C to 150°C

Major Component	Thermal Conductivity (W/m°C)
Carbohydrate	$k = 0.20141 + 1.3874 \times 10^{-3}\,T - 4.3312 \times 10^{-6}\,T^2$
Fiber	$k = 0.18331 + 1.2497 \times 10^{-3}\,T - 3.1683 \times 10^{-6}\,T^2$
Protein	$k = 0.17881 + 1.1958 \times 10^{-3}\,T - 2.7178 \times 10^{-6}\,T^2$
Fat	$k = 0.18071 + 2.7604 \times 10^{-3}\,T - 1.7749 \times 10^{-6}\,T^2$
Ash	$k = 0.32962 + 1.4011 \times 10^{-3}\,T - 2.9069 \times 10^{-6}\,T^2$
Liquid water	$k = 0.57109 + 1.7625 \times 10^{-3}\,T - 6.7036 \times 10^{-6}\,T^2$
Ice	$k = 2.2196 - 6.2489 \times 10^{-3}\,T + 1.0154 \times 10^{-4}\,T^2$
	Density (kg/m³)
Carbohydrate	$\rho = 1599.1 - 0.3105\,T$
Fiber	$\rho = 1311.5 - 0.3659\,T$
Protein	$\rho = 1330.0 - 0.518\,T$
Fat	$\rho = 925.6 - 0.417.6\,T$
Ash	$\rho = 2423.8 - 0.2806\,T$
Liquid water	$\rho = 997.2 + 3.1439 \times 10^{-3}\,T - 3.7574 \times 10^{-3}\,T^2$
Ice	$\rho = 916.9 - 0.1307\,T$
	Specific Heat (kJ/m²°C)
Carbohydrate	$c = 1.5488 + 1.9625 \times 10^{-3}\,T - 5.9399 \times 10^{-6}\,T^2$
Fiber	$c = 1.8459 + 1.8306 \times 10^{-3}\,T - 4.6509 \times 10^{-6}\,T^2$
Protein	$c = 2.0082 + 1.2089 \times 10^{-3}\,T - 1.3129 \times 10^{-6}\,T^2$
Fat	$c = 1.9842 + 1.4733 \times 10^{-3}\,T - 4.8008 \times 10^{-6}\,T^2$
Ash	$c = 1.0926 + 1.8896 \times 10^{-3}\,T - 3.6817 \times 10^{-6}\,T^2$
Water (−40°C to 0°C)	$c = 4.0817 - 5.3062 \times 10^{-3}\,T + 9.9516 \times 10^{-6}\,T^2$
Water (0°C to 150°C)	$c = 4.0817 - 5.3062 \times 10^{-3}\,T + 9.9516 \times 10^{-6}\,T^2$
Ice	$c = 2.0623 + 6.0769 \times 10^{-3}\,T$

Source: Adapted from Choi, Y. and Okos, M.R., in *Food Engineering and Process Applications*, Vol. 1, Elsevier, New York, 1986, 93–101. With permission.

$$q_1 = -kA\frac{\Delta T_1}{\Delta x} = -0.45\,[\mathrm{W/m^\circ C}] \times 0.02\,[\mathrm{m^2}] \times \frac{70^\circ\mathrm{C} - 55^\circ\mathrm{C}}{0.01\ \mathrm{m}/2} = -27\ \mathrm{W}$$

$$q_2 = -kA\frac{\Delta T_2}{\Delta x} = -0.45\,[\mathrm{W/m^\circ C}] \times 0.02\,[\mathrm{m^2}] \times \frac{65^\circ\mathrm{C} - 5^\circ\mathrm{C}}{0.01\ \mathrm{m}/2} = -108\ \mathrm{W}$$

$$q_3 = -kA\frac{\Delta T_3}{\Delta x} = -0.45\,[\mathrm{W/m^\circ C}] \times 0.02\,[\mathrm{m^2}] \times \frac{5^\circ\mathrm{C} - 2^\circ\mathrm{C}}{0.01\ \mathrm{m}/2} = -5.4\ \mathrm{W}$$

From the above example, we can see that the cooling load or energy consumption varies significantly during cooling. The transient cooling load or energy consumption profiles should be considered as an important factor for optimizing the operation of the refrigerator and minimizing its energy consumption.

To introduce the concept of thermal resistance, Equation 1.1 is rearranged as

$$q = -kA\frac{\Delta T}{\Delta x} = -\frac{\Delta T}{\Delta x / kA} \tag{1.7}$$

The thermal resistance is thus expressed as

$$R = \frac{\Delta x}{kA} \tag{1.8}$$

Substituting Equation 1.8 into Equation 1.7, we obtain

$$q = -\frac{\Delta T}{R} \tag{1.9}$$

The advantage of using the thermal resistance concept is significant when we study conduction in a multilayer wall made of several materials of different thermal conductivities and thicknesses. The conductive heat transfer rate through a composite wall can be expressed as

$$q = -\frac{\Delta T}{\sum_{i=1}^{n} \frac{\Delta x_i}{k_i A}} = -\frac{\Delta T}{\sum_{i=1}^{n} R_i} \tag{1.10}$$

where

i is the total number of layers

Δx_i, k_i, and R_i are the thickness, thermal conductivity, and thermal resistance, respectively, of the ith layer

ΔT is the temperature difference in the pathway of heat conduction

A is the total cross-sectional area perpendicular to the pathway of heat conduction

Example 1.2

A cold storage room has a 15 m^2 wall made of 5 cm thick concrete. The temperature on the cold side is 2°C, and that on the ambient side is 30°C. The thermal conductivity of concrete is 0.2 W/m°C. What is the rate of heat transferred through the wall? If 2 cm of polystyrene insulation with a thermal conductivity of 0.1 W/m°C is applied to one side of the wall, what is the rate of heat transfer through the composite wall? Assume that the temperatures have been maintained for a long period.

Solution 1.2

Using Equation 1.8, the thermal resistance of the concrete wall is

$$R_c = \frac{\Delta x_c}{k_c A} = \frac{0.05\,[\text{m}]}{0.2\ [\text{W/m°C}] \times 15\,[\text{m}^2]} = 0.0167\text{°C/W}$$

Using Equation 1.9, we can obtain the heat transfer rate as

$$q = -\frac{\Delta T}{R} = -\frac{2°\text{C} - 30°\text{C}}{0.0167°\text{C/W}} = 1677 \text{ W}$$

If an insulation layer of polystyrene is applied, the thermal resistance of the polystyrene is

$$R_\text{p} = \frac{\Delta x_\text{p}}{k_\text{p} A} = \frac{0.02 \text{ [m]}}{0.1 \text{ [W/m°C]} \times 15 \text{ [m}^2\text{]}} = 0.0133°\text{C/W}$$

The total resistance of the composite wall is

$$\begin{aligned} R_\text{t} &= R_\text{c} + R_\text{p} \\ &= 0.0167°\text{C/W} + 0.0133°\text{C/W} \\ &= 0.03°\text{C/W} \end{aligned}$$

Using Equation 1.9, we can obtain the heat transfer rate as

$$q = -\frac{\Delta T}{R_\text{t}} = -\frac{2°\text{C} - 30°\text{C}}{0.03°\text{C/W}} = 933 \text{ W}$$

Therefore, an insulation layer can significantly reduce the heat transfer rate and thus energy loss.

For calculating conductive heat transfer rate though a flat plate (or slab), the values of the thickness, Δx, and area, A, in Fourier's law can be determined easily. It is, however, difficult to determine the cross-sectional area and conductive heat transfer rate along the radial direction in a hollow cylinder. For calculating the conductive heat transfer rate along the radial direction in a hollow cylinder of inner radius, r_i outer radius, r_o, and length L, Fourier's law can be expressed as (Singh and Heldman, 2001)

$$q = -kA\frac{\text{d}T}{\text{d}r} = -k(2\pi rL)\frac{\text{d}T}{\text{d}r} \tag{1.11}$$

After integrating with boundary conditions ($T = T_\text{i}$ at $r = r_\text{i}$, and $T = T_\text{o}$ at $r = r_\text{o}$ on the inside and outside boundary of the hollow cylinder), Equation 1.11 can be expressed as

$$q = -\frac{T_\text{o} - T_\text{i}}{\dfrac{\ln(r_\text{o}/r_\text{i})}{2\pi Lk}} = -\frac{T_\text{o} - T_\text{i}}{R} \tag{1.12}$$

where the thermal resistance in the radial direction for a hollow cylinder is

$$R = \frac{\ln(r_\text{o}/r_\text{i})}{2\pi Lk} \tag{1.13}$$

Example 1.3

A 0.5 cm thick steel pipe with 5 cm inside diameter is used to convey steam from a boiler to a process equipment for a distance of 50 m. The inside pipe surface temperature is 115°C, and the outside pipe surface temperature is 114.5°C. The thermal conductivity of steel is 42 W/m°C. Calculate the total heat loss to the surroundings under a steady-state condition.

Solution 1.3

Using Equation 1.13, the thermal resistance of the steel pipe wall is

$$R = \frac{\ln(r_o/r_i)}{2\pi Lk} = \frac{\ln(0.03/0.025)}{2\pi \times 50\,[\mathrm{m}] \times 42\,[\mathrm{W/m°C}]} = 1.38 \times 10^{-5}\,°\mathrm{C/W}$$

From Equation 1.12, we can obtain the heat transfer rate as

$$q = -\frac{\Delta T}{R} = -\frac{115°\mathrm{C} - 114.5°\mathrm{C}}{1.38 \times 10^{-5}\,°\mathrm{C/W}} = -36\,\mathrm{kW}$$

1.2.3 Convective Heat Transfer

Convection uses the movement of fluids to transfer heat. For example, when cold air flows over the warm surface of food in a refrigerator, the cold air close to the surface is heated as it comes in contact with the surface. As the air flows, the air heated at the surface mixes and exchanges heat with the free stream air. The movement, which causes heat transfer, may occur in a natural or forced form. Natural convection creates the fluid movement by the difference between fluid densities due to the temperature difference. Forced convection uses external means such as agitators, pumps, and fans to produce fluid movement. Whenever a fluid moves past a solid surface, it may be observed that the fluid velocity varies from zero adjacent to the solid surface to a free fluid velocity at some distance away from the surface.

Convective heat transfer is the major mode of heat transfer between the surface of a solid material and the surrounding fluid. For analyzing heat transfer by convection, a boundary layer is normally assumed near the surface of the solid material as shown in Figure 1.1b. Heat is transferred by conduction through this layer. The layer contains almost all of the resistance to heat transfer because of the relatively low thermal conductivity of the layer and rapid heat transfer from the outer edge of the boundary layer into the bulk of the fluid. Using the boundary layer concept, the rate of convective heat transfer may be written as

$$q = kA\frac{(T_s - T_\infty)}{\delta} = A\frac{(T_s - T_\infty)}{\delta/k} \quad (1.14)$$

where

k and δ are the thermal conductivity and thickness of the boundary layer, respectively

A is the boundary surface area

T_s and T_∞ are the surface temperature and the temperature of bulk fluids, respectively

However, as the thickness of the boundary layer, δ, can neither be predicted nor measured easily, the thermal resistance of the boundary layer cannot be determined. δ/k is thus replaced with the term $1/h_c$, in which h_c is a convective heat transfer coefficient. Equation 1.14 can then be rewritten as

$$q = h_c A(T_s - T_\infty) \tag{1.15}$$

The convective heat transfer coefficient is a function of the fluid velocity, the difference in temperature between the fluid and the solid surface, the orientation of the surface, the roughness of the surface, and the properties of the fluid. The values of convective heat transfer are usually determined experimentally and they can also be calculated mathematically from empirical formulas. The typical values of convective heat transfer are given in Table 1.3 (Singh and Heldman, 2001).

In many heating/cooling applications, conductive and convective heat transfer may occur simultaneously. An example is heat loss through the wall of a cold storage room that circulates cold air at a temperature lower than the temperature of the environment surrounding the outside of the room. In this case, heat must be transferred first from the surrounding environment to the outer surface of the room by natural or free convection, by conduction through the wall material from the outer surface to the inside surface of the wall, and finally by forced convection from the inside surface of the wall to the inside cold air. The heat transfer is thus realized through three layers in series. The total thermal resistance in the pathway of heat transfer is

$$R_t = R_i + R_w + R_o \tag{1.16}$$

TABLE 1.3
Some Approximate Values of Convective Heat Transfer Coefficients

Fluid	Convective Heat Transfer Coefficient (W/m²°C)
Air natural convection	5–25
Air forced convection	10–200
Water natural convection	20–100
Water forced convection	50–10,000
Boiling water	3,000–100,000
Condensing water vapor	5,000–100,000

Source: From Singh, R.P. and Heldman, D.R., in *Introduction to Food Engineering*, Academic Press, San Diego, 2001. With permission.

where

$$R_i = \frac{1}{h_i A_i} \tag{1.17}$$

$$R_w = \frac{\Delta x}{k A_a} \tag{1.18}$$

$$R_o = \frac{1}{h_o A_o} \tag{1.19}$$

In the above equations,

h_i and h_o are the inside and outside convective heat transfer coefficients, respectively
A_i, A_o, and A_a are inside, outside, and average surface areas, respectively
Δx is the thickness of the wall
k is the thermal conductivity of wall material
R_i, R_w, and R_o are the thermal resistances to the inside convection, the wall conduction, and the outside convection, respectively

Example 1.4

A 0.5 cm thick steel pipe with 5 cm inside diameter is used to convey steam from a boiler to a process equipment for a distance of 50 m. The inside steam temperature is 115°C and the outside ambient temperature is 30°C. The inside and outside convective heat transfer coefficients are 100 W/m² °C and 10 W/m°C, respectively. The thermal conductivity of steel is 42 W/m°C. Calculate the total heat loss to the surroundings under a steady-state condition.

Solution 1.4

Using Equations 1.17 through 1.19, the thermal resistances of the inside convective layer, the steel pipe wall, and the outside convective layer are

$$R_i = \frac{1}{h_i A_i} = \frac{1}{100\,[\text{W/m}^2\,°\text{C}] \times \pi \times 5\,[\text{cm}] \times \left(\dfrac{1\,[\text{m}]}{100\,[\text{cm}]}\right) \times 50\,[\text{m}]} = 1.27 \times 10^{-3}\,°\text{C/W}$$

$$R_w = \frac{\ln(r_o/r_i)}{2\pi L k} = \frac{\ln(0.03/0.025)}{2\pi \times 50\,[\text{m}] \times 42\,[\text{W/m°C}]} = 1.38 \times 10^{-5}\,°\text{C/W}$$

$$R_o = \frac{1}{h_o A_o} = \frac{1}{10\,[\text{W/m}^2\,°\text{C}] \times \pi \times 6\,[\text{cm}] \times \left(\dfrac{1\,[\text{m}]}{100\,[\text{cm}]}\right) \times 50\,[\text{m}]} = 1.06 \times 10^{-2}\,°\text{C/W}$$

Therefore, the overall thermal resistance is

$$R_t = R_i + R_w + R_o = 1.27\times10^{-3} + 1.38\times10^{-5} + 1.06\times10^{-2} = 1.19\times10^{-2}\ ^\circ\text{C/W}$$

From Equation 1.10, we can obtain the overall heat transfer rate at

$$q = -\frac{\Delta T}{R} = -\frac{115^\circ\text{C} - 30^\circ\text{C}}{1.19\times10^{-2}\,^\circ\text{C/W}} = -7143\ \text{W}$$

From the above equation, we know that the outside convection is the dominant contribution to the overall thermal resistance. The wall conduction makes a negligible contribution to the overall thermal resistance. In order to decrease the overall heat transfer rate, insulation should be considered to increase the thermal resistance of the wall conduction. However, in order to increase the overall heat transfer rate, the outside convection should be enhanced to reduce its thermal resistance.

For convenience in calculation, the concept of overall heat transfer coefficient, U_i or U_o (based on inside area or outside area), is frequently used. The overall heat transfer coefficient based on inside area is defined as (Singh and Heldman, 2001)

$$\frac{1}{U_i A_i} = R_i + R_w + R_o = \frac{1}{h_i A_i} + \frac{\ln\left(r_o/r_i\right)}{2\pi L k} + \frac{1}{h_o A_o} \tag{1.20}$$

or

$$\frac{1}{U_i} = \frac{1}{h_i} + \frac{r_i \ln\left(r_o/r_i\right)}{k} + \frac{r_i}{h_o r_o} \tag{1.21}$$

The overall heat transfer rate is thus

$$q = U_i A_i \Delta T \tag{1.22}$$

The selection of the area over which to calculate the overall heat transfer is arbitrary. The overall heat transfer coefficient based on outside area is defined as

$$\frac{1}{U_o A_o} = R_i + R_w + R_o = \frac{1}{h_i A_i} + \frac{\ln\left(r_o/r_i\right)}{2\pi L k} + \frac{1}{h_o A_o} \tag{1.23}$$

or

$$\frac{1}{U_o} = \frac{r_o}{h_i r_i} + \frac{r_o \ln\left(r_o/r_i\right)}{k} + \frac{1}{h_o} \tag{1.24}$$

The overall heat transfer rate is thus

$$q = U_o A_o \Delta T \tag{1.25}$$

Example 1.5

What is the overall heat transfer coefficient based on inside area and outside area for Example 1.4?

Solution 1.5

From Example 1.4, we have

$$R_i = 1.27 \times 10^{-3}\ °\text{C/W}$$

$$R_w = 1.38 \times 10^{-5}\ °\text{C/W}$$

$$R_o = 1.06 \times 10^{-2}\ °\text{C/W}$$

Therefore, the overall heat transfer coefficient based on inside area is

$$\frac{1}{U_i A_i} = R_i + R_w + R_o = 1.27 \times 10^{-3} + 1.38 \times 10^{-5} + 1.06 \times 10^{-2} = 0.0119°\text{C/W}$$

$$U_i = 10.70\ \text{W/m}^2\,°\text{C}$$

Similarly, the overall heat transfer coefficient based on outside area is

$$U_o = 8.92\ \text{W/m}^2\,°\text{C}$$

1.2.4 Radiative Heat Transfer

Radiation does not require a medium for transferring heat but uses the electromagnetic waves emitted by an object for exchanging heat. The energy emitted from a surface depends on the temperature of the surface, which can be described using the Stefan–Boltzmann law (Wang and Sun, 2006):

$$q = \sigma A \varepsilon T_K^4 \tag{1.26}$$

where
σ is the Stefan–Boltzmann constant at $\sigma = 5.669 \times 10^{-8}$ W/m^{2}K^4
ε is the emissivity (at $\varepsilon = 1$ for a black body)
A is the area
T_K is the temperature in Kelvin

When two bodies exchange energy by radiation, the energy emitted by one is not completely absorbed by the other as it can only absorb the portion that it intercepts.

Therefore, a shape factor, F, is defined. The radiative energy exchange between the surface, 1, of a body and the surroundings, 2, of the body can be determined by

$$q = \sigma F_{12} A_1 \varepsilon (T_{K1}^4 - T_{K2}^4) \tag{1.27}$$

If the surface, 1, is enclosed by the surroundings, 2, as shown in Figure 1.1c, then $F_{12} = 1$. Similar to the convective heat transfer coefficient, a radiative heat transfer coefficient, h_r, may be expressed as

$$q = h_r A (T_{K1} - T_{K2}) \tag{1.28}$$

where

$$h_r = \sigma \varepsilon (T_{K1}^2 + T_{K2}^2)(T_{K1} + T_{K2}) \tag{1.29}$$

Example 1.6

A food item is heated in an oven. The oven temperature is set at 180°C. The average surface temperature of the food is 50°C. The food item has a spherical shape with a diameter of 20 cm. What is the radiative heat transfer rate? What is the radiative heat transfer coefficient?

Solution 1.6

The surface area of the food item is

$$A = 4\pi r^2 = 4 \times 3.14159 \times 0.1^2 = 0.1257 \text{ m}^2$$

Since the food item is placed in an oven, the shape factor is $F_{12} = 1$. The emissivity of the food surface is assumed to be 1. Using Equation 1.27, the radiative heat transfer rate is

$$\begin{aligned} q &= \sigma F_{12} A_1 \varepsilon (T_{K1}^4 - T_{K2}^4) \\ &= 5.669 \times 10^{-8} \times 1 \times 0.1257 \times 1 \times \left[(273.15 + 180)^4 - (273.15 + 50)^4\right] \\ &= 222.8 \text{ W} \end{aligned}$$

Using Equation 1.29, the radiative heat transfer coefficient is

$$h_r = \sigma \varepsilon (T_{K1}^2 + T_{K2}^2)(T_{K1} + T_{K2}) = 13.63 \text{ W/m}^2 \text{ °C}$$

1.2.5 Phase Change Heat Transfer

Most foods such as raw meats and vegetables have high moisture contents. Water itself is widely used as a processing medium in food processes. Water normally exists in one of three states: solid, liquid, and gas. The transition between two states is called a phase transformation or phase change. During phase transformation, the

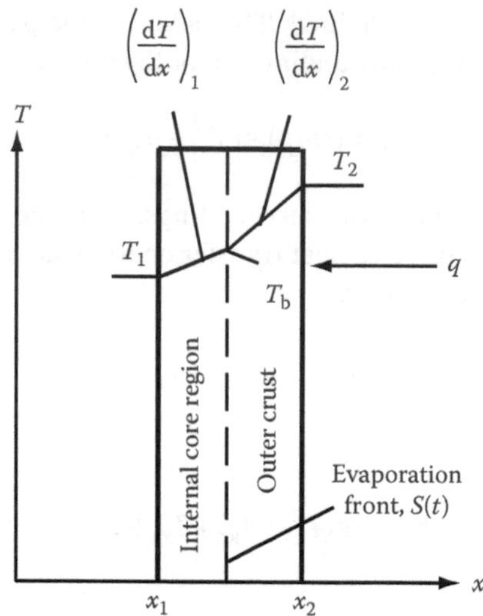

FIGURE 1.2 Schematic of heat transfer with phase changes (frying). (Reprinted from Wang, L.J. and Sun, D.W., *Thermal Food Processing: New Technologies and Quality Issues*, CRC Press, Boca Raton, 2006. With permission.)

temperature of pure water remains constant with added energy because all energy is used to transform water from one state to another. As water is widely present in foods and is used as a processing medium, it is necessary to discuss heat transfer with the phase changes of water in the food industry.

During thermal food processing, the water in the food may experience phase changes. Frying and grilling of foods involve a phase change from liquid water to vapor. There is an evaporation front, which divides the food body into two parts of the outer crust and internal core regions, as shown in Figure 1.2. The evaporation front moves toward the center as frying and grilling processes proceed. If a frozen food is used during frying and grilling, there will be two moving boundaries: the thawing front and the evaporation front. The heat transfer mechanisms across the moving boundary must account for the latent heat of phase changes of water. The moving front of phase change in the food can be tracked by the energy balance on the front, which is given by (Wang and Singh, 2004)

$$-k_1\left(\frac{\partial T}{\partial x}\right)_1 + k_2\left(\frac{\partial T}{\partial x}\right)_2 = \lambda \rho X_w \frac{dS(t)}{dt} \tag{1.30}$$

$$t > 0, x = S(t) \tag{1.31}$$

where

subscripts 1 and 2 denote phase 1 and phase 2, respectively
$\partial T/\partial x$ is the temperature gradient
k is thermal conductivity

λ is the latent heat of phase change
ρ is the density of the product
X_w is the moisture content
$S(t)$ is the position of the moving boundary at time t

Water is also widely used as a processing medium. Boiling and condensation involve a phase change between liquid water and vapor. Boiling heat transfer is particularly important in processing operations such as evaporation in which the boiling of liquids takes place either at submerged surfaces or on the inside surface of vertical tubes as in a climbing film evaporator. The heat flux changes dramatically as a function of the temperature difference between the surface and the boiling liquid, rising to a peak value and falling away sharply. This is caused by the strong dependence between the heat transfer coefficient and the temperature difference, which is shown in Figure 1.3. In order to avoid overheating and damaging the walls of the heater, equipment should ideally be operated in the nucleate boiling zone, just below the critical temperature difference as shown in Figure 1.3. Vapor condensation is also used in food thermal processes. Consider a sterilization process used for manufacturing canned foods. If steam is used as a heating medium, the condensation of steam on the metal surface of a can results in a significantly higher heat transfer than if hot water were used to heat the cans. Vapor condenses on a cold surface in one of two distinct ways: film condensation and drop condensation. The presence of noncondensable gases in steam affects the rate of condensation, and the film heat transfer coefficient may be reduced considerably (Wang and Sun, 2006).

The heat flux due to phase change of boiling and condensation can be expressed as

$$q = h_p A(T_s - T_\infty) \tag{1.32}$$

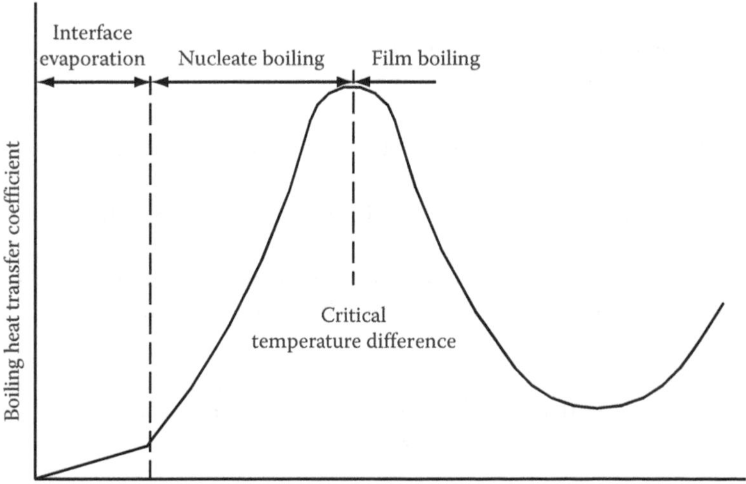

FIGURE 1.3 Relationship between boiling heat transfer coefficient and temperature difference. (Reprinted from Wang, L.J. and Sun, D.W., *Thermal Food Processing: New Technologies and Quality Issues*, CRC Press, Boca Raton, 2006. With permission.)

where
h_p is the heat transfer coefficient of phase change
T_s and T_∞ are the temperatures of solid surface and bulk fluid, respectively
A is the surface area where the phase change occurs

The heat transfer coefficients of a phase change when a liquid is vaporized or when a vapor is condensed are considerably greater than that for heat transfer without a phase change, as shown in Table 1.3. However, it is difficult to measure heat transfer coefficients of phase changes.

The heat transfer coefficient in nucleate boiling may be calculated by a correlation. Kutateladze's correlation is a commonly used one, which is given by (Wang and Sun, 2006):

$$\left(\frac{h_b}{k}\right)\psi^{0.5} = 0.0007\left[\frac{q_{max}}{\alpha\lambda\rho_v\sigma}\frac{P}{}\Psi\right]^{0.7}\text{Pr}^{-0.35} \tag{1.33}$$

where

$$\Psi = \frac{\sigma}{g(\rho_l - \rho_v)} \tag{1.34}$$

$$q_{max} = 0.16\lambda\rho_v\left(\frac{\sigma g(\rho_l - \rho_v)}{\rho_v^2}\right)^{0.25} \tag{1.35}$$

where
k is the thermal conductivity (W/m°C)
α is the thermal diffusivity (m²/s)
P is the pressure (Pa)
λ is the latent heat of phase change (J/kg)
ρ is the density (kg/m³)
Pr is the Prandtl number
σ is the Stefan–Boltzmann constant at $\sigma = 5.669 \times 10^{-8}$ W/m²K⁴
g is the acceleration due to gravity (m/s²)
Subscripts l and v denote the liquid and gaseous phases, respectively

The film heat transfer coefficient for condensation can be predicted from the Nusselt theory, which gives the mean film coefficient by (Wang and Sun, 2006)

$$h_{cd} = 0.943\left(\frac{\rho^2 k^3 g\lambda}{\mu L\Delta T}\right)^{0.25}, \quad \text{for a vertical surface} \tag{1.36}$$

$$h_{cd} = 0.725\left(\frac{\rho^2 k^3 g\lambda}{\mu d\Delta T}\right)^{0.25}, \quad \text{for a horizontal tube} \tag{1.37}$$

where

ρ is the density (kg/m^3)
k is the thermal conductivity (W/m°C)
g is the acceleration due to gravity (m/s^2)
λ is the latent heat of the phase change (J/kg)
μ is the viscosity (Pa s)
ΔT is the temperature difference for the phase change (°C)
L is the length (m)
d is the diameter of the tube (m)

1.2.6 Heat Transfer with Electromagnetic Waves

Microwave energy, which is an electromagnetic energy, is widely used in the food industry. Microwaves are transmitted as waves, which can penetrate foods and interact with the polar molecules in foods such as water to be converted to heat. An electromagnetic spectrum is normally characterized by wavelength (λ) and frequency (f). Microwaves are nonionizing electromagnetic waves, and commercial microwave heating applications use frequencies of 2450MHz, sometimes 915MHz in the United States and 896MHz in Europe. The depth of penetration into a food is directly related to frequency, and lower frequency microwaves penetrate more deeply. As a microwave can penetrate into foods, it can heat foods quicker than traditional thermal processing methods and transfer heat from the outer surface to the inside of foods by conduction. However, it should be noted that conduction or convection may also occur during microwave heating if a temperature difference exists in foods. In this case, the conversion rate of microwave energy per unit volume of foods can be considered as a source term in a heat transfer model.

The conversion of microwave energy to heat depends on the properties of the microwave energy source and the dielectric properties of the foodstuffs. The power dissipation or rate of energy conversion per unit volume of foods, P (W/m^3), is given by (Wang and Sun, 2006)

$$P = 5.56 \times 10^{-15} E^2 f \varepsilon'' \tag{1.38}$$

and

$$\varepsilon'' = \varepsilon' \tan \delta \tag{1.39}$$

where

E is the microwave electrical field strength (V/m)
f is the frequency of the microwave
ε' and ε'' are the dielectric constant and dielectric loss, respectively
δ is the loss angle

However, the suitability of a food for microwave heating is crucially dependant on the penetration characteristics of microwave into the food. The distribution of microwave energy within a material is determined by the attenuation factor, α, which is calculated by

$$\alpha = \frac{2\pi}{\lambda}\left[\frac{\varepsilon'}{2}\left(\sqrt{1+\tan^2\delta}-1\right)\right]^{0.5} \tag{1.40}$$

The microwave electrical field strength is a function of penetration depth, x, which can be given by

$$E = E_0 e^{-2\alpha x} \tag{1.41}$$

The microwave electrical field strength profile obtained by Equation 1.41 can be used to calculate the local conversion rate of microwave energy using Equation 1.38. The power absorption rate at the penetration depth, x, can also be determined by the Lambert's model, which is expressed as

$$P = P_0 e^{-2\alpha x} \tag{1.42}$$

where P_0 is the incident power on the surface (W/m^2).

1.2.7 Unsteady-State Heat Transfer

For steady-state heat transfer, the temperature of a food does not change with time. However, in the majority of thermal food processes, the temperature of a food product changes continuously and unsteady-state heat transfer is more commonly found. The generalized governing equation of unsteady-sate heat transfer can be expressed as

$$\rho c\left(\underbrace{\frac{\partial T}{\partial t}}_{\text{Accumulation}} + \underbrace{u_x\frac{\partial T}{\partial x} + u_y\frac{\partial T}{\partial y} + u_z\frac{\partial T}{\partial z}}_{\text{Convection}}\right) = \underbrace{\frac{\partial}{\partial x}\left(k\frac{\partial T}{\partial x}\right) + \frac{\partial}{\partial y}\left(k\frac{\partial T}{\partial y}\right) + \frac{\partial}{\partial z}\left(k\frac{\partial T}{\partial z}\right)}_{\text{Diffusion}} + \underbrace{S}_{\text{Source}} \tag{1.43}$$

In the above equation, the specific heat, c, of a food product can be calculated from the composition and temperature of the food. The temperature-dependent specific heat of major food components is given in Table 1.2. The power dissipation rate during microwave heating determined by Equation 1.38 or 1.42 is one example of source item, S, to be included in the above equation. In order to find the solution of Equation 1.43, it is necessary to know the initial and boundary conditions. The initial conditions give what happens at the start. The initial conditions may be the same initial temperature, $T|_{t=0} = T_0$. The initial conditions may also be an initial temperature profile, $T|_{t=0} = T_0(x,y,z)$. The boundary conditions give what happens at the boundaries of the material to be investigated. The boundary conditions may be (1) a constant, $T|_\Gamma = T_s$; (2) a flux, $T|_\Gamma = q_s$; (3) a convection, $T|_\Gamma = h(T_s - T_\infty)$; or (4) a combination of flux and convection, $T|_\Gamma = q_s + h(T_s - T_\infty)$.

In the modeling of an unsteady-state thermal process, the values of temperature depend on the time and position in the material. The equation governing the unsteady-state heat transfer is thus of a partial differential type, as shown in Equation 1.43. Numerical methods have been widely used to solve the partial differential equation.

Numerical methods can generate discretized solutions to the partial differential equation (Wang and Sun, 2002, 2003; Wang and Singh, 2004; Amezquita et al., 2005).

An analytical solution of the partial differential equation is continuous. The possibility of an analytical solution is restricted to rather simple forms of the governing equations, and the boundary and initial conditions. Sometimes, depending on the geometry of the product to be studied, it is useful to consider alternative coordinate systems such as the cylindrical coordinate and spherical coordinate systems. However, the intrinsic mechanisms and physical laws of heat transfer remain the same, irrespective of the system used. For one-dimensional heat transfer, Equation 1.43 can be rewritten in a general format as (Singh and Heldman, 2001)

$$\rho c\frac{\partial T}{\partial t}=\frac{1}{r^n}k\frac{\partial}{\partial r}\left(r^n\frac{\partial T}{\partial r}\right) \tag{1.44}$$

where

r is the distance from the center (m)
$n = 0$ for an infinite slab
$n = 1$ for an infinite cylinder
$n = 2$ for a sphere

Suppose the initial condition and boundary conditions are

$$T\big|_{t=0}=T_0 \tag{1.45}$$

$$-k\frac{\partial T}{\partial r}\bigg|_{r=0}=0 \tag{1.46}$$

$$-k\frac{\partial T}{\partial r}\bigg|_{r=R}=h(T_s-T_\infty) \tag{1.47}$$

To simplify the analysis, two parameters are defined as

$$\alpha=\frac{k}{\rho c}\ \text{(thermal diffusivity)} \tag{1.48}$$

$$\text{Bio}=\frac{hD}{k}\text{(Biot number)} \tag{1.49}$$

Biot number is the ratio of the internal resistance to heat transfer in the solid to the external resistance to heat transfer in the fluid. There are

- $\text{Bio} \le 0.1$, negligible internal resistance to heat transfer
- $0.1 < \text{Bio} < 40$, finite internal and external resistance to heat transfer
- $\text{Bio} \ge 40$, negligible external resistance to heat transfer (Singh and Heldman, 2001)

If Bio ≤ 0.1, the internal resistance can be neglected, and the heat balance equation becomes

$$\rho c V \frac{\partial T}{\partial t} = hA\left(T_{\mathrm{a}} - T\right) \tag{1.50}$$

Integrating Equation 1.44, we obtain

$$\frac{T_{\mathrm{a}} - T}{T_{\mathrm{a}} - T_{\mathrm{i}}} = \mathrm{e}^{-(hA/\rho c V)t} \tag{1.51}$$

If 0.1 < Bio < 40, both internal and external resistances should be considered. The solutions to Equation 1.44 with initial and boundary conditions are

Sphere:

$$\frac{T_{\mathrm{a}} - T}{T_{\mathrm{a}} - T_{\mathrm{i}}} = \frac{2}{\pi}\left(\frac{D}{r}\right)\sum_{n-1}^{\infty}\frac{(-1)^{n+1}}{n}\,\mathrm{e}^{\left(-n^2\pi^2\alpha t/D^2\right)}\sin\left(n\pi r/D\right) \tag{1.52}$$

Root equation,

$$\mathrm{Bio} = 1 - \beta_1 \cot\beta_1 \tag{1.53}$$

Infinite cylinder:

$$\frac{T_{\mathrm{a}} - T}{T_{\mathrm{a}} - T_{\mathrm{i}}} = 2\sum_{n-1}^{\infty}\frac{\mathrm{e}^{\left(-\lambda_n^2\alpha t/D^2\right)}J_0\left(\lambda_n r/D\right)}{\lambda_n J_1\left(\lambda_n\right)} \tag{1.54}$$

Root equation,

$$\mathrm{Bio} = \frac{\beta_1 J_1\left(\beta_1\right)}{J_0\left(\beta_1\right)} \tag{1.55}$$

Infinite slab:

$$\frac{T_{\mathrm{a}} - T}{T_{\mathrm{a}} - T_{\mathrm{i}}} = 2\sum_{n-1}^{\infty}\frac{(-1)^{n+1}}{\lambda_n}\,\mathrm{e}^{\left(-\lambda_n^2\alpha D^2\right)}\cos\left(\lambda_n r/D\right) \tag{1.56}$$

Root equation,

$$\mathrm{Bio} = \beta_1 \tan\beta_1 \tag{1.57}$$

If Bio ≥ 40, the external resistance can be neglected, and the boundary condition becomes constant ambient temperature or it is assumed that the surface heat transfer coefficient is infinitely big.

1.3 FLUID MECHANICS IN FOOD PROCESSING

1.3.1 Viscosity of Fluids

A fluid is a substance that is capable of flowing. A fluid may be a gas or liquid. Gases are readily compressible while liquids are only slightly compressible. There are many liquid food products such as milk and juice. Furthermore, fluids such as water and air are widely used as processing media in food processing facilities. Fluid statics deals with fluids at rest while fluid dynamics deals with fluids in motion. A fluid begins to move when a force such as pressure, gravity, or friction acts upon it.

Viscosity is a measure of the ease with which a fluid flows. Viscosity is a property of a fluid that causes resistance to relative motion within the fluid. The dynamic viscosity of a fluid is a measure of the resistance to internal deformation of shear. The unit of dynamic viscosity in the SI system is Pascal-second (Pa s). In the literature, dynamic viscosity of liquids is often expressed in centipoise (1 cP = 1 mPa s = 0.001 Pa s). Table 1.4 gives the dynamic viscosity of some liquid food products and processing media used in food processing facilities. Kinematic viscosity is the ratio of dynamic viscosity to the density. The SI unit of kinematic viscosity is m^2/s.

1.3.2 Laws of Fluid Dynamics

The mass and energy conservation of fluid flow is governed by physical laws. The conservation of mass for fluid flow is expressed as the continuity equation. According to the continuity equation, no fluid is created or destroyed as fluid flows in a closed conduit such as a pipe and duct. The continuity equation can be expressed on a basis of either mass flow or volumetric flow rate. The continuity equation can be expressed as

$$\dot{m}_1 = \dot{m}_2 \tag{1.58}$$

TABLE 1.4
Dynamic Viscosity of Some Liquid Food Products and Processing Medium

Fluid	Temperature (°C)	Viscosity (mPa s)
Air	25	0.0185
Water	25	0.993
Corn oil	25	56.5
Soybean oil	30	40
Raw milk	20	1.99
Apple juice (20°Brix)	27	2.1
Apple juice (60°Brix)	27	30
Liquid honey (18.6% total solid)	25	3860
Golden syrup (48.5% total solid)	27	53

The mass flow rate can be calculated from the volumetric flow rate or the fluid velocity:

$$\dot{m} = \rho \dot{V} = \rho A \bar{u} \tag{1.59}$$

Therefore, Equation 1.58 can be re-written as

$$\rho_1 A_1 \bar{u}_1 = \rho_2 A_2 \bar{u}_2 \tag{1.60}$$

In Equations 1.58 through 1.60, $\dot{m}$ is the mass flow rate, (kg/s), $\dot{V}$ is the volumetric flow rate, (m^3/s), ρ is the density, (kg/m^3), A is the cross-sectional area of the conduit, and $\bar{u}$ is the average velocity, (m/s).

Equation 1.60 is applicable to any fluid flow. If the fluid is incompressible ($\rho_1 = \rho_2$), Equation 1.60 can be simplified as

$$A_1 \bar{u}_1 = A_2 \bar{u}_2 \tag{1.61}$$

Fluid flow also follows the law of energy conservation. Energy within a system is neither created nor destroyed. However, it can be transformed from one form to another. There are three forms of energy that are considered: (1) pressure energy or flow work due to pressure imposed on the fluid, (2) kinetic energy due to the velocity or momentum of the fluid, and (3) potential energy due to the elevation of the fluid. Bernoulli's equation is used to describe the energy conservation of fluid flow in a closed system without (1) work done on or by the fluid, (2) heat transfer between the fluid and its surroundings, and (3) frictional energy loss, which is expressed as

$$P_1 + \frac{1}{2}\rho_1 u_1^2 + \rho_1 g z_1 = P_2 + \frac{1}{2}\rho_2 u_2^2 + \rho_2 g z_2 \tag{1.62}$$

where

P is the pressure
u is the velocity
z is the elevation
ρ is the density of the fluid
g is the acceleration due to gravity

According to Bernoulli's equation, if there is no heat added to the fluid, no work done on the fluid, and no frictional energy loss, the change in energy is zero. That is

$$\Delta E = \left(\frac{P_2}{\rho_2} + \frac{1}{2}u_2^2 + g z_2\right) - \left(\frac{P_1}{\rho_1} + \frac{1}{2}u_1^2 + g z_1\right) = 0 \tag{1.63}$$

where ΔE is the energy change in the SI unit of J/kg.

However, in case of a real fluid flow, we cannot ignore the frictional energy loss. A pump or fan is used to increase the energy of the fluid and overcome the frictional energy loss in a transport system. In this case, the energy change can be determined by

$$\Delta E = E_p - E_f = \left(\frac{P_2}{\rho_2} + \frac{1}{2}u_2^2 + gz_2\right) - \left(\frac{P_1}{\rho_1} + \frac{1}{2}u_1^2 + gz_1\right) \quad (1.64)$$

where

E_p is the energy imposed on the fluid by a pump or fan (J/kg)

E_f is the total frictional energy loss (J/kg)

Therefore, the energy requirement of a pump or fan per unit mass of fluid is determined by

$$E_p = \left(\frac{P_2}{\rho_2} + \frac{1}{2}u_2^2 + gz_2\right) - \left(\frac{P_1}{\rho_1} + \frac{1}{2}u_1^2 + gz_1\right) + E_f \quad (1.65)$$

The head of the pump, h_{pump} (m), which is usually used in the specifications of a pump, is defined as

$$h_{pump} = \frac{E_p}{g} \quad (1.66)$$

where g is the acceleration due to gravity (m/s^2).

The required power output of a pump or fan is calculated by

$$P = \dot{m}E_p \quad (1.67)$$

where $\dot{m}$ is the mass flow rate (kg/s).

A pump is usually located in the pipeline used to deliver a fluid. Careful attention is required to prevent vaporization of the liquid inside the pump. Therefore, the liquid pressure on the suction side of a pump should be higher than the saturation vapor pressure of the liquid at the operating temperature. That is

$$\text{NPSH}_R = \left(\frac{P_s}{\rho g} + \frac{1}{2g}u_s^2\right) - \frac{P_v}{\rho g} > 0 \quad (1.68)$$

Pump manufacturers specify the required net positive suction head (NPSH_R). The designed net positive suction head should be higher than the specified value.

1.3.3 Energy Loss of Fluid Flow

Energy is lost because of the friction in the flowing fluid due to viscosity. The magnitude of the energy loss for a given fluid is directly dependent on the characteristics of the fluid flow. The flow characteristics could be laminar, turbulent, or in between (transitional). In a laminar flow, the fluid appears to flow in layers in a smooth and regular manner while under the turbulent condition, the fluid flows in random directions within the flow stream. The flow characteristics of a fluid are quantitatively described by the Reynolds number, which is defined as

$$N_{\mathrm{Re}} = \frac{\rho \bar{u} D}{\mu} \tag{1.69}$$

where

- D is the characteristic dimension (e.g., tube diameter) (m)
- ρ is the density (kg/m^3)
- $\bar{u}$ is the average velocity (m/s)
- μ is the dynamic viscosity (Pa s)

The flow characteristics can be determined by the Reynolds number:

- $N_{\mathrm{Re}} < 2100$, laminar flow
- $2100 \leq N_{\mathrm{Re}} \leq 4000$, transitional flow
- $N_{\mathrm{Re}} > 4000$, turbulent flow

During fluid flow, a velocity gradient exists between the fluid at the wall and the center of a conduit due to friction between the individual molecules of the fluid and between the fluid and the wall. The major energy loss due to the friction is usually expressed as

$$E_{\mathrm{f,major}} = 2f \frac{L}{D} \frac{\bar{u}^2}{2} \tag{1.70}$$

where

- E_{f} is the energy loss due to friction
- f is the loss coefficient or friction factor
- $\bar{u}$ is the velocity of the fluid
- L is the length of the conduit
- D is the diameter of the conduit

The friction factor, f, varies with the flow characteristics described with the Reynolds number and surface roughness. The Moody chart shown in Figure 1.4 presents the friction factor as a function of the Reynolds number for various magnitudes of relative roughness of different pipes. The relative roughness of a pipe is the ratio of roughness, ε (m), and diameter, D (m), of the pipe. The equivalent roughness, ε, for new pipes made of different materials is (Singh and Heldman, 2001):

- Cast iron: 259×10^{-6} m
- Galvanized iron: 152×10^{-6} m
- Steel or wrought iron: 45.7×10^{-6} m

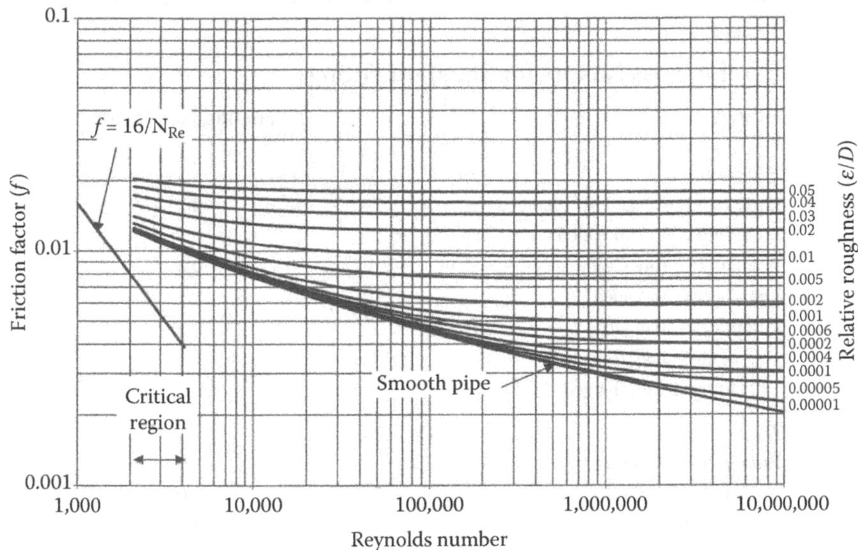

FIGURE 1.4 Moody diagram for the Fanning friction factor. (Reprinted from Singh, R.P. and Heldman, D.R., *Introduction to Food Engineering*, Academic Press, San Diego 2001. With permission.)

The various components used in pipelines, such as valves, tees, and elbows, and contraction and expansion of fluid when it enters into and exits from a pipe cause minor frictional losses. Minor frictional energy loss can be calculated by

$$E_{\text{f,minor}} = C_{\text{f,minor}} \frac{\bar{u}^2}{2} \tag{1.71}$$

The friction loss factor for sudden contraction is determined by (Singh and Heldman, 2001)

$$C_{\text{f,concentration}} = 0.4\left(1.25 - \frac{A_2}{A_1}\right) \quad \text{if } \frac{A_2}{A_1} < 0.715 \tag{1.72}$$

$$C_{\text{f,concentration}} = 0.75\left(1 - \frac{A_2}{A_1}\right) \quad \text{if } \frac{A_2}{A_1} > 0.715 \tag{1.73}$$

The friction loss factor for sudden expansion is determined by (Singh and Heldman, 2001)

$$C_{\text{f,expansion}} = \left(1 - \frac{A_2}{A_1}\right)^2 \tag{1.74}$$

In Equations 1.72 through 1.74, A_2 and A_1 are small and large cross-sectional areas, respectively. Friction loss factors for standard fittings are given in Table 1.5 (Singh and Heldman, 2001).

TABLE 1.5
Friction Loss Factors for Standard Fittings

Fittings	Friction Loss Factor
Elbows	
Long radius 45° and 90°, flanged	0.2
Long radius 90°, threaded	0.7
Regular 45°, threaded	0.4
Regular 90°, threaded	1.5
Regular 90°, flanged	0.3
180° return bend, threaded	1.5
180° return bend, flanged	0.2
Tees	
Branch flow, flanged	1.0
Branch flow, threaded	2.0
Line flow, flanged	0.2
Line flow, threaded	0.9
Union, threaded	0.8
Valve	
Angle, fully open	2
Ball valve, 1/3 closed	5.5
Ball valve, 2/3 closed	210
Ball valve, fully open	0.05
Diaphragm valve, ¼ closed	2.6
Diaphragm valve, ½ closed	4.3
Diaphragm valve, open	2.3
Gate, ¾ closed	17
Gate, ¼ closed	0.26
Gate, ½ closed	2.1
Gate, fully open	0.15
Globe, fully open	10
Swing check, backward flow	∞
Swing check, forward flow	2

Source: From Singh, R.P. and Heldman, D.R., in *Introduction to Food Engineering*, Academic Press, San Diego, 2001. With permission.

Example 1.7

A pumping system as shown in Figure 1.5 is designed to pump milk at 20°C from an open tank through a steel pipe 20 m in length and 2.5 cm inside diameter to a second tank at a higher level. The mass flow rate of milk is 2 kg/s. The supply tank maintains a liquid level of 2 m from the floor. The milk leaves the system at an elevation of 15 m above the floor (reference elevation). Determine the major and minor energy losses due to friction. Determine the required pump head and power requirement of the

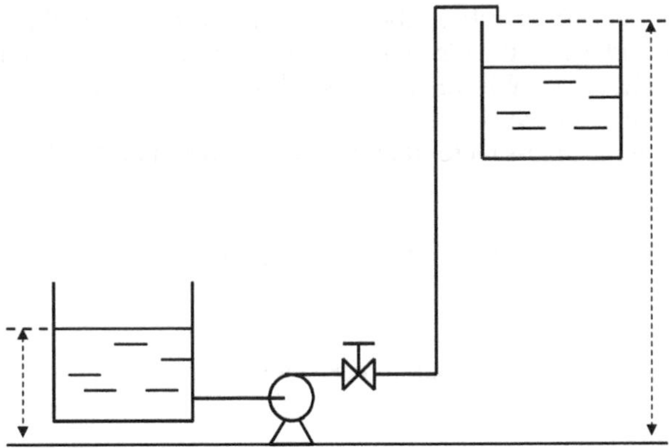

FIGURE 1.5 Typical pumping system.

pump. (Suppose that the density and viscosity of milk are 1000 kg/m^3 and 1.99 m Pas, respectively.)

Solution 1.7

1. From Equation 1.59, the mean velocity of milk in the pipe is

$$\bar{u} = \frac{\dot{m}}{\rho A} = \frac{2\,[\text{kg/s}]}{1000\,[\text{kg/m}^3] \times \frac{\pi(0.025)^2}{4}} = 4.08\text{ m/s}$$

2. From Equation 1.69, the Reynolds number is

$$N_{\text{Re}} = \frac{\rho \bar{u} D}{\mu} = \frac{1{,}000\,[\text{kg/m}^3] \times 4.08\,[\text{m/s}] \times 0.025\,[\text{m}]}{0.00199\,[\text{Pa s}]} = 51{,}256$$

The relative roughness of the pipe is

$$\frac{\varepsilon}{D} = \frac{45.7 \times 10^{-6}\,[\text{m}]}{0.025\,[\text{m}]} = 1.828 \times 10^{-3}$$

Using the Reynolds number and relative roughness, the frictional factor can be found from Figure 1.4: f = 0.0065. Using Equation 1.71, the major frictional energy loss is

$$E_{\text{f,major}} = 2f\frac{L}{D}\frac{\bar{u}^2}{2} = 2 \times 0.0065 \times \frac{20}{0.025} \times \frac{4.08^2}{2} = 86.6\text{ J/kg}$$

3. There are two 90° standard elbows and one angle valve in the pipeline. The entrance and exit of the pipeline are the concentration and expansion types, respectively. From Table 1.5, the energy loss factor for elbow and angle valve is 1.5 and 2, respectively.

   The entrance loss factor due to sudden contraction ($A_2/A_1 \approx 0$) is

$$C_{f,\text{entrance}} = 0.4\left(1.25 - \frac{A_2}{A_1}\right) = 0.4 \times (1.25 - 0) = 0.5$$

The exit loss factor due to sudden expansion ($A_1/A_2 \approx 0$) is

$$C_{f,\text{entrance}} = \left(1 - \frac{A_1}{A_2}\right)^2 = (1-0)^2 = 1$$

The total minor energy loss factor is thus

$$C_{f,\text{minor}} = C_{f,\text{entrance}} + 2C_{f,\text{elbow}} + C_{f,\text{valve}} + C_{f,\text{exit}} = 0.5 + 2 \times 1.5 + 2 + 1 = 6.5$$

The total minor energy loss is

$$E_{f,\text{minor}} = C_{f,\text{minor}} \frac{\overline{u}^2}{2} = 6.5 \times \frac{4.08^2}{2} = 54.1\ \text{J/kg}$$

4. From Equation 1.65, the energy requirement of the pump is

$$\begin{aligned} E_p &= \left(\frac{P_2}{\rho_2} + \frac{1}{2}u_2^2 + gz_2\right) - \left(\frac{P_1}{\rho_1} + \frac{1}{2}u_1^2 + gz_1\right) + E_f \\ &= \left(\frac{101325}{1000} + \frac{1}{2} \times 0^2 + 9.81 \times 15\right) - \left(\frac{101325}{1000} + \frac{1}{2} \times 0^2 + 9.81 \times 2\right) + (86.6 + 54.1) \\ &= 268.2\ \text{J/kg} \end{aligned}$$

5. From Equation 1.66, the pump head is

$$h_{\text{pump}} = \frac{E_p}{g} = \frac{268.2}{9.8} = 27.4\ \text{m}$$

6. From Equation 1.67, the power requirement of the pump is

$$P = \dot{m}E_p = 2 \times 268.2 = 536.5\ \text{W}$$

1.3.4 Fluid Handing Equipment in Food Processing Facilities

Fluids in food processing facilities are transported mostly in closed conduits such as pipes if they are round or ducts if they are not round. A typical fluid transporting system consists of four basic components: tanks, pipeline, pump or fan, and fittings (Singh and Heldman, 2001). Stainless steel is widely used to construct these four components because it provides smoothness, cleanability, and corrosion resistance.

Fluid flow occurs with the application of a force. Pumps are frequently used to provide mechanical energy to overcome the forces opposing transport of the liquid. There are several types of pumps used in food processing facilities. These pumps can be classified as positive displacement pumps and centrifugal pumps. Centrifugal pumps are most efficient with low-viscosity liquids such as milk and fruit juices at a high flow rate and moderate pressure. Positive displacement pumps such as reciprocating pumps can be used to deliver low-viscosity liquids at a low flow rate and high pressure. Blowers or fans are used to move gases.

The pump characteristic diagram, as shown in Figure 1.6, gives the plots of a pump head, and efficiency and power as functions of volumetric flow rate. It is usually provided by pump manufacturers and obtained for water. The peak of efficiency curve represents the volumetric flow rate where the pump is the most efficient. The flow rate at peak efficiency is thus the designed flow rate. In order to achieve a high efficiency, various impeller speeds or pumps in series or parallel may be used in a process. Affinity laws govern the performance of a centrifugal pump at various impeller speeds, which include

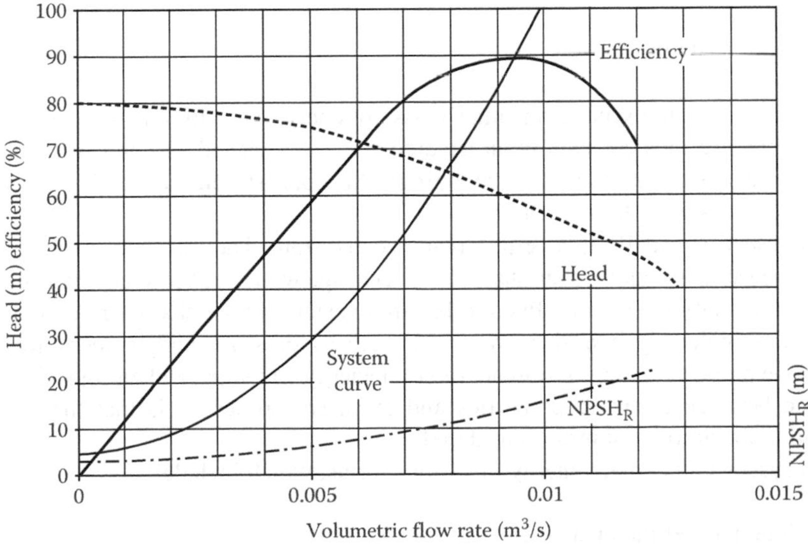

FIGURE 1.6 Performance characteristic curve of a pump. (Reprinted from Singh, R.P. and Heldman, D.R., *Introduction to Food Engineering*, Academic Press, San Diego, 2001. With permission.)

$$\frac{\dot{V}_2}{\dot{V}_1} = \frac{N_2}{N_1} \text{ (volumetric flow rate)} \tag{1.75}$$

$$\frac{h_2}{h_1} = \left(\frac{N_2}{N_1}\right)^2 \text{ (pump head)} \tag{1.76}$$

$$\frac{P_2}{P_1} = \left(\frac{N_2}{N_1}\right)^3 \text{ (power)} \tag{1.77}$$

1.4 THERMODYNAMICS IN FOOD PROCESSING

1.4.1 System and Its States

Thermodynamics deals with the transformation of energy and the accompanying changes in the state of the involved matter. To study thermodynamics, we need to define the systems and their surroundings to which the thermodynamic laws are applied. A system is the part of the whole that we wish to study, such as the contents of a reactor or a room. The space outside the chosen system is defined as the environment. After we define the system and its environment, we can examine the flow of mass and energy into and out of it. Based on the flow of mass and energy, the systems could be categorized as

- Isolated systems
- Closed systems
- Open systems

An isolated system exchanges neither mass nor energy with its surroundings. Mass and energy are thus constant in an isolated system. A closed system can exchange energy but not mass with its surroundings. An open system can exchange either energy or mass with its surroundings.

A system is usually in a certain macroscopic state. The transient state changes with time while the equilibrium state does not change with time but it may change with location in a flowing system. In equilibrium, the state can be characterized by state properties including intensive properties such as pressure and temperature and extensive properties such as volume, internal energy, enthalpy, and entropy. The difference between intensive properties and extensive properties is that the later is related to the amount of mass considered.

A system can contain energy in various forms, which include

- Internal thermal energy, U
- Kinetic energy, KE
- Potential energy, PE
- Other energy forms such as electrical, magnetic, and chemical energy

Internal thermal energy is stored in the molecules without regard external fields. Kinetic energy is related to the motion of the system, $KE = \frac{1}{2}mu^2$. Potential energy is caused due to elevation above a reference plane, $PE = \rho gz$.

1.4.2 Processes and State Changes

A system may be engaged in a process, which changes the states of the system. The isothermal, isobaric, isochoric, isentropic, or isenthalpic processes take place at a constant temperature, pressure, volume, entropy, or enthalpy, respectively. Furthermore, if no heat exchange takes place between the system and its environment, the process is adiabatic. If the frictional forces are zero, a process is reversible. A reversible process takes place when all driving forces go to zero. However, a real process is irreversible. The frictional forces have to be overcome, requiring driving forces such as temperature difference, ΔT, and pressure difference, ΔP.

Classical thermodynamics is based on the development of a set of general physical laws of macroscopic behavior without considering the microscopic structure of a matter. Classical thermodynamics is based on two physical laws: the first law deals with the quantities involved in energy changes and the second law deals with the direction in which the changes take place. Statistical thermodynamics is based on the laws of averaging to predict the macroscopic behavior of a collection of microscopic atoms and molecules (Witte et al., 1988).

1.4.3 First Law of Thermodynamics

A system, which contains energy, can exchange its energy with its surroundings in two ways:

- Work, W
- Heat, Q

Work and heat can be identified only as they cross a system boundary. The first law is a conservation law of energy. The first law of thermodynamics is used to perform an energy balance on a system. It states that energy is conserved during various types of changes in the states of matter. It says the net energy flowing into and out of a system via heat and work is equal to the energy stored within the system. Therefore, the first law for a closed system, which receives heat from the environment and performs work on the environment, is

$$\Delta Q_{in} - \Delta W_{out} = \Delta U \tag{1.78}$$

Internal energy, U, is the energy associated with motions and configurations of interior particles such as molecules and atoms of a substance, primarily due to temperature as the result of the addition or removal of heat. If more heat is received by the system than the work the system performs on the environment, the difference is stored as an addition to the internal energy, U, of the system.

For an open system where streams flow through a fixed control volume in a steady state, the first law is

$$\dot{Q}_{in} - \dot{W}_{out} = \dot{m}\left(\Delta h + \frac{\Delta u^2}{2} + g\Delta z\right) \tag{1.79}$$

where

$\dot{m}$ is the mass flow rate
h is the enthalpy
u is the velocity of the flowing system
z is the elevation of the system
g is the acceleration due to gravity

The enthalpy is a thermal property of a substance, which is a measure of the total energy content of the substance. It consists of the internal energy of the substance plus energy in the form of work due to flow into or out of the system, $h = U + PV$. Internal energy is the major component of enthalpy. The energy due to flow is the production of pressures times the volume. Although the enthalpy is actually the total energy content of a substance, the flow of work component, PV, is usually ignored since it is relatively small compared to the total heat of the substance.

If there is no work and fluid flow involved in a process, Equation 1.79 becomes

$$\dot{Q}_{in} = \dot{m}(\Delta h) = \Delta H \tag{1.80}$$

This is a typical heat transfer process as discussed in Section 1.2. Heat may be stored in a substance in two forms: sensible heat and latent heat. Sensible heat is the heat that changes the temperature of a substance without any phase change while latent heat is the heat that causes a change in phase of a substance without any change in temperature.

Constant pressure processes are most commonly encountered in food processing. If a process involves only a change of temperature from T_2 to T_1 at a constant pressure, the sensible heat can be calculated by

$$\Delta H = Q = mc_p(T_2 - T_1) \tag{1.81}$$

where

ΔH is the change in enthalpy of the system (J)
Q is quantity of heat added or removed from the substance (J)
m is the mass of the substance (kg)
c_p is the specific heat of the substance at a constant pressure (J/kg°C)
$T_2 - T_1$ is the temperature change (°C)

Specific heat is the quantity of heat required to change one unit mass of a substance by a degree. Specific heat varies slightly with temperature. In most calculations, it may be assumed to be constant.

Example 1.8

Ham has a specific heat of 2.5 kJ/kg°C. How much heat is required to raise the temperature of 5 kg ham from 4°C to 70°C?

Solution 1.8

Using Equation 1.81, the total heat required is

$$q = mc(T_2 - T_1) = 5\,[\text{kg}] \times 2.5\,[\text{kJ/kg°C}] \times (70°\text{C} - 4°\text{C}) = 825\,\text{kJ}$$

During heating/cooling of foods, phase changes may occur. The latent heat of fusion for water at 0°C is 333.2 kJ/kg. The latent heat of vaporization of water varies with temperature and pressure. At 100°C and atmospheric pressure, the latent heat of vaporization of water is 2257.06 kJ/kg.

If there is no heat transfer involved in a process, Equation 1.79 becomes

$$\dot{W}_{\text{out}} = \dot{m}\left(\Delta h + \frac{\Delta u^2}{2} + g\Delta z\right) \tag{1.82}$$

This is a typical fluid dynamic process as discussed in Section 1.3.

1.4.4 Second Law of Thermodynamics

The second law is associated with the direction of a process. Energy will always flow from a region with a higher energy state to a region with a lower energy state. The second law of thermodynamics is also known as the law of degradation of energy. It states that in any real process the direction of the process corresponds to the direction in which the total entropy increases. Entropy, *S*, is an extensive property. According to the second law of thermodynamics, the total change of entropies of both the system and environment should be positive, that is

$$\Delta S_{\text{total}} = \Delta S_{\text{system}} + \Delta S_{\text{environment}} \tag{1.83}$$

and

$$\Delta S_{\text{total}} > 0 \tag{1.84}$$

The second law of thermodynamics implies that it is impossible to manufacture any device that, operating in a cycle, has no effect other than cooling of one thermal reservoir and heating a second reservoir at a higher temperature.

1.4.5 Energy and Exergy

The energy content of a substance can be defined by

$$\text{En} = \dot{m}h \tag{1.85}$$

where

$\dot{m}$ is the mass or flow rate of the substance

h is the specific enthalpy of the substance

However, not all energy stored in a substance can be converted to work. Exergy is defined as the maximum amount of work that can be produced by a system or a flow of matter or energy as it comes to equilibrium with a reference environment (Dincer, 2000; Dincer and Cengel, 2001). Like enthalpy, exergy values depend on the reference state. The reference temperature and pressure are usually $T_0 = 25°C$ and $P_0 = 101,325\,\text{Pa}$ used in the chemical industry.

For a fluid, the exergy rate is determined as (Kuzgunkaya and Hepbasli, 2007)

$$\text{Ex} = \dot{m}\psi \tag{1.86}$$

The specific flow exergy, ψ, is determined by

$$\psi = (h - h_0) - T_0(s - s_0) \tag{1.87}$$

where

$\dot{m}$ is the mass flow rate
h is the enthalpy
s is the entropy
subscript zero denotes the reference state

In order to use the above equation to determine specific flow exergy, we need to determine enthalpy, h, and entropy, s. The specific enthalpy of an incompressible substance can be determined by (Kuzgunkaya and Hepbasli, 2007)

$$h = c(T - T_0) \tag{1.88}$$

The specific entropy of an incompressible substance can be determined by (Kuzgunkaya and Hepbasli, 2007)

$$s = c\ln\left(\frac{T}{T_0}\right) \tag{1.89}$$

where the unit of temperature is Kelvin.

Therefore, the exergy of an incompressible substance may be written as

$$\psi = c\left(T - T_0 - T_0\ln\frac{T}{T_0}\right) \tag{1.90}$$

where c is the specific heat of the substance.

Example 1.9

What are the specific energy and exergy contents of water at 100°C? The specific heat of water is 4.18 kJ/kg°C. Assume that water is incompressible and the reference temperature is $T_0 = 25°C$.

Solution 1.9

From Equation 1.88, the specific energy content is

$$h = c(T - T_0) = 4.18 \times (100 - 25) = 313.5 \text{ kJ/kg}$$

From Equation 1.90, the specific exergy is

$$\begin{aligned}\psi &= c\left(T - T_0 - T_0 \ln\frac{T}{T_0}\right)\\ &= 4.18 \times \left[(100 + 273.15) - (25 + 273.15) - (25 + 273.15)\ln\left(\frac{100 + 273.15}{25 + 273.15}\right)\right]\\ &= 33.9 \text{ kJ/kg}\end{aligned}$$

1.4.6 Loss of Work and Exergy

According to the first law, energy can neither be created nor be destroyed. Energy consumption does not mean that energy is consumed but the quality of energy or available work decreases (de Swaan Arons et al., 2004). Heat in free fall from a higher to a lower temperature incurs a loss in its quality. Its quality vanishes at the temperature of the environment, T_0. The loss of available work can be identified as the product of the entropy generated and the absolute temperature of the environment, T_0. The maximum amount of available work for the heat, Q, at a temperature $T > T_0$ is given by

$$\text{Ex} = W_{\text{max}} = Q\left(1 - \frac{T_0}{T}\right) \tag{1.91}$$

The factor $1 - (T_0/T)$ is called the thermodynamic efficiency or Carnot factor. The quality of heat supplied at temperature $T > T_0$ is the maximum fraction available for useful work with respect to a defined environment. Sometimes, the heat available for useful work is called the exergy of heat and the remaining part of the heat is called anergy of the heat, which is the minimal part of the original heat transferred to the environment. Therefore,

$$Q = Q\left(1 - \frac{T_0}{T}\right) + Q\frac{T_0}{T} = E_{\text{ex}} + E_{\text{an}} \tag{1.92}$$

where the first item on the right side is the exergy and the second item is anergy.

The amount of lost work or exergy from a process is

$$W_{\text{lost}} = E_{\text{ex,in}} - E_{\text{ex,out}} = QT_0\left(\frac{1}{T_{\text{low}}} - \frac{1}{T_{\text{high}}}\right) \tag{1.93}$$

From Equations 1.92 and 1.93, we can find

- Sum of exergy and anergy is always constant
- Anergy can never be converted into exergy

1.4.7 Energy and Exergy Efficiencies

The first law of thermodynamics, or the principle of energy conservation, is usually used to trace the flow of energy through the relevant industrial system to determine the primary energy inputs needed to produce a given amount of a product. The first law based on energy and the second law based on exergy give different efficiencies. Energy efficiency (the first law efficiency) can be defined as

$$\eta_{en} = \frac{\sum \text{En}_{out}}{\sum \text{En}_{in}} \times 100\% \tag{1.94}$$

The analysis based on the first law of thermodynamics takes no account of the type of an energy source in terms of its thermodynamic quality. The second law of thermodynamics facilitates the assessment of the maximum amount of work achievable in a given system with different energy sources. Exergy is the available energy for conversion from an energy source with a reference environmental condition. Therefore, exergy represents the thermodynamic quality of an energy carrier and heat or energy loss in the waste stream. Electricity, for instance, has a high quality or exergy while low-temperature hot water has a low quality. The exergy efficiency (the second law efficiency) is an important parameter, which is defined as

$$\eta_{ex} = \frac{\sum \text{Ex}_{out}}{\sum \text{Ex}_{in}} \times 100\% \tag{1.95}$$

Example 1.10

A saturation steam at 100°C is used to vaporize ethanol at 78.4°C in a shell-tube heat exchanger. The heat transfer rate from steam to ethanol is 1000 W. The heat exchanger is well insulated and the heat loss from the heat exchanger to the environment is negligible. What is the exergy loss and exergy efficiency? The reference temperature for calculation is 25°C.

Solution 1.10

Since there is no energy loss from the heat exchanger, the energy loss is zero and energy efficiency of the heat exchanger is 100%. However, the degradation of the energy source (steam) occurs during heat exchange.

From Equation 1.91, the available work or the exergy of the saturation steam is

$$\text{Ex}_{high} = Q\left(1 - \frac{T_0}{T_{high}}\right) = 1000\ \text{W} \times \left(1 - \frac{273.15 + 25}{273.15 + 100}\right) = 201\ \text{W}$$

After the energy in the steam is transferred to the ethanol at a lower temperature, the energy quality decreases and the available work becomes

$$\mathrm{Ex}_{\mathrm{low}} = Q\left(1-\frac{T_0}{T_{\mathrm{low}}}\right) = 1000\ \mathrm{W} \times \left(1-\frac{273.15+25}{273.15+78.4}\right) = 152\ \mathrm{W}$$

Therefore, the exchange of heat has taken place with the rate of loss of available work

$$W_{\mathrm{lost}} = E_{\mathrm{ex,in}} - E_{\mathrm{ex,out}} = 201-152 = 49\mathrm{W}$$

From Equation 1.95, the exergy efficiency of the heat exchanger is

$$\eta_{\mathrm{ex}} = \frac{E_{\mathrm{ex,out}}}{E_{\mathrm{ex,in}}} \times 100\% = \frac{152}{201} \times 100\% = 75.6\%$$

Energy has two aspects: its energy content and quality. As seen from the above example, there is no change in the energy content of the heat exchanger system. However, the energy quality is degraded. That is, the quality of the same amount of energy in the ethanol is lower than that in the steam.

Example 1.11

This example is adapted from de Swaan Arons et al., 2004. A stream of 1 kg/s of liquid water at 100°C is adiabatically mixed with a second stream of 1 kg/s of liquid water at 0°C to produce a stream of 2 kg/s of liquid water at 50°C as shown in Figure 1.7. Suppose that the temperature of the environment is 25°C. Determine the exergy loss of the mixing process and the exergy efficiency of the mixing process. (Suppose that the specific heat of water is 4.18 kJ/kg°C and the reference temperature is 25°C).

Solution 1.11

1. From Equation 1.90, the specific exergies of the three streams are

$$\psi_1 = c\left(T_1 - T_0 - T_0 \ln\frac{T_1}{T_0}\right) = 4180 \times \left[(100-25)-(273.15+25)\times\ln\left(\frac{273.15+100}{273.15+25}\right)\right]$$
$$= 33.86\ \mathrm{kJ/kg}$$

$$\psi_2 = c\left(T_2 - T_0 - T_0 \ln\frac{T_2}{T_0}\right) = 4180 \times \left[(0-25)-(273.15+25)\times\ln\left(\frac{273.15+0}{273.15+25}\right)\right]$$
$$= 4.64\ \mathrm{kJ/kg}$$

$$\psi_3 = c\left(T_3 - T_0 - T_0 \ln\frac{T_3}{T_0}\right) = 4180 \times \left[(50-25)-(273.15+25)\times\ln\left(\frac{273.15+50}{273.15+25}\right)\right]$$
$$= 4.15\ \mathrm{kJ/kg}$$

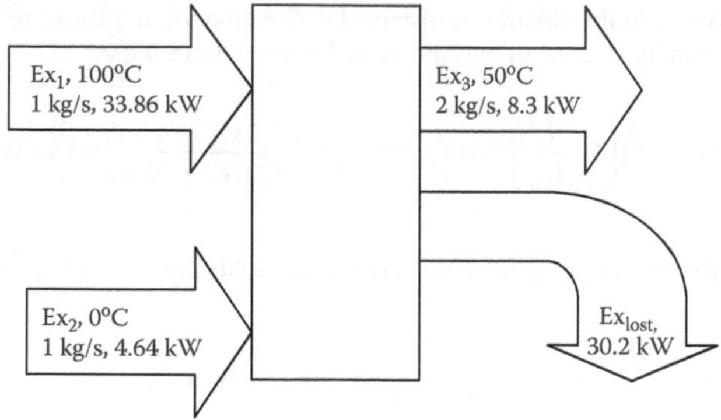

FIGURE 1.7 Exergy flow diagram. (Adapted from de Swaan Arons, J., van der Kooi, H., and Sankaranarayanan, K., *Efficiency and Sustainability in the Energy and Chemical Industries,* Marcel Dekker Inc., New York, 2004. With permission.)

2. From Equation 1.86, the exergies of the three streams are

$$\text{Ex}_1 = \dot{m}_1\psi_1 = 1\ [\text{kg/s}] \times 33.86\ [\text{kJ/kg}] = 33.86\ \text{kW}$$

$$\text{Ex}_2 = \dot{m}_2\psi_2 = 1\ [\text{kg/s}] \times 4.64\ [\text{kJ/kg}] = 4.64\ \text{kW}$$

$$\text{Ex}_3 = \dot{m}_3\psi_3 = 2\ [\text{kg/s}] \times 4.15\ [\text{kJ/kg}] = 8.3\ \text{kW}$$

3. From Equation 1.93, the exergy loss of the mixing process is

$$\text{Ex}_{\text{lost}} = E_{\text{ex,in}} - E_{\text{ex,out}} = (33.86 + 4.64) - 8.3 = 30.2\ \text{W}$$

From Equation 1.95, the exergy efficiency is

$$\eta_{\text{ex}} = \frac{E_{\text{ex,out}}}{E_{\text{ex,in}}} \times 100\% = \frac{8.3\ [\text{kW}]}{33.86\ [\text{kW}] + 4.64\ [\text{kW}]} \times 100\% = 21.6\%$$

Therefore, the mixing process causes about 80% of the available work in the feedstock to be lost.

1.5 SUMMARY

Analyses of energy consumption by food processing equipment and systems such as heat exchangers, boilers, pumps, fans, refrigerators, and dryers involve the application of basic scientific and engineering principles of heat transfer, fluid mechanics, and thermodynamics. In this chapter, these engineering principles were summarized. Heat transfer models can be used to quantify transient energy consumption during different food processes. Transient energy consumption profiles can be used to optimize unit operations and to minimize energy consumption. Heat transfer calculations are also essential in selecting an insulation material and determining the thickness of the insulation layer. Pumps and fans are widely used to deliver liquid foods and processing media in food processing facilities. The power requirement and energy loss of a pumping system can be calculated by engineering principles of fluid mechanics. Energy loss can be interpreted by the loss of energy content using the first law of thermodynamics and by the loss of energy quality using the second law of thermodynamics. Thermodynamic and exergy analyses of a process are used to determine the thermodynamic and exergy efficiencies of the process. Examples were given to demonstrate the application of these engineering principles in energy analyses.

REFERENCES

Amezquita, A., L.J. Wang, and C.L. Weller. 2005. Finite element modeling and experimental validation of cooling rates of large ready-to-eat meat products in small meat-processing facilities. *Transaction of the ASAE* 48: 287–303.

Choi, Y. and M.R. Okos. 1986. Effects of temperature and composition on the thermal properties of foods. In *Food Engineering and Process Applications, Vol. 1, Transport Phenomena*, Maguer, M.L. and P. Jelen, (Eds.), pp. 93–101. New York: Elsevier.

Davey, L.M. and Q.T. Pham. 1997. Predicting the dynamic product heat load and weight loss during beef chilling using a multi-region finite difference approach. *International Journal of Refrigeration* 20: 470–482.

de Swaan Arons, J., H. van der Kooi, and K. Sankaranarayanan. 2004. *Efficiency and Sustainability in the Energy and Chemical Industries.* New York: Marcel Dekker, Inc.

Dincer, I. 2000. Thermodynamic, exergy and environmental impact. *Energy Sources* 22: 723–732.

Dincer, I. and A.Y. Cengel. 2001. Energy, entropy and exergy concepts and their roles in thermal engineering. *Entropy International Journal* 2: 87–98.

Fryer, P.J., D.L. Pyle, and C.D. Rielly. 1997. *Chemical Engineering for the Food Industry.* London: Blackie Academic & Professional.

Gould, W.A. 1996. *Unit Operations for the Food Industries.* Maryland: CTI Publications, Inc.

Kuzgunkaya, E.H. and A. Hepbasli. 2007. Exergetic performance assessment of a ground-source heat pump drying system. *International Journal of Energy Research* 31: 760–777.

Mannaperuma, J.D. and R.P. Singh. 1990. Developments in food freezing. In *Biotechnology and Food Process Engineering*, Schwartzberg, H. and A. Rao, (Eds.), New York: Marcel Dekker, Inc.

Mont, J.A. and M.R. Harrison. 2006. Industrial insulation. In *Energy Management Handbook* (6th ed.), Turner, W.C. and S. Doty, (Eds.), pp. 437–470. GA. The Fairmont Press, Inc.

Murakami, E.G. and M.R., Okos. 1989. Measurement and prediction of thermal properties of foods. In *Food Properties and Computer Aided Engineering of Food Processing Systems*, Singh, R.P. and A.G. Medina, (Eds.), pp. 3–48. Amsterdam: Kluwer Academic Publishing.

Simpson, R., C. Cortes, and A. Teixeira. 2006. Energy consumption in batch thermal processing: Model development and validation. *Journal of Food Engineering* 73: 217–224.

Singh, R.P. and D.R. Heldman. 2001. *Introduction to Food Engineering* (3rd ed.), San Diego: Academic Press.

Sweat, V.E. 1995. Thermal properties of foods. In *Engineering Properties of Foods*, Rao, M.A. and S.S.H. Rizvi, (Eds.), pp. 99–138. New York: Marcel Dekker, Inc.

Wang, L.J. and R.P. Singh. 2004. Mathematical modeling and sensitivity analysis of double-sided contact-cooking process for initially frozen hamburger patties. *Transaction of the ASAE* 47: 147–157.

Wang, L.J. and D.W. Sun. 2002. Modeling vacuum cooling process of cooked meat-Part 2: Mass and heat transfer of cooked meat under vacuum pressure. *International Journal of Refrigeration* 25: 861–872.

Wang, L.J. and D.W. Sun. 2003. Recent developments in numerical modeling of heating and cooling processes in the food industry: A review. *Trends in Food Science and Technology* 14: 408–423.

Wang, L.J. and D.W. Sun. 2006. Heat and mass transfer in thermal food processing, Chapter 2. In *Thermal Food Processing: New Technologies and Quality Issues,* Sun, D.W. (Ed.), pp. 35–72. CRC Press, Boca Raton.

Wang, L.J. and C.L. Weller. 2006. Thermophysical properties of frozen foods, Chapter 5. In *Handbook of Frozen Food Processing and Packaging*, Sun, D.-W. (Ed.), pp. 101–125. New York: Marcel Dekker, Inc.

Witte, L.C., P.S. Schmidt, and D.R. Brown. 1988. *Industrial Energy Management and Utilization*. Washington DC: Hemisphere Publishing Corporation.

2 Fundamentals of Energy Auditing

2.1 INTRODUCTION

No-cost or very low-cost operational changes are considered to save a customer or an industry 10%–20% on utility bills. Capital investment programs with payback times of two years or less can save an additional 20%–30% (Capehart et al., 2006). In most cases, the reduction of energy cost will also reduce emissions of environmental pollutants. This may further reduce the costs for the control of environmental pollution.

Energy audit is the first step to identify energy saving opportunities in a food processing facility. Energy audit is a system used to keep track of energy consumption and costs through the whole facility. Energy audit is also called energy survey, energy analysis, or energy evaluation. Energy auditing is the study of a facility (1) to determine how and where energy is used or converted from one form to another, (2) to identify opportunities to reduce energy usage, (3) to evaluate the economics and technical practicability of implementing these reductions, and (4) to formulate prioritized recommendations for implementing process improvements to save energy (Witte et al., 1988; Capehart et al., 2006).

Electric and gas utilities in the United States offer free residential energy audits. Commercial or industrial customers may hire an engineering consulting company to perform a complete energy audit for a process with complex equipment and operations. The U.S. Department of Energy financially supports about 30 Industrial Assessment Centers operated by universities around the United States to provide free energy audits for small- and medium-sized manufacturing companies. Some large commercial and industrial companies may hire an energy manager or set up a team to conduct the energy audit and keep trace of the latest available energy efficiency technologies. In this chapter, the procedure of an energy audit is introduced. Measurements and instruments used in an energy audit are discussed,

2.2 PROCEDURES FOR ENERGY AUDIT

2.2.1 Preparing for an Energy Audit

An energy audit starts from the review of energy-related records to establish a baseline with which progress can be measured. Specific data that should be gathered for the review include

- Energy bills
- Descriptive information about the facility such as a plant layout

- Geographic location and weather data
- A list of each piece of equipment that significantly affects the energy consumption
- Operating hours of each piece of equipment

Electricity and fuel bills for at least 12 months before the audit should be used. Since the bills normally include several cost components in addition to direct expenditures for electricity or fuel, the figures appearing on the bills must be interpreted appropriately to determine the actual energy consumption. For electricity bills, the most common components of rate schedules include (Capehart et al., 2006)

- Administrative/customer charge to cover the utility's fixed costs to serve the customer such as providing a meter, reading the meter, and sending a bill
- Energy charge to cover the actual amount of electricity consumption in kWh
- Fuel cost adjustment to cover the increased cost of primary fuel for electricity generation
- Demand charge to allocate the cost of the capital facilities for electricity supply
- Demand ratchet to charge for creating a large power demand in only a few months of the year
- Power factor to charge for a machine with a poor power factor

Popular fuels used in food processing facilities include natural gas, fuel oil, and coal. Natural gas rate schedules are similar in structure to electrical rate schedules. However, natural gas companies normally do not charge for peak demand. They usually place customers into interruptible priority classes with different gas charge rates. Customers with a high priority and high charge rate will not be interrupted while customers with the lowest priority and lowest charge rate will be interrupted whenever a shortage exists. Fuel oils and coal vary in grades and sulfur content. Their billing schedules vary widely among geographical areas. In the United States, natural gas is priced on a per million cubic feet (MCF) basis; fuel oils are priced on a per gallon basis; and coal is priced on a per ton basis. Coal does not burn as completely as fuel oil and natural gas. If a combustion process is properly controlled, natural gas can be completely burned; fuel oil has only a small amount of unburned residue; and it is difficult to burn coal completely. To determine the costs of operating individual pieces of equipment or individual unit operations accurately in a food processing facility, the energy bills must be broken down into their components such as demand charge and energy charges for electrical bills.

Information about the facility layout should also be obtained and reviewed to determine the facility size, floor plan, and construction features. For the energy audit of a building, the heating degree days (HDD) and cooling degree days (CDD) are important concepts. The base for computing HDD and CDD is 65°F based on an assumption that the average building has a desired indoor temperature of 70°F and that 5°F of this is supplied by internal heat sources such as lights, appliances, equipment, and people. For example, if there were a period of 20 days when the outside temperature averaged 50°F each day, the number of HDD for these 3 days would be HDD = (65° – 50°) × 5 days = 75 degree days (Capehart et al., 2006).

Finally, an equipment list and the operating hours of each pieces of equipment should be obtained for a good understanding of major energy-consuming equipment at the facility. Knowing the operating hours allows determining whether any loads could be shifted to off-peak times.

2.2.2 Normalizing Energy Consumption Data

Energy consumption is typically related to the production rate and other possible variables such as seasonal weather conditions. The change in energy consumption may not be caused by the change in energy efficiency itself. It is sometimes necessary to normalize the energy consumption data to reflect changes in production. Normalizing energy consumption data becomes more complicated when energy use in the process is diversified and more than a single product is produced. Example 2.1 shows how to normalize energy consumption data.

Example 2.1

This example is modified from Witte et al., 1988. A plant making cereal products from grain produces two products, *A* and *B*. Product A is a meal produced by grinding the grain and then drying it to a specified moisture content. Product B is made by further drying the meal and extruding. The schematic process and the energy consumption of each step are shown in Figure 2.1. Please normalize the energy consumption to accurately reflect the trends of energy use in the plant. Table 2.1 gives the production figures and total electricity consumption for a 3 month period.

Solution 2.1

The energy consumption per kilogram of products *A* and *B* is E_A = 6 kWh/kg and E_B = 12 kWh/kg, respectively. The total monthly energy consumption, which is related to the volumes of products *A* and *B* produced, is

$$E = AE_A + BE_B$$

or

$$E_A = \frac{E}{A + (E_B/E_A)B}$$

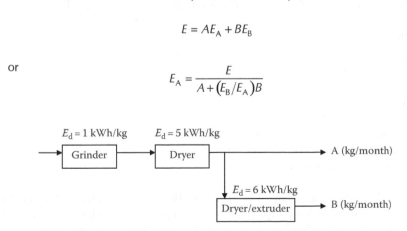

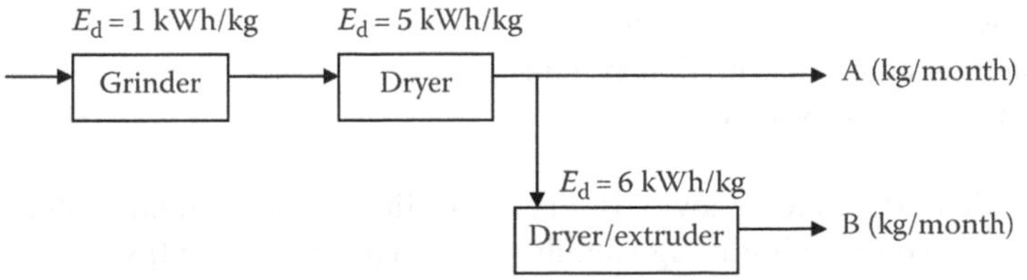

E_d = 1 kWh/kg
Grinder
E_d = 5 kWh/kg
Dryer
A (kg/month)
E_d = 6 kWh/kg
Dryer/extruder
B (kg/month)

FIGURE 2.1 Schematic of the process and the energy consumption of each step in a cereal processing facility.

Table 2.1
Production Figures and Total Electricity Consumption for a 3 Month Period

Month	A (kg)	B (kg)	Total Energy (kWh)	Normalized by A + B (kWh/kg)	Normalized by A + 2B (kWh/kg)
January	20,000	0	122,000	6.10	6.10
February	10,000	10,000	178,000	8.93	5.93
March	6,000	12,000	182,000	10.11	6.07

Therefore, the energy consumption for each unit of *B* produced is equivalent to E_A/E_B units of *A*. Specifically, in terms of energy consumption, each kilogram of *B* is equivalent to 2 kg of *A*. The monthly energy can thus be normalized in terms of energy per equivalent *A* produced by dividing *E* by the quantity of $A+2B$. The energies normalized to simple total production $A+2B$ and to equivalent A production are given in Table 2.1. Table 2.1 shows that

- Energy consumption shows a sharp increase from January to February
- If this energy consumption is normalized simply on the basis of total kilogram of production A and B, the energy consumption trend is the same
- If the difference in energy requirements of the two products is taken into account, the actual per unit energy consumption remained approximately constant during the period

2.2.3 Facility Inspection

It is necessary to conduct tours at different times to the entire facility to examine the operational patterns and equipment usage. Capehart et al. (2005) list nine major systems within a facility that should be examined for understanding and managing energy utilization within the facility. These nine major systems include

- Building envelope
- Heating, ventilating, and air conditioning (HVAC) system
- Electrical supply system
- Lighting system
- Motors
- Boiler and steam distribution system
- Hot water distribution system
- Compressed air distribution system
- Manufacturing system

When utilities offer free or low-cost energy audits to commercial customers, they usually provide only walk-through audits rather than detailed audits. They generally consider the lighting, HVAC system, water heating, insulation, and some motors (Capehart et al., 2006).

The facility inspection may have six main steps, which include (Capehart et al., 2005)

- Introductory meeting with the facility manager and the maintenance supervisor to explain the purpose of the energy audit and the information needed
- Audit interviews with the general manager, chief operating officer, facility/plant manager, financial officer, floor supervisors, equipment operators, and maintenance supervisor for correct information on facility equipment and operation
- Initial walk-through tour with facility/plant manager to obtain a general understanding of the facility's operation
- Gathering detailed data by examining the nine major energy-using systems in the facility
- Preliminary identification of energy saving opportunities based on the audit and on the knowledge of available energy efficiency technologies
- Preparation for an energy audit report to provide final results of the energy analyses and energy cost saving recommendations

Among the above six steps, process measurements will be required to gather detailed data on the nine major energy-using systems in the facility. The definition of data is thus a matter of critical importance. These data may include the following:

- Accurate flow sheets are necessary for the process as a guide to the flows of mass and energy.
- Past records of fuel and electricity usage and related production data for normalization and establishment of a baseline for plant energy consumption are needed.
- Nameplate specifications of major energy-using equipment are needed to determine the operating efficiency.
- Material properties including thermodynamic, physical, and chemical properties are needed for energy balance calculations and analyses of technical and economical feasibility.
- Dimensional data such as the length and diameter of pipes and thickness of insulation and available space for retrofitting equipment are important in estimating energy losses and evaluating the feasibility of implementing energy conservation measures.
- Equipment operating profiles are required to determine the accurate load profiles of equipment.
- Current and projected fuel and electricity costs are important to estimate the potential savings to be realized from the implementation of energy conservation technologies.

Safety is a critical part of an energy audit. The audit person or team should be thoroughly briefed on safety equipment and procedures. Adequate safety equipment should be worn at all appropriate times. Capehart et al. (2006) gives a safety checklist:

1. Electrical

 - Avoid working on live circuits, if possible.
 - Securely lock off circuits and switches before working on a piece of equipment.
 - Always keep one hand in your pocket while making measurements on live circuits to help prevent cardiac arrest.

2. Respiratory

 - When necessary, wear a full-face respirator mask with adequate filtration particle size.
 - Use activated carbon cartridges in the mask when working around low concentrations of noxious gases.
 - Use a self-contained breathing apparatus for work in a toxic environment.

3. Hearing

 - Use foam insert plugs while working around loud machinery to reduce sound levels up to 20 decibels.

2.2.4 Energy Analysis and Energy Action Plan

After the audit visit to the facility, the data collected should be examined, organized, and reviewed for completeness. In some cases, not all of the data required for comprehensive evaluations of the energy conservation opportunities can be obtained. The audit team will have to rely on their judgment to fill in missing information. In many cases, it is necessary to make indirect estimations of some quantities. These estimations are based on figures and tables presented in engineering handbooks, manufacturer's literature, and technical periodicals.

The preliminary energy saving opportunities identified during the audit visit should be reviewed. Each energy conservation opportunity must be reviewed to determine whether it is technically applicable to the process. If so, it should further determine what the associated energy savings would be in the particular operation being evaluated. The evaluation procedure consists of calculating energy and mass balances for each item if the energy conservation modification is made. The actual analysis of the equipment or operational changes should be conducted. The cost of the modification must be considered and lifetime economic evaluation must be carried out to determine the profitability of the measure. Economics of the potential energy saving opportunities should be determined. Economical and financial analyses such as simple payback period and discounted benefit–cost ratio are discussed in Chapter 3.

Witte et al. (1987) provide the following guidelines for evaluation:

- Energy savings should be cited separately from the economic saving because it may be desirable to consider energy savings alone.

- It is necessary to consider fuel costs over the lifetime of the project not simply at present levels.
- Various measures of economic and financial performance may be used.

Example 2.2

A food processing plant needs a 10 kW motor for a new process, which will run at a full load for two shifts a day or 4800 h per year. The company has two choices: a standard motor at an efficiency of 85% and a high-efficiency motor at an efficiency of 90%. The high-efficiency motor costs about $150 more than the standard efficiency motor. The electricity price is $0.05/kWh. Determine the simple payback period for the high-efficiency motor.

Solution 2.2

The simple payback period is the initial cost divided by the annual saving. The initial cost for this energy opportunity is the price difference of two motors: $150. The annual saving is 10 kW × 4800 h/year × (1/0.85 – 1/0.9) × $0.05/kWh = $157/year. The simple payback period is thus $150/$157/year = 0.96 years. This is a very attractive energy saving opportunity. More sophisticated economic financial analyses are discussed in Chapter 3.

2.2.5 Energy Audit Report

After the energy consumption data have been collected and analyzed, and the energy action plan for energy savings has been recommended for the facility, an energy management program should be set up to implement the audit recommendations. The report should begin with an executive summary that provides the manager of the facility with a brief of the total savings available and the highlights of each energy saving opportunity. The report should describe the facility that has been audited and provide information on the operation of the facility that is related to energy costs. The energy costs should be analyzed. The recommended energy saving opportunities are then recommended along with the analyses of costs and benefits and the cost effectiveness criterion.

Witte et al. (1988) and Capehart et al. (2006) give an outline for a typical energy audit report:

- Executive summary (the audit procedures, primary results, and a table of energy conservation recommendations)
- Introduction (the concept of the audit, the energy systems of the facility, and the major points of energy use within the facility)
- Energy audit procedures (the general procedures carried out in the audit)
- Plant energy distribution
- Evaluation of energy conservation opportunities
- Recommendation of an energy action plan for project implementation

Appendixes may also be provided to include data compilation, example calculations, and equipment cost estimates and quotations.

2.3 MEASUREMENTS, INSTRUMENTATION, AND DATA COLLECTION

Measurements during audit visit are essential to collect information for identifying energy saving opportunities. Instrumentation plays a crucial role in an energy audit to provide operating data such as temperature, pressure, and flow rate.

2.3.1 Dimension Measurement

A tape is the most basic measuring device to check the dimensions of the walls, ceilings, doors, and windows of a building and the distance between pieces of equipment for determining the length of a pipe for waste heat recovery.

2.3.2 Temperature Measurement

Temperature measuring devices are invaluable in providing information for analyzing energy balances and estimating heat loss. Temperature is also an important parameter to identify waste heat sources for potential heat recovery. A handheld temperature indicator with surface or immersion probes, as shown in Figure 2.2a, can be used to determine temperatures (Omega Engineering, Inc., Stamford, Connecticut). For surfaces that are inaccessible, the radiation-type (e.g., infrared) thermometer (Omega Engineering, Inc., Stamford, Connecticut) shown in Figure 2.2b is very useful. However, radiation thermometers tend to be less accurate at relatively low temperatures. Their measurement depends on the radiant emissivity of the surface, which may vary widely with surface conditions and composition. It is necessary to calibrate radiation thermometers by measuring the temperature of an easily accessible and similar surface with both the radiation thermometer and a handheld surface probe.

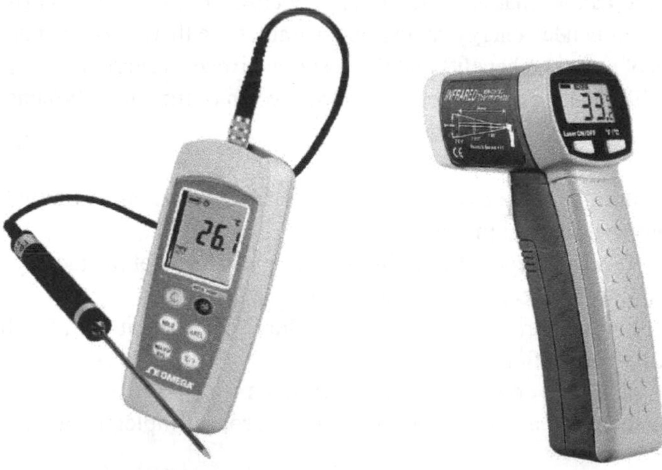

(a) Handheld temperature indicator and probes

(b) Handheld infrared thermometer

FIGURE 2.2 Temperature measuring devices.

FIGURE 2.3 Bourdon pressure gauge.

2.3.3 Pressure Measurement

Pressure in steam and compressed air lines is an important parameter to determine their energy contents. The most common means of measuring pressure is the Bourdon gauge (Omega Engineering, Inc., Stamford, CT) as shown in Figure 2.3. This type of pressure gauge is usually inexpensive and available for pressures ranging from low vacuum to high positive pressure. Appropriate mounting ports must be provided in the lines to be measured.

2.3.4 Fluid and Fuel Flow Measurement

Air, steam, and water are the three main processing fluids used in food processing facilities. Measuring fluid flow rate and fluid velocity is one of the difficult challenges in auditing and monitoring energy consumption because most fluid flow meters are relatively expensive and require rather elaborate installation. Airflow measurement devices include a velometer, an anemometer, and an airflow hood. A typical anemometer is shown in Figure 2.4(a). Steam flow and water flow to the processing equipment can be measured with the use of orifice plates mounted in the fluid lines. The orifice plate as shown in Figure 2.4(b) (Flowline Manufacturing Ltd, Herts, United Kingdom) measures the flow rate based on a pressure differential created across a built-in calibrated plate.

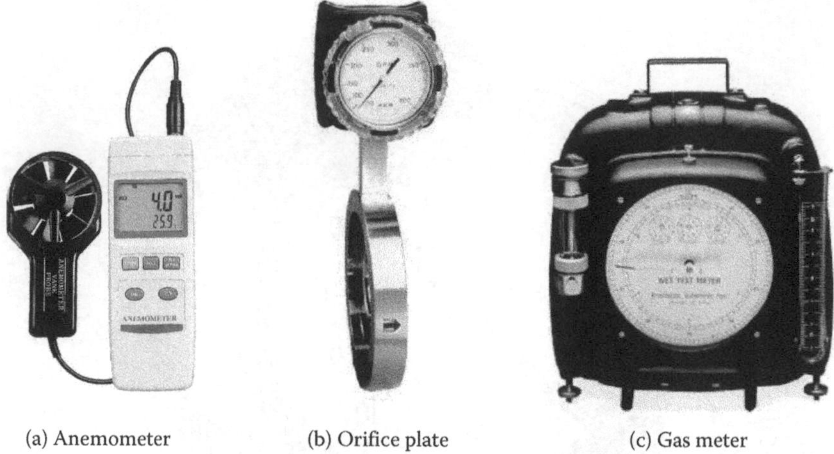

(a) Anemometer (b) Orifice plate (c) Gas meter

FIGURE 2.4 Fluid and fuel flow measuring devices.

Because of the hazardous nature of gaseous and liquid fuels, meters to measure these fuels must be constructed and installed with particular attention to safety. Positive displacement meters as shown in Figure 2.4c (Fisher Scientific, Pittsburg, Pennsylvania) are most commonly used. The working principle is that at each rotation of the lobe or vane, a fixed volume of liquid or gas is passed through the meter. This volume can be related to the energy content of the fuel if its heating value and density are known.

2.3.5 Combustion Gas Composition Measurement

In the food industry, combustion equipment such as boilers accounts for the use of high levels of energy. In evaluating the efficiency of combustion devices, it is necessary to determine the composition of the exhaust gases, particularly their oxygen, carbon dioxide, and carbon monoxide content. The Orsat analyzer has traditionally been used for determining the constituents of combustion flue gas. An electronic combustion gas analyzer as shown in Figure 2.5 (E. Instruments Group, LLC, Langhorne, Pennsylvania) is much more convenient and provides a direct meter reading of the percentage of oxygen and combustibles, usually calibrated in terms of carbon monoxide, in the gas.

2.3.6 Electrical Measurement

Electrical measurements in an energy audit are to determine the energy consumption in electrical devices and the operating characteristics such as voltage and power factors of energy-using equipment. Voltmeters and wattmeters/power factor meters are widely used in electrical measurement. A voltmeter is useful for determining the operating voltages of electrical equipment. The most versatile instrument is a combined volt-ohm-ammeter as shown in Figure 2.6(a) (Fisher Scientific, Pittsburg, Pennsylvania). The electrical consumption by various processing equipment can be measured with the use of electric wattmeters or transducers. A portable handheld

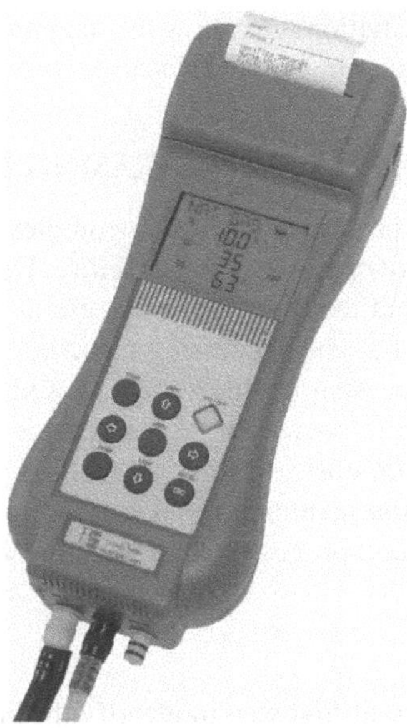

FIGURE 2.5 Flue gas analyzer.

wattmeter and power factor meter as shown in Figure 2.6(b) (ATEQUIP.com, San Diego, California) are very useful for determining the power consumption and power factor of individual motors and other inductive devices and the load factors of motors.

2.3.7 Light Measurement

A light meter is used to measure illumination levels in facilities. A light meter reads in foot candles. Many areas in buildings and plants are significantly overlighted.

(a) Volt-ohm-ammeter

(b) Wattmeter

FIGURE 2.6 Electrical measurement devices.

Measuring the excess illumination will allow the recommendation for a reduction in lighting levels through lamp removal or replacement.

2.4 ENERGY AUDIT IN FOOD PROCESSING FACILITIES

The energy audit for a food processing facility is complex because of the tremendous variety of equipment and unit operations in the facility. Therefore, the auditors should have the ability to understand the specialized processing equipment and operations. Examples of energy audits in food processing plants are provided by Singh (1986). Different existing food processing facilities are reviewed in Chapters 12 through 16. In order to come up with an improvement to the existing process and equipment for saving energy and production costs, the auditors should also have knowledge about available energy conservation technologies, discussed in Chapters 4 through 9 and emerging energy efficient food processes, discussed in Chapters 17 through 22.

2.5 SUMMARY

Energy audits are an important first step to identify the energy saving opportunities in a food processing facility. An energy audit starts with the review of past energy consumption data. Audit tours are necessary to collect detailed data on the facility. The measurement can identify the most significant sources of energy loss in a food processing facility or unit operation. After data review and measurement during energy auditing, it is important to identify measures to decrease the energy loss and to evaluate the energy savings and profitability potential in implementing changes. Finally, an attractive energy saving program or an energy action plan, which comes out through the systematic analyses of the data and energy savings potentials, is recommended and implemented. To conduct an energy audit and come up with an effective energy action plan for a food processing facility, it is necessary to have some knowledge about the food processing equipment and unit operations, energy conservation technologies, and emerging energy efficient processes.

REFERENCES

Capehart, B.L., M.B. Spiller, and S. Frazier. 2006. Energy auditing, Chapter 3, In *Energy Management Handbook* (6th ed.), Turner, W.C. and S. Doty, (Eds.), pp. 23–39. Lilburn, GA: The Fairmont Press Inc.

Capehart, B.L., W.C. Turner, and W.J. Kennedy. 2005. *Guide to Energy Management* (5th ed.), Lilburn, GA: The Fairmont Press.

Singh, R.P. 1986. Energy accounting of food processing operations, Chapter 3, In *Energy in Food Processing*. Singh, R.P. (Ed.), pp. 19–68. New York: Elsevier Science Publishing Company Inc.

Witte, L.C., P.S. Schmidt, and D.R. Brown. 1988. *Industrial Energy Management and Utilization*. Washington, DC: Hemisphere Publishing Corporation.

3 Energy Project Management in Food Processing Facilities

3.1 INTRODUCTION

The food processing industry is a large consumer of raw materials, which are used both as feedstock to produce numerous food products and as an energy source to derive its various processes. An energy conservation and management project can lower operating or production costs by reducing the energy consumption and improve the sustainability of processing systems by reducing energy consumption and using the waste streams as an alternative energy source.

Improvement of energy efficiency is based on better management and upgrading of equipment and production processes. Initiation of an energy project starts with energy analysis, which usually requires data from energy audit as discussed in Chapter 2. We should also have some degree of knowledge of new equipment and technologies as discussed in Chapters 4 through 9, and novel processes as discussed in Chapters 17 through 22 that can replace existing ones to save energy and reduce energy costs.

An energy conservation project usually requires a capital investment. Therefore, it is important to conduct economical analysis and quantify the returns of the project in order to make decisions to implement the project. Financing can be a key success factor for an energy conservation project. The investment analyses thus should include both the economic analysis and financial analysis. Adaptation of an energy conservation technology into a food processing facility requires a thorough understanding of the technical and economical performance of the technology. The aim of this chapter is to bring together in a concise form the basic principles of energy and sustainability analyses, economical analysis, and financial arrangements for an energy conservation project.

3.2 ENERGY ANALYSIS

3.2.1 Mass and Energy Flows in a Food Processing Facility

Typical mass and energy flows in a food processing facility are given in Figure 3.1 (Muller et al., 2007). Horizontal flows represent the transformation of raw materials into food products and by-products. Vertical flows represent energy and water. Maximizing the horizontal flows and minimizing the vertical flows will minimize the production costs and environmental impact.

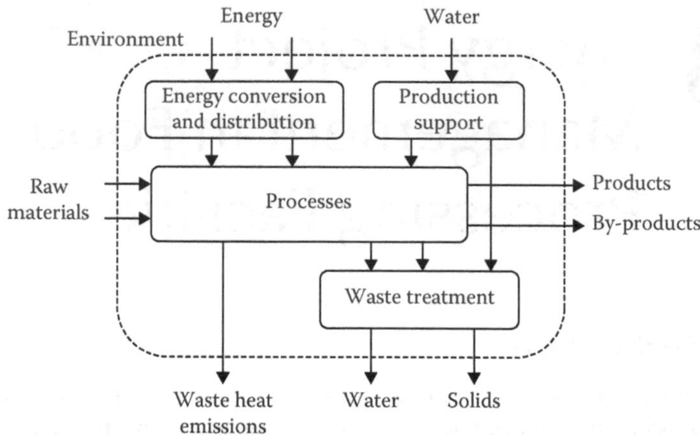

FIGURE 3.1 Typical mass and energy flows in a food processing facility. (Reprinted from Muller, D.C.A., Marechal, F.M.A., Wolewinski, T., and Roux, P.J., *Appl. Thermal Eng.*, 27, 2677, 2007. With permission.)

Two methods are usually used to analyze the energy consumption of a food processing facility: top–down and bottom–up approaches. Using the top–down approach, the total energy consumption on energy bills is allocated among the different users in the facility to identify the main energy users. In this framework, multiple linear regressions can be used to define the relationship between dependent variables, such as energy consumption, and independent variables, such as production volumes and ambient temperature (Vogt, 2004). Unlike the top–down approach, the bottom–up approach is to model the energy consumption of different unit operations in the facility to recalculate the global energy consumption by summing up the energy consumption of individual unit operations (Marechal et al., 1997; Hostrup et al., 2001).

3.2.2 Energy Sources and Their Quality

Electricity, steam, compressed air, and fuels are common energy carriers in food processing facilities. Energy carriers can be characterized by their energy content and quality. With respect to energy content, 1 kWh of electricity has an energy equivalent of 3.6 MJ of steam and fuels. For a thermal energy source, the energy content can be defined by

$$\mathrm{En} = H = \dot{m}h \tag{3.1}$$

where

h is the specific enthalpy of the substance (J/kg)
$\dot{m}$ is the mass or mass flow rate (kg/s)

Energy consumption is an energy conversion process. For example, a fossil power generation plant involves the energy conversion from chemical energy in fossil fuels to thermal energy in steam, to mechanical energy via steam turbines, and to electricity via electrical generators. During each conversion step, part of the energy is lost. Exergy is defined as the maximum amount of work that can be produced by a system as it reaches an equilibrium with a reference environment (Dincer and Cengel, 2001; Dincer, 2000). Exergy values depend on the reference state or the dead state of the energy carrier. The reference temperature and pressure are usually T_0 = 25°C and P_0 = 101,325 Pa used in the chemical industry.

Energy sources can be graded in quality. The thermodynamic quality of an energy carrier is defined as the ratio of exergy to enthalpy:

$$\Theta = \frac{\text{Ex}}{\text{En}} \tag{3.2}$$

where Ex and En are exergy and energy contents of the energy carrier, J.

For electricity, since the values of energy and exergy inputs are the same, the quality of electricity is thus

$$\Theta = 1 \tag{3.3}$$

That is, all energy stored in electricity can be used to generate work or power if the efficiency of equipment such as a motor is 100%.

When processing heat flows from a high temperature to a lower temperature, part of the energy stored in the processing heat is converted to work. The amount of work produced depends on the temperature of the heat source and the environmental temperature of the heat source at its dead state (or reference temperature). The amount of work produced is determined by

$$\Theta = 1 - \frac{T_0}{T} \tag{3.4}$$

where

T is the temperature of the heat source in Kelvin scale

T_0 is the reference temperature in Kelvin scale

Example 3.1

Steam at 350°C is used to produce mechanical power (or work) via a steam turbine. How much energy in the steam can be converted to power if the final temperature of the steam exiting from the turbine is 100°C? If the final temperature is the same but the initial steam temperature is 200°C, how much energy in the steam can be converted to power?

Solution 3.1

From Equation 3.4, the quality or the available work in the steam at 350°C is

$$\Theta = 1 - \frac{T_0}{T} = 1 - \frac{273.15 + 100}{273.15 + 350} = 0.40$$

This means only 40% of the energy stored in the steam at 350°C can be converted into mechanical power if the mechanical efficiency of the turbine is 100%.

If the steam decreases to 200°C, the quality or the available work in the steam is

$$\Theta = 1 - \frac{T_0}{T} = 1 - \frac{273.15 + 100}{273.15 + 200} = 0.21$$

This means only 21% of the energy stored in the steam at 200°C can be converted into mechanical power if the mechanical efficiency of the turbine is 100%.

High-grade or high-quality energy sources provide more organized forms of energy than low-grade sources. The energy sources that have a high grade include kinetic energy of moving matter, gravitational potential energy, and electrical energy, which can be converted into another energy form with small energy losses. Chemical energy has an intermediate grade, followed by high-temperature heat and low-temperature heat. From Example 3.1, if the electrical efficiency of an electrical motor and the mechanical efficiency of a steam turbine are 100%, 1 J of electricity, 2.5 J of steam at 350°C, and 4.76 J of steam at 200°C are required to provide 1 J of mechanical power or work, respectively. However, if they are used to provide processing heat, they have the same amount of energy equivalent in terms of energy content. Therefore, direct use of high-quality electricity to generate heat does not coincide with the concept of sustainability. Electrical energy is measured by its ability to do work. The traditional unit of measure of electrical energy is kilowatt-hour. One kilowatt-hour (kWh) is equivalent to 3412 Btu or 3.6 MJ. Electricity is usually generated by steam turbines with steam generated by boilers fired by fossil fuels. During electricity generation, a large amount of the energy stored in fuels is lost to produce electricity and losses also occur during transmission of electricity from one location to another. Thus about three times of equivalent fuel energy is required to produce electricity. This means that about 10.8 MJ of fuel energy is required to produce 3.6 MJ or 1 kWh of electricity.

3.2.3 Energy and Exergy Flow in a Food Processing Facility

Traditionally, the increase in energy efficiency is accomplished mainly by the integration of complex heat flows within processing facilities. Integration requires considerable capital investment. However, the production routes and the selection of raw materials and energy sources also largely determine the overall thermodynamic efficiency of a processing facility. Thus making a wise choice of production routes, raw materials, and energy carriers is required to significantly improve energy efficiency and process economics simultaneously.

According to the first law of thermodynamics, the sum of anergy (lost energy) and exergy (available energy) is constant. Exergy eventually ends up in the desired products (Baehr, 1988). Exergy can always be converted into anergy while anergy can never be converted into exergy. In real processes, part of exergy is lost. More exergy enters the process than leaves it. Therefore, it is important to delay the degradation of exergy as long as possible in a processing facility. Exergy may be lost in two ways: internal energy losses due to the irreversibility of the process itself and external energy losses in the waste streams that do not reach equilibrium with the environment. The exergy destruction due to irreversibility is

$$\text{Ex}_{\text{destruction}} = \text{Ex}_{\text{in}} - \text{Ex}_{\text{out}} = T_0 \Delta S \tag{3.5}$$

where
T_0 is the environmental temperature in Kelvin
ΔS is the generated entropy during conversion

The exergy stored in the waste steams is determined by

$$Ex_{waste} = Ex_{out} - Ex_{product} \quad (3.6)$$

Example 3.2

Hot air at 70°C, relative humidity of 15%, and mass flow rate of 0.01 kg/s is used to dry a food product. What are the energy rate and exergy rate of air into the dryer? (Note: the equations used to calculate energy content and exergy are given in Chapter 1)

Solution 3.2

1. Using the psychrometric chart (Singh and Heldman, 2001), the specific humidity ratio at the bulb temperature of 70°C and relative humidity of 15% is found at $w = 0.03$ kg moisture/kg dry air.
2. The specific heat of dry air and water vapor at 70°C are $c_a = 1.018$ kJ/kgK and $c_v = 1.958$ kJ/kgK. Therefore, the specific heat of dry air is

$$c = c_a + \omega c_v = 1.018 + 0.03 \times 1.958 = 1.0767 \text{ kJ/kgK}$$

3. Using $T_0 = 25$°C as a reference temperature, the enthalpy of the air is

$$h = c(T - T_0) = 1.0767 \times (70 - 25) = 48.45 \text{ kJ/kg}$$

and the specific exergy of the air is

$$\psi = c\left(T - T_0 - T_0 \ln\frac{T}{T_0}\right) = 1.0767 \times \left[(70 - 25) - (273.15 + 25)\ln\frac{273.15 + 70}{273.15 + 25}\right]$$
$$= 3.326 \text{ kJ/kg}$$

The energy and exergy rates into the dryer are

$$\text{En} = \dot{m}h = 0.01 \times 48.45 = 0.4845 \text{ kW}$$

and

$$\text{Ex} = \dot{m}\psi = 0.01 \times 3.326 = 0.03326 \text{ kW}$$

From the above example, we can see that although the energy rate into the dryer is high, the exergy rate into the dryer is very low because the quality of drying air is low. However, if electricity is used to dry the same product, the energy and exergy rates into the dryer should be the same since the quality of electricity is high ($\Theta = 1$). Since exergy eventually ends up in the products, hot drying air is more suitable for use as an energy carrier in a dryer in terms of the lower exergy to achieve the same drying effect.

3.2.4 Energy and Exergy Efficiencies in a Food Processing Facility

When energy is converted from one form to another, the useful output is usually not equal to the input and some energy is always lost. The ratio of the useful output to the input is called the efficiency of the process. The first law of thermodynamics, or the principle of energy conservation, is usually used to trace the flow of energy through a processing system to determine the primary energy input needed to produce a given amount of product. Energy efficiency (the first law efficiency) can be defined as

$$\eta_{en} = \frac{\sum En_{out}}{\sum En_{in}} \times 100\% \tag{3.7}$$

Since part of the energy output is in the form of waste heat, a more practical definition for energy efficiency is

$$\eta_{en} = \frac{\sum En_{product,\ or\ useful}}{\sum En_{in}} \times 100\% \tag{3.8}$$

The energy efficiency of the whole U.K. energy system was about 69% during 1965–1995 (Hammond, 2004).

The analysis of energy efficiency based on the first law of thermodynamics does not take into account the thermodynamic quality of energy sources. The second law of thermodynamics facilitates the assessment of the maximum amount of work achievable in a given system with different energy sources. Exergy is the available energy for the conversion from an energy source to work with a reference environmental condition. Therefore, exergy represents the thermodynamic qualities of an energy carrier and heat or energy lost in waste streams. Electricity, for instance, has a high quality or exergy while low-temperature hot water has a low quality. Exergy efficiency (the second law efficiency) is an important parameter, which is given by

$$\eta_{ex} = \frac{\sum Ex_{out}}{\sum Ex_{in}} \times 100\% \tag{3.9}$$

Exergy analysis is a very useful tool, which can be used to provide useful information for the improvement of the design and operation of an energy system. Exergy analysis can identify the magnitudes and locations of exergy destruction in the whole system. In recent years, exergy analysis has been widely used for performance evaluation of a thermal system (Dincer et al., 2005; Utlu and Hepbasli, 2006; Kuzgunkaya and Hepbasli, 2007a,b). Fuels and electricity are two main sources used in food processing facilities. The energy and exergy efficiencies for heating, cooling, and work production processes using electricity and fuels are listed in Table 3.1 (Dincer et al., 2005).

TABLE 3.1
Energy and Exergy Efficiencies for Heating, Cooling, Work Production, and Kinetic Energy Production Processes

Process	Energy Efficiency	Exergy Efficiency
Electrical heating	$\eta_{en} = \frac{Q_p}{E_e}$	$\eta_{ex} = \frac{Ex_{Q_p}}{Ex_{E_e}}$
		$\eta_{ex} = \left[1 - \frac{T_0}{T_p}\right]\frac{Q_p}{E_e}$
		$\eta_{ex} = \left[1 - \frac{T_0}{T_p}\right]\eta_{en}$
Fuel heating	$\eta_{en} = \frac{Q_p}{m_f H_f}$	$\eta_{ex} = \frac{Ex_{Q_p}}{m_f H_f \lambda_f}$
		$\eta_{ex} = \left[1 - \frac{T_0}{T_p}\right]\frac{Q_p}{m_f H_f \lambda_f}$
		$\eta_{ex} = \left[1 - \frac{T_0}{T_p}\right]\eta_{en}$
Electrical cooling	$\eta_{en} = \frac{Q_p}{E_e}$	$\eta_{ex} = \frac{Ex_{Q_p}}{Ex_{E_e}}$
		$\eta_{ex} = \left[1 - \frac{T_0}{T_p}\right]\frac{Q_p}{E_e}$
		$\eta_{ex} = \left[1 - \frac{T_0}{T_p}\right]\eta_{en}$
Shaft work production via electricity	$\eta_{en} = \frac{W_p}{E_e}$	$\eta_{ex} = \frac{Ex_{W_p}}{Ex_{E_e}} = \frac{W_p}{E_e} = \eta_{en}$
Shaft work production via fuel	$\eta_{en} = \frac{W_p}{m_f H_f}$	$\eta_{ex} = \frac{Ex_{W_p}}{m_f H_f \lambda_f} = \frac{\eta_{en}}{\lambda_f}$

Source: Adapted from Dincer, I., Hussain, M.M., and Al-Zaharnah, I., *Energy Policy*, 33, 1461, 2005. With permission.

Note: λ_f, the exergy grade function; H_f, higher heating value of fuels (kJ/kg); Q_p, heat transfer to product (kJ); W_p, shaft work (kJ); E_e, electricity (kJ); m_f, mass of fuels (kg); T_0, temperature of reference environment (K); T, temperature for energy conversion (K).

Example 3.3

Following Example 3.2, hot air at 70°C, relative humidity of 15%, and mass flow rate of 0.01 kg/s is used to dry 0.5 kg/h of food product from an initial moisture content of 65% to a final moisture content of 15% on a wet basis at 50°C. What are the energy and exergy efficiencies of the drying process?

Solution 3.3

1. Latent heat of water evaporation at 50°C is $h_{fg} = 2382.8$ kJ/kg.
2. Calculate the moisture removal rate, $\dot{m}_w$:

$$0.5 \times 65\% = \dot{m}_w + (0.5 - \dot{m}_w) \times 15\%$$

There is

$$\dot{m}_w = 0.294 \text{ kg/h}$$

3. Determine the heat transfer rate and exergy transfer rate for moisture evaporation:

$$q_{ev} = \dot{m}_w h_{fg} = 0.294 \text{ [kg/h]} \times \left(\frac{1 \text{ [h]}}{3600 \text{ [s]}}\right) \times 2382.8 \text{ [kJ/kg]} = 0.195 \text{ kW}$$

and

$$\begin{aligned} \text{Ex}_{ev} &= \left(1 - \frac{T_0}{T}\right) q_{ev} = \left(1 - \frac{273.15 + 25}{273.15 + 50}\right) \times 0.195 \\ &= 0.0151 \text{ kW} \end{aligned}$$

4. From Example 3.2, the energy and exergy rates into the dryer are $\text{En}_{in} = 0.4845$ kW and $\text{Ex}_{in} = 0.03326$ kW. Therefore, the energy and exergy efficiencies of the drying process are

$$\begin{aligned} \eta_{en} &= \frac{\sum \text{En}_{out}}{\sum \text{En}_{in}} \times 100\% \\ &= \frac{0.195 \text{ [kW]}}{0.4845 \text{ [kW]}} \times 100\% \\ &= 40.25\% \end{aligned}$$

$$\begin{aligned} \eta_{ex} &= \frac{\sum \text{Ex}_{out}}{\sum \text{Ex}_{in}} \times 100\% \\ &= \frac{0.015 \text{ [kW]}}{0.03326 \text{ [kW]}} \times 100\% \\ &= 45.1\% \end{aligned}$$

3.3 SUSTAINABILITY IN THE FOOD INDUSTRY

3.3.1 Sustainable Food Industry

The food processing industry depends heavily on raw materials and energy. Most of the energy sources used in food processing facilities have a limited availability. The food industry is under increasing pressure from governments and environmental groups to improve the sustainability of its process and create sustainable business practices. Gerbens-Leenes et al. (2003) used three indicators that address global environmental sustainability of a food production system: (1) the total energy from both fossil and renewable sources, (2) the water, and (3) land requirements per kilogram of available foods. This method can be used to compare the trends of production sustainability over time.

Energy resources and their utilization are intimately related to the sustainable development of the food industry. Hundreds of research projects have been conducted around the world with the general goals of (1) improving the thermodynamic efficiency of energy systems and (2) developing alternatives to the fossil sources used in the food industry. One of the major factors undermining the sustainability of a food production process is the depletion of resources it uses. The difference between renewable and nonrenewable resources is that renewable resources are created at least as fast as they are consumed while nonrenewable resources are consumed faster than they are created (de Swaan Arons et al., 2004).

Three parameters can be used to characterize the sustainability of a process (de Swaan Arons et al., 2004). These three parameters are

- Thermodynamic efficiency of the process
- Extent of the use of renewable resources
- Extent of a process cycle to be closed

3.3.2 Improving Energy Efficiency for Sustainability

Energy conservation is vital for sustainable development. Reduced energy consumption through conservation can benefit not only energy consumers by reducing their energy costs, but also the society. The most direct benefit that the society has from the improvement of energy efficiency is reduction in the use of energy resources and the emission of many air pollutants such as CO_2. To develop a sustainable society, much effort must be devoted not only to discovering renewable energy resources, but also to increasing the energy efficiency of devices and processes utilizing these resources.

A significant reduction in consumer energy costs can occur if conservation measures are adopted appropriately. The payback period for many energy conservation projects is less than 2 years. Energy conservation involves the improvement of energy efficiency, the formulation of pricing policies, good industrial practices, and good load management strategies. Among these measures, pricing policies play a key role in sustainable development. The improvement of energy efficiency will not automatically reduce energy use and thus improve sustainability. The improvement of energy efficiency may lower the price of energy sources and make them more affordable, which could increase the use of the energy resources and decrease the sustainability (Herring, 2006).

The following factors may limit the efforts of energy conservation projects (Dincer and Rosen, 1999):

- Lack of availability, reliability, and knowledge of energy efficient technologies
- Lack of expertise in appropriate technical input, financial support, and proper program design and monitoring
- Lack of explicit financing mechanisms
- Inappropriate program management practices and staff training
- Inappropriate pricing of electricity and other energy commodities
- Lack of appropriate information

Various energy conservation technologies are discussed in Chapters 4 through 9.

3.3.3 Combining Energy Efficiency and Renewables for Sustainability

All unit operations associated with food processing facilities are ultimately limited by our ability to supply useful energy to the processes. It is not enough to operate processes efficiently. Improvement of thermodynamic efficiency is not the only contribution to sustainability. The improvement of thermodynamic efficiency may lower the consumption rate of nonrenewable natural resources. In terms of the sustainability of the process, the net exergy entering the process to make it proceed has to come from renewable resources. This means the technology itself should be sustainable. The food processing facilities driven by nonrenewable resources should be transformed into those based on renewable resources. The food industry can adapt the zero-waste concept in its processing facilities. The main idea of the zero-waste concept is that the waste generated in one industry could become a raw material for other industries (Pauli, 1998). Each food processing facility produces a large mass of waste streams, which can be re-used or recycled. The conversion of food processing wastes into energy products is discussed in Chapters 23 through 28.

3.4 ECONOMIC ANALYSIS

3.4.1 Capital Investment Characteristics

Investments can be broadly categorized into capital costs and operating costs. Capital investments are generally more strategic and have long-term effects. Decisions regarding capital investments are usually made at higher management levels within a company with the consideration of additional tax consequences as compared to the operating costs. Capital investments have four unique characteristics

- Large initial costs
- Long recovery period for the investment
- Irreversible procedure
- Significant tax implications depending on the choice of financing methods

A capital investment usually requires a relatively large amount of money. Sometimes, the funds available for the capital investment of an energy conservation project in a company are limited. That is, the initial capital costs exceed the total available funds. This creates a situation known as capital rationing. A capital investment is recovered over a period of years. The period between the initial installation and the last future cash flow is the life cycle or the life of the investment. Since the benefits of the investment must be realized over the time when the energy conservation project is implemented, it is necessary to consider the time values of money to properly evaluate the feasibility of the investment. Furthermore, once the capital investment is made, it is very hard to retract. Therefore, careful attention should be paid to the economic analysis of a potential energy conservation project.

Capital investment is usually a function of the capacity or the scale of the project. The relationship between a capital cost and plant size is usually estimated by two methods: economic estimation using actual survey data and direct calculation from synthetic engineering data. The power function is the standard estimation function to determine capital investment at different scales. The main advantage of the power function is that estimates can be provided where no operating process exists (Gallagher et al., 2005). The power function is convenient for use in estimation because plant construction costs, K, can increase more or less proportionately with plant capacity, Q (Henderson and Quandt, 1980):

$$K = AQ^n \tag{3.10}$$

The power function is often referred to as the "0.6 factor rule," which says that a 1% expansion in processing capacity yields a little less than 0.6% increase in capital costs (Ladd, 1998). In the chemical processing industry, the factor ranges from 0.4 to 0.9 (Peters and Timmerhaus, 1991). In the dry mill ethanol plant, the estimated power factor is 0.836 and the capital costs of a dry mill ethanol plant, therefore, increase more rapidly than the average for all processing plants with an average power factor of 0.6 (Gallagher et al., 2005).

3.4.2 Time Value of Money

Interest and inflation are two primary factors associated with the time value of money. Interest is the ability to earn a return on money if it is loaned rather than consumed. There are two kinds of interest: simple interest and compound interest. Under simple interest, interest is earned only on the original amount loaned. Under compound interest, interest is earned on the original amount loaned and any interest accumulated from the previous periods.

The formulas used to calculate future value from present value based on simple interest and compound interest are given by, respectively,

$$F_n = P(1+ni) \tag{3.11}$$

$$F_n = P(1+i)^n \tag{3.12}$$

where

F_n is the future amount of money at the end of the nth year
P is the present amount of money
i is the interest
n is the number of years between P and F_n

Compound interest is more commonly used in practice than simple interest. The factor $(1 + i)^n$ is known as the single sum, future worth factor or the single payment, compound amount factor. Table 3.2 summarizes the time value of money factors.

Example 3.4

Determine the balance that will accumulate at the end of year 30 in an account that pays a 6%/year compound interest if a monthly deposit of $675 is made today.

Solution 3.4

This is a uniform series cash flow problem. From Table 3.2, we have

$$(F|A,i,n) = \frac{(1+i)^n - 1}{i}$$

where

n = 30 years × 12 month/year = 360 month
i = 6%/12 = 0.5%

TABLE 3.2
Summary of Discrete Compounding Time Value of Money Factors

To Find	Given	Factor	Symbol	Name
P	F	$(1 + i)^{-n}$	$(P\|F,i,n)$	Single payment, present worth factor
F	P	$(1 + i)^n$	$(F\|P,i,n)$	Single payment, compound amount factor
P	A	$\frac{(1+i)^n - 1}{i(1+i)^n}$	$(P\|A,i,n)$	Uniform series, present worth factor
A	P	$\frac{i(1+i)^n}{(1+i)^n - 1}$	$(A\|P,i,n)$	Uniform series, capital recovery factor
F	A	$\frac{(1+i)^n - 1}{i}$	$(F\|A,i,n)$	Uniform series, compound amount factor
A	F	$\frac{i}{(1+i)^n - 1}$	$(A\|F,i,n)$	Uniform series, sinking fund factor
P	G	$\frac{1-(1+ni)(1+i)^{-n}}{i^2}$	$(P\|G,i,n)$	Gradient series, present worth factor
A	G	$\frac{(1+i)^{-n} - (1+ni)}{i[(1+i)^{-n} - 1]}$	$(A\|G,i,n)$	Gradient series, uniform series factor

Source: Reprinted from Pratt, D. *Energy Management Handbook* (6th ed.), The Fairmont Press Inc., Georgia, 2006. With permission.

$$F = A \times (F|A,i,n) = 675 \times \frac{(1+0.005)^{360} - 1}{0.005} = 0.678 \text{ millions dollars}$$

Example 3.5

Determine the equal monthly withdrawals that can be made for 30 years from an account with 0.678 million dollars available at the end of year 30 given in Example 3.4. The account continues to be paid at a compound interest of 6%/year. The first withdrawal is to be made 1 month after the end of year 30.

Solution 3.5

From Table 3.2, we have

$$(A|P,i,n) = \frac{i(1+i)^n}{(1+i)^n - 1}$$

where

n = 30 [years] × 12 [month/year] = 360 month
i = 6%/12 = 0.5%
P = $678,000

$$A = P \times (A|P,i,n) = 678{,}000 \times \frac{0.005(1+0.005)^{360}}{(1+0.005)^{360} - 1} = \$4{,}065$$

Example 3.6

Determine the equal monthly payment for a 15 year mortgage loan of $170,000 at an annual fixed interest of 5.25%.

Solution 3.6

From Table 3.2, we have

$$(A|P,i,n) = \frac{i(1+i)^n}{(1+i)^n - 1}$$

where

n = 15 [years] × 12 [month/year] = 180 month
i = 5.25%/12 = 0.4375%

$$A = P \times (A|P,i,n) = 170{,}000 \times \frac{0.004375(1+0.004375)^{180}}{(1+0.004375)^{180} - 1} = \$1{,}366.6$$

Inflation is a complex subject. Inflation is a rise in the general price level. It can be described as a decrease in the purchasing power of money in general. An inflation rate is published by the government based on the Consumer Price Index (CPI).

Taking into account the inflation effect, the current cash value can be converted to constant cash value by

$$\text{Constant } \$ = (\text{current } \$)/(1+f)^n \tag{3.13}$$

The constant cash value can also be converted to current cash value by

$$\text{Current } \$ = (\text{constant } \$)(1+f)^n \tag{3.14}$$

Example 3.7

Determine the constant cash value for $5000 at the end of 30 years from now if the inflation rate is 3%.

Solution 3.7

Using Equation 3.13, Constant $\$ = (\text{current } \$)/(1 + f)^n = 5000/(1 + 0.03)^{30} = \$2060$

3.4.3 Depreciation and Taxes

Depreciation is a recognition that most assets decrease in value over time. The U.S. federal income tax law permits deductions from taxable income to allow for this value loss of assets. These deductions are called depreciation allowances. A depreciable asset must be held by the business for producing income, wear out, or be consumed in its use and have a life longer than a year. Many methods of depreciation including straight line, sum-of-the-years digits, declining balance, and the accelerated cost recovery system have been allowed under the U.S. tax law over the years. The method currently used for calculating the depreciation of assets placed in service after 1986 is the Modified Accelerated Cost Recovery System (MACRS). The allowable MACRS depreciation deduction for an asset is a function of the property class, the basis value, and the recovery period of the asset. Table 3.3 gives the MACRS property classes and

TABLE 3.3
MACRS Property Classes and Percentages by Recovery Year

Property Class	3 year	5 year	7 year	10 year	15 year
Example assets	Special handling devices for food Special tools for motor vehicle manufacturing	Computers and office machines General purpose trucks	Office furniture Manufacturing machine tools	Assets used in manufacturing of certain food products, Tugs and water transport equipment Petroleum refining assets	Fencing and landscaping Cement manufacturing assets Telephone distribution equipment

TABLE 3.3 (continued)
MACRS Property Classes and Percentages by Recovery Year

Recovery	1	33.33%	20.00%	14.29%	10.00%	5.00%
Year	2	44.45%	32.00%	24.49%	18.00%	9.50%
	3	14.81%	19.20%	17.49%	14.40%	8.55%
	4	7.41%	11.52%	12.49%	11.52%	7.70%
	5		11.52%	8.93%	9.22%	6.93%
	6		5.76%	8.92%	7.37%	6.23%
	7			8.93%	6.55%	5.90%
	8			4.46%	6.55%	5.90%
	9				6.56%	5.91%
	10				6.55%	5.90%
	11				3.28%	5.91%
	12					5.90%
	13					5.91%
	14					5.90%
	15					5.91%
	16					2.95%

percentages by recovery year. The basis of an asset is the cost of placing the asset in service. In most cases, the basis includes the purchase cost of the asset plus the costs, such as installation costs, necessary to place the asset in service.

Taxes are primarily designed to generate revenues for governmental entities. Cash flows used for economic analysis should always be adjusted for the combined impact of all relevant taxes. Tax laws and regulations are too complex and intricate to be discussed in detail in this book. The amount of taxes is usually determined based on the tax rate multiplied by the taxable income. In the United States, depending on income ranges, the marginal tax rates vary from 15% to 39% of taxable income as shown in Table 3.4.

Taxable income is calculated by subtracting allowable deduction from gross income. Gross income is generated when a company sells products or services. Allowable deductions include salaries and wages, materials, interest payments, depreciation, and other costs associated with the business as listed in tax regulations. Given the availability of the above information and the Before Tax Cash Flows (BTCF), the procedure to determine the After Tax Cash Flow (ATCF) on a year-by-year basis is given below:

$$\text{Taxable income} = \text{BTCF} - \text{loan and bond interest} - \text{deprecation} \\ - \text{other allowable deductions} \quad (3.15)$$

TABLE 3.4
Federal Tax Rates based on the Omnibus Reconciliation Act of 1993

Taxable Income, X($)	Marginal Tax Rate	Taxes Due
0–50,000	0.15	$0.15X$
50,000–75,000	0.25	$7{,}500 + 0.25(X - 50{,}000)$
75,000–100,000	0.34	$13{,}750 + 0.34(X - 75{,}000)$
100,000–335,000	0.39	$22{,}250 + 0.39(X - 100{,}000)$
335,000–10,000,000	0.34	$113{,}900 + 0.34(X - 335{,}000)$
10,000,000–15,000,000	0.35	$3{,}400{,}000 + 0.35(X - 10{,}000{,}000)$
15,000,000–18,333,333	0.38	$5{,}150{,}000 + 0.38(X - 15{,}000{,}000)$
>18,333,333	0.35	$6{,}416{,}667 + 0.35(X - 18{,}333{,}333)$

Source: Adapted from Pratt, D. *Energy Management Handbook* (6th ed.), The Fairmont Press Inc., Georgia, 2006. With permission.

$$\text{Taxes} = \text{Taxable incomes} \times \text{Tax rate} \tag{3.16}$$

$$\text{ATCF} = \text{BTCF} - \text{total loan and bond payments} - \text{taxes} \tag{3.17}$$

It worth noting that depreciation reduces taxable income and thus taxes but does not directly enter into the calculation of cash flows. Depreciation is an accounting concept to stimulate business by reducing taxes over the life of an asset and it does not involve any cash changes.

3.4.4 Cash Flow Diagrams

For a quantitative analysis of an investment project, the project is usually separated into several elements including

- Initial investment
- Returns on investment
- Economic life
- Salvage value

The initial investment includes the price of equipment and materials and the ancillary costs of the finished project such as transportation, installation, and licensing fees. After an initial investment is made, returns will be achieved over the lifetime of the project. Taxes, operating costs, and maintenance costs must be subtracted from the gross project income. Returns are typically credited at discrete times on a monthly or yearly basis rather than on a continuous day-to-day basis. The economic return on an investment will not last indefinitely. Equipment used for energy conservation and conversion has a physical lifetime. An estimate of the expected economical life is required. Economic life used in economic analysis of an investment project is usually the best estimate of the length of time that equipment can be economically used. The physical lifetime for equipment may be used as the economic life of the equipment.

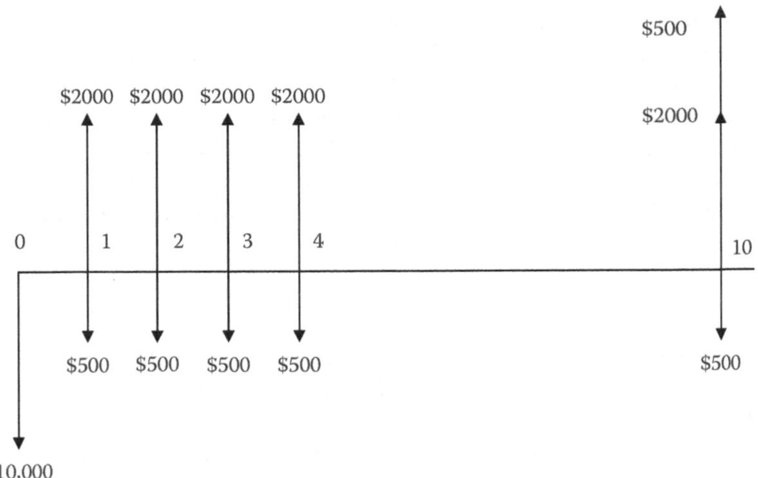

FIGURE 3.2 Cash flow diagram for uniform and discrete returns with a salvage value.

Economic life may be arbitrarily agreed on by contract. The Internal Revenue Service in the United States gives some guidelines for the economic life of assets as shown in Table 3.3. The economic value of the equipment at the end of its economic life must be included in the returns on investment.

The bottom line of an energy conservation project is that the revenues or savings generated by the investment must be greater than the costs involved. The number of years over which the revenues accumulate, the future revenues, the initial and future costs, and the future money value relative to present money value are important factors in making an investment decision for an energy conservation project. A convenient way to display the revenues (or savings) and costs associated with an investment is a cash flow diagram. A cash flow diagram, as shown in Figure 3.2, is to show all cash inflows and outflows plotted along a horizontal time line.

A graphic projection of the details of an investment project or a cash flow diagram can be used to represent the initial investment and the returns on the investment along with the economic life. Although cash flow diagrams are simply graphical representations of revenues and costs, good cash flow diagrams are complete, accurate, and legible. It is usually advantageous to first determine the time frame over which the cash flows occur. The lifetime of the investment is divided into periods, which are frequently, but not always, years. Individual revenues and costs are indicated by drawing vertical lines appropriately placed along the time scale. Upward directed lines indicate cash inflow of revenues or savings while downward directed lines indicate cash outflow of costs. The relative magnitudes of cash inflow and outflow are indicated by the height of the vertical lines.

3.4.5 Economic Evaluation Methods

There are a number of methods for evaluating economic performance. These methods include the simple payback period, life cycle cost method, net benefit or net savings method, benefit/cost ratio method, internal rate of return method, overall rate of return method, and discounted payback method (Singh, 1986; Kreith and West, 1997; Mull, 2001). Usually, several methods are used to provide better understanding of an investment's worth. The simple payback period is commonly used by businesses. However, the primary criterion mandated for assessing the effectiveness of energy conservation investments in federal facilities and many state government facilities is the minimization of life cycle costs. Therefore, it is important to understand the life cycle cost analysis for federal and state facility projects.

3.4.5.1 Simple Payback Period

Simple payback period analysis is also called payback period analysis. It determines the number of years required to recover an initial investment through project returns. The advantages of the simple payback period analysis are that it is simple and easily understood. However, it does not consider the time value of money, costs, or benefits of the investment following the payback period. It is assumed that the lifetime of the project is longer than the simple payback period.

Example 3.8

A heat pump has an initial cost of \$10,000, energy savings of \$25,000 per year, and a maintenance cost of \$500 per year. Determine its simple payback period.

Solution 3.8

The heat pump's simple payback period is

$$\text{SPP} = \frac{\$10{,}000}{\$2{,}500 - \$500} = 5 \text{ years}$$

3.4.5.2 Discounted Payback Method

The discounted payback evaluation method measures the elapsed time between the point of an initial investment and the point at which accumulated savings and net of other accumulated costs are sufficient to offset the initial investment. The shorter the length of time until the investment pays off, the more desirable the investment is. However, a shorter payback time does not always mean a more economically efficient investment. Discounted payback method is often used as a supplementary measure when the project life is uncertain. It is used to identify feasible projects when the investor's time horizon is constrained. It is not a reliable guide for most complex investment decisions where the objective is to choose the most profitable investment alternatives. To determine the discounted payback period, find the minimum solution value of $N_{\min}$ in the following equation:

$$\sum_{n=1}^{N_{\min}} \frac{(B_n - C_n)}{(1+d)^n} = I_0 \tag{3.18}$$

where

B_n is the benefit in period n
C_n is the cost in period n

I_0 is the initial investment cost of an alternative
d is the discount rate
N_{min} is the discounted payback period

3.4.5.3 Benefit to Cost Ratio Method

The benefit to cost ratio method divides the benefits by costs and thus this method gives the measure as a dimensionless number. The higher the ratio is the more dollar savings are realized for every dollar of investment. The ratio method can be used to determine whether or not to accept or reject a given investment on economic grounds. A formula for calculating the ratio of savings to investment costs is given by

$$\text{SCR} = \sum_{n=0}^{N} \frac{S_n / C_n}{(1+d)^n} \tag{3.19}$$

where
SCR is the savings to cost ratio
S_n is the cost saving in year n
C_n is the cost in year n
d is the discount rate

3.4.5.4 Net Benefits or Savings Method

The net benefits or savings method is used to determine the excess of benefits or savings over costs. The time values of all amounts are used. The net benefits method is particularly suitable for decisions made on the basis of long-run profitability. A formula for determining the net benefits from an investment is given by

$$\text{NB} = \sum_{n=0}^{N} \frac{(B_n - C_n)}{(1+d)^n} \tag{3.20}$$

where
NB is the net benefit
B_n is the benefit in year n
C_n is the cost in year n
d is the discount rate

3.4.5.5 Internal Rate of Return Method

The internal rate of return method solves for the interest rate for which dollar savings are just equal to dollar costs over the relevant period. It is the rate for which net savings are zero. Unlike other methods, the internal rate of return method does not require a prespecified discount rate in the computation but solves for a rate. It is a widely used method. The shortcomings include the possibility of no solution or multiple solution values.

The rate of return is typically calculated by a process of trial and error, by which various compound rates of interest are used to discount cash flows until a rate is found for which the net value of the investment is zero. For determining the internal rate of return, a trial interest rate is substituted for the discount rate, d, in Equation

3.20. A positive NB means the rate of return is less than the trial interest while a negative NB means the rate of return is larger than the trial interest. Based on the information, try another rate until you find the rate at which NB approaches zero.

3.4.5.6 Life Cycle Cost Method

A life cycle cost analysis is to quantify costs over the entire life cycle of the project investment. The life cycle cost method sums the costs of acquisition, maintenance, repair, replacement, energy, and other costs such as salvage value that are affected by the investment decision. All amounts are usually measured either in a present value or in an annual value. The time value of money must be taken into account for all amounts over the relevant period. Two important parameters for a life cycle cost analysis are the lifetimes of equipment and the interest rate. Different pieces of equipment may have different lifetimes. The lifetime is usually either found through vendors or estimated based on experience. A common life period must be chosen so that all alternatives are considered over the same time line. If the shortest life is used, a salvage value must be estimated for the longer-lived alternatives. If the longest life is used, shorter-lived alternatives are assumed to be repeatable (Capehart et al., 2006). The interest rate is the value that a company or organization uses for evaluating its investments. This interest is often known as the minimum attractive rate of return (MARR). The value is company or organization specific and should be supplied by the company or organization.

The life cycle cost method is particularly useful for decisions that are made primarily on the basis of cost effectiveness to determine whether a given conservation or renewable energy investment will lower total cost. However, it cannot be used to find the best investment in general. Numerous alternatives may be compared. The alternative with the lowest life cycle cost that meets the investor's objectives and constraints is the preferred investment. A formula for calculating the life cycle costs of each alternative is given by

$$\mathrm{LCC}_n = I_n + E_n + M_n + R_n - S_n \tag{3.21}$$

where

LCC_n is the life cycle cost of alternative n
I_n is the present value of investment costs of alternative n
E_n is the present value of energy costs associated with alternative n
M_n is the present value of nonfuel operating and maintenance cost of alternative n
R_n is the present value of repairing and replacement costs of alternative n
S_n is the present value of resale or salvage value associated with alternative n

Example 3.9

An energy efficient air compressor is proposed by a vendor. The compressor and installation will cost $30,000. It will require $1000 worth of maintenance each year for its life of 10 years. Energy costs will be $6000 per year. A standard air compressor will cost $25,000 and will require $500 worth of maintenance each year. Its energy costs will be $10,000 per year. Suppose the MARR is 10%, would you invest in the energy efficient air compressor based on a life cycle cost analysis?

Solution 3.9

Alternative 1:

$$LCC_1 = \$30,000 + \$6,000(P/A,10,10\%) + \$1,000((P/A,10,10\%) = \$73,012$$

Alternative 2:

$$LCC_2 = \$25,000 + \$10,000(P/A,10,10\%) + \$500((P/A,10,10\%) = \$89,518$$

The decision is to choose the energy efficient air compressor due to its lowest LCC.

3.5 FINANCIAL ANALYSIS

Capital investment requires a source of funds. The process of obtaining funds for capital investment is called financing. Equipment can be purchased with

- Cash on hand
- Loan and government bond
- Capital lease, true lease, or sale of stock
- Performance contract

With performance contracting, a company is not to pay for the equipment itself, but the benefits provided by the equipment. The decision to purchase or utilize equipment partly depends on the company's strategic focus. If the company itself wants to be intricately involved with the energy project, purchasing the equipment could yield the greatest profits. If the company wants to delegate some or all of the responsibility of energy project management, it should use a true lease or a performance contract.

There are two broad sources of financing for purchases: debt financing and equity financing. In many cases the financing for a set of capital investments is obtained by packaging a combination of the capital sources to achieve a desired level of available funds.

Debt financing involves borrowing and utilizing money from a lender, which is to be repaid at a later point in time. The lender does not hold an ownership position within the borrowing organization. The borrower will pay the borrowed funds and accrued interest to the lender for using the money according to a repayment schedule. The interest rate is called the cost of capital. The cost of capital influences the return on investment. If the cost of capital increases, the return on investment decreases. Mortgage loans are an example of debt financing. Loans and bonds are two primary sources of debt capital. One benefit of debt financing is the cost of the debt capital is relatively easy to calculate since interest rates and repayment schedules are usually clearly documented in a legal financing arrangement. Another benefit of debt financing under current U.S. tax law is that the interest payments made by borrowers on debt capital are tax deductible.

Equity financing creates an ownership or equity position for the lender within the borrowing organization. Stocks and retained earnings are two primary sources of equity financing. The cost of the debt capital associated with shares of stocks is calculated with difficulty. In addition, the stock dividend payments are not tax deductible. Retained earnings are the accumulation of annual earnings that a company retains within the company rather than pays out to the stockholders as dividends. Although these earnings are held by the company, they belong to the stockholders.

Example 3.10

Suppose a food processing company needs a new drying system for a specific process in its manufacturing plant. The capital cost including installation of the new system is $1.5 million. The expected equipment life is 10 years. However, the process is only planned to be used for 5 years, after which the drying system will be sold at an estimated market value of $700,000. The drying system is expected to save the company about $500,000/year in energy costs. The company tax rate is 35%. The equipment's annual maintenance and insurance cost is $20,000. The company minimum attractive rate of return (MARR) is 15%. The company does not have $1.5 million to pay for the new system so it considers having a 5 year loan with 20% down payment and an annual interest rate of 12%. Conduct a financial analysis for this company.

Solution 3.10

The net annual savings are the energy saving minus the maintenance and insurance cost. That is $500,000 – $20,000 = $480,000. The recovery period is 10 years and the annual depression is calculated using the MACRS method given in Table 3.4. The annual total payment for the loan is calculated by the formula given in Table 3.3, which is $A = P\times(A|P,i,n) = P\times\frac{i(1+i)^n}{(1+i)^n-1}$. In this formula, P is the present value of the loan, P = 1,200,000; i is the annual interest rate of the load, and i = 12%; n is the period of the load, n = 5 years. The depression and paid interest are tax deductible so the taxable income is the net savings minus the depression and paid interest. The ATCF is the net savings minus the total payments and tax. The financial analysis is given in Table 3.5. The net present value of the project is $295,588.

3.6 ENERGY MANAGEMENT PROJECT PLANNING, IMPLEMENTATION, AND EVALUATION

3.6.1 Energy Management Programs

Fischer et al. (2007) reported that around 57% of industry's primary energy inputs are lost or diverted before reaching the intended process activities. Estimates from several studies indicate that on average, savings of 20% to 30% can be achieved without capital investment, using only procedural and behavioral changes. Industry's energy consumption can be cost-effectively reduced by 10% to 20% through well-structured energy management programs that combine technology, operations management practices, and energy management systems.

A strategic approach to energy management can result in significant energy savings for all types of businesses, including food processors. The "Energy Star" program developed jointly by the U.S. Department of Energy and the Environmental

TABLE 3.5
Financial Analysis for a Loan with a 20% Down-Payment at an APR of 12%

End of Year	Savings	Depression	Payment Principal	Payment Interest	Total	Principal Outstanding	Taxable Income	Tax	ATCF
0					300,000	1,200,000			–300,000
1	480,000	150,000	188,892	144,000	332,892	1,011,108	186,000	65,100	82,008
2	480,000	270,000	211,559	121,333	332,892	799,550	88,667	31,033	116,075
3	480,000	216,000	236,946	95,946	332,892	562,604	168,054	58,819	88,289
4	480,000	172,800	265,379	67,512	332,892	297,225	239,688	83,891	63,217
5	480,000	138,300	297,225	35,667	332,892	0	306,033	107,112	39,996
5′	700,000	552,900					147,100	51,485	648,515
		1,500,000	Net present value at 15%						**295,588**

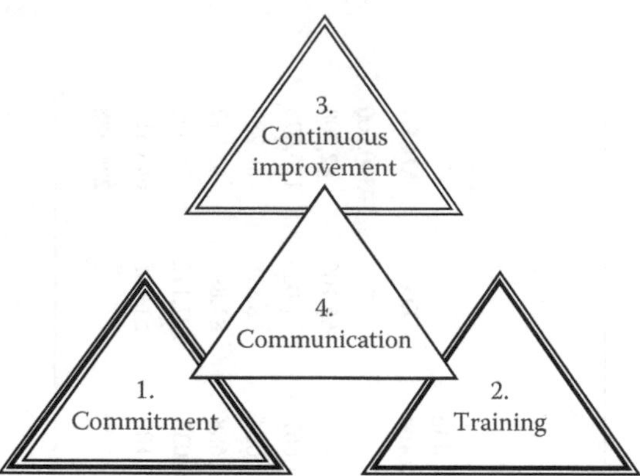

FIGURE 3.3 Energy management diagram. (Reprinted from Center for Industrial Research and Services (CIRAS), Iowa State University. 2005. *Energy-Related Best Practices: A Sourcebook for the Food Industry.* Available at http://www.ciras.iastate.edu/publications/EnergyBP%2DFoodIndustry/. With permission.)

Protection Agency is one of the existing energy management programs. The Energy Star Program has more than 450 industrial participants including Anheuser-Busch, Ben and Jerry's, Cargill, McCain Foods, Sargento Foods, and Weaver Potato Chip Company. Management principles like Lean, Total Quality (TQ), Six Sigma, ISO 9000, and Theory of Constraints have been widely used to achieve high performance in different industries. The energy management program will complement these management principles. Lean management program deals with the issues of overproduction, unnecessary transportation, inappropriate processing, production of defective products, and waiting. However, these issues also involve energy waste. Efforts to manage energy are consistent with reducing waste. TQ, Six Sigma, ISO 9000, and ISO 14000 have been widely used in the food processing industry to improve the quality of all processes. These programs provide many tools to build an effective and high-quality energy management program. The Theory of Constraints is based on the concept that the improvement of a few capacity constraint resources in a production system will have the greatest impact on the bottom line (Caffall, 1988; Witte et al., 1988; Wulfinghoff, 1999; Mull, 2001).

An energy management program requires (1) commitment from leadership, (2) training, (3) continuous improvement through strategic goals and action plans, and (4) communication as shown in Figure 3.3 (CIRAS, 2005).

3.6.2 Commitment from Leadership

An effective energy management program starts with the support and participation of the company's leaders. The leaders should make the energy management program to be part of the corporate strategic plan and associate the program with corporate financial environmental goals. The best way to convince leaders and obtain their commitment is to show them facts and statistics. An initial measurement and assessment of energy performance are necessary to understand energy cost structure and current energy usage and trends including end users and fuel types. An important step in energy management is to determine the exact sources of energy consumption. The end-use patterns can be determined by an energy audit as discussed in Chapter 2. The data, which are collected and tracked, can be used to establish energy standards. Energy productivity index, energy cost index, and productivity standards should be developed to measure whether the energy purchased is productively utilized. Once energy accounting is established and standards are set, management should begin to compare actual performance to the standards.

Company leaders must commit resources including time, talent, and money on an ongoing basis. Since one person may not have all the talent necessary for a successful energy management program, an energy management team should be formed. Accountability should be clearly established. Energy management is an agenda item at all regular leadership meetings. The energy policy is evaluated regularly and updated as needed. Adequate budget is provided annually for effective energy management (CIRAS, 2005).

3.6.3 Training

After top leaders in the organization have made their commitment to an energy management program and after current energy performance has been assessed and standards are set, it is time to provide training for all company personnel. Sharing information and increasing the knowledge level of employees are prerequisites to a successful energy management program. Every employee will need some training on topics such as awareness of the corporate energy policy, current usage and trends, basic energy management terminology, and energy measures. More specific topics in energy management should be provided to smaller identified groups. A training calendar should be established, and all staff should be scheduled for the training they need. Follow-through is critical in establishing awareness of employees to energy management. It is important to determine the effectiveness of the training. Learning objectives in each session should be evaluated. The feedback should be used to make changes to training sessions.

3.6.4 Continuous Improvement

There are two important factors of potential energy projects: (1) the potential impact of a successful project on company finances; and (2) the investment required for implementation. Goals need to be set. These goals should be achievable, measurable, and specific. A good strategy is to start with no- or low-investment energy conservation projects that have moderate or high potential for energy savings. Every food processing facility has a few good energy saving opportunities. Examples may include (Capehart et al., 2006) the following:

- Repairing steam leaks
- Insulating steam, hot water, and other heated fluid lines and tanks
- Installing highly efficient motors

A significant portion of the savings generated by these projects should then be budgeted to finance the investment in more costly projects. No matter what method of financial analysis is used, it is critical to carefully account for not only the savings that come directly from a project, but also any measurable returns that are caused by the project or made possible because of it. We should also consider all impacts including the potential negative impacts when making a strategic choice of projects. The steps needed to achieve improvement should be carefully planned and, at a minimum, should include the following (Capehart et al., 2006; CIRAS, 2005):

- Clear statement of desired outcomes and success measures
- List of resources that are and are not available
- Sequential list of steps involved
- List of key milestones or intermediate indicators of success
- Expected completion date
- Clear explanation of reporting requirements (frequency and scope)
- Rewards if successful (if applicable)

The success of implementation against the established should be evaluated. The evaluation will indicate problems and adjustments should be made to the action plan.

3.6.5 Communication

Company leaders expect and need data that will help them make better decisions. Employees need to see that their efforts are appreciated and that they make a difference. The centerpiece of communication is reporting energy performance. Everyone in the organization should be continuously aware of the current facts and figures on energy performance. It is thus important to provide the appropriate information in the most understandable format to each level throughout the company. Information on energy performance should also be shared with those responsible for planning training. This will help them identify the training that is needed during the ongoing process (CIRAS, 2005).

3.7 SUMMARY

This chapter has discussed energy efficiency analysis, economic analysis, financial analysis and project management for an energy conservation and management program in food processing facilities. An energy management program starts with the energy efficiency and sustainability analysis. Any measure that improves profits or enhances competitive positions through economic and financial analyses is considered effective energy management. A successful energy management program requires (1) commitment from leadership, (2) training at all levels, (3) continuous improvement through strategic goals and action plans, and (4) communication through sharing information on energy performance.

REFERENCES

Baehr, H.D. 1988. *Thermodynamics* (6th ed.), New York: Springer-Verlag.

Caffall, C. 1995. *Learning from Experiences with Energy Management in Industry*. Sittard, the Netherlands: CADDET.

Capehart, B.L., W.C. Turner, and W.J. Kennedy. 2006. *Guide to Energy Management* (5th ed.), Lilburn, GA: The Fairmont Press.

Center for Industrial Research and Services (CIRAS), Iowa State University. 2005. *Energy-Related Best Practices: A Sourcebook for the Food Industry*. Available at http://www.ciras.iastate.edu/publications/EnergyBP%2DFoodIndustry/

Corzo, O., N. Bracho, A. Vasquez, and A. Pereira. 2008. Energy and exergy analyses of thin layer drying of coroba slices. *Journal of Food Engineering* 86: 151–161.

De Swaan Arons, J., H. van der Kooi, and K. Sankaranarayanan. 2004. *Efficiency and Sustainability in the Energy and Chemical Industries*. New York: Marcel Dekker, Inc.

Dincer, I. 2000. Thermodynamic, exergy and environmental impact. *Energy Sources* 22: 723–732.

Dincer, I. and A.Y. Cengel. 2001. Energy, entropy and exergy concepts and their roles in thermal engineering. *Entropy International Journal* 2: 87–98.

Dincer, I. and M.A. Rosen. 1999. Energy, environment and sustainable development. *Applied Energy* 64: 427–440.

Dincer, I., M.M. Hussain, and I. Al-Zaharnah. 2005. Energy and exergy utilization in agricultural sector of Saudi Arabia. *Energy Policy* 33: 1461–1467.

Fischer, J.R., J.E. Blackman, and J.A. Finnell. 2007. Industry and energy: Challenges and opportunities. *Resource: Engineering & Technology for a Sustainable World* 4: 8–9.

Gallagher, P.W., H. Brubaker, and H. Shapouri. 2005. Plant size: Capital cost relationships in the dry mill ethanol industry. *Biomass and Bioenergy* 28: 565–571.

Gerbens-Leenes, P.W., H.C. Moll, and A.J.M. Schoot Uiterkamp. 2003. Design and development of a measuring method for environmental sustainability in food production systems. *Ecological Economics* 46: 231–248.

Hammond, G.O. 2004. Engineering sustainability: Thermodynamics, energy systems, and the environment. *International Journal of Energy Research* 28: 613–639.

Henderson, J.M. and R.E. Quandt. 1980. *Microeconomic Theory: A Mathematical Approach*. New York: McGraw-Hill.

Herring, H. 2006. Energy efficiency—A critical view. *Energy* 31: 10–20.

Hostrup, M., R. Gani, Z. Kravanja, A. Sorsak, and I. Grossmann, I. 2001. Integration of thermodynamic insights and MINLP optimization for the synthesis, design and analysis of process flowsheets. *Computers and Chemical Engineering* 25: 73–83.

Kreith, F. and R.E. West. 1997. *CRC Handbook of Energy Efficiency*. Boca Raton: Taylor & Francis.

Kuzgunkaya, E.H. and A. Hepbasli. 2007a. Exergetic evaluation of drying of laurel leaves in a vertical ground-source heat pump drying cabinet. *International Journal of Energy Research* 31: 245–258.

Kuzgunkaya, E.H. and A. Hepbasli. 2007b. Exergetic performance assessment of a ground-source heat pump drying system. *International Journal of Energy Research* 31: 760–777.

Ladd, G. 1998. Grain transport and industry structure. In *Structural Change and Performance of the US Grain Marketing System* Larson D. et al., (Ed.), pp. 51–53. Urbana, IL: Sherer Co.

Marechal, F., G. Heyen, and B. Kalitventzeff. 1997. Energy savings in methanol synthesis: Use of heat integration techniques and simulation tools. *Computers and Chemical Engineering* 21: S511–S516.

Mull, T.E. 2001. *Practical Guide to Energy Management for Facilities Engineers and Managers*. New York: ASME.

Muller, D.C.A., F.M.A. Marechal, T. Wolewinski, and P.J. Roux. 2007. An energy management method for the food industry. *Applied Thermal Engineering* 27: 2677–2686.

Ozdogan, S. and M. Arikol. 1995. Energy and exergy analyses of selected Turkish Industries. *Energy* 20: 73–80.

Pauli, G. 1998. Upsizing: The road to zero emissions. *More jobs, More Income and No Pollution*. London: GreenLeaf Publishing.

Peters, M.S. and K.D. Timmerhaus. 1991. *Plant Design and Economics for Chemical Engineers*. New York: McGraw-Hill.

Pratt, D. 2006. Economic analysis, Chapter 4. In *Energy Management Handbook* (6th ed.), Turner, W.C. and S. Doty (Eds.), pp. 41–86, Lilburn, GA: The Fairmont Press Inc.

Rosen, M.A., N. Pedinelli, and I. Dincer. 1999. Energy and exergy analyses of cold thermal storage systems. *International Journal of Energy Research* 23: 1029–1038.

Singh, R.P. 1986. *Energy in Food Processing*. New York: Elsevier Science Publishing Company Inc.

Singh, R.P. and D.R. Heldman. 2001. *Introduction to Food Engineering* (3rd ed.), San Diego: Academic Press.

Stegou-Sagia, A. and N. Paignigiannis. 2003. Exergy losses in refrigerating systems. A study for performance comparison in compressor and condenser. *International Journal of Energy Research* 27: 1067–1078.

Utlu, Z. and A. Hepbasli. 2006. Assessment of the energy and exergy utilization efficiencies in the Turkish agricultural sector. *International Journal of Energy Research* 30: 659–670.

Vogt, Y. 2004. Top-down energy modeling. *Strategic Planning for Energy and the Environment* 24: 66–80.

Witte, L.C., P.S. Schmidt, and D.R. Brown. 1988. *Industrial Energy Management and Utilization*. New York: Hemisphere Publishing Corporation.

Wulfinghoff, D.R. 1999. *Energy Efficiency Manual*. MD: Energy Institute Press.

Part II

Energy Conservation Technologies Applied to Food Processing Facilities

4 Energy Conservation in Steam Generation and Consumption System

4.1 INTRODUCTION

Thirty-seven percent of fossil fuels burned in the United States industry is used to generate steam in general. In the food industry, about 57% of fossil fuel consumption is to generate steam (Einstein et al., 2001). The steam is used to provide space heat and process heat for cooking, concentrating liquid foods, drying, and sterilizing, and generate mechanical power and electricity in some food processing facilities. A boiler is a pressure vessel designed to transfer heat from a combustion process to a fluid, usually steam or hot water. In this chapter, typical equipment, its energy efficiency, and energy losses in food processing facilities for generation and distribution of steam are examined. Energy conservation technologies for a steam generation and distribution system are discussed.

4.2 COMPONENTS OF A STEAM GENERATION AND DISTRIBUTION SYSTEM

Different manufacturing industries have their major steam generation and distribution systems in common. Figure 4.1 shows a typical steam generation and distribution system used in food processing facilities. Treated cold feed water is fed to the boiler, where it receives heat from fuel combustion and is heated to generate steam. Although the feed water is treated, some impurities still remain in the treated water and can build up in the boiler. Therefore, an amount of water is periodically drained from the bottom of the boiler in a process called as blowdown. The generated steam is delivered along the pipes of a steam distribution system to reach the processes where its heat is used. When the steam passes through the pipes, heat loss from the steam occurs and some of the steam may condense. A steam trap, which allows condensate to pass through but blocks the passage of steam, is usually used to remove the condensate from steam. The condensate can be recirculated to the boiler for recovering some heat and reducing the need for fresh treated feed water. If a process requires low-pressure steam, the high-pressure steam from the boiler is passed through a pressure reduction valve.

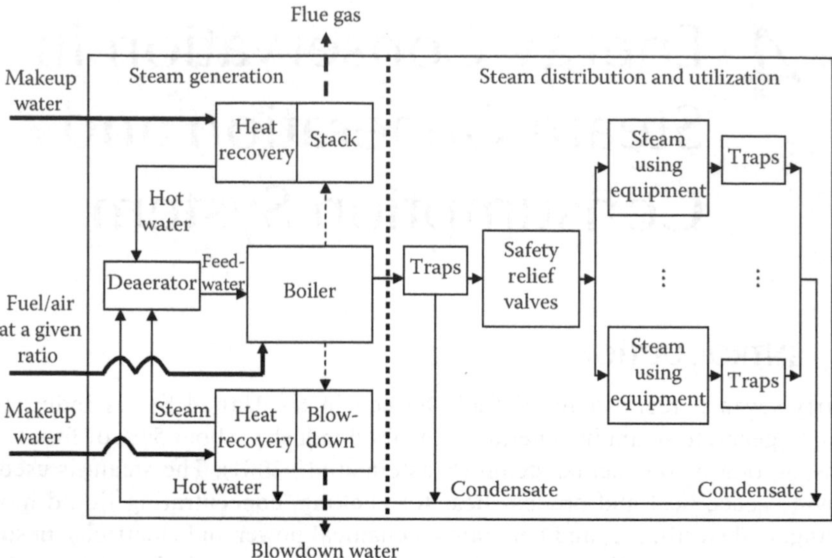

FIGURE 4.1 Typical steam generation and distribution system in a food processing plant.

4.3 STEAM GENERATION SYSTEM

4.3.1 Components of a Boiler System

A boiler is used to change the energy in a fuel, usually fossil fuel, into steam. It can be either a fire tube design or a water tube design. Whatever the design is, a boiler has several main components:

- Burner
- Tubes
- Chamber between tubes
- Flue

Most boilers also have an economizer, water treatment, and some environmental protection ancillaries with the flue. Fuel and air are mixed and burnt in a burner to release its heat. The kind of burner depends on the fuel. In a fire tube boiler, the flame from the burning of fuel flows inside the tubes. The heat from the flame transfers to water surrounded outside the tubes. In a water tube boiler, water flows inside the tubes and flame flows outside the tubes. The flue is the stack that carries off combustion products such as carbon dioxide and excess air.

An economizer is a heat exchanger used to recover some of the heat from the flue gas for increasing the temperature of makeup water to the boiler. However, the stack temperature is usually kept above 150°C to avoid condensation of gases that can produce acids inside the stack and damage the stack. Environmental protection equipment such as wet scrubbers, electrostatic precipitators, cyclone separators, and filters may also be used to treat the flue gas depending on the boiler design, fuel, and regulations.

When steam and water flow through a closed piping system, they may pick up some dissolved solids, which degrade the efficiency and operation of the boiler. The dissolved solids eventually become sludge and settle on the bottom of the boiler, which decreases the heat transfer capability of the boiler and causes significant damage to the boiler. Large amounts of dissolved solids may also lead to foaming and carry liquid water into the stream, causing a water hammer phenomenon. To reduce the concentration of these solids, a certain amount of water is removed from the boiler frequently, which is called blowdown. The hot blowdown water can be used to heat the boiler water. There are two types of blowdown: bottom and surface. Bottom blowdown is a manual process to remove the dissolved solids that have accumulated on the bottom of the boiler. The procedure is performed at regular intervals according to the type of boiler, and steam and water usage. Surface blowdown, also known as top blowdown, removes solids that are floating on or near the surface of the water in the boiler. Boilers have a metered opening just below the water's surface. High pressure inside the boiler blows hot water and the dissolved solids intermittently, continuously, or automatically through the opening. The quantity of total dissolved solids (TDS) in the water or the amount of makeup water used determine whether the blowdown should be intermittent, continuous, or automated.

A water treatment unit is needed to control the acidity and remove dissolved gases such as air in the boiler water. Water enters the boiler from different sources including makeup water, hot water from blowdown heat recovery, and steam condensate return. Before the water enters the boiler, all dissolved gases are removed to minimize corrosion. This process requires energy, which is usually supplied by injecting some of the steam produced by the boiler or low-pressure steam flashed from blowdown heat recovery. Deaerators can use an atmospheric tank or a pressurized tank. One advantage of a pressurized tank is that the condensate can be returned at a higher pressure and temperature, which reduces the amount of energy needed at the boiler to generate steam. The higher temperature also reduces the amount of dissolved gases in the water.

When a pressurized tank is used, deaeration takes place in two stages: the first stage is a spray assembly; the second stage will use either an atomizer or trays. Atomizer or spray-type deaerators use a high-velocity steam jet to remove gases. Tray-type deaerators use the agitation created by spilling the water over several stacked plates usually arranged in a staggered pattern.

4.3.2 Fuels Used in Boilers

Popular fuels used in boilers can be natural gas, fuel oil, and coal. Some organic processing wastes and biomass can also be used as fuel in a boiler. However, a boiler is always designed for a particular fuel. In a boiler, the air/fuel ratio is optimal when the amount of air mixed with the fuel is neither too little nor too much. If the air/fuel mixture does not have enough air (i.e., oxygen), incomplete combustion occurs. On the other hand, if the air/fuel mixture has too much air, the combustion efficiency decreases because part of the energy that could be used to heat the water in the boiler is used instead to heat the excess air. Since it is impossible to supply the precise volume of air needed for complete combustion, it is generally recommended that a small amount of

TABLE 4.1
Advantages and Disadvantages of Different Industrial Fuels

Fuel	Heating Value (MJ/kg)	Price	Cost ($/GJ)	Availability	Combustion Technology	Emission Control	Handling
Coal	30	35–70 $/ton	1.5–3	Moderately abundant	Mature	Difficult (NOx and SOx)	Complex
Fuel oil NO 6	42	1–2 $/gallon	6–13	Moderately abundant	Mature	Difficult (NOx and SOx)	Simple
Natural gas	53	5–10 $/GJ	5–10	Moderately abundant	Mature	Easy	Simple
Organic wastes	20	—	—	Abundant	Immature	Easy	Complex

excess air should be available to the boiler. The amount of excess air depends on fuels. The advantages and disadvantages of industrial fuels are given in Table 4.1.

Example 4.1

Natural gas is burned in a boiler. The volumetric analysis of natural gas at 20°C is given in Table 4.2. What is the weight of the oxygen and what is the weight of the air required for stoichiometric combustion?

Solution 4.1

1. Change the volumetric fraction to mass fraction given in Table 4.2
2. Write combustion equations for CH_4 and C_2H_6. Since both N_2 and CO_2 are inert, they do not participate in reactions.

$$CH_4 + 2O_2 \rightarrow CO_2 + 2H_2O$$

$$C_2H_6 + \frac{7}{2}O_2 \rightarrow 2CO_2 + 3H_2O$$

3. Determine the mass of oxygen required for combustion of CH_4 and C_2H_6.

$$CH_4 : 0.8823\left(\frac{64}{16}\right) = 3.5292$$

$$C_2H_6 : 0.0527\left(\frac{112}{30}\right) = 0.1967$$

Total O_2 required per kilogram of fuel is 3.5292 + 0.1967 = 6.7259 kg

4. Determine the mass of air required for combustion of CH_4 and C_2H_6

Air contains about 78% nitrogen and 21% oxygen by volume or 76.8% nitrogen and 23.2% oxygen by mass. Therefore, the required air mass is

TABLE 4.2
Mass Fraction of Components in Natural Gas for Example 4.1

Constituent Product	Volume (%)	Density (kg/m³)	Mass of Component (kg/m³)	Mass Fraction (%)
CH_2	94.2	0.665	0.6264	88.23
C_2H_6	3.0	1.247	0.0374	5.27
CO_2	2.4	1.829	0.0439	6.18
N_2	0.4	0.582	0.0023	0.32
Total	100		0.71	100

$$6.7259 + 6.7259\left(\frac{0.768}{0.232}\right) = 28.99 \text{ kg air/kg fuel}$$

In actual combustion processes, more air than the theoretical amount of air determined in Example 4.1 must be supplied to ensure complete combustion. The amount of air required above the theoretical requirement is called excess air. The optimum levels of excess air depend on the fuel type and firing method.

Different fuels have different hydrogen content. When a fuel is burned, the hydrogen becomes water, which is released as part of the flue gas. Since the temperature of flue gas is higher than the boiling point of water, energy is lost through the steam in the flue gas. Therefore, the higher the hydrogen content of the fuel, the more energy is lost to the flue gas via steam.

4.3.3 Energy Analysis for a Steam Generation System

All matter is conserved in a boiler. First, the total mass of fuel plus air entering the system is the same as the total mass of flue gas exiting the stack. Second, individual atomic species, such as C, O, and H, are conserved. Although the energy stored in the fuel is changed from one form to another, energy is conserved, which is neither lost nor destroyed. Determination of boiler efficiency is largely a matter of careful energy accounting based on the principle of energy conservation.

The heat content or enthalpy of steam includes two parts: latent heat required to vaporize water into steam and sensible heat required to increase the feed water temperature to the water boiling point and increase the steam temperature if the steam is superheated. For saturated steam, the steam temperature is directly related to its pressure. The heat content or enthalpy of saturated steam can be determined from Table 4.3 if either temperature or pressure of the steam is known. However, both temperature and pressure should be known to determine the heat content of superheated steam from Table 4.4.

The boiler efficiency defines the ratio of useful heat in the product's steam to the energy content of the fuel. The efficiency can be determined by measuring the flow and thermodynamic properties of the incoming fuel and outgoing product. The boiler efficiency, η_b, can be considered as the product of two components: the combustion efficiency, η_c, of fuels and the heat exchanger efficiency, η_h:

TABLE 4.3
Thermodynamic Properties of Saturated Steam

Temperature (°C)	Pressure (kPa)	Specific Volume (m³/kg)		Enthalpy (kJ/kg)		Entropy (kJ/kgK)	
		Liquid	Vapor	Liquid	Vapor	Liquid	Vapor
0.01	0.6113	0.001	206.14	0.01	2501.4	0.000	9.1562
5	0.8721	0.001	147.12	20.98	2510.6	0.0761	9.0257
10	1.2276	0.001	106.38	42.01	2519.8	0.1510	8.9008
15	1.7051	0.001001	77.93	62.99	2528.9	0.2245	8.7814
20	2.339	0.001002	57.79	83.96	2538.1	0.2966	8.6672
25	3.169	0.001003	43.36	104.89	2547.2	0.3674	8.5580
30	4.246	0.001004	32.89	125.79	2556.3	0.4369	8.4533
35	5.628	0.001006	25.22	146.68	2565.3	0.5053	8.3531
40	7.384	0.001008	19.52	167.57	2574.3	0.5725	8.2570
45	9.593	0.001010	15.26	188.45	2583.2	0.6387	8.1648
50	12.349	0.001012	12.03	209.33	2592.1	0.7038	8.0763
55	15.758	0.001015	9.568	230.23	2600.9	0.7679	7.9913
60	19.940	0.001017	7.671	251.13	2609.6	0.8312	7.9096
65	25.03	0.001020	6.197	272.06	2618.3	0.8935	7.8310
70	31.19	0.001023	5.042	292.98	2626.8	0.9549	7.7553
75	38.58	0.001026	4.131	313.93	2635.3	1.0155	7.6824
80	47.39	0.001029	3.407	334.91	2643.7	1.0753	7.6122
85	57.83	0.001033	2.828	355.90	2651.9	1.1343	7.5445
90	70.14	0.001036	2.361	376.92	2660.1	1.1925	7.4791
95	84.55	0.001040	1.982	397.96	2668.1	1.2500	7.4159
100	101.35	0.001044	1.6729	419.04	2676.1	1.3069	7.3549
110	143.27	0.001052	1.2102	461.30	2691.5	1.4185	7.2387
120	198.53	0.001060	0.8919	503.71	2706.3	1.5276	7.1296
130	270.1	0.001070	0.6685	546.31	2720.5	1.6344	7.0269
140	361.3	0.001080	0.5089	589.13	2733.9	1.7391	6.9299
150	475.8	0.001091	0.3928	632.20	2746.5	1.8418	6.8379
160	617.8	0.001102	0.3071	675.55	2758.1	1.9427	6.7502
170	791.7	0.001114	0.2428	719.21	2768.7	2.0419	6.6663
180	1002.1	0.001127	0.19405	763.22	2778.2	2.1396	6.5857
190	1254.4	0.001141	0.15654	807.62	2786.4	2.2359	6.5079
200	1553.8	0.001157	0.12736	852.45	2793.2	2.3309	6.4323
210	1906.2	0.001173	0.10441	897.76	2798.5	2.4248	6.3585
220	2318	0.001190	0.08619	943.62	2802.1	2.5178	6.2861
230	2795	0.001209	0.07158	990.12	2804.0	2.6099	6.2146
240	3344	0.001229	0.05976	1037.32	2803.8	2.7015	6.1437
250	3973	0.001251	0.05013	1085.36	2801.5	2.7927	6.0730
260	4688	0.001276	0.04221	1134.37	2796.9	2.8838	6.0019
270	5499	0.001302	0.03564	1184.51	2789.7	2.9751	5.9301
280	6412	0.001332	0.03017	1235.99	2779.6	3.0668	5.8571
290	7436	0.001366	0.02557	1289.07	2766.2	3.1594	5.7821
300	8581	0.001404	0.02167	1344.0	2749.0	3.2534	5.7045

TABLE 4.4
Thermodynamic Properties of Superheated Steam

Pressure (kPa)		Temperature (°C) 100	200	300	400	500	600	700	800	900	1000	1100	1200	1300
10	v	17.196	21.825	26.445	31.063	35.679	40.295	44.911	49.526	54.141	58.757	63.372	67.987	72.602
	h	2687.5	2879.5	3076.5	3279.6	3489.1	3705.4	3928.7	4159.0	4396.4	4640.6	4891.2	5147.8	5409.7
	s	8.4479	8.9038	9.2813	9.6077	9.8978	10.1608	10.4028	10.6281	10.8396	11.0393	11.2287	11.4091	11.5811
50	v	3.418	4.356	5.284	6.209	7.134	8.057	8.981	9.904	10.828	11.751	12.674	13.597	14.521
	h	2682.5	2877.7	3075.5	3278.9	3488.7	3705.1	3928.5	4158.9	4396.3	4640.5	4891.1	5147.7	5409.6
	s	7.6947	8.1580	8.5373	8.8642	9.1546	9.4178	9.6599	9.8852	10.0967	10.2964	10.4859	10.6662	10.8382
100	v	1.6958	2.172	2.639	3.103	3.565	4.028	4.490	4.952	5.414	5.875	6.337	6.799	7.260
	h	2676.2	2875.3	3074.3	3278.2	3488.1	3704.4	3928.2	4158.6	4396.1	4640.3	4891.0	5147.6	5409.5
	s	7.3614	7.8343	8.2158	8.5435	8.8342	9.0976	9.3398	9.5652	9.7767	9.9764	10.1659	10.3463	10.5183
200	v		1.0803	1.3162	1.5493	1.7814	2.013	2.244	2.475	2.705	2.937	3.168	3.399	3.630
	h		2870.5	3071.8	3276.6	3487.1	3704.0	3927.6	4158.2	4395.8	4640.0	4890.7	5147.5	5409.3
	s		7.5066	7.8926	8.2218	8.5133	8.7770	9.0194	9.2449	9.4566	9.6563	9.8458	10.0262	10.1982
400	v		0.5342	0.6548	0.7726	0.8893	1.0055	1.1215	1.2372	1.3529	1.4685	1.5840	1.6996	1.8151
	h		2860.5	3066.8	3273.4	3484.9	3702.4	3926.5	4157.3	4395.1	4639.4	4890.2	5146.8	5408.8
	s		7.1706	7.5662	7.8985	8.1913	8.4558	8.6987	8.9244	9.1362	9.3360	9.5256	9.7060	9.8780
600	v		0.3520	0.4344	0.5137	0.5920	0.6697	0.7472	0.8245	0.9017	0.9788	1.0559	1.1330	1.2101
	h		2850.1	3061.6	3270.3	3482.8	3700.9	3925.3	4156.5	4394.4	4638.8	4889.6	5146.3	5408.3
	s		6.9665	7.3724	7.7079	8.0021	8.2674	8.5107	8.7367	8.9486	9.1485	9.3381	9.5185	9.6906
800	v		0.2608	0.3421	0.3843	0.4433	0.5018	0.5601	0.6181	0.6761	0.7340	0.7919	0.8497	0.9076
	h		2839.3	3056.5	3267.1	3480.6	3699.4	3924.2	4155.6	4393.7	4638.2	4889.1	5145.9	5407.9
	s		6.8158	7.2328	7.5716	7.8673	8.1333	8.3700	8.6033	8.8153	9.0153	9.2050	9.3855	9.5575
1000	v		0.2060	0.2579	0.3066	0.3541	0.4011	0.4478	0.4943	0.5407	0.5871	0.6335	0.6798	0.7261
	h		2827.9	3051.2	3263.9	3478.5	3697.9	3923.1	4154.7	4392.9	4637.6	4888.6	5145.4	5407.4
	s		6.6940	7.1229	7.4651	7.7622	8.0290	8.2731	8.4996	8.7118	8.9119	9.1017	9.2822	9.4543

$$\eta_b = \eta_c \times \eta_h \tag{4.1}$$

The combustion efficiency reflects the extent to which the energy content of the fuel is converted to heat in the combustion gas, which is usually close to 100%. The efficiency of heat exchangers represents the fraction of the heat content in the combustion gas actually transferred to the steam, which is usually 70%–80% for an industrial boiler. CEC (2008) reports that the efficiency of a 746kW fire tube boiler without economizers used in a food processing facility is 86%.

To determine the relative operating efficiency and to establish energy conservation benefits for an excess-air-control program, the following information should be available:

- Percent oxygen by volume in the flue gas
- Stack temperature rise (the difference between the flue gas temperature and the combustion air inlet temperature)
- Fuel type

4.3.4 Heat Loss from a Boiler System

The physical and chemical properties of fuels have a significant influence on the efficiency of boilers. Usually, boilers with gaseous or liquid fuels have a higher energy efficiency than those with solid fuels because complete combustion of solid fuels requires a larger amount of excess air and combustion of solid fuels discharges hot solid refuse. Both the design and operating parameters of boilers also affect the performance of boilers. The energy losses of a steam generation system or boiler could include

- Flue gas including the sensible heat in the dry flue gas, water vapor formed by the combustion of hydrogen in fuels, and the moisture in fuels and combustion air
- Incomplete combustion
- Boiler blowdown water
- Fouling of heat transfer surface
- Heat convection and radiation loss from the hot boiler surface

Heat is lost from the stack through the hot flue gas and the radiation of the hot stack surface. Over 20% of the total heat is normally exhausted from the boiler through the stack without a waste-heat recovery unit. The energy loss from flue gas increases with the increase in excess air. Depending on the fuel, the temperature of leaving gases can be in the range of 205°C to 325°C. The heat loss in the flue gas can be estimated in three ways (Carpehart et al., 2006):

- Calculation from the amount and enthalpy of each flue gas component
- Determination from a combustion efficiency curve as given in Figure 4.2 (Parker et al., 2006)
- Calculation from the mass and heat balances of the boiler

A combustion efficiency curve, as shown in Figure 4.2, is a simple and useful method to determine the boiler combustion efficiency. For the stack temperature rise of 260°C (or 500°F) and excess oxygen 2% by volume (in the flue gas), it can be found from Figure 4.2 that the excess air is about 10% and the combustion efficiency is 81.4%.

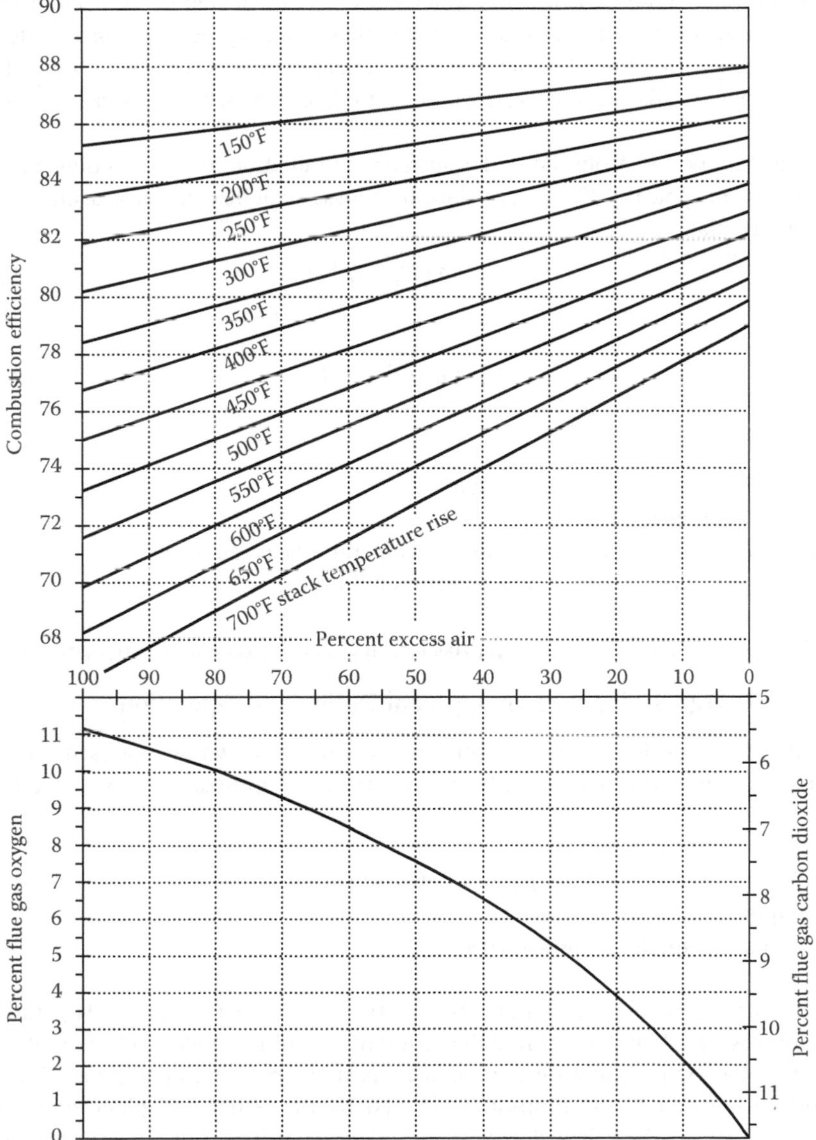

FIGURE 4.2 Combustion efficiency chart for natural gas. (Reprinted from Parker, S.A., Scollon, R.B., and Smith, R.D., *Energy Management Handbook*, The Fairmont Press Inc., Lilburn, 2006. With permission.)

Blowdown is used to remove dissolved solids in the boiler system. Heat loss in boiler blowdown depends on the temperature and amount of blowdown water. The amount and frequency of blowdowns are dependent on the quantity and condition of the boiler makeup water. The blowdown rate normally should not be over 1%–3% of the steam output. However, the blowdown rate for some boilers used in food processing facilities can be as high as 15% (CEC, 2008). The reverse osmosis system can be used to improve the feed water quality if the quality of makeup water is too low.

Inside a boiler, both the fire and water sides can be affected by the accumulation of dirt. The accumulation acts as an insulator, thus reducing the heat transfer rate and wasting energy. In a fouled boiler, the wasted heat increases the stack temperature. Heat exchangers like economizers can accumulate dirt, which also affects boiler efficiency.

Heat may be lost from the boiler surface via radiation and convection. Surface heat losses are typically 1%–2%. Surface heat losses can be calculated using the following formula:

$$q_r = \sigma A\left(T_{s,R}^4 - T_{a,R}^4\right) \tag{4.2}$$

$$q_c = hA\left(T_s - T_a\right) \tag{4.3}$$

where

σ is the Stefan–Boltzmann constant
A is the surface area
T_s and T_a are the surface and ambient temperatures, respectively
h is the surface convective heat transfer coefficient

4.3.5 Energy Conservation Technologies for Steam Generation System

4.3.5.1 Energy Saving through Optimal Design and Operation

The energy savings for a boiler system can be divided into two categories: design and operation optimization and waste-heat recovery. The optimization may include

- Proper size of a boiler
- Proper pressure and temperature of steam
- Optimal amount of excess air
- Optimal amount of blowdowns

The efficiencies of boilers differ due to the difference in their designs and operations. Food processors should determine the required size of the boilers, identify the efficiency of the boilers, and determine the optimal operating conditions. Pressure should be reduced to the minimum allowed by the boiler and plant equipment. At low pressures, there is a decrease in leakage and losses due to transportation resistance. The reduction in steam leakage saves energy and makeup water. Makeup water costs can be significant because of the requirement of water treatment. In addition, low-pressure steam is also at a lower temperature and thus there is direct saving of energy stored in the steam. Steam at a low pressure and temperature also reduces heat transfer losses during steam distribution. Automatic controls can be installed to monitor the temperature and pressure of the system and control the boilers. Operating boilers at a very low load reduces their efficiency since some losses are constant independent of the amount of steam produced.

Air slightly in excess of the ideal stoichiometric fuel/air ratio is required for safety and to reduce NO_x emission. The more the air used to burn the fuel, the more the heat wasted in heating air. Approximately, 10%–15% excess air is adequate depending on fuels. Poorly maintained boilers can have up to 140% excess air. If the excess air is reduced from 140% to 15%, it will save 8% of energy (Einstein et al., 2001). The amount of excess air can be determined by the measurement of oxygen or carbon dioxide in the flue gas with inexpensive equipment. Equipment that is used to measure oxygen is more precise than carbon dioxide measuring devices. Information from the combustion gas analysis is used to calibrate the settings on the air and fuel supply systems. Flue gas monitors can be used to maintain an optimum flame temperature, monitor carbon monoxide oxygen and smoke, and detect any air leaks. It is possible to optimize the fuel/air mixture for a high flame temperature and low emissions. The monitoring measure can save about 3% of the energy but it may be too expensive for small boilers (Einstein et al., 2001).

The blowdown water itself cannot be recirculated back to the boiler and is usually sent to the sewer because of its high dissolved solid content. The blowdown should be performed in such a way that the total dissolved solid level is kept to its maximum allowable value.

4.3.5.2 Energy Recovery from Flue Gas

Stack heat recovery systems can improve boiler efficiency by as much as 15%. When possible, heat from the stack should be recovered. An economizer can efficiently recover wasted stack heat and transfer it to boiler makeup water. The limiting factor for flue gas heat recovery is that the economizer wall temperature should not drop below the dew point of acids in the flue gas. Since the high heat transfer coefficient is present on the water side of the economizer, it is reasonable to preheat the feedwater to close to the acid dew point before it enters the economizer to recover the waste heat from flue gas. This allows the temperature of flue gas exiting the economizer to be just above the acid dew point. An increase in water temperature every 6°C or decrease in flue gas temperature every 25°C can save approximately 1% of the boiler fuel. For air pre-heaters used to recover the stack waste heat, an increase in air temperature every 4.5°C will increase the boiler efficiency by 1% (Ganapathy, 1994). A waste-heat recovery unit used to save 1% energy for the boiler has a payback period of 2 years (Einstein et al., 2001).

4.3.5.3 Energy Recovery from Blowdown Water

If the right equipment is used, up to 78% of the heat stored in blowdown water can be recovered. To reduce energy loss through the high-temperature and high-pressure blowdown water, the heat content in the blowdown water can be recovered via either a heat exchanger or a flash steam generator. The heat content of the blowdown water can be transferred to the boiler makeup water through a heat exchanger. Alternatively, if the blowdown water, which exits from a boiler at a high temperature and pressure, is released into a flash unit at a lower pressure, part of the blowdown water is immediately converted into low-pressure steam without solids. The heat/energy stored in the blowdown is thus captured in the form of the low-pressure steam. The steam can be returned to the boiler via a deaerator or used as low-pressure steam in the processing facilities. The heat recovery from blowdown water is considered to save 1.3% of boiler fuel use for small boilers with a payback of 2.7 years (Einstein et al., 2001).

4.3.5.4 Maintenance of Boiler

The poor maintenance of boilers can reduce the boiler efficiency up to 20%–30% of initial efficiency within 2–3 years. The maintenance may include

- Improvement of insulation
- Repairing of air leaks
- Reduction of heat transfer resistance
- Replacement of refractory lining

Air leaks into the boiler and the flue gas reduce the heat transferred to the steam and increase the energy consumption of air fans. We can save 2%–5% of the energy through repairing air leaks. Improved boiler insulation can save 6%–26% of the energy. Heat transfer surfaces should be cleaned periodically. It is recommended that the heat transfer surfaces should be cleaned at least once a year depending on the boiler type and the amount of makeup water used. Correct blowdown and water treatment will help keeping the water side surface clean. Insufficient blowdown will increase the amount of dirt in the steam system. Boiler water sampling can detect fouling on the water side. (Einstein et al., 2001).

4.4 STEAM DISTRIBUTION SYSTEM

The steam distribution system is responsible for carrying the steam from the boiler to the end-use equipment in a plant. The distribution system must be properly designed to account for condensate drainage, system pressure, and flow control.

4.4.1 Components of a Steam Distribution System

As steam moves throughout the system, part of its heat may be lost, generating some condensate. Condensate may cause pipe hammering and damage equipment with rust. Steam traps are automatic valves that separate air and condensate from the steam in a pipeline. A leaky trap wastes energy by allowing steam to enter the condensate return or by not expelling the condensate from the steam line, which reduces the efficiency of the system. Traps are classified into three main groups—mechanical, thermostatic, and thermodynamic. There are several different types of traps in each group. Because different traps operate in different ways, sizing, positioning, and installation procedures vary. Safety relief valves are placed on a steam system to protect the system from damage caused by buildup of excessive pressure in the system. These valves are often placed near the end-use equipment to be protected (Mull, 2001). Information about proper installation of pipes and relief valves can be found in ASME B31.1-Power Piping Code.

4.4.2 Heat Loss and Energy Efficiency of a Steam Distribution System

Pipelines that carry steam lose heat to the ambient. The amount of heat loss depends on the steam temperature, the ambient temperature, and the amount of insulation of the pipes. Although heat loss through a steam distribution system is unavoidable, effective insulation of vessels, steam lines, valves, and condensate return lines can reduce energy losses. If steam lines have leaks, large amounts of steam loss can be a major waste of energy from a steam distribution system. After the heat carried with steam is converted to process heat to be used in the food processing facilities, the steam may be condensed into hot water. The condensate can be returned to the boiler and converted into steam again. Therefore, condensate leaks represent both the energy loss and treated water loss.

4.4.3 Energy Conservation Technologies for a Steam Distribution System

4.4.3.1 Steam Trap Maintenance and Condensate Recovery

Energy savings for regular steam trap checks and follow-up maintenance are estimated to be 10% with a payback period of 0.5 years. Using an automatic monitoring system to immediately notice any malfunctioning or failure of steam traps can save an additional 5% of energy with a payback of 1 year (Einstein et al., 2001).

When a steam trap purges condensate from a pressurized steam distribution system to an ambient pressure, flash steam is produced. This steam can be used for food processing and space heating. The hot condensate can be re-used in the boiler to save energy and reduce the need for treated boiler feed water. The condensate recovery system is used to capture the condensate present in the steam system and return it to the deaerator system. The water quality of condensate is usually high enough for it to be re-used in the boiler without the requirement of additional treatment. Because the temperature of condensate is relatively high compared to cold makeup water, it requires much less energy to be reconverted into steam. Therefore, condensate should be returned to the deaerator system as soon as possible.

4.4.3.2 Repairing of Steam Leaks

Table 4.5 shows the cost of leaked steam at 0.6895 MPa gauge pressure (around 170°C for saturation steam) as a function of a leak diameter. At a higher steam pressure, the leakage is even greater. The loss rate of steam increases at a rate proportional to the square root of the pressure, which is given by (Schmidt, 2006)

$$\frac{\dot{m}_1}{\dot{m}_2} = \left(\frac{P_1}{P_2}\right)^2 \tag{4.4}$$

TABLE 4.5
Annual Cost of Steam Leaks at a Pressure Difference of 0.6895 MPa

Leak Diameter (mm)	Steam Wasted per Month (kg/month)	Cost per Month ($/month)[a]	Cost per Year ($/year)
1.588	6,038	60	720
3.175	23,699	237	2,844
6.350	94,886	949	11,388
12.700	378,182	3,782	45,384

Source: Adapted from Schmidt, P.S., in *Energy Management Handbook*, The Fairmont Press, Inc., Lilburn, 2006. With permission.

[a] Based on steam price at 10 $/ton (about 3.8 $/GJ steam).

Example 4.2

Steam at 0.6895 MPa leaks from a 1.588 mm leak at the rate of 6038 kg/month. What is the leak rate if the steam pressure increases to 1 MPa?

Solution 4.2

From Equation 4.4, $\dot{m}_2 = \left(\frac{P_2}{P_1}\right)^2 \times \dot{m}_1 = \left(\frac{1}{0.6895}\right)^2 \times 6038 = 12{,}701$ kg/month

If natural gas at 3.8 $/GJ is used to generate steam at 100% energy efficiency, the energy cost of steam is about 10 $/ton since the enthalpy of the saturated steam at 170°C is 2769 kJ/kg (from Table 4.3). The economic loss from the leaked steam is thus 1524 $/year.

4.4.3.3 Insulation Improvement

Improving the insulation of steam distribution systems may save an average of 3%–13% of energy (Einstein et al., 2001). Proper insulation materials should have low thermal conductivity, dimensional stability under temperature change, resistance to water absorption, and resistance to combustion. Some popular insulating materials are given in Table 4.6 (Capehart et al., 2005).

Example 4.3

Steam at 150°C is delivered from a boiler to a food processing unit through a 30 m stainless steel pipe. The diameter of the pipe is 5 cm and its thickness is 0.5 cm. The ambient temperature is 20°C. The thermal conductivity of stainless steel is 43 W/m°C. The inside steam convective heat transfer coefficient and the outside air convective heat transfer coefficient are 1000 W/m^2°C and 10 W/m^2°C, respectively. What is the energy loss from the pipe? To reduce the energy loss from the pipe, the pipe is insulated by 5 cm of aluminum-jacketed fiberglass. The thermal conductivity of the fiberglass is 0.035 W/m^2°C. What is the energy loss from the pipe after insulation? If the boiler efficiency is 80% and the natural gas at $5.00/GJ is used in the boiler for the steam generation, what are the energy savings per year through the insulation (8000 h)?

TABLE 4.6
Industrial Insulation Materials

Material	Temperature (°C)	Thermal Conductivity (W/m°C) (24°C–260°C)	Compressive Strength (kPa) at % Deformation	Cell Structure
Calcium silicate blocks	815	0.0534–0.0764	690–1725 at 5%	Open
Glass fiber blankets	650	0.0346–0.1053	0.14–24 at 10%	Open
Glass fiber boards	540	0.0317–0.0880	0.14–24 at 10%	Open
Glass fiber pipe covering	455	0.0332–0.0894	0.14–24 at 10%	Open
Mineral fiber blocks	1050	0.0332–0.1183	7–125 at 10%	Open
Cellular glass blocks	–270 to 480	0.0548–0.1038	700 at 5%	Closed
Expanded perlite blocks	815	–0.0909	620 at 5%	Open
Urethane foam blocks	–270 to 110	0.0231	110–520 at 10%	95% closed
Isocyanurate foam blocks	175	0.0216	120–170 at 10%	93% closed
Phenolic foam	–40 to 120	0.0317	90–150 at 10%	Open
Elastometric closed cell sheets	–40 to 105	0.0361	275 at 10%	Closed
MIN-K blocks and blanks	980	0.0274–0.0346	700–1310 at 8%	Open
Ceramic fiber blankets	1425	–0.0779	3.5–7 at 10%	Open

Source: Adapted from Capehart, B.L., Turner, W.C., and Kennedy, W.J., *Guide to Energy Management*, The Fairmont Press, Inc., Lilburn, 2005. With permission.

Solution 4.3

The energy loss from the pipe can be determined by the heat transfer rate, which is discussed in Chapter 1.

1. The energy loss from the pipe before insulation is

$$\frac{1}{U_i} = \frac{1}{h_i} + \frac{r_i \ln\left(r_o/r_i\right)}{k_w} + \frac{r_i}{h_o r_o}$$

$$= \frac{1}{1000} + \frac{0.025 \ln\left(0.03/0.025\right)}{43} + \frac{0.025}{10 \times 0.03}$$

$$U_i = 11.84\ \text{W/m}^2\,^\circ\text{C}$$

$$A_i = 2\pi r_i L$$

$$= 2 \times 3.14159 \times 0.025 \times 30$$

$$= 4.712\ \text{m}^2$$

$$\begin{aligned}\Delta T &= T_i - T_o \\ &= 150 - 20 \\ &= 130^\circ\text{C}\end{aligned}$$

$$\begin{aligned}Q &= U_i A_i \Delta T \\ &= 11.84 \times 4.712 \times 130 \\ &= 7253\ \text{W}\end{aligned}$$

2. The overall coefficient of the pipe before insulation is

$$\frac{1}{U_i'} = \frac{1}{h_i} + \frac{r_i \ln\left(r_{wo}/r_i\right)}{k_w} + \frac{r_i \ln\left(r_{so}/r_{wo}\right)}{k_s} + \frac{r_i}{h_o r_{so}}$$

$$= \frac{1}{1000} + \frac{0.025\ln\left(0.03/0.025\right)}{43} + \frac{0.025\ln\left(0.08/0.03\right)}{0.035} + \frac{0.025}{10 \times 0.08}$$

$$U_i' = 1.36\ \text{W/m}^2\ {}^\circ\text{C}$$

$$\begin{aligned}A_i &= 2\pi r_i L \\ &= 2 \times 3.14159 \times 0.025 \times 30 \\ &= 4.712\ \text{m}^2\end{aligned}$$

$$\begin{aligned}\Delta T &= T_i - T_o \\ &= 150 - 20 \\ &= 130^\circ\text{C}\end{aligned}$$

$$\begin{aligned}Q' &= U_i' A_i \Delta T \\ &= 1.36 \times 4.712 \times 130 \\ &= 833\ \text{W}\end{aligned}$$

3. Energy savings

$$\begin{aligned}\Delta Q &= Q - Q' \\ &= 7253 - 833 \\ &= 6420\ \text{W}\end{aligned}$$

Total energy savings are

$$\begin{aligned}E &= \Delta Q \times \Delta t \\ &= 6420 \times 8000 \times 3600 \\ &= 185\ \text{GJ}\end{aligned}$$

The economic saving is

$$\begin{aligned}\$ &= \frac{E}{\eta} \times c \\ &= \frac{185}{0.8} \times 5 \\ &= 1156 \text{ \$/year}\end{aligned}$$

Since both energy savings and insulation investment increase with the insulation thickness, there is an optimum insulation thickness. The implementation cost of an energy saving project is discussed below.

4.5 ECONOMIC ANALYSIS OF ENERGY EFFICIENCY IMPROVEMENT FOR A STEAM SYSTEM

The implementation cost of an energy saving potential should be lower than fuel cost saving. The annual implementation costs (AIC) include the annual capital investment cost and operating and maintenance costs. It can be defined as

$$\text{AIC} = \frac{\text{Annualized investment } + \text{Annual change in operating and maintenance costs}}{\text{Annual energy savings}} \tag{4.5}$$

The annual capital investment is calculated by

$$\text{Annualized investment } = \text{ Capital cost } \times \frac{i}{1-(1+i)^{-n}} \tag{4.6}$$

where

i is the discount rate
n is the expected lifetime of the capital investment

Einstein et al. (2001) studied the economic potential of energy efficiency improvement of industrial boilers in the United States, which is shown in Table 4.7. Conservation technologies can save about 20% of energy consumption in the steam system. The payback periods for these conservation measures are less than 3 years at a discount rate of 30%.

4.6 COGENERATION

Cogeneration is the process of sequentially producing both electricity and steam from a single fuel source. A cogeneration facility uses some of the thermal energy for food processing and space heating from an electric power plant. Otherwise, this thermal energy is rejected to the environment. Furthermore, if industrial steam is generated at a pressure and temperature above that required for end use, steam can be brought down to the desired pressure and temperature through a turbine generator for additional electricity (Teixeira, 1986). Cogeneration can produce a given amount

TABLE 4.7
Energy Efficiency Measures for the Industrial Steam Systems in the United States

Measure	AIC ($/GJ)	Payback Period (year)
Boiler maintenance	0.01	0
Leak repair	0.26	0.4
Steam trap maintenance	0.36	0.5
Improved process control	0.43	0.6
Condensate return	0.69	1.1
Automatic steam trap monitoring	0.74	1.0
Improved insulation	1.08	1.1
Flue gas heat recovery	1.53	2.0
Blowdown steam recovery	1.86	2.7

Source: Reprinted from Einstein, D., Worrell, E., and Khrushch, M., 2001. Steam systems in industry: Energy use and energy efficiency improvement potentials. Lawrence Berkeley National Laboratory. Paper LBNL-49081. Available at http://repositories.cdlib.org/lbnl/LBNL-49081. With permission.

of electric power and thermal energy for 10% to 30% less fuel than a power plant, which produces the same amount of electricity alone (Capehart et al., 2005). Cogeneration cycles are discussed in Chapter 9. For many food processing facilities, cogeneration offers a way to provide both low-cost electric power and large amounts of thermal energy needed for processing heat.

4.7 SUMMARY

There is a large potential for energy savings in steam generation and distribution systems through the application of different energy conservation technologies. The savings can easily represent approximately 20% of final energy use in steam systems. These savings can also substantially reduce carbon dioxide emissions. Optimization of design and operation, waste-heat recovery, and reduction of heat loss are three main measures to improve the energy efficiency of a steam generation and distribution system. Cogeneration cycles can provide both steam and electricity simultaneously and thus the efficiency of the whole process is high.

REFERENCES

California Energy Commission (CEC). 2008. *California's Food Processing Industry Energy Efficiency Initiative: Adoption of Industrial Best Practices.* Available at: http://www.energy.ca.gov/2008publications/CEC-400-2008-006/CEC-400-2008-006.PDF.

Capehart, B.L., W.C. Turner, and W.J. Kennedy. 2006. *Guide to Energy Management* (5th ed.), Lilburn, GA: The Fairmont Press, Inc.

Einstein, D., E. Worrell, and M. Khrushch. 2001. *Steam systems in industry: Energy use and energy efficiency improvement potentials.* Lawrence Berkeley National Laboratory. Paper LBNL-49081. Available at http://repositories.cdlib.org/lbnl/LBNL-49081.

Ganapathy, V. 1994. Understand steam generator performance. *Chemical Engineering Progress.*

Mull, T.E. 2001. *Practical Guide to Energy Management for Facilities Engineers and Plant Managers*, New York: ASME Press.

Parker, S.A., R.B. Scollon, and R.D. Smith. 2006. Boilers and fired systems, Chapter 5. In *Energy Management Handbook* (6th ed.), Turner, W.C. and S. Doty (Eds.), Lilburn, GA: The Fairmont Press, Inc.

Schmidt, P.S. 2006. Steam and condensate systems, Chapter 6. In *Energy Management Handbook* (6th ed.), Turner, W.C. and S. Doty, (Eds.) Lilburn, GA: The Fairmont Press, Inc.

Teixeira, A.A. 1986. Cogeneration in food processing plants, Chapter 18. In *Energy in Food Processing*, Singh, R.P., (Ed.), pp. 283–302, New York: Elsevier Science Publishing Company Inc.

5 Energy Conservation in Compressed Air System

5.1 INTRODUCTION

Compressed air is used in different operations such as aerating, conveying, and pneumatic control in a food processing facility. Compressed air is used in food processing facilities in cooking retorts, wrapping machines, color sorters, lid machines, open blowing for water removal and drying of cans prior to labeling, and pneumatic controls.

The total operating cost of compressed air is four to five times that of electrical energy used to generate the compressed air. Proper improvements to compressors and compressed air systems can save 20%–50% of the energy consumed by the systems (Mull, 2001). Usually, the devices that can produce air at pressure up to 80 kPa (gauge, or 12 psig) are considered blowers and fans. Compressors are usually used to produce air in the range of 500–1000 kPa (gauge, or 80–150 psig).

Compressed air is one of the major utilities in food processing facilities. The production of compressed air can be one of the most expensive processes in manufacturing facilities. Therefore, energy conservation for compressed air systems should be an important part of the energy conservation project in food processing facilities. In this chapter, the main components and energy loss sources of compressed air systems are reviewed. Several energy conservation technologies for compressed air systems are discussed.

5.2 MAIN COMPONENTS OF COMPRESSED AIR SYSTEMS

A diagram of a typical compressed air system is given in Figure 5.1. A compressed air system may include

- Compressors
- Air distribution line
- Air storage units
- Air filtration and drying units

Compressed air is usually supplied by several air compressors through a pipe system. In order to meet the varying air demands, air storage systems are usually installed. Extra filters and air driers are also required to supply food grade quality air. There are two types of air compressors: positive displacement compressors such as reciprocating air compressors and rotary air compressors, and dynamic compressors such as centrifugal air compressors. Reciprocating air compressors have a piston that moves back and forth inside a cylinder to compress air. Rotary air compressors have

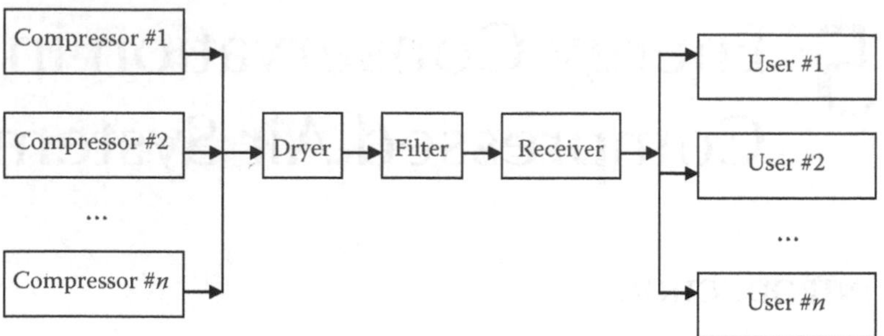

FIGURE 5.1 Diagram of a compressed air system.

a rotary vane or rotating screw housed in a cylinder to compress air. Centrifugal air compressors are machines in which air is compressed by the mechanical action of rotating impellers. Centrifugal air compressors usually have large airflow capacities from 30 to 850 m^3/min at pressure in the range from 690 to 860 kPa (gauge). Atmospheric air contains small amounts of moisture, dirt, and impurities. There are three basic methods to remove moisture from compressed air: chemical drying, adsorption, and refrigeration. Filters are usually installed to remove the impurities. Air receivers are large volume storage tanks that are used to store compressed air to supplement compressor output when temporary demand for air is greater than the compressor output capacity. In positive displacement compressors, the receivers can also dampen pulsations from the compressors (Mull, 2001).

The compressed air is characterized by air quality, quantity, and level of pressure. Compressed air quality is determined by the dryness and contaminant level required by end-users, which is accomplished with filtering and drying equipment. The higher the air quality, the more the costs required to produce the air. The required compressed air quantity can be determined by summing the requirements of compressors by the tools and process operations in the facility. High short-term demands should be met by the air stored in air receivers. Over-sized air compressors are extremely inefficient because air compressors use more energy per unit volume of air produced when operating at a part load. Multiple compressors with sequencing controls may provide an economic operation if the processing facility has a wide variation in air demands over time. For example, Del Monte Foods in California, which mainly processes canned fruits and vegetables, used nine compressors ranging in size from 56 to 93 kW to supply compressed air at a total rate of 120 m^3/min (standard) and pressure from 690 to 860 kPa through a single pipe system. The average plant air demand ranges from 35 to 102 m^3/min (CEC, 2008).

5.3 SOURCES OF ENERGY LOSSES FROM A COMPRESSED AIR SYSTEM

The energy losses from a compressed air system may come from

- Partially loaded compressors
- Air leaks
- Improper pressure and pressure drop

Partial load decreases the energy efficiency of a compressor significantly. All compressed air systems have some uncontrolled air leakage and inappropriate uses of compressed air. The supplying pressure from the air compressors is usually higher than the target pressure at the utilization sites to overcome pressure drop. Pressure drop occurs by insufficient storage volume, extra filtration to remove oil contaminants, and drying to remove carryover water.

5.4 ENERGY CONSERVATION TECHNOLOGIES FOR COMPRESSED AIR SYSTEMS

5.4.1 Energy Conservation in Compressed Air Systems

Compressed air is one of the major utilities in food processing facilities. The production of compressed air can be one of the most expensive processes in manufacturing facilities. There are several publications that describe the energy saving potential of compressed air systems (Talbott, 1993; Cerci et al., 1995; Risi, 1995; Terrell, 1999; Kaya et al., 2002). The energy saving opportunities for compressed air can come from three main areas:

- Compressors themselves
- Compressed air distribution systems
- Compressed air utilization units

Generally, energy saving for a compressed air system can be achieved by (Kaya et al., 2002)

- Installing high-efficiency motors (for compressors)
- Reducing the average air inlet temperature by using outside air (for compressors),
- Repairing air leaks (for distribution systems)
- Reducing compressor air pressure (for utilization units)

5.4.2 High-Efficiency Motors

Most industrial equipment in manufacturing facilities is powered by electric motors. The electrical energy that a motor consumes to generate a specified power output is inversely proportional to its efficiency. Electric motors cannot completely convert the electrical energy consumed into mechanical energy. The ratio of the mechanical power supplied by a motor to the electrical power consumed during operation is called the efficiency of the motor. Therefore, high-efficiency motors cost less to operate than their standard counterparts. Motor efficiencies range from about 70% to over 96%.

Replacement of a standard motor with an energy efficient motor can result in a decrease in energy consumption. Potential savings, S, through the improved efficiency can be calculated by

$$S = p \times L \times C \times N \times \left(\frac{100}{\eta_l} - \frac{100}{\eta_h} \right) \tag{5.1}$$

where

p is the horsepower of the motor (kW)
L is the percentage of full-specified load
C is the electricity price ($/kWh)
N is the life expectancy (h)
η_l and η_h are the efficiencies of a low-efficiency motor and a high-efficiency motor, respectively

Example 5.1

At a food processing facility, there is an 850 kW standard motor associated with a compressor to generate compressed air. The average price of electricity is 0.05 $/kWh. The fraction of rated load is 0.9. At this load, the energy efficiency of the motor is 77%. The annual operating time is 7200 h/year. If a high-efficiency motor at an efficiency of 82% is used to replace the standard motor, what is the annual energy saving?

Solution 5.1

Using Equation 5.1, the annual energy saving with the high-efficiency motor is

$$S = p \times L \times C \times N \times \left(\frac{100}{\eta_l} - \frac{100}{\eta_h}\right) = 850 \times 0.9 \times 0.05 \times (7200 \times 1) \times \left(\frac{100}{77} - \frac{100}{82}\right)$$
$$= \$21{,}809/\text{year}$$

The efficiency of a motor depends on its load. To obtain optimal energy efficiency, compressors should run at their full-specified load. In addition, variable-speed motors can be used to meet varying air demands.

5.4.3 Repairing of Air Leaks

Air leaks are the greatest single cause of energy loss from a compressed air system in manufacturing facilities. The cost of compressed air leaks is the cost of the energy required to compress the lost air from the atmospheric pressure to the compressor operating pressure. Leaks often represent as much as 25% of the output of an industrial compressed air system (Terrell, 1999). Eliminating air leaks totally is impractical, and a leakage rate of 10% is considered acceptable in practice (Cerci et al., 1995). The cost of compressed air leaks increases exponentially with the increase in leak diameters, as shown in Figure 5.2. When the ratio of the atmospheric pressure to the line pressure of compressed air is less than 0.5283, the air flow is considered to be choked (i.e., the air travels at the speed of sound). The volumetric flow rate of free air exiting the holes is dependent on the extent to which the flow is choked. The volumetric flow rate of free air under the choked condition, V_f, m³/h, exiting all the leaks of a given size can be calculated by (Cerci et al., 1995):

$$V_f = \frac{\text{NL} \times (273.15 + T_i) \times P_1/P_i \times C_1 \times C_2 \times C_d \times \pi D^2/4}{C_3 \times \sqrt{273.15 + T_1}} \tag{5.2}$$

where

D is the leak diameter (mm)
NL is the number of air leaks

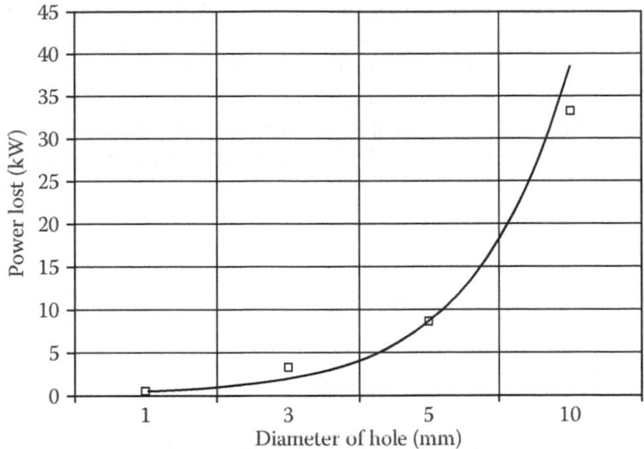

FIGURE 5.2 Power loss as a function of leak diameter at 600 kPa pressure. (Adapted from Kaya, D., Phelan, P., Chan, D., and Sarac, H.I., *Int. J. Energy Res.*, 26, 837, 2002. Copyright John Wiley & Son, Ltd. With permission.)

T_l and T_i are the average temperatures of the air in the pipeline and inlet point of the compressor (°C)
P_l and P_i are the pressures of the air in the pipeline and at the ambient (kPa)
C_1 is the isentropic sonic volumetric flow constant, $C_1 = 7.3587\,\text{m/sK}^{0.5}$
C_2 is the conversion constant, $C_2 = 3600\,\text{s/h}$
C_3 is the conversion constant, $C_3 = 10^6\,\text{mm}^2/\text{m}^2$
C_d is the isentropic coefficient of discharge for square edges orifice, $C_d = 0.8$

The power loss from leaks can be estimated as the power required to compress the volume of air lost from the compressor inlet pressure, P_i, to the compressor discharge pressure, P_l. The power loss from a leak can be calculated by (Cerci et al., 1995)

$$\text{PL} = \frac{P_i \times (1/C_2) \times V_f \times k/(k-1) \times N \times \left[(P_l/P_i)^{(k-1)/(k \times N)} - 1 \right]}{E_a \times E_m} \tag{5.3}$$

where
PL is the power loss from a given air leak (kW)
k is the specific heat ratio of air, $k = 1.4$
N is the number of compression stages
P_l and P_i are the compressor operating pressure and inlet air pressure, respectively (kPa)
E_a is the compressor isentropic efficiency
E_m is the compressor motor efficiency

Example 5.2

In a food processing facility, five air leaks are found and are estimated as 5 mm in diameter. The air compressor used in this facility operates at 800 kPa (absolute pressure) with 1 compression stage. The atmospheric pressure is 101.3 kPa (absolute pressure). The average ambient temperature and compressed air temperature are 25°C and 32°C. Rotary screw compressor isentropic (adiabatic) efficiency is 82% (E_a) and compressor motor efficiency is 90%. Estimate the volumetric flow rate of air from the leaks and determine the power loss. If the compressor operates 24 h/day for 300 days per year and the electricity price is \$0.05/kWh, what will be the energy saving if all leaks are repaired?

Solution 5.2

1. From Equation 5.2, the volumetric flow rate of air from all five leaks is

$$V_f = \frac{NL \times (273.15 + T_l) \times P_l/P_i \times C_1 \times C_2 \times C_d \times \pi D^2/4}{C_3 \times \sqrt{273.15 + T_i}}$$

$$= \frac{5 \times (273.15 + 32) \times (800/101.3) \times 7.3587 \times 3600 \times 0.8 \times 3.14159 \times 5^2/4}{10^6 \times \sqrt{273.15 + 25}}$$

$$= 290.38 \text{ m}^3/\text{h}$$

2. From Equation 5.3, the power loss is

$$PL = \frac{P_i \times \left(1/C_2\right) \times V_f \times k/(k-1) \times N \times \left[\left(P_l/P_i\right)^{(k-1)/(k \times N)} - 1\right]}{E_a \times E_m}$$

$$= \frac{101.3 \times (1/3600) \times 290.38 \times 1.4/(1.4-1) \times 1 \times \left[(800/101.3)^{(1.4-1)/(1.4 \times 1)} - 1\right]}{0.82 \times 0.9}$$

$$= 31.2 \text{ kW}$$

3. The energy savings per year is

$$S = 31.2 \text{ [kW]} \times 24 \text{ [h/day]} \times 300 \text{ [days/year]} \times 0.05 \text{ \$/kWh}$$
$$= 11{,}232 \text{ \$/year}$$

Repairing of air leaks may involve replacement of couplings or hoses, replacement of seals around filters, shutting off air flow during break periods, or repairing breaks in lines. All these costs should be very low (e.g., \$20/leak). Therefore, the payback period for the implementation cost is very short (Cerci et al., 1995).

5.4.4 Reduced Air Pressure

The Affinity law is used to determine the performance of fans and blowers at different rotational speeds. There are three rules of the Affinity law to determine the fan capacity, pressure produced by the fan, and horsepower required to drive the fan with a change in the fan rotational speed, respectively:

$$\frac{\dot{V}_2}{\dot{V}_1} = \frac{RPM_2}{RPM_1} \tag{5.4}$$

$$\frac{P_2}{P_1} = \left(\frac{\text{RPM}_2}{\text{RPM}_1}\right)^2 \tag{5.5}$$

$$\frac{W_2}{W_1} = \left(\frac{\text{RPM}_2}{\text{RPM}_1}\right)^3 \tag{5.6}$$

A small change in motor speed can cause a significant change in energy consumption according to Equation 5.6. Different tools and process unit operations may require compressed air at different pressures. Therefore, energy conservation can be achieved with energy efficient motor retrofits.

Example 5.3

An 850 kW compressor is used to supply compressed air. Suppose the operating time is 7200 h/year and the electricity price is $0.05/kWh. If the pressure of the air can be reduced by 10% of the initial value, what would be the energy saving for 1 year?

Solution 5.3

According to Equation 5.5, if the air pressure is reduced by 10%, the rotational speed will be reduced to

$$\frac{\text{RPM}_2}{\text{RPM}_1} = \left(\frac{P_2}{P_1}\right)^{\frac{1}{2}} = (0.9)^{0.5} = 0.95$$

From Equation 5.6, the power will be reduced to

$$\frac{W_2}{W_1} = \left(\frac{\text{RPM}}{\text{RPM}_1}\right)^3 = 0.95^3 = 85.4\%$$

Therefore, the total electricity savings would be

$$S = (1 - 85.4\%) \times 850 \times 7200 \times 0.05 = 44{,}676 \text{ \$/year}$$

5.4.5 Reduced Air Inlet Temperature

Compressors generate heat during operation. If they are located inside production facilities, the temperature of intake air drawn from the inside of a building is higher than that of outside air. The energy consumption increases with the increase in air intake temperature. Therefore, the air supplied directly from the outside of a building can reduce the energy consumption. The compressor work, W, is proportional to the absolute temperature of the intake air temperature, which can be calculated by (Mull, 2001)

$$\frac{W_2}{W_1} = \frac{T_2}{T_1} \tag{5.7}$$

According the above equation, if the air intake temperature increases by 10°C from the outside temperature of 25°C, the energy consumption of the compressor will increase by 3.35%.

For multi-stage compression, the heat generated by compression work may increase the air temperature to be as high as 205°C if no cooling unit is installed. Cooling the air between stages can increase the density of the air and reduce the power required for compression (Mull, 2001).

5.4.6 Waste-Heat Recovery

Air compressors generate large amounts of heat as they compress air. Adiabatic compression of air to 690 kPa (gauge) results in air temperature between 175°C and 260°C. Approximately 80% of the energy used to compress air finally becomes heat stored in the compressed air. Heat can be recovered from the high-temperature compressed air. Waste-heat recovery is discussed in Chapter 8.

5.5 LOCALIZED AIR DELIVERY SYSTEM

Cold air is widely used in food processing facilities to prevent the growth of airborne microorganisms. Chilled food products must be manufactured in a factory operating at low temperatures to prevent the growth of microorganisms on the foods. Localized air delivery provides the opportunity to reduce the airborne contamination of foods, to improve operator comfort, and to reduce energy use. The concept of localized air delivery is considered in which cold air is supplied close to foods to maintain their low required temperature and higher temperatures are allowed elsewhere in the production area (Burfoot et al., 2000; Burfoot et al., 2004).

For a localized air delivery system as shown in Figure 5.3, cold air from the supply ducts passes through air slots, where the air flow rate and temperature are

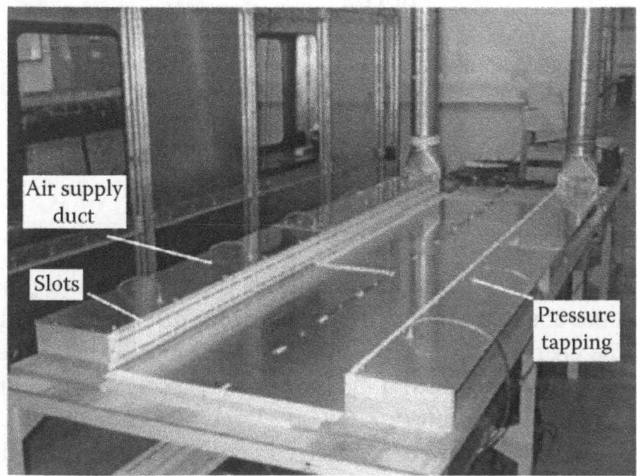

FIGURE 5.3 Localized air supply system. (Reprinted from Burfoot, D., Reavell, S., Wilkinson, D., and Duke, N., *J. Food Eng.*, 62, 23, 2004. With permission.)

adjusted to provide the temperature required by the food ducts. Using a localized air delivery system to supply cold air at 5°C in food ducts operating in a room at 16°C, energy savings are up to 26% and 32% compared to conventional production areas operated at 5°C or 10°C. More energy savings can be achieved by reducing the air flow rate, optimizing the slots in the ducts, and using localized air delivery in areas such as transfer zones in the production area (Burfoot et al., 2004).

5.6 SUMMARY

Compressed air is widely used as an important utility for conveying food products, pneumatic process control, and as a processing medium of many unit operations in a food processing facility. Energy conservation measures can save 20%–50% of energy consumed by a compressed air generation and distribution system. These energy conservation technologies include the use of high-efficiency and variable-speed motors, reduction of inlet air temperature, use of cooling or a waste-heat recovery unit for compressors, reduction of air leaks along the air distribution line, reduction of air pressure, and use of a localized air delivery system.

REFERENCES

Burfoot, D., K. Brown, Y. Xu, S.V. Reavell, and K. Hall. 2000. Localized air delivery systems in the food industry. *Trends in Food Science and Technology* 11: 410–418.

Burfoot, D., S. Reavell, D. Wilkinson, and N. Duke. 2004. Localized air delivery to reduce energy use in the food industry. *Journal of Food Engineering* 62: 23–28.

California Energy Commission (CEC). 2008. *California's Food Processing Industry Energy Efficiency Initiative: Adoption of Industrial Best Practices.* Available at http://www.energy.ca.gov/2008publications/CEC-400-2008-006/CEC-400-2008-006.PDF

Cerci, Y., Y.A. Cengel, and H.T. Turner. 1995. Reducing the cost of compressed air in industrial facilities. *Thermodynamics and the Design, Analysis, and Improvement of Energy Systems, ASME, AES* 35: 175–86.

Kaya, D., P. Phelan, D. Chau, and H.I. Sarac. 2002. Energy conservation in compressed air systems. *International Journal of Energy Research* 26: 837–849.

Mull, T.E. 2001. *Practical Guide to Energy Management for Facilities Engineers and Plant Managers.* New York: ASME Press.

Risi, J.D. 1995. Energy savings with compressed air. *Energy Engineering* 92: 49–58.

Talbott, E.M. 1993. *Compressed Air Systems: A Guidebook on Energy and Cost Savings* (2nd ed.), Lilburn, GA: The Fairmont Press, Inc.

Terrell, R.E. 1999. Improving compressed air system efficiency: Know what you really need. *Energy Engineering* 96: 7–15.

6 Energy Conservation in Power and Electrical Systems

6.1 INTRODUCTION

Electricity is a main energy source used in the food industry, which is 20% of the total energy consumed in the food industry of the United States. The food manufacturing industry produces about 9% of the electricity with its onsite power generation systems, 95% of which are cogeneration systems for simultaneous supply of power and steam (U.S. EPA, 2007). In the food industry, about 25% of the electricity is used for process cooling and food refrigeration and 48% for machine drive (Okos et al., 1998). However, in the meat sector, refrigeration constitutes between 40% and 90% of total electricity use during production time and almost 100% during nonproduction time (Ramiez et al., 2006). The mover of a mechanical compression refrigerator is a motor or compressor. Pumps and pumping represent one of the most important aspects of nearly all food processing operations. Pumps are a mechanical system of moving liquid and semi-liquid products in or around a food processing facility. Positive displacement pumps and centrifugal pumps are two main types of pumps. Motors are needed to drive a pump. Non-process uses for space heating, venting, air conditioning, lighting, and onsite transport consume about 16% of electricity (Okos et al., 1998).

Since motors, which are used to drive refrigerators and pumps, are the main consumer of electricity in the food industry, the focus of this chapter is energy conservation technologies for electrical motors. Electric motors were primarily chosen on the basis of their price. With the increased price of electrical energy, it is important to consider the energy efficiency of motors along with their price when buying electrical motors. The electricity bill may be determined by power demand, electricity consumption, power factor, and ratchet charge. In this chapter, typical electrical loads and electricity billing structure are reviewed. The sources of energy loss from electrical equipment are then discussed. Finally, energy conservation technologies for power and electrical systems are discussed.

6.2 TYPICAL ELECTRICAL EQUIPMENT

6.2.1 Types of Electrical Loads

There are three types of electrical loads:

- Resistance
- Inductance
- Capacitance

Standard utility power systems supply alternating current (AC) and voltage. In a load with an electrical resistance, the AC current and voltage perform work during both the positive and the negative cycles. During both positive and negative cycles, the power, which is the product of current and voltage, is always positive. The power load of resistance can be calculated by

$$P_{\mathrm{R}}(\mathrm{W}) = I(\mathrm{A}) \times E(\mathrm{V}) = \frac{E^2(\mathrm{V})}{R(\Omega)} \tag{6.1}$$

Inductance is the property of a coil that opposes a change of current in a circuit. A pure or ideal inductor is one without any electrical resistance. The property of an inductor has a significant effect on a power line. However, in an electrical load with an inductor, the current and voltage are out of phase by 90° of their product, and the power thus goes from positive to negative each half cycle. This means that the inductor takes power from the source and returns the power to the source each half cycle, resulting in zero power dissipation or consumption. An AC wattmeter connected to the inductive load will measure zero watts. However, a VAR meter can measure the product of current and voltage, or VAR, of an inductor load. Since there is no real power load for an inductor, a reactive power load, which is the product of voltage and current, is defined. The reactive power load of an inductor can be calculated by

$$P_{\mathrm{L}}(\mathrm{VARs}) = I(\mathrm{A}) \times E(\mathrm{V}) = \frac{E^2(\mathrm{V})}{X_{\mathrm{L}}(\Omega)} \tag{6.2}$$

Capacitance is a property that opposes any change of voltage in a circuit. The current in a capacitor leads the voltage changes across the capacitor by 90°. Like an inductor, the power dissipation or consumption of a capacitor is also zero. The reactive power load of a capacitor is given by

$$P_{\mathrm{C}}(\mathrm{VARs}) = I(\mathrm{A}) \times E(\mathrm{V}) = \frac{E^2(\mathrm{V})}{X_{\mathrm{C}}(\Omega)} \tag{6.3}$$

The product of current and voltage is defined as apparent power while the power supplied to the resistance loads is real power. Although there is current and voltage through a resistor, inductor, and capacitor, there is real power in the resistor but no real power in the inductor and capacitor. In a circuit with resistance load, and inductance or capacitance, the apparent power should be bigger than the real power.

6.2.2 Electric Motors

Motors are designed for rated voltage, frequency, and number of phases. Motor efficiency, which is the shaft power divided by the electrical input power, is expressed as

$$\eta = \frac{746 \times \mathrm{HP\ output}}{\mathrm{Watts\ input}} \times 100\% \tag{6.4}$$

The efficiency of a motor is usually stamped on the nameplate of the motor.

6.3 ELECTRICITY BILL STRUCTURE

An electricity bill may be determined by electricity consumption, power demand, power factor, and ratchet charge. Utilities usually use a declining block rate of electricity consumption for very large users and assesses a rather heavy charge for power demand.

Utility companies sometimes choose to charge for the power factor by modifying the power demand charge. Electricity customers should maintain a power factor of at least 80%; otherwise, they may be penalized for the low power factors of their electrical power systems. Usually, there is no charge or reward if the power factor is above 80%. For the power factor lower than 80%, the billed demand may be determined by

$$\text{Billed demand} = \text{Actual demand} \times \frac{\text{Base power factor (0.8)}}{\text{Actual power factor}} \tag{6.5}$$

Utility companies may have a ratchet clause. A sample one is that the billed demand for any month is 65% of the highest on-peak season maximum demand corrected for the power factor of the previous 12 months or the actual demand corrected for an actual power factor, whichever is greater.

6.4 SOURCES OF ENERGY LOSSES IN POWER AND ELECTRICAL SYSTEMS

6.4.1 Low Power Factor

Three are three types of currents produced in a circuit: working current, reactive current, and total current. Working current is a power-producing current, which is finally converted into useful work by the equipment. The unit of the real power produced from the working current is kilowatt (kW). Reactive current is the current that is required to produce the flux necessary to operate electrical devices. The unit of the reactive power produced from the reactive current is kilovar (kVAR). The third current is total current, which produces the apparent power or total power. The unit of the apparent power is kilovoltampere (kVA). The relationship of the three types of powers is

$$\text{kVA} = \sqrt{(\text{kW})^2 + (\text{kVAR})^2} \tag{6.6}$$

The power factor, pf, is the ratio of real power to apparent power, which is expressed as

$$\text{pf} = \frac{\text{Real power } (P)}{\text{Apparent power } (I \times E)} = \frac{\text{kW}}{\text{kVA}} = \cos\theta \tag{6.7}$$

where

pf is an angle in degrees

Real power is the power supplied to resistance loads

Apparent power includes the real power and reactive power supplied to reactive loads including inductance and capacitance

The power factor of pure resistance load is 1. A power factor may lead or lag the real power depending on inductance and capacitance loads in a circuit. In most cases, a low power factor is caused by the use of inductive or magnetic devices, which include (Lobodovsky, 2006)

- Induction motors (55%–90% pf)
- Lifting magnets (20%–50% pf)
- Induction heating equipment (60%–90% pf)
- Small dry pack transformers (30%–95% pf)
- Non-power factor corrected fluorescent and high-intensity discharge lighting fixture ballasts (40%–80% pf)
- Arc welders (50%–70% pf)
- Solenoids (20%–50% pf)

Among those inductive devices, induction motors are generally the main cause of a low power factor due to the large number of motors used in food processing facilities. Capacitors can be used to improve the power factor of a circuit with a large inductive load. The current through a capacitor leads the voltage by 90° and has an opposing effect on the 90° lagging current caused by an inductive load. The increase in power factors may reduce the penalty charges by the electric utility companies and reduce power losses in feeders, transformers, and distribution equipment. Capacitors can be placed on each piece of equipment, ahead of the equipment group and at the main service line (Lobodovsky, 2006).

Example 6.1

A clamp-on ammeter measured the current and voltage of a motor as 20 A and 120 V and a wattmeter measured the real load of the motor as 1250 W. What were the power factor and the phase angle?

Solution 6.1

The apparent power of the circuit is calculated from the measured current and voltage, which is 20 A × 120 V = 2200 VARs and the measured real power is 1250 W. Therefore, from Equation 6.7, the power factor is

$$\text{pf} = 1250\,\text{W}/2200\ \text{VARs} = 0.568$$

The phase angle between the current and voltage in the motor is $\cos^{-1}(0.568) = 56°$.

6.4.2 Improper Motor Load

Most electric motors are designed to operate at 50% to 100% of their rated load. Optimum efficiency is generally achieved at 75% of the rated load since motors are usually designed for their starting requirement of high load. Underloaded motors operate inefficiently and have low power factors. Typical part-load efficiency and power factor characteristics of motors are given in Figure 6.1. The motor load is defined as 1 (Lobodovsky, 2006)

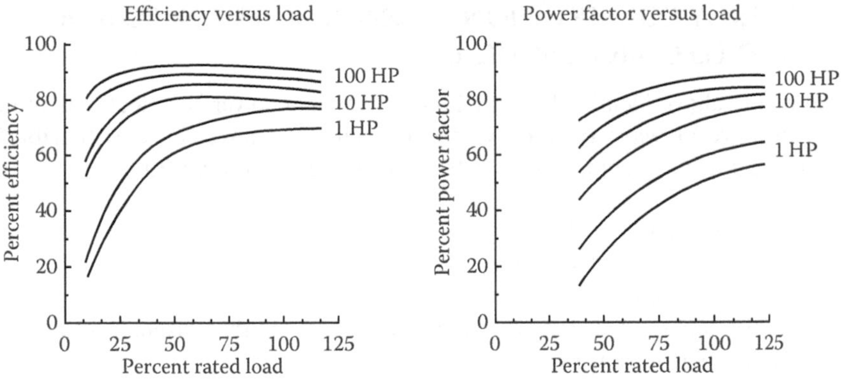

FIGURE 6.1 Typical part-load efficiency and power factor characteristics. (Reprinted from Lobodovsky, K.K., *Energy Management Handbook*, The Fairmont Press, Inc., Georgia, 2006. With permission.)

$$(\%\text{ motor load}) = \frac{\text{NLS} - \text{OLS}}{\text{NLS} - \text{FLS}} \times 100\% \tag{6.8}$$

where

NLS is the no load or synchronous speed
OLS is the operating load speed
FLS is the full load speed

Example 6.2

For a 50 horsepower (HP) and 1800 RPM no load speed (NLS) motor, if the full load speed (FLS) and operating load speed (OLS) are 1775 RPM and 1786 RPM, respectively, what is the motor load?

Solution 6.2

From Equation 6.8, the motor load is

$$(\%\text{ motor load}) = \frac{\text{NLS} - \text{OLS}}{\text{NLS} - \text{FLS}} \times 100\% = \frac{1800 - 1786}{1800 - 1775} \times 100\% = 56\%$$

6.4.3 Poor Control

Many pump and fan applications involve the control of flow or pressure by means of throttling or bypass devices. Throttling and bypass valves are in series or parallel to power regulators that perform their function by dissipating the difference between the source energy supplied and the desired sink energy. These losses can be significantly reduced by controlling the flow rate or pressure via control of the pump or fan speed with a variable speed drive.

6.5 ENERGY CONSERVATION TECHNOLOGIES FOR POWER AND ELECTRICAL SYSTEMS

Based on the above identified energy loss sources in a power and electrical system, there are several ways to improve the efficiency of an electrical distribution system and a drive system, some of which include (Lobodovsky, 2006)

- Maintain voltage levels
- Minimize phase imbalance
- Maintain a high power factor
- Identify and fix distribution system losses, particularly resistance loss
- Select efficient transformers
- Use adjustable speed motors
- Select efficient motors
- Match motor operating speeds
- Size motor for high energy efficiency

6.5.1 Power Management for Demand Control

Reduction of peak power demand is very important when a utility company charges the peak demand in electrical bills. The utility company is concerned about the peak power demand because the size of the equipment it uses to generate and transmit electricity must be sufficient to handle the peak demand of its customers. Therefore, many utility companies may charge the peak demand. The electrical demand is measured in kilowatts (kW) over a short time, usually 15 to 30 min. Billing rates for the peak demand vary greatly by billing plans. The highest recorded electrical demand during the month may be used to determine the demand charge on a monthly electrical bill. Sometimes the peak demand charges are based on the highest average peak demand in one demand interval during the previous year. This is known as a "ratchet" clause. In such cases, electric charges will be affected every month for a single 15 or 30 min period of the highest demand during the previous year.

Power management devices, which regulate the on and off times of loads such as fans, pumps, and motors, can reduce the electrical demand and regulate electricity consumption. In multiple shift operations, it may be possible to reduce peak demand by performing energy intensive processes on different shifts. For example, using dryers on one shift and cookers on another will reduce the peak load, compared to running both at the same time, and therefore, will reduce energy costs. Some power management devices can serve as load shedders. They can reduce the demand in critical demand periods by interrupting the electricity supply to motors, fans, and other loads for a short period to avoid the demand charge.

Example 6.3

A food company has four large machines in its processing facility. The machines have a demand of 100 kW each. According to the production requirement, four machines with a total demand of 400 kW should be used for one 8 h shift. The company has chosen to limit the use of the machines by changing one shift to two shifts. What could be the total cost change in the next bill? The electricity billing structure is

- Customer cost: $21.00/month
- Energy cost: $0.05/kWh
- Demand cost: $6.50/kW month
- Demand ratchet: 65%
- Power factor: 75%
- Taxes: 8%
- Fuel adjustment: 1.15 ¢/kWh
- The highest corrected demand in the previous year: 500 kW

Solution 6.3

1. Demand charge:

   The ratchet clause is Ratch = 500 × 0.65 = 325 kW
   From Equation 6.5, the billed demand prior to shift change is

$$\text{Billed demand before change} = \text{Actual demand} \times \frac{\text{Base power factor (0.8)}}{\text{Actual power factor}}$$

$$= 400 \times \frac{0.8}{0.75} = 427\,\text{kW}$$

   The actually billed demand before change = max (427 kW, 325 kW) = 427 kW

$$\text{Billed demand after change} = \text{Actual demand} \times \frac{\text{Base power factor (0.8)}}{\text{Actual power factor}}$$

$$= 200 \times \frac{0.8}{0.75} = 213\,\text{kW}$$

   The actually billed demand after change = max (213 kW, 325 kW) = 325 kW
   The saving due to demand charge is $S1$ = (427 − 325) × 6.50 = 663 $/month

2. Electricity consumption costs:

   There is no change in energy consumption. The total cost of electricity is
   400 kW × 8 h/day × 30 day/month × ($0.05/kWh + $0.0115/kWh) = 5904 $/month
   $S2 = 0$

3. Total savings are

   $S = S1 + S2$ = 663 $/month
   After tax
   S' = 663 × (1 + 0.08) = 716 $/month
   This shift change may also reduce the ratchet clause next year.

6.5.2 Power Factor Improvement

Induction motors are often the major cause of low power factor in food processing facilities as they are the main electrical loads. Power factor improvement can increase the facility capacity and reduce any power factor penalty from utility companies. Power factor improvement can improve the voltage characteristics of the supply and reduce power losses in distribution lines and transformers. A normal watt-hour meter shows only the real power delivered to the system, irrespective of power factor. If the electricity is billed by a watt-hour meter, improvement of power factor may not reduce the utility bill directly but it may reduce the bill indirectly because implicit energy losses due to a low power factor occur in utility lines and transformers.

Example 6.4

A 230 volt electrical distribution system has a lagging power factor of 0.7. If the power factor has to be increased to a lagging power factor of 0.9, what size capacitor should be used to correct a 10 kW load?

Solution 6.4

Using Equation 6.7, the apparent power before installing the capacitor is

$$\text{kVA} = \frac{\text{kW}}{\text{pf}} = \frac{10\,[\text{kW}]}{0.7} = 14.3\,\text{kVA}$$

Using Equation 6.6, the reactive power before installing the capacitor is

$$\text{kVAR} = \sqrt{(\text{kVA})^2 - (\text{kW})^2} = \sqrt{14.3^2 - 10^2} = 10.22\,\text{kVAR}$$

The apparent power after installing the capacitor is

$$\text{kVA}' = \frac{\text{kW}}{\text{pf}} = \frac{10\,[\text{kW}]}{0.9} = 11.1\,\text{kVA}$$

The reactive power after installing the capacitor is thus

$$\text{kVAR}' = \sqrt{(\text{kVA¢})^2 - (\text{kW})^2}\ \sqrt{11.1^2 - 10^2} = 4.82\,\text{kVAR}$$

The decrease in reactive power caused by the addition of a capacitor is the difference between the reactive powers before and after installing the capacitor, which is

$$\Delta\text{kVAR} = 10.22 - 4.82 = 5.4\,\text{kVAR}$$

Using Equation 6.3, the size of the capacitor should be

$$X_c = \frac{E^2}{\Delta\text{kVAR}} = \frac{(230\text{ V})^2}{5400\text{ VAR}} = 9.8\,\Omega$$

6.5.3 Replacement with High-Efficiency Motors

Replacement of a standard motor with an energy efficient motor can decrease energy consumption. Potential economic savings, S, through improved efficiency can be calculated by

$$S = 0.746 \times \text{hp} \times L \times C \times N \times \left(\frac{100}{\eta_l} - \frac{100}{\eta_h}\right) \tag{6.9}$$

where

hp is the horsepower of the motor (hp)
L is the percentage of rated load
C is the electricity price (\$/kWh)
N is the life expectancy (h)
η_l and η_h are efficiency of a low-efficiency motor and high-efficiency motor, respectively

Example 6.5

An old 5 hp electrical motor has an efficiency of 77% and 40,000 h life expectancy. If the electricity price is \$0.05/kWh and a high-efficiency motor at an efficiency of 85% is used as an alternative, what could be the cost saving through the whole life of the motor? Suppose that the motor operates at 100% of the rated load.

Solution 6.5

Using Equation 6.9, the replacement will save

$$S = 0.746 \times \text{hp} \times L \times C \times N \times \left(\frac{100}{\eta_l} - \frac{100}{\eta_h}\right)$$

$$= 0.746 \times 5 \times 100\% \times 0.05 \times 40{,}000 \times \left(\frac{100}{77} - \frac{100}{85}\right)$$

$$= \$912$$

If the loads are sensitive to the motor speed, a correction factor of $\left(\frac{\text{RPM}_{ee}}{\text{RPM}_{std}}\right)^3$ should be used to correct the expected savings.

6.5.4 Replacement with Electronic Adjustable Speed Motors

In most motor installations, motors are sized to provide the maximum power output required. If the rotational speed is constant at its maximum value to provide the maximum designed load, the power input to the motor remains constant at the maximum value. However, if the load decreases, significant energy savings can be achieved if the rotational speed of the motor is decreased to match the load requirement of the device driven by the motor. The Affinity law is used to determine the performance of fans and blowers at different rotational speeds. There are three rules of the Affinity law to determine the fan capacity, pressure produced by the fan, and the horsepower required to drive the fan with a change in the fan rotational speed:

$$\frac{\dot{V}_2}{\dot{V}_1} = \frac{\text{RPM}_2}{\text{RPM}_1} \tag{6.10}$$

$$\frac{P_2}{P_1} = \left(\frac{\text{RPM}_2}{\text{RPM}_1}\right)^2 \tag{6.11}$$

$$\frac{W_2}{W_1} = \left(\frac{\text{RPM}_2}{\text{RPM}_1}\right)^3 \tag{6.12}$$

It can be seen from Equation 6.12 that the power required to drive a load such as a fan and a pump is proportional to the cube of the rotational speed. A small change in motor speed can cause a significant change in energy consumption. Therefore, energy conservation can be achieved with energy efficient motor retrofits.

Variable speed drive can change the speed of the motor by changing the voltage and frequency of the electricity supplied to the motor based on the load requirement. This is accomplished by converting the AC to DC and then by inverting the DC to a synthetic AC output with controlled voltage and frequency based on various switching mechanisms. Variable frequency drives change the speed of a motor by changing the voltage and frequency of the power supplied to the motor (Hordeski, 2003). A variable frequency drive motor can result in energy savings by eliminating throttling and frictional losses affiliated with mechanical or electromechanical adjustable speed technologies.

Example 6.6

A fan, which has a power input of 15 kW at the rated speed of 1800 RPM, is used to supply 1000 m³/min air for an air blast chiller at an operating speed of 1750 RPM. A new energy efficient motor, which has a power output of 15 kW at the rated speed of 1800 RPM, is considered as an alternative. The new motor operates at 1790 RPM speed. Can this energy efficient motor save energy?

Solution 6.6

The new motor will increase the volumetric flow rate of air supply:

$$\begin{aligned}\dot{V}_2 &= \dot{V}_1 \times \frac{\text{RPM}_2}{\text{RPM}_1}\\ &= 1000 \times \frac{1790}{1750}\\ &= 1023\ \text{m}^3/\text{min}\end{aligned}$$

$$\begin{aligned} W_2 &= W_1 \times \left(\frac{\text{RPM}_2}{\text{RPM}_1}\right)^3 \\ &= 15 \times \left(\frac{1790}{1750}\right)^3 \\ &= 16.05 \text{ kW} \end{aligned}$$

Therefore, the replacement of a standard motor with an energy efficient motor in the fan can result in a 7% increase in energy consumption if the energy efficient motor operates at a higher RPM. However, the energy efficient motor at a higher RPM can supply more air at a flow rate of 1023 m^3/min than 1000 m^3/min for the old motor.

Example 6.7

A 400 kW compressor is used to supply air. If the pressure of the air can be reduced to one-half of the initial value, what could be the energy savings for 1 year? Suppose that the operating time of the compressor is 4000 h per year and the electricity price is $0.05/kWh.

Solution 6.7

According to Equation 6.11, the ratio of the rotational speed of the compressor before and after pressure reduction is

$$\frac{\text{RPM}_2}{\text{RPM}_1} = \left(\frac{P_2}{P_1}\right)^{\frac{1}{2}} = 0.5^{0.5} = 0.707$$

The power ration is thus

$$\frac{W_2}{W_1} = \left(\frac{\text{RPM}_2}{\text{RPM}_1}\right)^3 = 0.707^3 = 35.4\%$$

Therefore, the total electricity savings would be

$$S = 35.4\% \times 400 \times 4000 \times 0.05 = 2832 \text{ \$/year}$$

However, the actual savings would be much lower than this theoretical value.

6.6 SUMMARY

Electric motors are used extensively to drive fans, pumps, conveyors, and many other operations in food processing facilities. Electric motors are the largest electricity consumers in food processing facilities. Cost savings can be achieved by controlling the peak demand, improving the power factor, and reducing the electricity consumption. Since the energy loss in a motor is in the range of 5%–30% of the input power, it is important to consider energy conservation technologies for motors. High efficiency, selection of a high power factor, and a variable speed motor may result in significant savings when purchasing a new motor. In addition, the required load should closely match with the motor power so that the motor will be operated at a full load and high efficiency.

REFERENCES

Hordeski, M.F. 2003. *New Technologies for Energy Efficiency*. New York: Marcel Dekker, Inc.

Knutson, G.D. 1986. Selection of electric motors and electrical measurements on motors for maximum efficiency, Chapter 6. In *Energy in Food Processing*, Singh, R.P. (Ed.), New York: Elsevier Science Publishing Company Inc.

Okos, M., N. Rao, S. Drecher, M. Rode, and J. Kozak. 1998. *Energy Usage in the Food Industry*. American Council for an Energy-Efficient Economy. Available at http://www.aceee.org/pubs/ie981.htm

Ramirez, C.A., M. Patel, and K. Blok. 2006b. How much energy to process one pound of meat? A comparison of energy use and specify energy consumption in the meat industry of four European countries. *Energy* 31: 2047–63.

U.S. Environmental Protection Agency. 2007. *Energy Trends in Selected Manufacturing Sectors: Opportunities and Challenges for Environmentally Preferable Energy Outcomes*. Online: http://www.epa.gov/ispd/energy/index.html

Lobodovsky, K.K. 2006. Electric energy management, Chapter 11. In *Energy Management Handbook* (6th ed.), Turner, W.C. and S. Doty (Eds.), Lilburn, GA: The Fairmont Press, Inc.

7 Energy Conservation in Heat Exchangers

7.1 INTRODUCTION

In food processing facilities, many unit operations such as refrigeration, freezing, thermal sterilization, drying, and evaporation involve the transfer of heat between food products and heating or cooling medium. Heating and cooling of foods is achieved with heat exchangers. Heat exchangers also play a key role in waste heat recovery. There is a wide variety of heat exchangers available for food applications. The area or size of heat exchangers affects their effectiveness and initial and operating costs.

There are two fluid streams at different temperatures present in a heat exchanger. The fluids are usually separated by a wall through which the heat is transferred from the fluid on one side to the fluid on the other side. Several energy conservation technologies including heat transfer enhancement, fouling removal, and optimization of heat exchanger network have been used to improve the energy efficiency of heat exchangers. In this chapter, different types of heat exchangers used in food processing facilities are reviewed. Energy analysis of a heat exchanger is then discussed. Finally, several energy conservation technologies applied to heat exchangers are discussed.

7.2 TYPICAL HEAT EXCHANGERS

7.2.1 Double-Pipe Heat Exchanger and Shell and Tube Heat Exchanger

A double-pipe heat exchanger consists of a pipe located concentrically inside another pipe. Two fluids flow in the annular space and in the inner pipe, respectively. Shell and tube heat exchangers as shown in Figure 7.1 are most common in food processing facilities. The heat exchanger consists of a fixed head with inlet and outlet ports for the heating fluids. A tube bundle is placed within the shell. One end of the tubes is attached to the fixed head while the other end is often attached to a floating head that allows the tubes to expand and contract by the temperature change without damage. The fluid flow path on the shell side becomes complex by the presence of baffles to force the fluid to flow across the tubes (Rozzi et al., 2007).

7.2.2 Plate Heat Exchanger

Plate heat exchangers are widely used in the dairy industry. A plate heat exchanger is shown in Figure 7.2. A plate heat exchanger consists of a series of parallel, closely spaced plates pressed in a frame. Gaskets, made of natural or synthetic rubber, seal the plate edges and ports to prevent intermixing of liquids. The hot and cold streams

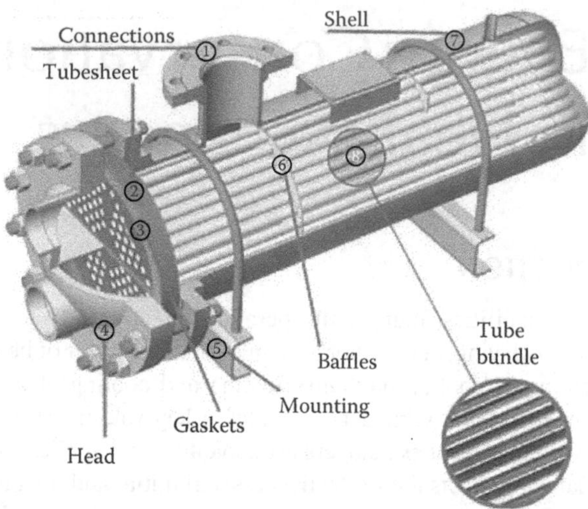

FIGURE 7.1 Shell and tube heat exchanger.

are directed into the alternate gaps formed by two plates by the gaskets. Plate heat exchangers can be easily dismantled for cleaning. The capacity of a plate heat exchanger can be adjusted by adding and removing plates to the frame. If the directions of the product stream and processing medium steam are in the same direction, it is a parallel flow plate heat exchanger. Otherwise, if the directions are opposite to each other, it is a counterflow heat exchanger. The plates used for food processing are usually constructed from stainless steel.

FIGURE 7.2 Plate heat exchanger.

Plate heat exchangers are extensively used for heating, cooling, and heat regeneration in food, chemical, and pharmaceutical industries because of their high thermal efficiency, flexibility, and ease of sanitation (Galeazzo et al., 2006). Plate heat exchangers are suitable for low-viscosity liquid foods such as dairy products and juices. Large solid particles present in food products can bridge across the plate contact points. Kim et al. (1999) determined that the heat transfer coefficients in a plate heat exchanger for heating orange juice and hot water were from 983 to 1,046 W/m^2°C and from 8,387 to 24,245 W/m^2°C, respectively. Fouling on the plate surface decreases the heat transfer between the processing medium and the product. For dairy products, usually ultrahigh temperature applications are required and the process time is often limited to 3–4 h (Singh and Heldman, 2001). The temperature difference between two streams separated by a plate can be within 1°C. Plate heat exchangers offer opportunities for energy conservation by regeneration.

7.2.3 Scraped-Surface Heat Exchanger

A scraped-surface heat exchanger is shown in Figure 7.3. The cylinder containing food products and a rotor is enclosed in an outside jacket. The inside rotor contains blades that are covered with plastic laminate or molded plastic. The rotor speed varies between 150 and 500 rpm. The heating/cooling medium is supplied to the outside jacket. Scraped surface heat exchangers are frequently used in the food and chemical industry for heating or cooling of high-viscosity products such as cream cheese, ice cream, and fruit concentrate. The main advantage of scraped surface heat exchangers is the prevention of fouling of the heat exchange surface by means of periodic scraping with blades (Mabit et al., 2003, 2004).

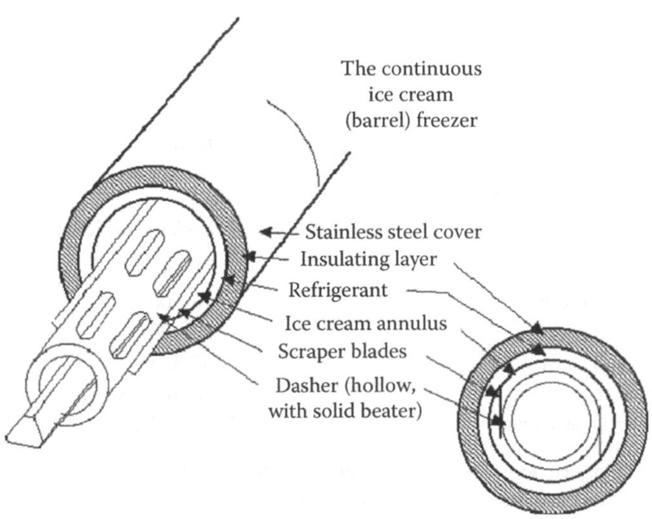

FIGURE 7.3 Scraped-surface heat exchanger.

7.3 ANALYSIS OF HEAT EXCHANGERS

7.3.1 Heat Transfer Coefficient

When a fluid comes into contact with a solid body, heat exchange occurs between the solid surface and the fluid by convection whenever there is a temperature difference between the two. Depending on whether the fluid flow is artificially induced or occurs naturally, there are two types of convective heat transfer: natural (or free) convection and forced convection. Heat transfer coefficient is an important parameter in the design and analysis of a heat exchanger. Convective heat transfer coefficient, h, depends on a number of properties of a fluid (i.e., density, specific heat, viscosity, and thermal conductivity), the velocity of the fluid, geometry, and roughness of the surface of the solid in contact with the fluid. Table 7.1 gives some approximate values of h at different conditions.

The heat transfer coefficients of surface convection are mostly calculated using a correlation between a set of dimensionless numbers:

$$\text{Nusselt number: Nu} = hL/k \tag{7.1}$$

$$\text{Prandtl number: Pr} = c\mu/k \tag{7.2}$$

$$\text{Reynolds number: Re} = \rho L u/\mu \tag{7.3}$$

$$\text{Grashof number: Gr} = L^3\rho^2 g\beta\Delta T/\mu^2 \tag{7.4}$$

In the above formulas, h is the convective heat transfer coefficient (W/m^2 °C); L is the characteristic dimension (e.g., diameter of a pipe) (m); k is the thermal conductivity of the fluid (W/m°C); c is the specific heat of the fluid (J/kg°C); μ is the viscosity of the fluid (Pa s); u is the velocity of the fluid (m/s); g is the acceleration due to gravity

TABLE 7.1
Some Approximate Values of Convective Heat Transfer Coefficients

Fluid	Convective Heat Transfer Coefficient (W/m^2°C)
Air free convection	5–25
Air forced convection	10–200
Water free convection	20–100
Water forced convection	50–10,000
Boiling water	3,000–100,000
Condensing water vapor	5,000–100,000

Source: Adapted from Singh, R.P. and Heldman, D.R., *Introduction to Food Engineering*, Academic Press, San Diego, 2001. With permission.

(m/s^2); β is the thermal expansion of the fluid (°C^{-1}); and ΔT is the temperature difference between the solid surface and the bulk flow.

Using empirical correlation to calculate convective heat transfer coefficient can follow four steps (Singh and Heldman, 2001):

- Identify flow geometry
- Determine the properties of the fluid
- Calculate the Reynolds number
- Select an appropriate empirical correlation

It should be noted that such correlations are normally restricted to a given range of operating conditions and reasonable accuracy can only be ensured under the given range of operating conditions. Air, water, and steam are the three main processing fluids used in food processing facilities. Their properties are given in Table 7.2. Some empirical correlations are given in Table 7.3.

Example 7.1

Hot air at 70°C and 0.5 m/s is passed through a bed of green peas. Assume that the surface temperature of peas is 30°C and the diameter of peas is 0.5 cm. Determine the convective heat transfer coefficient on the pea surface.

Solution 7.1

1. Identify the flow geometry:
   The air flows around a spherically immersed object, green peas. The characteristic dimension is $d_s = 0.5\,\text{cm}$
2. Identify the fluid and determine its properties:
   The properties of air are evaluated at

$$T_f = \frac{T_f + T_\infty}{2} = \frac{30+70}{2} = 50°\text{C}$$

From Table 7.2, the properties of air at 50°C are

$$\rho = 1.057\ \text{kg/m}^3$$

$$c_p = 1.016\ \text{kJ/kgK}$$

$$k = 0.0272\ \text{W/mK}$$

$$\mu = 19.515 \times 10^{-6}\ \text{Pas}$$

3. Calculate the Reynolds number:

$$\text{Re} = \frac{\rho d u}{\mu} = \frac{1.057 \times 0.005 \times 0.5}{19.515 \times 10^{-6}} = 135.41$$

TABLE 7.2
Properties of Air, Water, and Saturation Steam

Temperature (°C)	Density (kg/m^3)	Specific Heat (kJ/kg K)	Thermal Conductivity (W/mK)	Viscosity ($\times 10^{-6}$ Pa s)	Expansion Ratio ($\times 10^{-4}$ K^{-1})
Air					
–20	1.365	1.005	0.0226	16.279	39.7
0	1.252	1.011	0.0237	17.456	36.5
10	1.206	1.010	0.0244	17.848	35.3
20	1.164	1.012	0.0251	18.240	34.1
30	1.127	1.013	0.0258	18.682	33.0
40	1.092	1.014	0.0265	19.123	32.0
50	1.057	1.016	0.0272	19.515	31.0
60	1.025	1.017	0.0279	19.907	30.0
70	0.996	1.018	0.0286	20.398	29.1
80	0.968	1.019	0.0293	20.790	28.3
90	0.942	1.021	0.0300	21.231	27.6
100	0.916	1.022	0.0307	21.673	26.9
Water					
0	999.9	4.226	0.558	1793.636	–0.7
10	999.7	4.195	0.577	1296.439	0.95
20	998.2	4.182	0.597	993.414	2.1
30	995.7	4.176	0.615	792.377	3.0
40	992.2	4.175	0.633	658.026	3.9
50	988.1	4.178	0.647	555.056	4.6
60	983.2	4.181	0.658	471.650	5.3
70	977.8	4.187	0.668	404.034	5.8
80	971.8	4.194	0.673	352.059	6.3
90	965.3	4.202	0.678	308.909	7.0
100	958.4	4.211	0.682	277.528	7.5

Source: Adapted from Singh, R.P. and Heldman, D.R., *Introduction to Food Engineering*, Academic Press, San Diego, 2001. With permission.

$$\mathrm{Pr} = \frac{c_p \mu}{k} = \frac{1.016 \times 1000 \times 19.515 \times 10^{-6}}{0.0272} = 0.73$$

4. From Table 7.3, we can select a correlation to determine the Nusselt number:

$$\begin{aligned} \mathrm{Nu} &= 2 + 0.60\mathrm{Re}^{0.5}\,\mathrm{Pr}^{0.33} \\ &= 2 + 0.60 \times 135.41^{0.5}\, 0.73^{0.33} \\ &= 8.29 \end{aligned}$$

TABLE 7.3
Empirical Correlations for Calculation of Convective Heat Transfer Coefficients

Convection Type	Utilization Conditions	Correlations
Free convection on a vertical plate	Characteristic length, L, properties at film temperature, $10^4 < \text{Re}\,Gr < 10^9$	$\text{Nu} = 0.59(Gr\,\text{Re})^{0.25}$
Free convection on a horizontal plate (upper surface of a hot plate or lower surface of a cold plate)	Characteristic length, A/p, properties at film temperature, $10^4 < \text{Re}\,Gr < 10^7$	$\text{Nu} = 0.54(Gr\,\text{Re})^{0.25}$
Free convection on a horizontal plate (upper surface of a cold plate or lower surface of a hot plate)	Characteristic length, A/p, properties at film temperature, $10^5 < \text{Re}\,Gr < 10^{11}$	$\text{Nu} = 0.27(Gr\,\text{Re})^{0.25}$
Laminar flow in pipes	Re < 2100, constant surface temperature, fully developed flow, properties at average fluid temperature	$\text{Nu} = 3.66$
Laminar flow in pipes	Re < 2100, uniform surface heat flux, fully developed flow, properties at average fluid temperature	$\text{Nu} = 4.36$
Transition flow in pipes	2100 < Re < 10,000, properties at average fluid temperature	$\text{Nu} = \dfrac{(f/8)(\text{Re}-1000)\text{Pr}}{1+1.27(f/8)^{1/2}(\text{Pr}^{2/3}-1)}$, $f = \dfrac{1}{(0.790\ln\text{Re}-1.64)^2}$
Turbulence in pipes	Re > 10,000, properties at average fluid temperature but μ_w at surface temperature	$\text{Nu} = 0.023\,\text{Re}^{0.8}\,\text{Pr}^{0.33}\left(\dfrac{\mu_b}{\mu_w}\right)^{0.14}$
Flow past immersed spherical objects	1 < Re < 70,000 0.6 < Pr < 400, properties at film temperature	$\text{Nu} = 2 + 0.60\,\text{Re}^{0.5}\,\text{Pr}^{0.33}$

The convective heat transfer coefficient is thus

$$
\begin{aligned}
h_c &= \frac{\text{Nu} \times k}{d_s} \\
&= \frac{8.29 \times 0.0272}{0.005} \\
&= 45\,\text{W/m}^2\,°\text{C}
\end{aligned}
$$

In a heat exchanger, conductive and convective heat transfers occur simultaneously. Heat may first be transferred from the inside fluid by convection to the inside surface of a pipe, then through the pipe wall by conduction, and finally from the outer surface of the pipe to the surrounding environment by convection. Therefore, heat transfer is through three layers of resistance in a series. The overall resistance is determined by

$$R_t = R_i + R_w + R_o \tag{7.5}$$

The resistance of a convective layer can be calculated by

$$R_{\text{convection}} = \frac{1}{hA} \tag{7.6}$$

and the resistance of a conduction layer can be calculated by

$$R_{\text{conduction}} = \frac{\ln\left(r_0 / r_i\right)}{2\pi kL} \tag{7.7}$$

The overall heat transfer coefficient (based on the inside area) can be calculated using

$$\frac{1}{U_i A_i} = \frac{1}{h_i A_i} + \frac{\ln\left(r_0 / r_i\right)}{2\pi kL} + \frac{1}{h_0 A_0} \tag{7.8}$$

or

$$\frac{1}{U_i} = \frac{1}{h_i} + \frac{r_i \ln\left(r_0 / r_i\right)}{2\pi kL} + \frac{r_i}{h_0 r_0} \tag{7.9}$$

Example 7.2

Milk is pasteurized in a shell and tube heat exchanger, made of stainless steel. The inside diameter of the pipe is 2.5 cm and the thickness of the pipe is 0.5 cm. The thermal conductivity of stainless steel is 45 W/m°C. The inside convective heat transfer coefficient of milk and outer convective heat transfer coefficient of steam are 100 W/m^2 °C and 1000 W/m^2 °C, respectively. Determine the overall heat transfer coefficient.

Solution 7.2

The overall heat transfer coefficient (based on the inside area) can be calculated using Equation 7.9:

$$\frac{1}{U_i} = \frac{1}{100} + \frac{0.0125 \times \ln\left(0.0175/0.0125\right)}{45} + \frac{0.0125}{1000 \times 0.0175}$$

Therefore, we have $U_i = 92.53$ W/m² °C.

The flow behavior becomes more complex for non-Newtonian fluid foods with added particulates flowing in a heat exchanger. Many non-Newtonian fluids such as modified starch solution are pseudoplastic fluids. The flow characteristics can be described by the power-law model, which may be expressed as

$$\tau = K\gamma^n \tag{7.10}$$

where

K and n are constants for the particular fluid
$n < 1$ for pseudoplastic fluids
γ is the shear rate
τ is the shear stress

For a power-law fluid with a fully developed velocity profile, the following equation may be used to determine the heat transfer coefficient (Singh, 1992):

$$\text{Nu} = 1.75\left(\frac{3n+1}{4n}\right)^{1/3} Gz^{1/3} \tag{7.11}$$

and

$$Gz = {}^{mc}\!/_{Lk} \tag{7.12}$$

where

m is the mass flow rate (kg/h)
c is the specific heat (kJ/kg°C)
L is the length (m)
k is the thermal conductivity (W/m°C)

Experiments have shown that the presence of particles has a strong positive effect on heat transfer coefficient and heat transfer rate in a tubular heat exchanger. When increasing the particle concentration from 0% to 10%, the inner heat transfer coefficient is approximately doubled. An increase in particle concentration from 10% to 20% and 20% to 30% causes an increase of 25% for heating and 35% for cooling (Sannervik et al., 1996).

7.3.2 Temperature Difference

The amount of heat that is transferred between two fluids passing through a heat exchanger is a function of the temperature difference between the two fluid streams.

The temperature difference changes from one end of the heat exchanger to the other end. The log mean temperature difference ΔT_{LMTD} is the effective temperature difference between the two fluid streams flowing through a heat exchanger. ΔT_{LMTD} is the temperature difference at one end of the heat exchanger minus the temperature difference at the other end, divided by the natural logarithm of the ratio of these two temperature differences. ΔT_{LMTD} is expressed as

$$\Delta T_{LMTD} = \frac{(T_{h2} - T_{c2}) - (T_{h1} - T_{c1})}{\ln\left[(T_{h2} - T_{c2}) / (T_{h1} - T_{c1})\right]} F \tag{7.13}$$

where

T_{h1} is the temperature of hot fluid at end 1 of the heat exchanger
T_{h2} is the temperature of hot fluid at end 2 of the heat exchanger
T_{c1} is the temperature of cold fluid at end 1 of the heat exchanger
T_{c2} is the temperature of cold fluid at end 2 of the heat exchanger

Example 7.3

Milk enters a heat exchanger at 25°C and is heated to 121°C by steam at a temperature of 130°C. Assume that the saturation steam flows in the heat exchanger and saturation water flows out of the heat exchanger. What is the effective temperature difference between the milk and steam?

Solution 7.3

Using Equation 7.13, ΔT_{LMTD} is

$$\Delta T_{LMTD} = \frac{(T_{h2} - T_{c2}) - (T_{h1} - T_{c1})}{\ln\left[(T_{h2} - T_{c2}) / (T_{h1} - T_{c1})\right]}$$

$$= \frac{(130 - 121) - (130 - 25)}{\ln\left[(130 - 121) / (130 - 25)\right]}$$

$$= 39.08°\text{C}$$

The temperature difference is 130–25 = 105°C at one end and 130–121 = 9°C at the other end of the heat exchanger. If the temperature difference decreases from one end to the other end linearly, the average temperature difference is (105 + 9)/2 = 57°C. However, the decrease in temperature difference does not follow a linear relationship, so the concept of log mean temperature difference was introduced.

7.3.3 Energy Balances in Heat Exchangers

In a heat exchanger, the heat loss from the hot flow should equal the heat gain by the cold flow. The heat loss from the hot flow and heat gain by the cold flow should be the heat transferred between the two flows. The overall heat balance and the rate equation can be expressed as

$$q = \dot{m}_h c_{p,h} \left(T_{1,h} - T_{2,h}\right) = \dot{m}_c c_{p,c} \left(T_{2,c} - T_{1,c}\right) = U_i A_i \Delta T_{LMTD} \tag{7.14}$$

where

q is the rate of heat transfer (W)
$\dot{m}$ is the mass flow rate of cold or hot flow (kg/s)
c_p is the specific heat of a fluid (J/kg°C)
T is the temperature (°C)
U is the overall heat transfer coefficient (W/m^2°C)
A is the area for heat transfer (m^2)
ΔT_{LMTD} is the log mean temperature difference between two flows (°C)
subscripts h and c denote the hot and cold flows
subscripts 1 and 2 denote the entrance and exit
subscript i denotes the inside

Example 7.4

Following Examples 7.2 and 7.3, if the mass flow rate of milk is 1000 kg/h and the specific heat of milk is 4.18 kJ/kg°C, determine the required heat transfer area of the heat exchanger. The latent heat of saturation steam at 130°C is 2174 kJ/kg; determine the mass flow rate of steam.

Solution 7.4

From Examples 7.2 and 7.3, we have

$$U_i = 92.53\,\text{W/m}^2\,°\text{C}$$

and

$$\Delta T_{LMTD} = 39.08°\text{C}$$

The total heat required for the increase in milk temperature is

$$q = \dot{m}_c c_{p,c} \left(T_{2,c} - T_{1,c}\right) = \frac{1{,}000\ [\text{kg/h}]}{3{,}600\ [\text{s/h}]} \times 4{,}180\ [\text{J/kg°C}] \times (121 - 25)[°\text{C}] = 111{,}467\ \text{W}$$

The required heat transfer area of the heat exchanger

$$A_i = \frac{q}{U_i \Delta T_{LMTD}} = \frac{111{,}467\ [\text{W}]}{92.53\ [\text{W/m}^2\,°\text{C}] \times 39.08\,°\text{C}} = 30.8\ \text{m}^2$$

The mass flow of steam through the heat exchanger is

$$\dot{m}_h = \frac{q}{\Delta h_v} = \frac{111{,}467\ [\text{W}]}{2{,}174\ [\text{kJ/kg}]} = 184.6\ \text{kg/h}$$

7.3.4 Effectiveness of Heat Exchangers

Heat exchanger effectiveness is used to predict how well a heat exchanger will perform its job. It is defined as the ratio of the actual amount of heat transferred to the maximum possible amount of heat that could be transferred with an infinite area, which is expressed as

$$\varepsilon = \frac{\left[\dot{m}c\left(T_{\text{in}} - T_{\text{out}}\right)\right]_{\text{hot or cold}}}{\left(\dot{m}c\right)_{\text{min}}\left(T_{\text{hot in}} - T_{\text{cold in}}\right)} \tag{7.15}$$

The top and bottom items on the right side of the above equation are the actual amount of heat transferred and the maximum possible amount of heat that could be transferred with an infinite area, respectively. If the heat transfer area is infinite, the exit temperature of the cold fluid will approach the entrance temperature of the hot fluid while the exit temperature of the hot fluid will approach the entrance temperature of the cold fluid. Liquid–liquid heat exchangers usually operate at 75% effectiveness.

Once the effectiveness of a heat exchanger is known, the exit temperatures of hot and cold flow can be predicted. The actual heat load can be calculated by

$$Q = \varepsilon\left(\dot{m}c\right)_{\text{min}}\left(T_{\text{hot in}} - T_{\text{cold in}}\right) \tag{7.16}$$

The exit temperatures of the hot fluid and cold fluid are thus determined by

$$T_{\text{hot out}} = T_{\text{hot in}} - \frac{Q}{\left(\dot{m}c\right)_{\text{hot}}} \tag{7.17}$$

and

$$T_{\text{cold out}} = T_{\text{cold in}} + \frac{Q}{\left(\dot{m}c\right)_{\text{cold}}} \tag{7.18}$$

7.3.5 Exergy Analysis of a Heat Exchanger

The exergy rate of a fluid is a function of the mass flow rate of the fluid, $\dot{m}$, and the specific flow exergy, ψ, which is calculated by

$$\text{Ex} = \dot{m}\psi \tag{7.19}$$

The specific flow exergy is determined from enthalpy, h, and entropy, s, which is expressed as

$$\psi = \left(h - h_0\right) - T_0\left(s - s_0\right) \tag{7.20}$$

where the subscript 0 denotes the reference state of the fluid for the enthalpy and entropy.

The energy and exergy efficiency of a heat exchanger are defined by

$$\eta_{En} = \frac{\dot{m}_c \left(h_{c,o} - h_{c,i}\right)}{\dot{m}_h \left(h_{h,i} - h_{h,o}\right)} \times 100\% \tag{7.21}$$

$$\eta_{Ex} = \frac{\dot{m}_c \left(\psi_{c,o} - \psi_{c,i}\right)}{\dot{m}_h \left(\psi_{h,i} - \psi_{h,o}\right)} \times 100\% \tag{7.22}$$

where the subscripts c, h, i, and o denote the cold fluid, hot fluid, inlet, and outlet, respectively.

Example 7.5

Following Examples 7.2–7.4, what are the energy and exergy efficiencies of the heat exchanger? The properties of the fluids are given in Table 7.4. The reference temperature $T_0 = 25°C$.

Solution 7.5

The specific flow exergies of saturated steam into the heat exchanger and saturated water out of the heat exchanger at 130°C are

$$\begin{aligned}\psi_{h,i} &= \left(h_s - h_{s0}\right) - T_{s0}\left(s_s - s_{s0}\right)\\ &= (2720.5 - 2545.4)\ [\text{kJ/kg}] - (25 + 273)\ [\text{K}] \times (7.0269 - 8.5794)\ [\text{kJ/kgK}]\\ &= 637.7\ \text{kJ/kg}\end{aligned}$$

$$\begin{aligned}\psi_{h,o} &= \left(h_w - h_{w0}\right) - T_{w0}\left(s_w - s_{w0}\right)\\ &= (546.31 - 100.7)\ [\text{kJ/kg}] - (25 + 273)\ [\text{K}] \times (1.6344 - 0.3534)\ [\text{kJ/kgK}]\\ &= 63.87\ \text{kJ/kg}\end{aligned}$$

TABLE 7.4
Properties of Fluids for Example 7.5

	Temperature	Enthalpy (kJ/kg)	Entropy (kJ/kgK)
Saturated steam	130	2720.5	7.0269
Saturated steam	25	2545.4	8.5794
Saturated water	130	546.31	1.6344
Saturated water	25	100.7	0.3534
Milk	121	503.7	1.5276
Milk	25	100.7	0.3534

The specific flow exergies of milk into and out of the heat exchanger are

$$\begin{aligned}\psi_{c,i} &= (h_{m,i} - h_{m0}) - T_{m0}(s_{m,i} - s_{m0}) \\ &= (503.7 - 100.7)\,[\text{kJ/kg}] - (25 + 273)[\text{K}] \times (1.5276 - 0.3534)\ [\text{kJ/kgK}] \\ &= 53.1\ \text{kJ/kg}\end{aligned}$$

$\psi_{c,o} = \psi_{m0} = 0$ (The reference temperature is the outlet temperature of the milk.)

During heat exchange, the heat loss from the steam equals the heat gain by the milk. That is,

$$\dot{m}_c(h_{c,o} - h_{c,i}) = \dot{m}_h(h_{h,i} - h_{h,o})$$

From Equation 7.21, the energy efficiency of the heat exchanger is

$$\eta_{En} = \frac{\dot{m}_c(h_{c,o} - h_{c,i})}{\dot{m}_h(h_{h,i} - h_{h,o})} \times 100\% = 100\%$$

From Equation 7.22, the exergy efficiency of the heat exchanger is

$$\begin{aligned}\eta_{Ex} &= \frac{\dot{m}_c(\psi_{c,o} - \psi_{c,i})}{\dot{m}_h(\psi_{h,i} - \psi_{h,o})} \times 100\% \\ &= \frac{1000\ [\text{kg/h}]\ \times (53.1 - 0)\ [\text{kJ/kg}]}{184.6\ [\text{kg/h}]\ \times (637.7 - 63.87)\ [\text{kJ/kg}]} \times 100\% \\ &= 50.13\%\end{aligned}$$

It can be seen from the above example that although the energy efficiency of a heat exchanger is 100% if the heat loss is negligible, the exergy efficiency is only 50.13%. This means that the quality of the energy into the heat exchanger is degraded through the heat exchanger.

7.4 ENERGY CONSERVATION TECHNOLOGIES FOR HEAT EXCHANGERS

7.4.1 Energy Conservation through Heat Transfer Enhancement Techniques

Conservation of the useful part of energy (or exergy) can be achieved through enhanced heat transfer surfaces and flow configurations. Heat transfer enhancement can be achieved by either passive techniques without the direct application of external power such as rough surfaces and extended surfaces, or active techniques with the application of external power such as mechanical agitation, surface vibration, and fluid vibration. In some cases, two or more techniques may be used simultaneously (Zimparov, 2002). Heat transfer enhancement techniques can be used to

- Improve the performance of an existing heat exchanger
- Reduce the size and cost of a new heat exchanger
- Increase the energy and exergy efficiencies of the heat exchangers, thus reducing operating cost

The heat transfer rate of a pipe may be limited by the inside convective heat transfer coefficient, outside convective heat transfer coefficient, or the heat conduction of the pipe wall. Enhancement of heat transfer means to increase the heat transfer coefficients. Applications of heat transfer techniques should assess the limiting factor on heat transfer.

Example 7.6

Following Examples 7.2 and 7.5, if the inside convective heat transfer coefficient or the outside convective heat transfer coefficient is doubled, what will be the increase in the overall heat transfer coefficient?

Solution 7.6

From Example 7.2, the overall heat transfer coefficient before enhancement is $U_i = 92.53\ \text{W/m}^2\ °\text{C}$.

1. If the inside convective heat transfer coefficient is doubled from 100 to 200 W/m² °C, the overall heat transfer coefficient is determined by

$$\frac{1}{U_i} = \frac{1}{200} + \frac{0.0125\ln\left(0.0175/0.0125\right)}{45} + \frac{0.0125}{1000 \times 0.0175}$$

   Thus, the new overall heat transfer coefficient is $U_i = 172.18\ \text{W/m}^2\ °\text{C}$. The overall heat transfer coefficient is increased by 86% from 92.53 to 172.18 W/m² °C.

2. If the outside convective heat transfer coefficient is doubled from 1000 W/m² °C to 2000 W/m² °C, the overall heat transfer coefficient is determined by

$$\frac{1}{U_i} = \frac{1}{100} + \frac{0.0125\ln\left(0.0175/0.0125\right)}{45} + \frac{0.0125}{2000 \times 0.0175}$$

   Thus, the new overall heat transfer coefficient is $U_i = 95.69\ \text{W/m}^2\ °\text{C}$. The overall heat transfer coefficient is increased by 3.4% from 92.53 to 95.69 W/m² °C.

From the above example, we can see that the enhancement of heat transfer should be applied to the side with dominant thermal resistance or a smaller heat transfer coefficient.

Example 7.7

Following Examples 7.2 through 7.6, enhanced heat transfer pipes with double inside convective heat transfer coefficient are used to replace the pipes in the heat exchanger, and other conditions remain the same. Exergy efficiency can be improved by reducing steam temperature. What could be the steam temperature and exergy efficiency for the new heat exchanger? The properties of the fluids are given in Table 7.5. The reference temperature, T_0 = 25°C.

Solution 7.7

From Example 7.4, the total inside area of the pipes of the heat exchanger is $A_i = 30.8\,m^2$.

From Example 7.5, the overall heat transfer coefficient of the heat exchanger with the enhanced heat transfer pipes is $U_i = 172.18\,W11/m^2\,°C$.

Using Equation 7.14, for the same heating load using the heat exchanger with the enhanced heat transfer pipes, the temperature difference between steam and milk can be reduced to

$$\Delta T_{LMTD} = \frac{q}{U_i' A_i} = \frac{111{,}467\ [W]}{172.18\ [W/m^2\,°C] \times 30.8\ [m^2]} = 21.02°C$$

Using the heat exchanger with the enhanced heat transfer pipes, the temperature difference decreases from 39.08°C to 21.02 °C.

Using Equation 7.1, we can find the steam temperature for the heat exchanger with the enhanced heat transfer pipes:

$$T_{h1} = T_{h2} = 122°C$$

The specific flow exergies of saturated steam into the heat exchanger and saturated water out of the heat exchanger at 122°C are

$$\begin{aligned}\psi_s &= (h_s - h_{s0}) - T_{s0}(s_s - s_{s0}) \\ &= (2709.6 - 2545.4)\ [kJ/kg] - (25 + 273)\ [K] \times (7.1088 - 8.5794)\ [kJ/kgK] \\ &= 602.4\ kJ/kg\end{aligned}$$

TABLE 7.5
Properties of Fluids for Example 7.7

	Temperature	Enthalpy (kJ/kg)	Entropy (kJ/kgK)
Saturated steam	122	2709.6	7.1088
Saturated steam	25	2545.4	8.5794
Saturated water	122	490.97	1.4405
Saturated water	25	100.7	0.3534
Milk	121	503.7	1.5276
Milk	25	100.7	0.3534

$$\psi_w = (h_w - h_{w0}) - T_{w0}(s_w - s_{w0})$$
$$= (490.97 - 100.7)\ [\text{kJ/kg}] - (25 + 273)\ [\text{K}] \times (1.4405 - 0.3534)\ [\text{kJ/kgK}]$$
$$= 66.31\ \text{kJ/kg}$$

At 122°C, the latent heat of steam increases from 2174 kJ/kg to 2219 kJ/kg. The mass flow of steam through the heat exchanger is decreased to

$$\dot{m}_h = \frac{q}{\Delta h_v} = \frac{111{,}467\ [\text{W}]}{2{,}219\ [\text{kJ/kg}]} = 180.8\ \text{kg/h}$$

From Example 7.5, the specific flow exergies of milk into and out of the heat exchanger are

$$\psi_{m1} = 53.1\ \text{kJ/kg}$$

and

$$\psi_{m2} = \psi_{m0} = 0$$

$$\eta_{Ex} = \frac{1000\ [\text{kg/h}]\ \times (53.1 - 0)\ [\text{kJ/kg}]}{180.8\ [\text{kg/h}]\ \times (602.4 - 66.31)\ [\text{kJ/kg}]} = 54.78\%$$

Using the heat exchanger with the enhanced heat transfer pipes, the exergy efficiency will be increased by 9.3% from 50.13% to 54.78%.

Enhanced heat transfer surfaces have been successfully used to obtain more compact and efficient heat exchangers (Wang et al., 2000a,b). Several enhanced heat transfer rough surface configurations such as spirally fluted pipes, as shown in Figure 7.4, have been widely used to enhance the convective heat transfer of a single phase flow, with an increase of about 50% in the inner convective heat transfer coefficient (Wang et al., 2000a). Swirl flow devices such as a twisted tape insert are also used to enhance the surface protuberances (Zimparov, 2002). A gas usually has lower convective heat transfer coefficients than a liquid. Therefore,

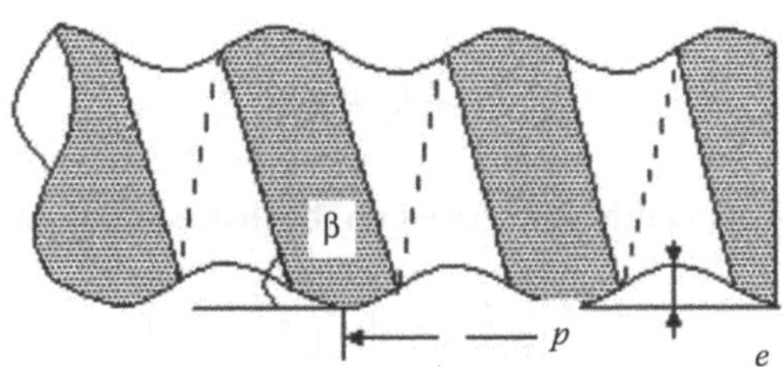

FIGURE 7.4 Schematic structure of the spirally fluted tube. (Reprinted from Wang et al., *Energ. Convers. Manage.*, 41: 993–1005, 2000a. With permission.)

methods should be developed to reduce the thermal resistance or increase the convective heat transfer coefficient on the gas side.

Fluid foods including both Newtonian and non-Newtonian fluids such as fruit and vegetable juices and milk are often subjected to thermal treatment inside heat exchangers. The heat exchangers must have a high heat transfer rate, low friction losses, and easy cleaning and sanitizing. The shell and tube heat exchanger equipped with helically corrugated walls as shown in Figure 7.1 can meet these requirements (Rozzi et al., 2007). Helically corrugated tubes are particularly effective in enhancing convective heat transfer for Reynolds number ranging from about 800 to the limit of the transitional flow regime (Rozzi et al., 2007).

Enhanced heat transfer surface configurations have also been developed to improve the heat transfer with phase changes of condensation and evaporation (Wang et al., 2000a,b). The heat transfer coefficient of a turbulent flow in rough tubes is as high as 250% of that in smooth tubes. The performance of a condenser with enhanced heat transfer tubes can increase up to 400%. However, the pressure drop of a fluid flowing through the rough surface of the enhanced heat transfer tubes may be significantly increased depending on the configuration of the enhanced heat transfer surface (Zimparov, 2002).

7.4.2 Energy Conservation through Cleaning of Fouling Layer

Fouling and cleaning of heat exchangers are serious industrial problems. It was reported that fouling caused an increase of up to 8% in the energy consumption in fluid milk plants and about 21% of the total energy was used to clean milk pasteurization plants (Ramirez et al., 2006). Although the chemistry of the fouling process is still not understood, the complex interaction between the chemistry and the fluid mechanics of heat exchangers makes it difficult to deal with the fouling problem (Fryer and Belmar-Beiny, 1991; Visser and Jeurnink, 1997). The dairy industry has been confronted with fouling on the metal surface of plate heat exchangers. In the dairy industry, fouling deposits are mainly caused by heat-sensitive whey proteins and heat-induced precipitation of calcium phosphate salts. Fouling may cause an increased pressure drop, heat transfer resistance, and microbial growth at the fouling places (Visser and Jeurnink, 1997; Changani et al., 1997).

If a fouling layer is generated on the inside wall of pipes, heat transfer will need to overcome four layers of resistance in a series: inside boundary layer, fouling layer, pipe wall, and outside boundary layer. The overall resistance is thus determined by

$$R_t = R_i + R_f + R_w + R_o \tag{7.23}$$

The overall heat transfer coefficient (based on the inside area) can be calculated using

$$\frac{1}{U_{fi}A_{fi}} = \frac{1}{h_{fi}A_{fi}} + \frac{\ln\left(r_{f0}/r_{fi}\right)}{2\pi k_f L} + \frac{\ln\left(r_{w0}/r_{wi}\right)}{2\pi k_w L} + \frac{1}{h_{w0}A_{w0}} \tag{7.24}$$

or

$$\frac{1}{U_{fi}} = \frac{1}{h_{fi}} + \frac{r_{fi}\ln\left(r_{f0}/r_{fi}\right)}{k_f} + \frac{r_{fi}\ln\left(r_{w0}/r_{wi}\right)}{k_w} + \frac{r_{fi}}{h_{w0}r_{w0}} \tag{7.25}$$

Example 7.8

Following Examples 7.2 through 7.7, if a 1 mm thick fouling layer at a thermal conductivity of 0.5 W/m°C is formed on the inside wall of the pipe, what will be the overall heat transfer coefficient? What would be the steam temperature and exergy efficiency of the heat exchanger after fouling occurs?

Solution 7.8

From Example 7.4, the total inside area of the pipes of the heat exchanger is A_i = 30.8 m².

The overall heat transfer coefficient (based on the inside area) after fouling can be determined by Equation 7.25:

$$\frac{1}{U_{fi}} = \frac{1}{100} + \frac{0.0115\ln\left(0.0125/0.0115\right)}{0.5} + \frac{0.0115\ln\left(0.0175/0.0125\right)}{45} + \frac{0.0115}{1000\times 0.0175}$$

The overall heat transfer coefficient is thus

$$U_{fi} = 78.98 \text{ W/m}^2\,^\circ\text{C}$$

Therefore, due to the fouling, the overall heat transfer coefficient is decreased from 92.53 W/m² °C (from Example 7.2) to 78.98 W/m² °C.

Due to the formation of a fouling layer, the inside surface area is reduced to

$$\begin{aligned} A_{fi} &= A_{wi}\left(\frac{r_{fi}}{r_{wi}}\right)^2 \\ &= 30.8 \text{ m}^2 \times \left(\frac{0.0115 \text{ m}}{0.0125 \text{ m}}\right)^2 \\ &= 26.1 \text{ m}^2 \end{aligned}$$

Due to the decrease in overall heat transfer coefficient and inside surface area, in order to supply the same amount of heat to milk, q = 111,467 W, the temperature difference between steam and milk should be increased to

$$\Delta T'_{LMTD} = \frac{q}{U'_i A_i} = \frac{111{,}467 \text{ [W]}}{78.98 \text{ [W/m}^2\,^\circ\text{C]} \times 26.1 \text{ [m}^2\text{]}} = 54.07^\circ\text{C}$$

The temperature difference is increased from 39.08°C (from Example 7.3) to 54.07°C.

TABLE 7.6
Properties of Fluids for Example 7.8

	Temperature	Enthalpy (kJ/kg)	Entropy (kJ/kgK)
Saturated steam	140.5	2733.9	6.9299
Saturated steam	25	2545.4	8.5794
Saturated water	140.5	589.13	1.7391
Saturated water	25	100.7	0.3534
Milk	121	503.7	1.5276
Milk	25	100.7	0.3534

Using Equation 7.1, the steam temperature for the heat exchanger with the enhanced heat transfer pipes is $T'_{h1} = T'_{h2} = 140.5°C$.

The enthalpy and entropy of steam at 140.5°C are given in Table 7.6. Using Equation 7.17, the specific flow exergies of saturated steam into the heat exchanger and saturated water out of the heat exchanger at 140.5°C are

$$\begin{aligned}\psi_s &= (h_s - h_{s0}) - T_{s0}(s_s - s_{s0}) \\ &= (2733.9 - 2545.4)\ [\text{kJ/kg}] - (25 + 273)\ [\text{K}] \times (6.9299 - 8.5794)\ [\text{kJ/kgK}] \\ &= 680.05\ \text{kJ/kg}\end{aligned}$$

$$\begin{aligned}\psi_w &= (h_w - h_{w0}) - T_{w0}(s_w - s_{w0}) \\ &= (589.13 - 100.7)\ [\text{kJ/kg}] - (25 + 273)\ [\text{K}] \times (1.7391 - 0.3534)\ [\text{kJ/kgK}] \\ &= 75.49\ \text{kJ/kg}\end{aligned}$$

At 140.5°C, the latent heat of steam is 2145 kJ/kg. The mass flow of steam through the heat exchanger is

$$\dot{m}_h = \frac{q}{\Delta h_v} = \frac{111{,}467\ [\text{W}]}{2{,}145\ [\text{kJ/kg}]} = 187.1\ \text{kg/h}$$

From Example 7.5, the specific flow exergies of milk into and out of the heat exchanger are

$$\psi_{m1} = 53.1\ \text{kJ/kg}$$

and

$$\psi_{m2} = \psi_{m0} = 0$$

$$\eta_{Ex} = \frac{1000\ [\text{kg/h}] \times (53.1 - 0)\ [\text{kJ/kg}]}{187.1\ [\text{kg/h}] \times (680.05 - 75.49)[\text{kJ/kg}]} = 46.94\%$$

The exergy efficiency decreases by 6.4% from 50.13% to 46.94%.

7.4.3 Energy Conservation through Optimization of Heat Exchanger Design

Traditional design of heat exchangers involves many trials by changing one variable at a time and using a trial–error or a graphical method to meet design specifications. The design variables may include

- Inside heat transfer coefficient
- Outside heat transfer coefficient
- Temperature difference
- Tube surface area

With the combination of exergy analysis and life cycle analysis, an exergy optimization of a heat exchanger can be obtained. There is a trade-off between exergy saving during operation and exergy consumption during construction of a heat exchanger (Cornelissen and Hirs, 1999; Unuvar and Kargici, 2004).

7.4.4 Energy Conservation through Heat Exchanger Network Retrofit

Process integration technology for improving energy efficiency has been widely used around the world. The financial benefit comes from both reduced energy costs and debottlenecking for increase in throughput. It can also reduce flue gas emission. Systematic methods for the design of heat exchanger networks have been developed (Wang et al., 1990; Silva and Zemp, 2000; Smith, 2000). The network retrofit usually starts by identifying the bottlenecking exchangers within an existing heat exchanger network structure using thermodynamic methods. To overcome the network pinch, a modification in the network structure is required by relocation of an existing heat exchanger to a new duty or addition of a new exchanger or change of stream splitting arrangement.

However, it is not straightforward to identify the most appropriate structural modification. The relationship among heat transfer coefficient, pressure drop, and exchanger area is complex. The retrofit area target should be implemented as a nonlinear optimization problem to minimize the requirement for additional area. Mathematical models are usually needed to identify the most beneficial structural changes (Smith, 2000). The minimum temperature difference in heat exchangers can be optimized, and trade-off between the capital costs and the energy saving revenue can be determined by mathematical models (Wang et al., 1990). In addition, besides the increased heat transfer coefficients, pressure drop constraints due to the additional heat exchanger area during retrofit of heat exchanger networks should also be considered (Silva and Zemp, 2000).

7.5 SUMMARY

Heat exchangers are widely used as a main component of different unit operations such as cooling, freezing, drying, sterilization, and pasteurization in food processing facilities. Heat transfer enhancement, fouling removal, and optimization of heat exchanger network can improve the energy and exergy efficiencies of heat exchangers.

On one hand, heat transfer enhancement technology can increase the overall heat transfer coefficient of a heat exchanger, thus reducing the temperature difference required for a given heat exchange load and increasing the exergy efficiency. On the other hand, fouling increases heat transfer resistance and decreases the overall heat transfer coefficient, thus increasing the temperature difference required for a given heat exchange load and decreasing the exergy efficiency. Therefore, it is necessary to frequently remove the fouling layers on the metal surface of a heat exchanger. Energy can also be saved by retrofitting the heat exchanger network.

REFERENCES

Changani, S.D., M.T. Belmar-Beiny, and P.J. Fryer. 1997. Engineering and chemical factors associated with fouling and cleaning in milk processing. *Experimental Thermal and Fluid Science* 14: 392–406.

Cornelissen, R.L. and G.G. Hirs. 1999. Thermo-dynamics optimization of a heat exchanger. *International Journal of Heat and Mass Transfer* 42: 951–959.

Fryer, P.J. and M.T. Belmar-Beiny. 1991. Fouling of heat exchanger in the food industry: A chemical engineering perspective. *Trends in Food Science & Technology* 2: 33–37.

Galeazzo, F.C.C., R.Y. Miura, J.A.W. Gut, and C.C. Tadini. 2006. Experimental and numerical heat transfer in a plate heat exchanger. *Chemical Engineering Science* 61: 7133–7138.

Kim, H.B., C.C. Tadini, and R.K. Singh. 1999. Heat transfer in a plate exchanger during pasteurization of orange juice. *Journal of Food Engineering* 42: 79–84.

Kuzgunkaya, E.H. and A. Hepbasli. 2007. Exergetic performance assessment of a ground-source heat pump drying system. *International Journal of Energy Research* 31: 760–777.

Mabit, J., F. Fayolle, and J. Legrand. 2003. Shear rates investigation in a scraped surface heat exchanger. *Chemical Engineering Science* 58: 4667–4679.

Mabit, J., C. Loisel, F. Fayolle, and J. Legrand. 2004. Relation between mechanical treatment of starch and flow conditions in a scraped surface heat exchanger. *Food Research International* 37: 505–515.

Ramirez, C.A., M. Patel, and K. Blok. 2006. From fluid milk to milk power: Energy use and energy efficiency in the European dairy industry. *Energy* 31: 1984–2004.

Rozzi, S., R. Massini, G. Paciello, G. Pagliarini, S. Rainieri, and A. Trifiro. 2007. Heat treatment of fluid foods in a shell and tube heat exchanger: Comparison between smooth and helically corrugated wall tubes. *Journal of Food Engineering* 79: 249–254.

Sannervik, J., U. Bolmstedt, and C. Tragardh. 1996. Heat transfer in tubular heat exchangers for particulate containing liquid foods. *Journal of Food Engineering* 29: 63–74.

Silva, M.L. and R.J. Zemp. 2000. Retrofit of pressure drop constrained heat exchanger networks. *Applied Thermal Engineering* 20: 1469–1480.

Singh, R.P. 1992. Heating and cooling processes for foods. In *Handbook of Food Engineering*, Heldman, D.R. and D.B. Lund, (Eds.), New York: Marcel Dekker Inc.

Singh, R.P. and D.R. Heldman. 2001. *Introduction to Food Engineering* (3rd ed.), San Diego: Academic Press.

Smith, R. 2000. State of the art in process integration. *Applied Thermal Engineering* 20: 1337–1345.

Soylemez, M.S. 2000. On the optimum heat exchanger sizing for heat recovery. *Energy Conversion and Management* 41: 1419–1427.

Unuvar, A. and S. Kargici. 2004. An approach for optimum design of heat exchangers. *International Journal of Energy Research* 28: 1379–1392.

Visser, J. and T.J.M. Jeurnink. 1997. Fouling of heat exchangers in the dairy industry. *Experimental Thermal and Fluid Science* 14: 407–424.

Wang, Y.P., Z.H. Chen, and M. Groll. 1990. A new approach to heat exchanger network synthesis. *Heat Recovery Systems and CHP* 10: 399–405.

Wang, L.J., D.S. Zhu, and Y.K. Tan. 1999. Heat transfer enhancement of the adsorber of an adsorption heat pump. *Adsorption* 5: 279–286.

Wang, L.J., D.W. Sun, P. Liang, L.X. Zhuang, and Y.K. Tan. 2000a. Heat transfer characteristics of the carbon steel spirally fluted tube for high-pressure preheaters. *Energy Conversion and Management*. 41: 993–1005.

Wang, L.J., D.W. Sun, P. Liang, L.X. Zhuang, and Y.K. Tan. 2000b. Experimental studies on heat transfer enhancement of the inside and outside spirally triangle finned tube with small spiral angles for high pressure preheaters. *International Journal of Energy Research* 24: 309–320.

Zimparov, V. 2002. Energy conservation through heat transfer enhancement techniques. *International Journal of Energy Research* 26: 675–696.

8 Waste-Heat Recovery and Thermal Energy Storage in Food Processing Facilities

8.1 INTRODUCTION

Waste heat may be the thermal energy stored in waste hot air, hot water, combustion flue gas, hot liquid foods, and any other fluids that leave processing units and facilities. It has been estimated that up to 50% of the energy consumed in the United States was discharged as waste heat to the environment. Typical process industries could save 20% of their fuel consumption through proper waste-heat recovery and management (Mull, 2001).

Recovery of waste heat can not only save money due to the reduced energy consumption and capacity requirement for energy conversion equipment, but also prevent thermal pollution to the environment. The economic benefits of waste-heat recovery include the reduction of energy costs for purchased fuels and capital costs for energy conversion equipment with less capacity. However, the economics of a waste-heat recovery system depends on utilization, quantity and quality of the recovered waste heat, and the heat transfer equipment for waste-heat recovery.

In the food industry, heat with end-use temperatures below 200°C is more than 90% of the total thermal energy demand (Ozdogan and Arikol, 1995). Part of the thermal energy demand can be easily supplied by the waste heat recovered from a higher-temperature source nearby if available. Also, renewable energy sources such as solar energy and geothermal energy have a potential for heat supply in food processing facilities. It is necessary that the users of recovered waste heat are available at the same time as the waste-heat sources. Otherwise, a thermal energy storage system is needed to temporarily store the surplus recovered waste heat. In this chapter, waste-heat recovery, utilization, and thermal storage are discussed.

8.2 RECOVERY OF WASTE HEAT IN FOOD PROCESSING FACILITIES

8.2.1 Quantity and Quality of Waste Heat in Food Processing Facilities

The quantity and quality of waste heat should be carefully evaluated when considering waste-heat recovery. The quality of waste heat is determined by its temperature. High-temperature waste heat not only has the highest quality but also is the most

TABLE 8.1
Quality of Waste-Heat Sources

Quality	Temperature Range (°C)	Type of Heat from Food Processes
High temperature	593–1650	—
Medium temperature	205–593	Steam boiler exhausts, gas turbine exhausts, drying and baking ovens
Low temperature	27–205	Process steam condensate, air compressors, pumps, internal combustion engines, drying and baking ovens

useful energy source. Usually, waste heat is divided into three categories according to its temperature, as given in Table 8.1. The potential for economic waste-heat recovery does not depend as much on the quantity of waste heat available as it does on whether its quality fits the requirement of potential heating loads or users and whether the waste heat is available at the time when thermal energy is required. The waste heat in a food processing facility may come from steam condensate, low-pressure steam, boiler flue gas, hot water, and hot air leaving from processing units and facilities. Exhausts from steam boilers, gas turbines, and some drying and baking ovens may generate medium-temperature waste-heat streams while process steam condensate, air compressors, pumps, and internal combustion engines, and some drying and baking ovens may generate low-temperature waste-heat streams.

Tables 8.2 through 8.4 give the quantities of waste heat in a canned fruit and vegetable processing facility, meat processing facility, and milk processing facility,

TABLE 8.2
Quantity of Waste Heat in a Canned Fruit and Vegetable Processing Facility

Unit Operation	Product	Quantity (L/t Product)	Temperature (°C)	Biochemical Oxygen Demand (kg/t)	Suspended Solids (kg/t)
Water blanching	Snap beans	124–335	90	0.69	0.13
	Lima beans	822	90	0.65	—
	Peas	240–385	90	1.4–3.0	—
Steam blanching	Snap beans	125–150	90	0.55	0.02
	Lima beans	113–238	90	3.5	—
	Peas	191–313	90	4.3	—
Vibratory spiral blancher	Snap beans	27	90	0.53	0.08
	Lima beans	25	90	0.90	0.54
	Brussels sprouts	15	90	0.43	0.08
	Broccoli	11	90	0.25	0.09
	Cauliflower	3	90	—	—

TABLE 8.2 (continued)
Quantity of Waste Heat in a Canned Fruit and Vegetable Processing Facility

Unit Operation	Product	Quantity (L/t Product)	Temperature (°C)	Biochemical Oxygen Demand (kg/t)	Suspended Solids (kg/t)
Steam blanching with water cooling	Snap beans	4937	90	1.6	—
	Lima beans	4967	90	3.4	—
	Peas	4967	90	2.9	—
Cooker condensate	—	117–210	120	—	—
Cooling water	—	250–415	55	—	—
Can topping water overflow	—	165–210	95	—	—
Boiler feed water	—	1590–1880	15	—	—

Source: Adapted from Singh, R. P., *Energy in Food Processing*. Elsevier Science Publishing Company Inc., New York, 1986. With permission.

respectively. Waste heat in food processing facilities is usually low- or medium-temperature energy sources. The temperature of waste water streams in food processing facilities is usually below 95°C as shown in Tables 8.2 through 8.4. These waste-heat sources provide low-temperature heat according to the categorization of waste-heat sources in Table 8.1.

Waste-heat sources are unique to each facility. To assess the potential for heat recovery from the waste streams in food processing facilities, it is necessary to

TABLE 8.3
Quantity of Waste Heat in a Meat Processing Facility

Processing Facility	Unit Operations	Quantity (L/t Live Weight)	Temperature (°C)
Hog production	Clean-up	36 liter/hog	60
Primary chicken processing	Scalding water overflow	613–430	60
	De-feathering	568	19
	Primary chilling	943	21
Slaughtering	Condensate from heating water	668	99
	Boiler feed water	250	15
Manufacturing	Condensate from heating water	1670–2100	99
	Boiler feed water	625–1045	15

Source: Adapted from Singh, R. P., *Energy in Food Processing*. Elsevier Science Publishing Company Inc., New York, 1986. With permission.

TABLE 8.4
Quantity of Waste Heat in a Dairy Processing Facility

Unit Operations	Quantity (L/t Live Weight)	Temperature (°C)
Sweet water—water cooled by ice	4590	5
Cooling glycol	2754	0
Cottage cheese clean-up	814	60
Ice cream room clean-up	1548	60
Pasteurizer overflow	12.5–17	70
Pasteurizer clean-up	210–250	65
Boiler feed water	33–42	15
Whey in cheese processing	993	38
Clean-up water in cheese processing	250–545	60
Pasteurizer overflow in cheese processing	13	70
Condensate curd/whey heating	29	95
Boiler feed water	150	15

Source: Adapted from Singh, R. P., *Energy in Food Processing*. Elsevier Science Publishing Company Inc., New York, 1986. With permission.

characterize the quantity and quality of each stream. A survey may be needed to identify and calculate the quantity of waste heat in a processing facility. The quantity of waste heat available is expressed in terms of the enthalpy (heat content) flow of the waste stream:

$$Q = \dot{H} = \dot{m}h \tag{8.1}$$

where

$\dot{H}$ is the total enthalpy flow rate of the waste stream (W or J/s)
$\dot{m}$ is the mass flow rate of the waste stream (kg/s)
h is the specific enthalpy of the waste stream (J/kg)

If only sensible heat is available in the waste stream, the specific enthalpy of the waste stream can be determined by

$$h = c\left(T_{\text{waste}} - T_{\text{reference}}\right) \tag{8.2}$$

where

c is the specific heat of the waste stream (J/kg°C)
T_{waste} is the waste stream temperature (°C)
$T_{\text{reference}}$ is the temperature at a reference state (or dead state) (°C)

Example 8.1

A waste hot water stream is available at a mass flow of 15,000 kg/h and a temperature of 50°C in a vegetable processing facility. What is the quantity of waste heat available from the waste hot water stream in terms of heat or energy content? Suppose that the reference temperature is 25°C and the specific heat of water is 4,180 J/kg°C.

Solution 8.1

From Equation 8.2, the specific enthalpy of the hot water stream is

$$h = c\left(T_{\text{waste}} - T_{\text{reference}}\right) = 4180 \times (50 - 25) = 1045 \text{ kJ/kg}$$

From Equation 8.1, the available heat is

$$Q = \dot{H} = \dot{m}h = 15{,}000 \times 1{,}045 = 1567 \text{ MJ/h} = 435 \text{ kW}$$

8.2.2 Waste-Heat Utilization

In food processing facilities, because the temperature of heating medium is relatively low, waste-heat streams are usually high-volume and low-temperature energy sources. The use of waste heat in the low-temperature range is more challenging. The use of waste heat may include heat for feed water, make-up water, air supply to boilers, heat for drying, concentration, evaporation, and other thermal processes, building heating, hot water supply, and generation of electricity. Practical applications are generally for preheating liquids such as boiler make-up water, which is usually at 15°C–25°C, or gases such as combustion air. Low-temperature waste-heat streams can be used for thawing, tempering, or defrosting. Low-temperature waste heat may also be recovered and used to power an absorption or adsorption refrigeration system as discussed in Chapter 9. Furthermore, low-temperature waste heat can be upgraded using a heat pump as discussed in Chapter 9.

Heat balances should be conducted to determine the available quantity of waste heat and potential use of recovered waste heat. To conduct a heat balance data on the mass flow rate, temperature, pressure, and specific heat of each waste stream are required. These data may be obtained from instruments already in the facility, specific measurements, and reasonably accurate estimates. Waste-heat recovery is closely related to the potential users of recovered heat. Maximum waste-heat recovery occurs when waste stream availability exactly coincides with the application demands. An example calculation of heat balance is given here.

Example 8.2

This example is adapted from Lund, 1986. The systems diagram of a canning facility for processing 13 t of raw peas per hour is given in Figure 8.1. The waste heat in the facility is recovered to preheat the boiler feed water at 15°C by indirect heat exchangers at an effectiveness of 75%. Analyze the potential of waste-heat recovery for the canning facility.

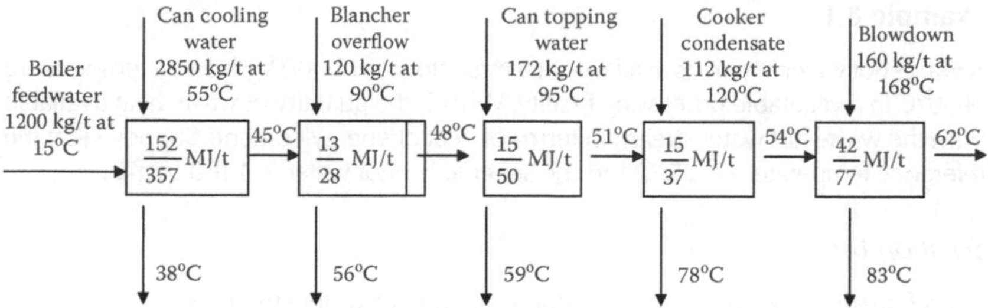

FIGURE 8.1 Systems diagram for a canning facility. (Adapted from Singh, R. P., *Energy in Food Processing*. Elsevier Science Publishing Company Inc., New York, 1986. With permission.)

Assumptions

- Condensate to the boiler is not recycled
- Blowdown is 10% of the boiler feed rate
- All waste streams are compatible with application streams
- No direct use is made of a waste stream as boiler feed water

Analysis

The waste-heat streams include

- Can cooling water at 55°C
- Blancher overflow at 90°C
- Can topping water at 95°C
- Cooker condensate at 120°C
- Boiler blowdown at 168°C

The recoverable waste heat is shown in Figure 8.1. The waste heat from the above sources is recovered and fed into the boiler feed water in sequence according to the increasing order of the temperatures of the waste-heat sources. The numerator in the ratio expressed in the boxes is the extracted heat from each waste stream while the denominator is the heat energy stored in each waste stream at a reference temperature of 15°C. As shown in Figure 8.1, the temperature of boiler feed water can be increased from 15°C to 62°C by indirect heat exchange with the waste-heat streams. The total recovered heat is about 45% based on the reference temperature of 15°C. If the boiler operates at 9.6×10^5 Pa and 170°C, the recovered energy is 7.3% of the total energy required by the boiler. More heat can be recovered if the condensate to the boiler is recycled.

8.2.3 Heat Exchangers for Waste-Heat Recovery

Waste heat can be recovered by heat exchangers directly. Heat exchangers are discussed in Chapter 7. For specific applications in waste-heat recovery, heat exchangers may be called as recuperators, regenerators, waste-heat boilers, condensers, tube and shell heat exchangers, plate-type heat exchangers, feed water heaters, economizers, and so on. There are many types of heat exchangers that can be used for waste-heat recovery. Some examples of heat exchangers are

- Shell and tube heat exchangers
- Gasketed plate heat exchangers
- Double pipe heat exchangers
- Heat pipe heat exchangers
- Lamella heat exchangers
- Spiral heat exchangers
- Rotary regenerative heat exchangers

Among the above heat exchangers, a heat pipe used in a heat exchanger as discussed in Chapter 9 can transfer up to 100 times more thermal energy than a copper rod at the same size. A heat pipe is the best known thermal conductor. A heat pipe heat recovery system is capable of operating at temperatures up to 300°C. Since a heat pipe can operate at a small temperature difference for heat transfer, the heat recovery capability of the heat pipe can be as high as 60%–80% (Akyurt et al., 1995; Lukitobudi et al., 1995).

The essential parameters that should be known and specified in order to make an optimum choice of waste-heat recovery devices include (Rohrer, 2006)

- Temperature of the waste-heat stream
- Flow rate of the waste-heat stream
- Chemical composition of the waste-heat stream
- Minimum allowable temperature of the waste-heat stream
- Amount and type of contaminants in the waste-heat stream
- Allowable pressure drop for the waste-heat stream
- Temperature of the heated fluid
- Chemical composition of the heated fluid
- Maximum allowable temperature of the heated fluid
- Allowable pressure drop in the heated fluid
- Control temperature if required

Potential fouling in heat exchangers caused by waste-heat streams should also be considered during waste-heat recovery in food processing facilities. Protein, sugars, other soluble organic compounds, and salt are readily deposited on the metal surface of heat exchangers, reducing the heat transfer efficiency of the heat exchangers. In some cases, a settling tank and skimmer may be used as a pretreatment method to reduce the fouling potential (Lund, 1986).

8.2.4 Heat Pumps for Waste-Heat Recovery

Waste heat in food processing facilities is usually a low- or medium-temperature energy source. It is usually impractical to extract work directly from the low-temperature waste-heat sources. For example, it is impractical to extract heat from 50°C to heat a fluid stream at 70°C. Heat pumps as discussed in Chapter 9 may be used to raise the temperature of a waste source so that the waste heat can be transferred to a fluid stream at a higher temperature. A heat pump is a device that absorbs and transfers energy from an energy source at a low temperature to an energy sink or receiver at a higher temperature at an expense of external work. The coefficient of performance (COP) of a heat pump cycle is the simple ratio of heat delivered to work consumed by the heat pump:

$$\text{COP} = \frac{Q_h}{W} = \frac{1}{1 - T_L/T_H} \tag{8.3}$$

where

T_L is the temperature of energy source (K)
T_H is the temperature of energy sink or potential waste-heat user (K)
Q_h is the heat recovered by a heat pump (kJ)
W is the work consumed by a heat pump (kJ)

Example 8.3

A heat pump is used to upgrade heat from a waste hot water stream at a temperature of 50°C to an energy sink at 90°C to be used by an absorption refrigeration system. What is the theoretical COP of the heat pump?

Solution 8.3

From Equation 8.3, the COP of the heat pump is

$$\text{COP} = \frac{Q_h}{W} = \frac{1}{1 - T_L/T_H} = \frac{1}{1 - (273.15 + 50)/(273.15 + 90)} = 9.1$$

The work required by a heat pump is usually either provided by an electrical motor or a liquid fuel engine. In order to be economically attractive, the COP of an electrical powered heat pump must be considerably higher than 3 since the generation efficiency of the electricity used by the heat pump is usually less than 35%. From Example 8.3, the theoretical COP of the heat pump is 9.1. However, the actual COP is much lower than the theoretical COP because

- Efficiency of the compressor used by the heat pump is not 100% but in the range of 65% to 85%.
- Energy loss occurs through the throttle valve and pipe line of the heat pump.
- Temperature differences are required to transfer heat from the waste stream into the evaporator of the heat pump and from the condenser of the heat pump to the energy load.

It is expected that an actual COP value ranging from 50% to 65% of the theoretical value can be achieved (Rohrer, 2006). Heat pumps offer only limited opportunities for waste-heat recovery because the capital and operating costs of a heat pump may exceed the value of the waste heat recovered.

8.3 THERMAL ENERGY STORAGE

8.3.1 Thermal Energy Storage System

If the available waste heat is more than the potential use of recovered waste heat at a time, thermal storage systems may be required to store the surplus recovered heat. Thermal storage systems store heat at a high temperature or cooling energy at a low temperature for use at a later time. Thermal storage systems are frequently used to handle short-term peak thermal demands in food processing facilities and store recovered waste heat. They offer greater flexibility in the use of thermal energy to shift peak demand periods to off-peak periods in a facility, reduce the size and cost of central equipment, and sometimes improve the energy efficiency of central equipment. Usually, the reduction in peak demand charge and cost of central equipment allows the installation of thermal energy storage systems at more or less the same cost.

Thermal energy storage is an important energy conservation technology. There are several advantages in using thermal energy storage systems, which may include (Dincer, 2002)

1. Consumption of purchased energy can be reduced by storing waste or surplus thermal energy available at certain times for use at other times.
2. Demand for purchased electrical energy can be reduced by storing electrically produced thermal energy during off-peak periods to meet the thermal loads that occur during high demand periods.
3. Purchase of additional equipment for heating, cooling, or air conditioning applications can be deferred and the equipment size in new facilities can be reduced.

The development of a thermal energy storage system involves a heat exchanger and thermal storage materials. Studies have been focused on the development of

1. Heat exchanger configurations such as shell and tube, double pipe, and plate heat exchangers
2. Sensible and latent heat storage materials
3. Methods to improve heat transfer in the thermal storage systems

8.3.2 Thermal Energy Storage Materials

A storage medium is needed to store the thermal energy. Thermal energy storage systems can be classified as either sensible heat storage (e.g., water, glycerol, and rock) or latent heat storage (water/ice, salt hydrates, and fatty acids).

Sensible heat storage is mainly used to store low-grade heat such as solar energy or waste heat. The sensible heat storage materials should be inexpensive, possess a good thermal capacity factor, and a high heat transfer rate. Table 8.5 gives some common sensible thermal energy storage materials and their properties (Dincer, 2002).

TABLE 8.5
Thermal Capacities of Some Common Sensible Thermal Energy Storage Materials at 20°C

Materials	Density (kg/m³)	Specific Heat (kJ/kg°C)	Volumetric Thermal Capacity (MJ/m³°C)
Clay	1458	0.879	1.28
Brick	1800	0.837	1.51
Sandstone	2200	0.712	1.57
Wood	700	2.39	1.67
Concrete	2000	0.880	1.76
Glass	2710	0.837	2.27
Aluminum	2710	0.896	2.43
Iron	7900	0.452	3.57
Steel	7840	0.465	3.68
Gravelly earth	2050	1.81	3.77
Magnetite	5177	0.752	3.89
Water	988	4.182	4.17

Source: Reproduced from Dincer, I., *Int. J. Energy Res.*, 26, 567, 2002. Copyright John Wiley & Son, Ltd. With permission.

The quantity of heat stored by a sensible heat storage material can be estimated by

$$Q = mc\Delta T = \rho Vc\Delta T \tag{8.4}$$

where

m is the mass of storage material (kg)
c is the specific heat of storage material (kJ/kg°C)
ΔT is the temperature rise (°C)
ρ is the density of storage material (kg/m³)
V is the volume of storage material (m³)

Iron, which has a high heat capacity and thermal conductivity, is an excellent thermal storage medium. Iron can be used as a high-temperature storage medium. Water, as a thermal energy storage material, has an excellent specific heat. It can be pumped. However, if water is used to store heat above 100°C, it should be pressurized. Some oils can be used at a temperature higher than 100°C without the requirement of pressurization. The specific heat of oils is only about 2.3 kJ/kg°C compared to 4.18 kJ/kg°C for water.

The latent heat storage system, which provides a much higher storage energy density with a smaller temperature difference between storing and releasing energy, is one of the most efficient ways to store thermal energy. Latent heat storage systems must have three key components: (1) a substance, also called phase change material, that undergoes a solid-to-liquid phase transition in the required operating temperature range, (2) a container for the storage substance, and (3) a heat exchanging surface to transfer heat between the hot or cold sources and the storage substance.

TABLE 8.6
Properties of Phase Change Materials

Storage Materials	Melting Temperature (°C)	Heat of Fusion (kJ/kg)	Liquid Density (kg/m³)	Solid Density (kg/m³)	Volume Expansion (%)	References
Ice at 0°C	0	340	1000	916	−8	Singh and Heldman, 2001
80% steric–20% myristic acid	61–65	191	870	940	10	Mazman et al., 2008
80% palmitic–20% lauric acid	55–58	183	850	950	11	Mazman et al., 2008

The quantity of heat stored by a latent heat storage material can be estimated by

$$Q = mh = \rho V h \tag{8.5}$$

where
m is the mass of storage material (kg)
h is the fusion heat of storage material (kJ/kg)
V is volume of storage material (m^3)
ρ is the density of storage material (kg/m^3)

The properties of some latent heat storage materials are given in Table 8.6. Latent heat storage materials typically include water/ice, inorganic salt hydrates, and organics such as fatty acids (Mazman, et al., 2008). The most popular method for thermal energy storage via latent heat is the conversion of water to ice. Low-volatile, anhydrous organic substances such as glycerol (Bakan et al., 2008), fatty acids (Mazman et al., 2008), and paraffins (Demirel and Ozturk, 2006) are also used as thermal energy storage materials.

The thermal conductivity of most phase change materials is too low to achieve a high heat transfer rate between the phase change material and the energy source. Therefore, the heat transfer has to be increased to efficiently use the phase change material as an energy storage medium. There are several methods used to enhance the overall heat transfer in a thermal energy storage system: the use of finned tubes, the combination of phase change materials with materials with high thermal conductivities such as metals, and the micro-encapsulation of the phase change materials to increase the heat exchanging surface (Mazman et al., 2008).

8.3.4 Hot Thermal Energy Storage

Hot thermal energy storage systems are used to store the recovered waste heat from processing facilities and solar heat collected during daytime (Andersen et al., 2008). Hot thermal energy storage systems can be used to store surplus recovered waste heat to adapt the temporary mismatch between heat loads and waste-heat sources. Since the availability of solar energy depends on the time of the day and varies in different seasons, a hot thermal energy storage system can be used to store excess solar energy and release it when the energy availability is inadequate or not available (Devahastin and Pitaksuriyarat, 2006).

8.3.5 Cooling Energy Storage

Cooling energy storage systems can be used to store cooling energy generated at off-peak demand times and supply cooling energy at peak demand times in food processing facilities. They can reduce the size of a refrigeration system and reduce the high demand charge of electricity. In cooling and freezing of foods, refrigeration loads vary significantly with the cooling/freezing time. Cooling storage systems can provide cooling effect at a constant temperature to meet the requirement of dynamic cooling/freezing loads. Cooling storage systems can be recharged using a small refrigeration system during off-peak demand time.

Cooling energy storage systems are a popular demand management tool for utilities and refrigeration. They can help avoid costly plant expansions to meet the requirement of increased cooling/freezing capacity and reduce peak electricity demand. The systems provide cooling energy that is produced using inexpensive electricity during off-peak hours and stored for utilization during on-peak hours when the demand is high and the electricity is more expensive. Cooling energy storage is an economically viable energy conserving technology for a food processing facility. Furthermore, cooling energy storage system is one of the most appropriate methods for correcting the mismatch between the supply and demand of cooling requirement (Cheralathan et al., 2007).

A cooling energy storage system with a glycol–water solution as its storage medium is shown in Figure 8.2. This system uses the temperature change of glycol to

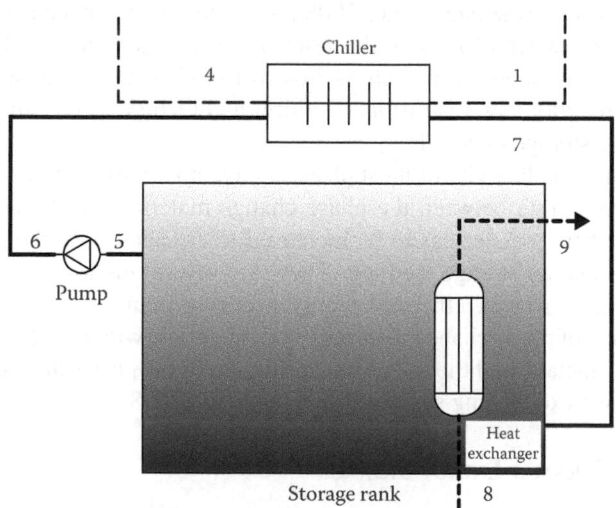

FIGURE 8.2 Glycol cooling energy storage system. (Reproduced from Bakan, K., Dincer, I., and Rosen, M.A., *Int. J. Energy Res.*, 32, 215, 2008. Copyright John Wiley & Son, Ltd. With permission.)

store cooling energy, usually generated from low-cost electricity (Bakan et al., 2008). It was found that the average energy and exergy efficiencies of the cooling energy storage system with a capacity of 350,000 kg of 45% glycol–water solution are 80% and 35%, respectively, at a storage temperature of 2.4°C –5.8°C and ambient temperature of 25°C–45°C (Bakan et al., 2008). Ice on coil, ice harvester or ice slurry, encapsulated ice, or other phase change materials are popular latent heat cooling storage media (Egolf et al., 2008; Yamaha et al., 2008). Integration of a cooling energy storage unit with a chiller can reduce the specific energy consumption by charging the system at lower condenser and optimal evaporator temperatures. Cheralathan, et al. (2007) found that a 1°C increase in evaporator temperature and decrease in condenser temperature can decrease the specific energy consumption by 3%–4% and 2.25%–3.25%, respectively. Cooling energy storage is used mainly in cooling of buildings. It uses chillers during off-peak hours at nights to produce a low-temperature medium, which can be stored for use during daytime (Rosen et al., 1999)

8.4 SUMMARY

In food processing facilities, there are large amounts of low-temperature waste-heat streams. Waste-heat streams can be used directly to increase the temperature of boiler feed water. It is possible to reduce 15%–50% to the fuel requirement of a boiler if all waste heat in a food processing facility can be extracted into the boiler feed water. It is technically feasible to recover heat energy from waste streams either directly or indirectly to reduce boiler energy requirements from 3% to 10%. Waste heat from many high-temperature unit operations can be recovered with heat exchangers for lower-temperature uses. Heat pumps can be used to upgrade low-temperature heat sources to higher-temperature energy sources. In some cases, thermal storage systems are required to store either high-temperature or low-temperature surplus recovered thermal energy or energy generated during off-peak demand periods for use during peak demand periods.

REFERENCES

Akyurt, M., N. J. Lamfon, Y. S. H. Najjar, M. H. Habeebullah, and T. Y. Alp. 1995. Modeling of waste heat recovery by looped water-in-steel heat pipes. *International Journal of Heat and Fluid Flow* 16: 263–271.

Andersen, E., S. Furbo, M. Hampel, W. Heidemann, and H. Muller-Steinhagen. 2008. Investigations on stratification devices for hot water heat stores. *International Journal of Energy Research* 32: 255–263.

Bakan, K., I. Dincer, and M. A. Rosen. 2008. Exergoeconomic analysis of glycerol cold thermal energy storage systems. *International Journal of Energy Research* 32: 215–225.

Cheralathan, M., R. Velraj, and S. Renganarayanan. 2007. Performance analysis on industrial refrigeration system integrated with encapsulated PCM-based cool thermal energy storage system. *International Journal of Energy Research* 31: 1398–413.

Demirel, Y. and H. H. Ozturk. 2006. Thermoeconomics of seasonal latent heat storage system. *International Journal of Energy Research* 30: 1001–12.

Devahastin, S. and S. Pitaksuriyarat. 2006. Use of latent heat storage to conserve energy during drying and its effect on drying kinetics of a food product. *Applied Thermal Engineering* 26: 1705–13.

Dincer, I. 2002. Thermal energy storage systems as a key technology in energy conservation. *International Journal of Energy Research* 26: 567–88.

Egolf, P. W., A. Kitanovski, D. Ata-Caesar, D. Vuarnoz, and F. Meili. 2008. Cold storage with ice slurries. *International Journal of Energy Research* 32: 187–203.

Lukitobudi, A. R., A. Akbarzadeh, P. W. Johnson, and P. Hendy. 1995. Design, construction and testing of a thermosyphon heat exchanges for medium temperature heat recovery in bakeries. *Heat Recovery System & CHP* 15: 481–491.

Lund, D. B. 1986. Low-temperature waste-heat recovery in the food industry, Chapter 17. In *Energy in Food Processing*, R. P. Singh, (Ed.), pp. 267–281. New York: Elsevier Science Publishing Company Inc.

Mazman, M., L. F. Cabeza, H. Mehling, H. O. Paksoy, and H. Evliya. 2008. Heat transfer enhancement of fatty acids when used as PCMs in thermal energy storage. *International Journal of Energy Research* 32: 135–143.

Mull, T. E. 2001. *Practical Guide to Energy Management for Facilities Engineers and Plant Managers*. New York: ASME Press.

Ozdogan, S. and M. Arikol. 1995. Energy and exergy analyses of selected Turkish Industries. *Energy* 20: 73–80.

Rohrer, W. M. 2006. Waste-heat recovery, Chapter 8. In *Energy Management Handbook* (6th ed.), Turner, W. C. and S. Doty, (Eds.), Lilburn, GA: The Fairmont Press, Inc.

Rosen, M. A., N. Pedinelli, and I. Dincer. 1999. Energy and exergy analyses of cold thermal storage systems. *International Journal of Energy Research* 23: 1029–38.

Singh, R. P. 1986. *Energy in Food Processing*. New York: Elsevier Science Publishing Company Inc.

Singh, R. P. and D. R. Heldman. 2001. *Introduction to Food Engineering* (3rd edition). San Diego: Academic Press.

Yamaha, M., N. Nakahara, and R. Chiba. 2008. Studies on thermal characteristics of ice thermal storage tank and a methodology for estimation of tank efficiency. *International Journal of Energy Research* 32: 226–244.

9 Novel Thermodynamic Cycles Applied to the Food Industry for Improved Energy Efficiency

9.1 INTRODUCTION

Several novel thermodynamic cycles such as refrigeration cycles, heat pumps, heat pipes, and combined heat and power generation cycles have been developed to improve the energy efficiency of existing processes or to use low-grade waste heat. Novel refrigeration cycles, such as absorption and adsorption refrigeration cycles, can be powered by waste heat or other renewable energy sources such as geothermal energy and solar energy at a low temperature (e.g., 70°C). A heat pump is used to transfer heat from a low-temperature source to a high-temperature sink if work is done on the heat pump cycle. Heat pumps can be used to upgrade a low-temperature heat source such as waste heat, solar heat, and groundwater to a higher-temperature one. Heat pipes can effectively transfer heat with a small temperature difference at a fast rate. Heat pipes can be used to enhance the heat transfer through a large food item such as a cooked ham with poor thermal conductivity. Heat pipes have also been used to recover waste heat such as the heat in the outlet air of an air conditioning system. Combined heat and power generation system can provide both thermal energy (i.e., steam) and power (i.e., electricity) simultaneously at a higher energy efficiency than traditional power generation systems.

In this chapter, the working principles of several novel refrigeration cycles, heat pumps, heat pipes, and cogeneration cycles are introduced. Their energy efficiencies are reviewed. Finally, the applications of these new energy conservation technologies in the food processing facilities are reviewed and discussed.

9.2 NOVEL REFRIGERATION CYCLES

9.2.1 Refrigeration Phenomena

There are many refrigeration phenomena. These refrigeration phenomena may be caused by (Sun and Wang, 2001)

- Evaporation of a low-pressure liquid at a low temperature
- Expansion of a high-pressure gas
- Flow of electric current through an interface of different semiconductors
- Chemical reactions

Based on the above refrigeration phenomena, a variety of refrigeration cycles have been developed. Energy shortage and the shortcomings of CFC refrigerants require the refrigeration industry not only to increase the efficiency of the traditional refrigeration cycles but also to accelerate the development of other suitable refrigerants and refrigeration cycles. According to the compensating energy used by the system (or driving power of the system), the refrigeration process can be divided into two categories: the mechanical energy-driven refrigeration system such as mechanical compression refrigeration, discussed in Section 9.2.4 and air cycle, discussed in Section 9.2.5, and the thermal energy-driven refrigeration system discussed in Sections 9.2.6 through 9.2.8.

9.2.2 Coefficient of Performance in a Mechanical Energy-Driving Refrigeration Cycle

The economic performance of a refrigeration system is defined as the refrigeration effect obtained per unit of compensating energy. The refrigeration coefficient ε is introduced to evaluate the performance of a mechanical energy-driven refrigeration cycle:

$$\varepsilon = \text{COP} = Q_e/W \tag{9.1}$$

where

Q_e is the refrigeration effect generated by the evaporator
W is the compensating energy consumed by the compressor

Sometimes, the refrigeration coefficient and the thermal coefficient are together named as coefficient of performance (COP). In a mechanical energy-driven refrigeration cycle, heat is transferred from a low-temperature heat source (i.e., Q_e absorbed in an evaporator) to a high-temperature one (i.e., Q_a released in a condenser) by consuming an amount of mechanic work (i.e., W done by a compressor). If the temperatures of the two heat sources are constant and the refrigeration cycle is reversible, according to the first law of thermodynamics, the energy balance equation of a mechanic compression refrigeration cycle is

$$Q_e + W = Q_a \tag{9.2}$$

For a reversible cycle running between two heat sources at constant temperatures of T_a and T_e (in Kelvin), the entropy of the cycle keeps constant according to the second law of thermodynamics. That is,

$$\frac{Q_a}{T_a} = \frac{Q_e}{T_e} \tag{9.3}$$

Substituting Equations 9.2 and 9.3, we obtain

$$\frac{T_a}{T_e} = 1 + \frac{W}{Q_e} \tag{9.4}$$

Therefore, Equation 9.1 can be rewritten as

$$\varepsilon_{max} = \text{COP}_{max} = \frac{Q_e}{W} = \frac{1}{T_a/T_e - 1} \tag{9.5}$$

where ε_{max} is the maximum value of the refrigeration coefficient or COP for a given mechanic compression refrigeration cycle.

The following conclusions can be drawn from Equation 9.5 (Sun and Wang, 2001):

- COP of a reversible refrigeration cycle operating between two constant-temperature heat sources is affected only by the temperatures of heat sources. The properties of the refrigerant in the cycle have no effect on its coefficient of performance.
- COP is associated with the temperature difference between the two heat sources. The smaller the temperature difference, the higher the coefficient.

Example 9.1

A mechanical compression refrigeration system performs as a reversible refrigeration cycle. The condenser and evaporator temperatures are 40°C and −20°C, respectively. What is the maximum value of COP of the system? If the evaporator temperature increases to 0°C, what is its COP?

Solution 9.1

1. From Equation 9.5, the COP of the refrigeration system operating at the condenser temperature of 40°C and evaporator temperature of −20°C is

$$\begin{aligned}\text{COP}_{max} &= \frac{Q_e}{W} = \frac{1}{T_a/T_e - 1} \\ &= \frac{1}{(273.15 + 40)/(273.15 - 20) - 1} \\ &= 4.22\end{aligned}$$

2. COP of the refrigeration system operating at the condenser temperature of 40°C and evaporator temperature of 0°C is

$$\begin{aligned}\text{COP}_{max} &= \frac{Q_e}{W} = \frac{1}{T_a/T_e - 1} \\ &= \frac{1}{(273.15 + 40)/(273.15 + 0) - 1} \\ &= 6.83\end{aligned}$$

9.2.3 Coefficient of Performance in a Thermal Energy-Driving Refrigeration Cycle

For a thermal energy-driven refrigeration cycle, the thermal coefficient ξ is defined as

$$\xi = \text{COP} = Q_e / Q_g \quad (9.6)$$

where

Q_e is the refrigeration effect generated in an evaporator
Q_g is the thermal energy consumed by the cycle

In the thermal energy-driven refrigeration cycle, the process of heat transferred from a low-temperature heat source (i.e., Q_e absorbed in an evaporator) to a high one (i.e., Q_a released in a condenser) is achieved by using a third thermal energy source (i.e., Q_g consumed in a regenerator). If the temperature of the third thermal energy source is also constant, according to the first law of thermodynamics, the energy balance equation of the thermal energy-driven refrigeration cycle is

$$Q_a = Q_e + Q_g \quad (9.7)$$

If the cycle is reversible, according to the second law of thermodynamics, there is

$$\oint dS = 0 \quad \text{or} \quad \frac{Q_a}{T_a} = \frac{Q_e}{T_e} + \frac{Q_g}{T_g} \quad (9.8)$$

Then, the thermal coefficient defined in Equation 9.6 can be rewritten as

$$\xi_{max} = \text{COP}_{max} = \frac{1}{T_a/T_e - 1}\left(1 - \frac{T_a}{T_g}\right) \quad (9.9)$$

The first element to the right of Equation 9.9 is similar to the COP of a mechanical energy-driven refrigeration cycle. The second element is the thermal efficiency of the reversed thermal engine running between T_g and T_a. The reversible thermal engine or regenerator can be regarded as an equal reversible mechanical energy-driven compressor, where a quantity of heat, Q_g, from the driving heat resource is first converted into a quantity of work, W, at a conversion factor of $(1 - T_a/T_g)$, and then the work drives the refrigeration cycle. Although the conversion factor of $(1 - T_a/T_g)$ is always smaller than one, it is incomparable between the COP of a mechanical compression refrigeration cycle, ε, and the COP of a thermal energy-driven refrigeration cycle, ξ, by quantity since the grade or quality of the compensating energies, W, used by a compressor and Q_g used by a regenerator is different.

The following conclusions can also be drawn from Equation 9.9 (Sun and Wang, 2001):

- The COP of a reversible refrigeration cycle powered by a thermal energy source running between heat sources at two constant temperatures is affected only by the temperatures of the three heat sources. The properties of the refrigerant in the cycle do not affect its COP value.

- The COP is associated with the temperature difference between two heat sources and the temperature of the driving heat source. The smaller the temperature difference between two heat resources and the higher the temperature of the driving heat source, the higher the coefficient of the refrigeration cycle.

Example 9.2

A thermal energy-driven refrigeration system performs as a reversible refrigeration cycle. The condenser and evaporator temperatures are 40°C and 0°C, respectively. The temperature of the thermal energy source is 80°C. What is the maximum value of the COP of the system?

Solution 9.2

From Equation 9.9, the COP of the refrigeration system is

$$\begin{aligned} COP_{max} &= \frac{Q_e}{Q_g} = \frac{1}{T_a/T_e - 1}\left(1 - \frac{T_a}{T_g}\right) \\ &= \frac{1}{(273.15+40)/(273.15+0)-1}\left(1 - \frac{273.15+40}{273.15+80}\right) \\ &= 0.77 \end{aligned}$$

The COP of the thermal energy-driven refrigeration system is much smaller than that of the mechanical compression refrigeration system running at between the same two constant temperatures given in Example 9.1 (0.77 vs. 6.83).

ε or ξ is the maximum value of the coefficients for a refrigeration cycle for given heat sources. However, the COP value for an actual refrigeration cycle is smaller than the theoretical value because there is an irreversible energy lost in the actual refrigeration cycle. There are two reasons to discuss the reversible refrigeration cycles though they are unattainable ideals: (1) they serve as a standard of comparison and (2) they provide a convenient guide to determine the temperatures that should be maintained to achieve maximum effectiveness (Sun and Wang, 2001). Therefore, the efficiency η is introduced to evaluate the perfecting extent of the thermal process in an actual refrigeration cycle, compared with that of an ideal one. It is defined as $\eta = \varepsilon/\varepsilon_{max}$ for the mechanical energy-driven system or $\eta = \xi/\xi_{max}$ for the thermal energy-driven system. There is $0 < \eta < 1$. The higher the η, the better the refrigeration cycle and the smaller the irreversible lost.

The values of COP and η are the criteria of the economic characteristics for a given refrigeration cycle. The COP gives a ratio of the desired energy to the required energy in a refrigeration cycle and its value can be bigger, smaller, or equal to one. However, the COP value cannot show the grade or quality of the energy sources. Although the value of COP is only associated with the temperatures of heat sources in an ideal refrigeration cycle, for an actual refrigeration cycle, its value is controlled by the temperatures of heat sources, the properties of refrigerants, and the efficiencies of components in the cycle. Therefore, the comparison of COP between two actual refrigeration cycles to evaluate their economic characteristics is reasonable only if the two cycles are of the same type and run between the same two temperatures. However, the thermal perfecting extent of η indicates the difference in thermal characteristics between the actual refrigeration cycle and its corresponding ideal refrigeration cycle. The comparison of the η between any two refrigeration cycles is meaningful (Sun and Wang, 2001).

9.2.4 Mechanical Compression Cycle

The mechanical vapor-compression cycle is the most widely used refrigeration cycle. As shown in Figure 9.1a, in this cycle, there are four main components: compressor,

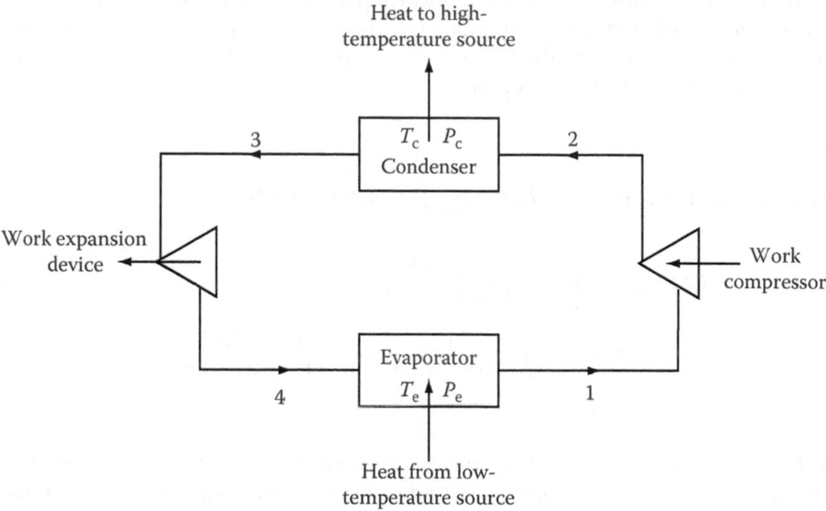

(a) Mechanic compression refrigeration cycle

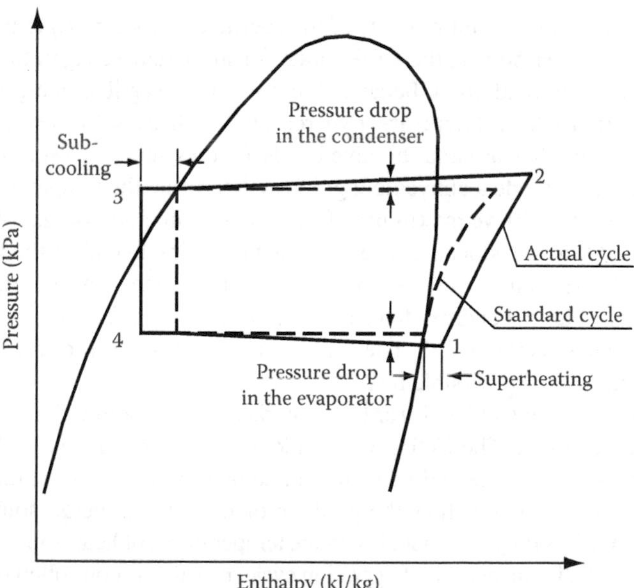

(b) Pressure-entropy diagram for the mechanical compression refrigeration cycle

FIGURE 9.1 Mechanical compression refrigeration cycle. (Adapted from Sun, D.W. and Wang, L.J., *Advances in Food Refrigeration*, Leatherhead Publishing, Leatherhead, 2001. With permission.)

expansion device, condenser, and evaporator. The refrigerant vapor exiting from the evaporator is compressed by the compressor, then condensed into liquid in the condenser, following which the high-pressure liquid flows through the expansion device and its pressure and temperature are dropped so that the liquid refrigerant can evaporate at a low pressure in the evaporator to complete the cycle.

A temperature-entropy for a standard vapor-compression cycle is shown in Figure 9.1b (Stoecker and Jones, 1982). The cycle constitutes

- 1–2 reversible adiabatic compression by the compressor
- 2–3 reversible rejection of heat at constant condensation pressure, desuperheating, and condensing the refrigerant vapor in the condenser
- 3–4 irreversible expansion through the expansion device
- 4–1 reversible addition of heat at constant evaporation pressure in the evaporator

A comparison of the pressure-enthalpy diagram between an actual cycle and the standard cycle is also given in Figure 9.1b. The essential difference between the actual cycle and the standard cycle is that there is a pressure drop in the condenser and the evaporator, subcooling of the liquid leaving the condenser, and superheating of the vapor leaving the evaporator for the actual cycle. Another difference is that the compression of an actual cycle is not isentropic due to friction and other losses.

9.2.5 Air Cycle Refrigeration

Air can also be used as a refrigerant in a vapor-compression cycle (Anon, 1998). An air cycle refrigeration system and its temperature-entropy diagram are given in Figure 9.2. An air cycle includes

- 1–2 isentropic compression
- 2–3 constant-pressure cooling
- 3–4 isentropic expansion
- 4–1 constant-pressure heating

The heat transfer in an air refrigeration cycle is a constant pressure process. A mechanical compression refrigeration cycle can operate between the two constant temperatures in its evaporator and condenser since the heat is transferred with phase changes of the refrigerant. However, in an air cycle, the heat transfer occurs only because of the temperature difference between the air and the surroundings. During heating (equal to the evaporation process of a mechanical compression refrigeration cycle), the air temperature must be lower than the temperature of the cooling medium so that the air can receive heat from the cooling medium until its temperature approaches the temperature of the cooling medium. Similarly, during cooling (equal to the condensation process), the air temperature must be above the atmospheric temperature so that the heat can be transferred from the hot air to the ambient. As shown in Figure 9.2b, the area *a* represents the increased amount of work required for the heat transfer during cooling and the area *b* represents the increased amount

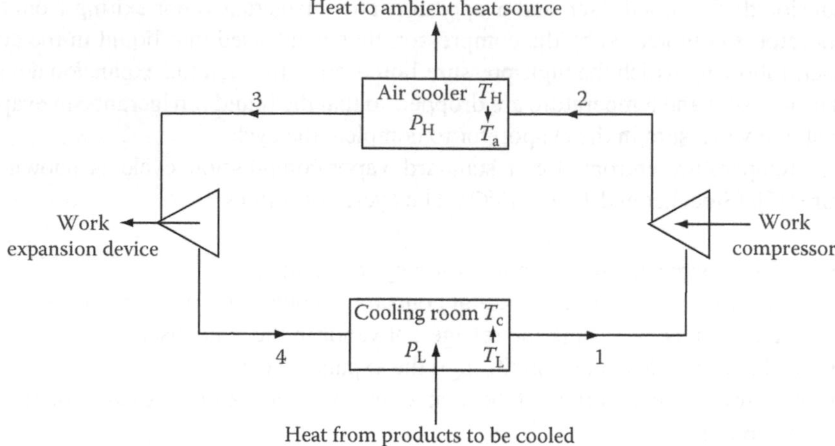

(a) Layout of air cycle refrigeration

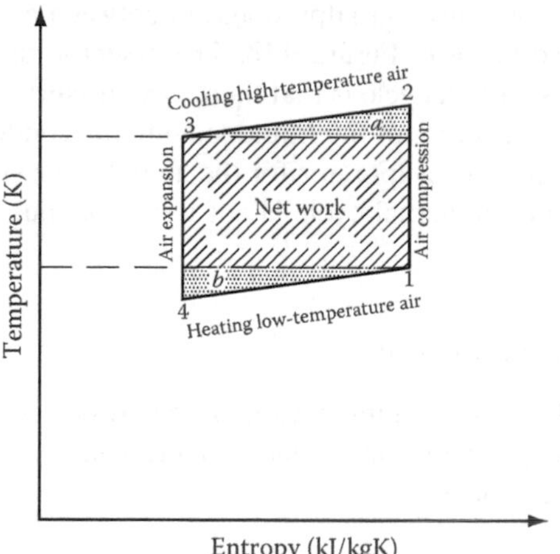

(b) Temperature-entropy diagram for the air cycle refrigeration

FIGURE 9.2 Air cycle refrigeration. (Reprinted from Sun, D.W. and Wang, L.J., *Advances in Food Refrigeration*, Leatherhead Publishing, Leatherhead, 2001. With permission.)

of work required for the heat transfer during heating. These non-isothermal effects of heat transfer reduce the COP of the air cycle. However, this does not mean that the energy consumption of air cycle refrigeration is greater than that of mechanical vapor-compression refrigeration. The larger the temperature difference between the high-temperature heat source and the low-temperature heat source, the nearer the COP of the air cycle approaches that of the vapor-compression cycle (Sun and Wang, 2001).

9.2.6 Absorption Refrigeration Cycle

An absorption–desorption cycle is based on the fact that the partial pressure of refrigerant vapor is a function of the temperature and concentration of a refrigerant solution. LiBr-H_2O and H_2O/NH_3 are two main working solutions used in absorption refrigeration cycles (Sun and Wang, 2001). In an absorption cycle as shown in Figure 9.3 (Sun and Wang, 2001), low-pressure refrigerant vapor from the evaporator is absorbed by a strong absorbent solution in the absorber. The pump increases the pressure of the weak solution absorber and delivers the high-pressure weak solution to the generator, where heat from a high-temperature source vaporizes the refrigerant vapor in the weak solution. The strong solution then returns to the absorber through a throttling valve to reduce the pressure of the strong solution. The high-pressure refrigerant vapor generated from the weak solution flows to the condenser where the vapor is condensed into liquid. The liquid refrigerant then flows into the evaporator through a throttling valve to reduce its pressure. In order to improve cycle performance, a solution heat exchanger is normally added to the cycle, as shown in Figure 9.3. The heat exchanger is an energy saving component but not essential to the successful operation of the cycle (Sun, 1998a). An absorption refrigeration cycle has four main heat exchangers: generator, evaporator, absorber, and condenser. A high-temperature heat source is added in the generator to change the weak absorbent-refrigerant solution to a strong absorbent-refrigerant solution, and low-temperature heat from the product to be chilled is added to the refrigerant in the evaporator. The heat of the cycle received is rejected to the outside of the cycle in the condenser and absorber.

The typical operating conditions for an H_2O/NH_3 absorption refrigeration cycle could be generator temperature of T_g = 60°C–90°C, condenser temperature of T_c = 20°C–40°C, absorber temperature of T_a = 20°C–40°C, evaporator temperature of T_e = 2.5°C–7.5°C, and mass flow rate of refrigerant of W_{re} = 1.0 kg/s (Sun, 1997b).

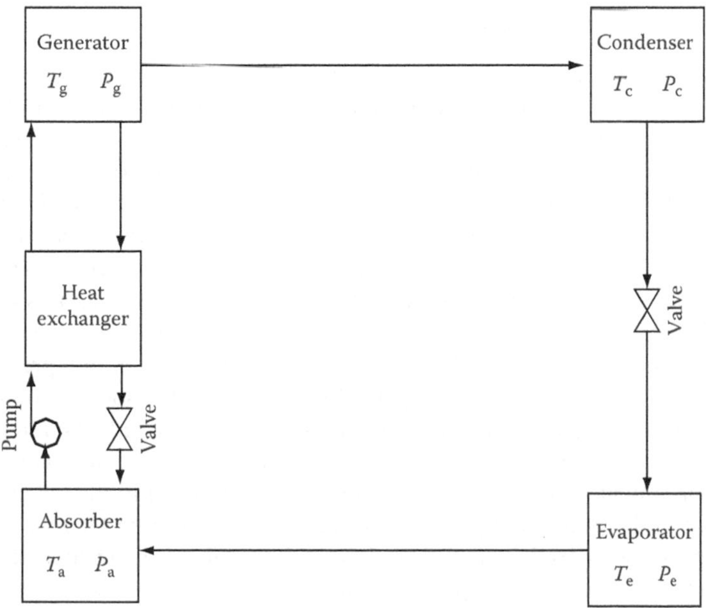

FIGURE 9.3 Absorption refrigeration cycle. (Reprinted from Sun, D.W. and Wang, L.J., *Advances in Food Refrigeration*, Leatherhead Publishing, Leatherhead, 2001. With permission.)

The COP for an absorption refrigeration machine is in the region of 0.6–0.9. These values are much smaller than that of a mechanical vapor-compression machine with a typical value of 3. However, it is difficult to compare two systems only in terms of COP values since electricity used in the vapor-compression cycle is much more expensive than the low-grade thermal energy used in the absorption cycle. The overall costs for absorption machines should be low. The economics of absorption refrigeration is largely determined by the heat exchangers of the cycle (Summerer, 1996). The heat exchanger area can be converted to the investment cost, C_i. The COP for a given heat exchanger area determines the necessary amount of driving heat for a specific cooling capacity, which is converted to the running cost, C_r. The total cost, C_t, is made up of the investment and running costs, which is given by

$$C_t = C_r + \frac{C_i}{D} \tag{9.10}$$

where D is a constant.

It has been reported that the lowest first cost is quite far from the lowest running cost (Berlitz et al., 1999). This means that the cheapest machine from a first cost argument will never be the cheapest machine from a total cost point of view. For a short running time, a relatively cheap machine is the most economical. For a long running time, a more expensive machine is optimal. However, the design of an actual machine depends on the thermal-physical properties of the working pairs as well as the external restrictions. There is much room for improving the economics of an absorption refrigeration cycle by improving its efficiency.

9.2.7 Ejector-Refrigeration Cycle

An ejector is a hydro-device with a special configuration and power by a high-pressure jet of a fluid such as steam. In the ejector, a high-pressure jet of a fluid is used to entrain a low-pressure fluid from the evaporator and then increase its pressure for the condenser (Sun and Eames, 1995). An ejector-refrigeration system is shown in Figure 9.4 (Sun, 1996b; Sun and Eames, 1997). This cycle is sometimes referred to as an ejector-compression refrigeration system. The ejector cycles may be powered by thermal energy. In an ejector-refrigeration cycle, low-grade thermal energy is supplied to the boiler, where the liquid refrigerant is vaporized. The vapor is used as the driving fluid of the ejector, where the refrigerant vapor is suctioned and mixed with the driving fluid to form a single stream. The stream is further compressed to the condenser pressure through another part of the ejector with a special configuration. Emerging from the ejector, the fluid undergoes a temperature reduction in the regenerator and then condenses in the condenser with the condensation heat being rejected to the environment. Finally, the condensate is partly pumped to the regenerator and flows back to the boiler and partly expands via a throttling valve and evaporates in the evaporator to produce the necessary cooling effect. The evaporated vapor is entrained by the ejector to complete the cycle.

The ejector performance is measured by the entrainment ratio, which is defined as the mass flow rate ratio of the secondary flow, refrigerant flow, over primary flow, and driving flow, which is given by

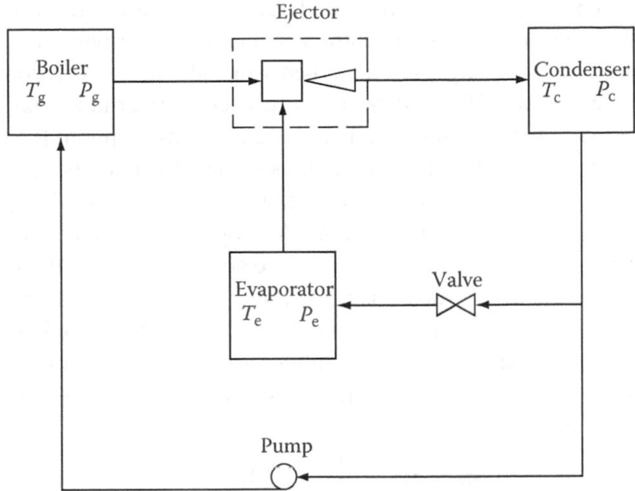

FIGURE 9.4 Ejector refrigeration cycle. (Reprinted from Sun, D.W. and Wang, L.J., *Advances in Food Refrigeration*, Leatherhead Publishing, Leatherhead, 2001. With permission.)

$$\omega = \frac{m_2}{m_1} \tag{9.11}$$

The system performance is measured by COP as discussed before. For an ejector cycle, the COP value can further be expressed as

$$\text{COP} = \frac{Q_e}{Q_b} = \omega \frac{h_{v,e} - h_{l,c}}{h_{v,b} - h_{l,c}} \tag{9.12}$$

where
- h is the enthalpy of the fluid
- Subscripts v and l denote the liquid and vapor
- Subscripts e, b, and c denote the evaporator, boiler, and condenser

From Equations 9.11 and 9.12, we can see that the system performance is closely related to the ejector performance and operating temperatures, and hence the ejector performance characteristics also represent the characteristics of the system.

For an ejector refrigeration cycle using HCFC-123 as ejection medium and operating at T_e = 5°C–10°C, T_c = 30°C, and T_g = 80°C–90°C, its COP is in the range 0.19–0.29. This COP can be improved by about 20% if a regenerator is introduced into the cycle. However, the addition of a pre-cooler has much less effect. In fact, its addition may not be economically justifiable (Sun and Eames, 1996).

9.2.8 Adsorption Refrigeration Cycle

Adsorption is the transfer of molecules from a fluid phase to rigid adsorbent particles. Desorption is the separation of molecules adsorbed from the solid adsorbent particle. Since adsorption involves the accumulation of refrigerant molecules into adsorbent particles, highly porous solids with very large internal area per unit volume are preferred. Adsorbents are natural or synthetic material of amorphous or microcrystalline structure. The commonly used adsorbents include activated carbon, silica gel, zeolite, and metal hydride. If the adsorption is caused by intermolecular forces rather than by formation of new chemical bonds, it is called phsisorption. If the adsorption is caused by making or breaking chemical bonds, the adsorption is chemisorption.

The key part of an adsorption refrigeration cycle is the working pairs such as zeolite–water or activated carbon–methanol (Critoph, 1986; Wang et al., 1999). The amount of refrigerant that can be charged by the adsorbent is usually determined by the properties of a refrigerant–adsorbent working pair, the temperature of the adsorbent, and the partial pressure of refrigerant vapor. The Dubinin–Askhov equation is one of the most widely used adsorption equations to calculate the amount of refrigerant charged by an adsorbent via a phsisorption process, which is expressed as

$$M = M_0 \exp\left[-K\left(T \ln\frac{P_s}{P}\right)^n\right] \tag{9.13}$$

where

M_0, K, and n are three constants
T is the temperature of the adsorbent in Kelvin
P is the vapor pressure of the refrigerant
P_s is the saturated pressure of the refrigerant at the adsorbent temperature

For a given absorbent, the value of M is related to the temperature of the adsorbent and the refrigeration partial pressure. The constants in Equation 9.13 can be determined by measuring the amount of refrigerant adsorbed under different refrigerant vapor pressures at a constant absorption temperature. Two isothermal adsorption curves at 150°C and 30°C for zeolite 13X-water working pair are given in Figure 9.5. The amount of refrigerant charged by the adsorbent increases with the decrease in temperature and the increase in pressure. For a zeolite 13X-water working pair, at a temperature of 150°C, pressure of 3 kPa, and equilibrium state, the maximum amount of water charged by zeolite 13X is 5% of the weight of zeolite 13X. If the temperature decreases to 30°C and the pressure increases to 1 kPa, the amount of water charged to zeolite 13X is 23% of the weight of zeolite 13X.

A typical adsorption refrigeration system is shown in Figure 9.6. Like a mechanical vapor-compression cycle, adsorption refrigeration cycle is also based the evaporation of a liquid refrigerant to generate a cooling effect. An adsorption refrigeration cycle consists of an evaporator, a condenser, and an adsorber. An adsorber filled with adsorbent replaces the compressor of a mechanical vapor-compression refrigeration cycle and the absorber of an absorption refrigeration cycle to accelerate the evaporation of a refrigerant. A whole adsorption refrigeration cycle can be divided into two stages. During the charging or adsorption period, the cold and fresh adsorbent bed is able to adsorb the refrigerant vapor from the evaporator to generate a cooling effect. During the discharging or desorption period, the saturated adsorber is heated by thermal energy to discharge the refrigerant vapor from the adsorber. The refrigerant vapor is delivered to the condenser and is condensed back to a liquid phase. The liquid refrigerant flows to the evaporator for the next cycle. The refrigeration occurs only during the charging period.

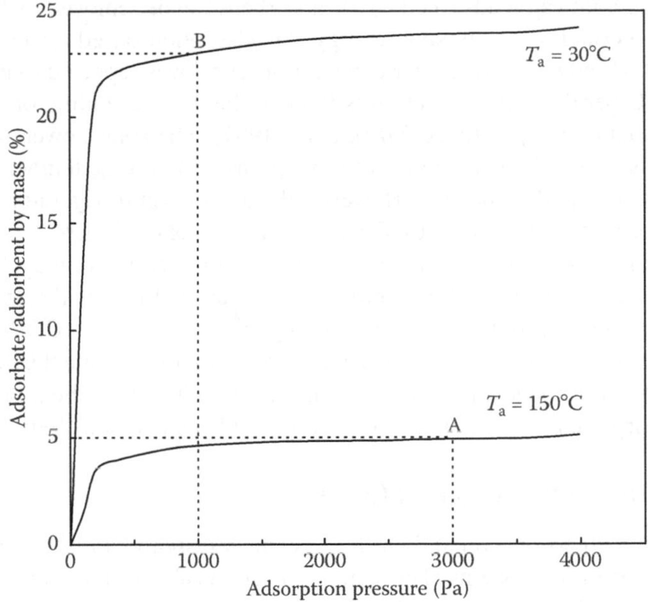

FIGURE 9.5 Isothermal adsorption curves for zeolite 13X. (Reprinted from Sun, D.W. and Wang, L.J., *Advances in Food Refrigeration*, Leatherhead Publishing, Leatherhead, 2001. With permission.)

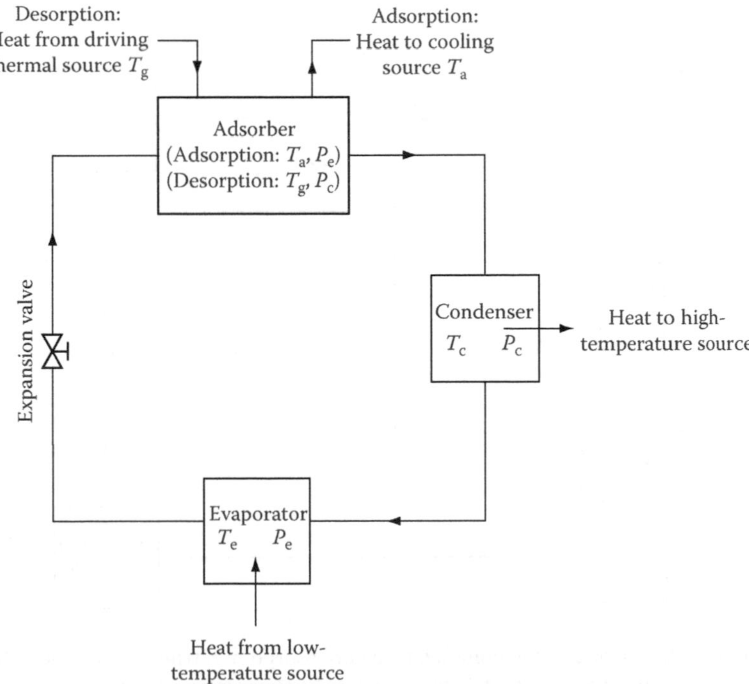

FIGURE 9.6 Adsorption refrigeration cycle. (Reprinted from Sun, D.W. and Wang, L.J., *Advances in Food Refrigeration*, Leatherhead Publishing, Leatherhead, 2001. With permission.)

The refrigerant vapor adsorbed by the adsorbent can be removed by heating with a low-grade thermal source or solar energy. An adsorption-based refrigerator with a volume of 103 L powered by a solar collector of 1.1 m^2 was reported to produce 6 kg of ice at −5°C per day and the COP was found to be 0.11 depending on the ambient and solar radiation temperatures (Wang et al., 1997). The solar-powered adsorption refrigerator is especially suitable for use in regions with a large temperature difference between day and night time. However, the adsorption refrigerator powered by natural gas or waste heat can be used everywhere (Wang et al., 1997).

The main technical issue related to an adsorption refrigeration cycle is the low heat transfer rate through a solid adsorbent bed because of its low thermal conductivity. The low heat transfer rate leads to a long cycling time, low COP, and a large size, compared to a mechanical vapor-compression unit to generate the same amount of cooling effect. Enhancement of heat transfer through an adsorbent bed is critical to make adsorption refrigerators economically viable (Zhu et al., 1996).

9.2.9 Combined Refrigeration Cycles

The COP for both absorption and ejector systems are significantly lower than those for vapor-compression systems. The combination of an absorption refrigeration cycle with an ejector refrigeration cycle brings together the advantages of absorption and ejector refrigeration systems and provides a higher COP for refrigeration (Sun, 1996a). This combined cycle is particularly suitable for utilizing waste thermal energy. As shown in Figure 9.7, an ejector integrated into the absorption cycle increases the refrigerant flow rate from the evaporator and therefore raises the cooling capacity of

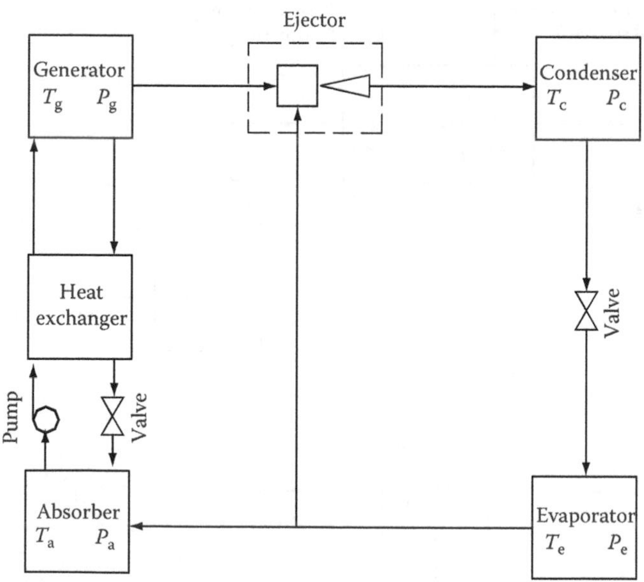

FIGURE 9.7 Schematic of the combined ejector-absorption refrigeration cycle. (Reprinted from Sun, D.W. and Wang, L.J., *Advances in Food Refrigeration*, Leatherhead Publishing, Leatherhead, 2001. With permission.)

the machine. The absorbent-refrigerant solution is heated at the generator by a low-grade heat source to produce high-pressure steam refrigerant. This steam (primary fluid) then flows through the ejector. A low-pressure region is caused by the expansion of the flow in the ejector, which induces vapor (secondary fluid) from the evaporator. The primary and secondary fluids then mix in the ejector. After emerging from the ejector, the mixed stream flows to the condenser where it condenses to liquid. The heat of condensation is rejected to the environment. The condensate is then expanded through a throttling valve to a low-pressure state and enters the evaporator where it evaporates to produce the necessary cooling effect. Then some of the evaporated vapor is entrained by the ejector and the remainder is absorbed by the strong solution coming from the generator via the solution heat exchanger and a throttling valve to form a weak solution. The heat produced by this absorption process is rejected to the environment. Finally, the weak solution is pumped back to the generator via the solution heat exchanger, thus gaining sensible heat from the strong solution returning to the absorber and hence completing the combined cycle. For the same evaporator temperature at 5°C and 10°C, the COP values of the combined machine are about 20% and 40% greater than those of the absorption machine, respectively (Sun, 1996).

A combined ejector-vapor-compression refrigeration cycle as shown in Figure 9.8 has also been investigated (Sun, 1997a; 1998b). It consists of an ejector refrigeration subcycle (ERSC) and a vapor-compression subcycle (VCSC). The combined cycle

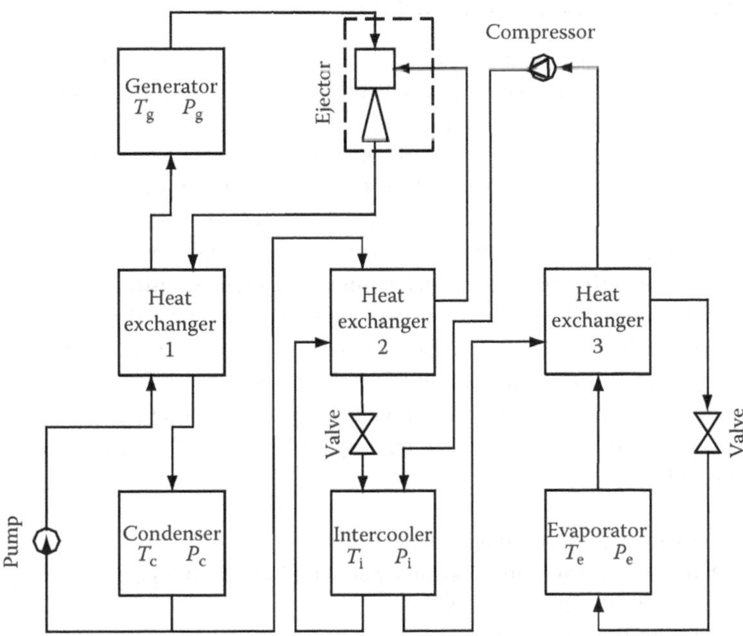

FIGURE 9.8 Schematic of the combined ejector-vapor-compression refrigeration cycle. (Reprinted from Sun, D.W. and Wang, L.J., *Advances in Food Refrigeration*, Leatherhead Publishing, Leatherhead, 2001. With permission.)

can apply a single refrigerant or dual refrigerants. The connection between the two subcycles is at the inter-cooler, which serves as the evaporator of the ERSC and condenser of the VCSC. Since the combined cycle brings the conventional ejector cycle and the vapor-compression cycle together, the following two coefficients are defined for analyzing the system performance characteristics.

$$(\text{COP})_W = \frac{Q_e}{W} \tag{9.14}$$

$$(\text{COP})_T = \frac{Q_e}{W + Q_g} \tag{9.15}$$

Equation 9.14 gives the power coefficient of performance $(\text{COP})_w$ which is the ratio between the useful cooling and the electric power consumption, while the thermal coefficient of performance $(\text{COP})_T$ is presented in Equation 9.15 (Sun, 1997a; 1998b). Sun (1997a; 1998b) found that the $(\text{COP})_w$ of the combined cycle was 50% higher than a vapor-compression cycle, depending on the condenser temperature. This means more than 50% saving in the electric energy for the same refrigeration capacity. The $(\text{COP})_T$ of the combined cycle is much lower than its $(\text{COP})_w$. However, the cost of low-grade waste heat is much cheaper than that of electric energy.

9.3 HEAT PUMPS

9.3.1 Working Principle of Heat Pumps

As discussed in Chapter 1, heat can only be transferred from a high-temperature region to a lower-temperature region in nature. If work is done on a thermodynamic cycle, the cycle can transfer heat in the opposite direction. A system, which transfers heat from a low-temperature region to a higher-temperature region, is called a heat pump. A heat pump operates like a refrigerator or an air conditioner. When the working substance changes from a liquid to a gas by evaporation or boiling, it absorbs heat. When it changes back from a gas to a liquid by condensation, it releases heat. The condition for the phase changes of the substance is determined by its temperature–pressure relationship. A liquid can be boiled at a low temperature to absorb heat if its pressure is low while a vapor can be condensed at a high temperature to release heat if its pressure is high. A heat pump uses these properties of a working substance such as a refrigerant to transfer heat from a low-temperature source to a higher-temperature sink.

Heat pumps can be used for both heating and cooling. There are four types of heat pump cycles:

- Closed vapor-compression cycle
- Mechanical vapor-recompression cycle with heat exchanger
- Open vapor-recompression cycle
- Heat-driven Rankine cycle

Closed vapor-compression cycle is a common design of a heat pump. In a closed vapor-compression cycle, a compressor is usually used to increase the vapor pressure while an expansion valve is used to decrease the liquid pressure. Two heat exchangers are used as an evaporator and condenser for phase changes of the working substance, respectively.

9.3.2 Efficiency of Heat Pumps

A heat pump extracts heat from a source at a low temperature and makes the extracted heat a useful energy source at a higher temperature. Since a heat pump transfers heat from a low-temperature source to a high-temperature source via a cycle, work should be done on the cycle according to the second law of thermodynamics. The COP is used to determine the efficiency of an ideal heat pump, which is given by

$$\text{COP} = \frac{Q_L}{W} = \frac{Q_L}{Q_H - Q_L} = \frac{T_H}{T_H - T_L} \tag{9.16}$$

where

Q is the heat
W is the work done on the heat pump cycle
T is the temperature in Kelvin
Subscripts H and L denote high temperature and low temperature

The COP is the ratio of the heat delivered to the sink to the power used by the compressor motor. If a heat pump is used to increase the temperature of a heat source from 70°C to 150°C, the COP of the ideal heat pump is 5.3 according to Equation 9.16. The COP of a heat pump decreases as the temperature of the heat source decreases and the temperature of the heat sink increases. In addition, due to the restriction of existing compressor design, the temperature of waste heat supplied to the heat pump should be lower than 110°C.

Since most compressors used in a heat pump cycle have only 65%–85% efficiency, there is frictional energy loss through the cycle, and since heat exchangers including the condenser and evaporator require a temperature difference for heat transfer, the COP of an actual heat pump is much lower than that of an ideal one. However, since the efficiency of electricity generation from fossil fuels is lower than 35%, the COP of a heat pump should be greater than 3 for economically attractive applications. However, it should be noted that the ratio of heat generated to the electricity consumed in an electric resistance heater is only 1.

9.3.3 Applications of Heat Pumps

The heat source and sink of a heat pump could be gas or liquid. Air-to-air, water-to-water, and water-to-air are the common types of heat pumps. Heat pumps can be used to recover low-temperature waste heat and upgrade it to a higher-temperature energy source to be used in food processing facilities. Heat pumps can also operate with groundwater and solar heating systems to upgrade the low-temperature heat sources to useful energy sources in food processing facilities.

Pasteurization and sterilization, evaporation and drying, cooking, refrigeration, and bottle washing are among the processes in food processing facilities where a heat pump can be used. Ozyurt et al. (2004) designed a liquid-to-liquid heat pump as

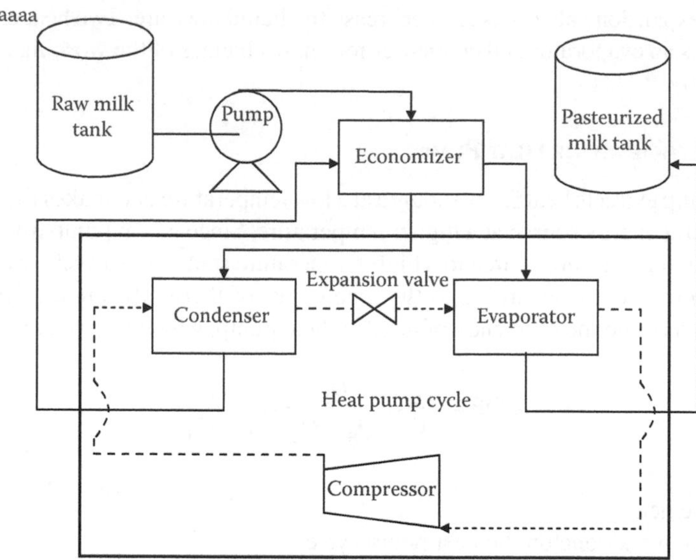

FIGURE 9.9 Schematic diagram of a liquid–liquid heat pump system for pasteurization of milk. (Reproduced from Ozyurt O., Comakli, O., Yilmaz, M., and Karsli, S., *Int. J. Energy Res.*, 28, 833, 2004. Copyright John Wiley & Son, Ltd. With permission.)

shown in Figure 9.9 for the pasteurization of milk. The heat source for the evaporator of the heat pump is hot pasteurized milk while the heat sink for the condenser of the heat pump is cold raw milk. The raw milk is pre-heated in an economizer by the hot milk from the condenser of the heat pump, and then sent to the condenser. When the milk reaches the set pasteurization temperature in the condenser, it is sent to the evaporator of the heat pump where the milk is cooled to the coagulation temperature. Finally, the milk is sent to the pasteurized milk tank. For the pasteurization temperature at 72°C and coagulation temperature at 32°C, the measured COP of the heat pump was found to be from 2.3 to 3.1. The heat pump system saved 66% of the primary energy compared to traditional plate and double jacket milk pasteurization systems. Specifically, the plate pasteurizer and double jacket boiler system consumed 1.9 and 2.8 times of the energy of the heat pump, respectively (Ozyurt et al., 2004).

Kuzgunkaya and Hepbasli (2007) investigated a ground-source heat pump drying system as shown in Figure 9.10. The system was formed by three circuits: ground water loop to extract geothermal heat, heat pump cycle, and drying air cycle. The measured COP of the heat pump was in the range between 1.63 and 2.88. Although the COP is the most commonly used measure for evaluating the efficiency of a heat pump, exergy analysis is needed to identify the possibilities for thermodynamic improvement. The exergy efficiency values of the heat pump were between 21.1% and 15.5% at a reference temperature of 27°C(Kuzgunkaya and Hepbasli, 2007).

Heat pumps have been used to increase the drying efficiency of convection drying (Jia et al., 1990; Jolly et al., 1990; Strommen and Kramer, 1994; Perera and Rahman, 1997; Adapa and Schoenau, 2005). As shown in Figure 9.11, a heat pump functions in a manner similar to a refrigerator. It consists of a condenser for

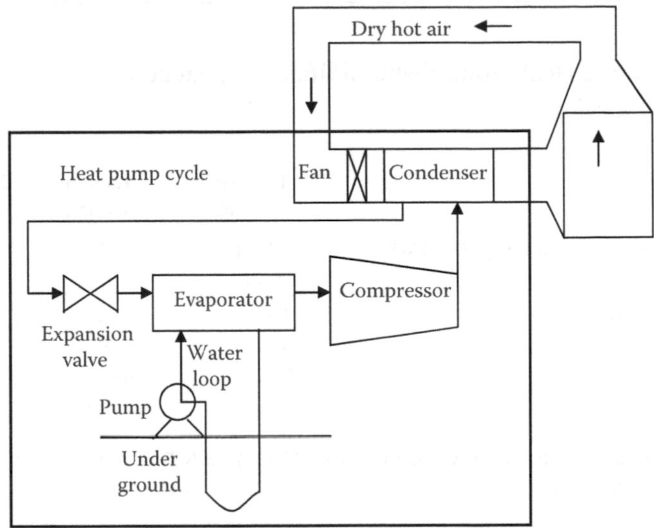

FIGURE 9.10 Schematic diagram of an air–water heat pump system for drying. (Reproduced from Kuzgunkaya, E.H. and Hepbasli, A., *Int. J. Energy Res.*, 31, 760, 2007. Copyright John Wiley & Son, Ltd. With permission.)

high-temperature heat exchange, a compressor, an evaporator for low-temperature heat exchange, and an expansion value to decrease the pressure of working liquid. A fan is usually used to provide air movement in the drying chamber. Dry and heated air is supplied continuously to the product tray to pick up moisture and is re-circulated. The moist air passes through the evaporator of the heat pump where the moisture is removed from the moist air by condensation and its latent heat of condensation is released. The cold dry air is reheated by the hot condenser of the heat pump. Too low drying temperatures limit drying rates. Some novel refrigerants such as R 134a can allow the use of a high drying temperature (e.g., 75°C for R 134a) with a requirement of supplementary heating (Perera and Rahman, 1997).

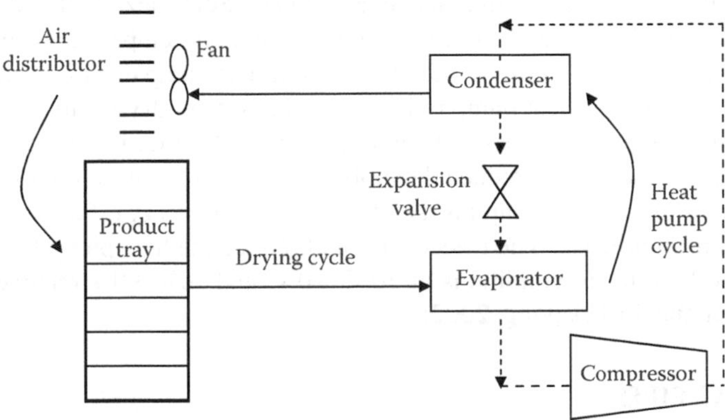

FIGURE 9.11 Schematic diagram of a typical heat pump dehumidifier dryer. (Adapted from Perera, C.O. and Rahman, M.S., *Trends Food Sci. Technol.*, 8, 75, 1997. With permission.)

TABLE 9.1
Comparison of a Heat Pump Dehumidifier with Vacuum and Hot-Air Drying

	Hot-Air Drying	Vacuum Drying	Heat Pump Dehumidifier Drying
Specific moisture removal rate (kg H2O/kWh)	0.12–1.28	0.72–1.2	1.0–4.0
Drying efficiency (%)	35–40	≤70	95
Operating temperature range (°C)	40–90	30–60	10–65
Operating relative humidity (%)	Variable	Low	10–65
Capital cost	Low	High	Moderate
Running cost	High	Very high	Low

Source: Reprinted from Perera, C.O. and Rahman, M.S., *Trends Food Sci. Technol.*, 8, 75, 1997. With permission.

Heat pump dehumidifier dryers are more energy efficient than conventional hot air dryers. Perera and Rahman (1997) compare the performance of heat pump dehumidifier dryers with hot air and vacuum dryers and their results are summarized in Table 9.1. The energy efficiency of a heat pump dehumidifier dryer can be as high as 95%. For a heat pump dehumidifier drying system, electrical energy is required to drive the compressor and air fan, and additional energy is required to heat the product and chamber and to replace any energy loss through the system and air leakage. The fan and compressor can be located in the drying chamber so that the heat produced by them can be used for the drying process.

Herbal and medicinal crops such as ginseng are usually dried at a low temperature of 30°C–35°C. Heat pump dryers can significantly reduce the energy consumption and the drying time compared to conventional dryers. It was reported that a re-circulating heat-pump-assisted continuous bed dryer can reduce energy consumption by 22% and drying time by 65% for drying herbal and medicinal crops compared to conventional dryers (Adapa and Schoenau, 2005). Soylemez (2006) developed a mathematical model to calculate the optimum operating temperatures and sizes of system components for minimum life cycle cost of heat pump drying systems.

The evaporator of a heat pump reduces the humidity of drying air while the condenser of the heat pump increases the temperature of the air. Silical gel from a staggered tube bank was used to reduce the moisture content of moist air before it entered the evaporator of the heat pump for the improvement of energy efficiency of the heat pump dehumidifier dryer. The desiccant could reduce the heat pump load by about 7%–20% at the inlet air conditions of 30°C–50°C and 70%–90% relative humidity (Kiatsiriroat and Tachajapong, 2002).

9.4 HEAT PIPES

9.4.1 Working Principle of Heat Pipes

A heat pipe is illustrated in Figure 9.12. One end of the tube serves as the evaporator and the other end is the condenser. Heating the evaporator end of the tube causes the

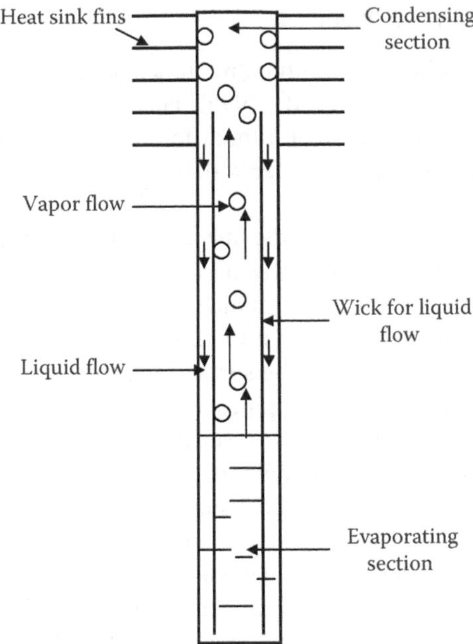

FIGURE 9.12 Schematic drawing of a heat pipe.

liquid to evaporate and the vapor travels along the tube to the other end. At the other end, which is the condenser, vapor condenses and releases its latent heat. The liquid travels back to the original end along the wick by capillary action to complete a cycle. Since a heat pipe uses latent heat rather than sensible heat to transfer heat from one location to another location, it is a highly efficient approach to transfer heat. A heat pipe can be used to quickly remove heat from a hot medium or transfer heat into a cold medium. The type of heat pipes is determined by the method by which the condensate returns to its hot end. The condensate return can be achieved by gravity or capillary action in a wick (Ketteringham and James, 2000).

A heat pipe is a pipe through which heat passes if there is a small temperature difference between its two ends. A typical heat pipe consists of an evacuated metal tube. A certain amount of working fluid is placed in the wick of the inside of the tube. For low-temperature applications, the working fluid may be helium, nitrogen, ammonia, water, acetone, and methanol while for high-temperature applications, the working fluid may be liquid metals such as sodium and potassium. Water is an excellent working fluid for heat pipes because of its high latent heat, availability, and high resistance to decomposition and degradation. Water has been used successfully in copper heat pipes for low-temperature applications. For water heat pipes working in the range of 150°C to 300°C, plain carbon steel would be considered (Akyurt et al., 1995).

The performance of a heat pipe may be affected by material–fluid combination, the channel arrangement, the cross-sectional shapes of the channel, and the operating parameters such as the working temperatures, the heat transfer rate, and external heat transfer coefficient at the condenser section (Sobhan et al., 2007).

9.4.2 Efficiency of Heat Pipes

A heat pipe can transfer heat several hundred times better than a solid copper rod of the same diameter (Lukitobudi et al., 1995). Heat pipes can dissipate substantial quantities of heat with a small temperature difference. This is achieved as a result of good heat transfer through mass transfer with the phase change of the working fluid inside the pipe.

During thermal processing of foods, only a small percentage of the energy consumed by the processing units is really used to change the temperature of foods while the rest of the energy is used to maintain the units at the set points. Cooking and cooling times for large food items or foods in large containers are limited by the rate at which heat can flow between the surface and the thermal center of the products or containers. Application of heat pipes to enhance the heat conduction through foods can reduce the processing time and increase the temperature uniformity inside the foods. The reduction in processing time and improvement of the temperature uniformity can thus reduce the energy consumption of processing facilities, such as a cooker and cooler, to maintain the operations at their set point and to change the temperature of foods (Ketteringham and James, 2000). James and Rhodes (1978) reported that between 74% and 85% of the energy consumed by a domestic electric oven was used to maintain the oven temperature at its set point while only between 15% and 26% of the consumed energy was used to increase the temperature of a meat joint from 5°C to a final temperature around 70°C during cooking. Therefore, a 50% reduction in cooking time due to the application of heat pipes would reduce the overall energy consumption by between 37% and 43%.

9.4.3 Applications of Heat Pipes

Heat pipes have been considered potential aids to improve cooking and cooling operations by providing a fast path for heat transfer between the surface and the thermal center of foods (Ketteringham and James, 2000; James et al., 2005). Use of heat pipes in cooking and cooling of foods can significantly save processing time and thus energy. James et al. (2005) found that the cooking and cooling times were reduced up to 50% and 25% if a heat pipe with a 12–19 mm diameter and 330 mm length was inserted in a 1 kg cylindrical meat joint with a 120–135 mm diameter and 77–106 mm length. Ketteringham and James (2000) found that heat pipes can reduce the cooling time of an 88 mm diameter cylinder filled with heated mashed potato from 70°C to 3°C by an average of 29%.

Heat pipes have been used to recover waste heat from flue or exhaust gas and outlet air of air conditioning systems (Akyurt et al., 1995; Lukitobudi et al., 1995; El-Baky and Mohamed, 2007). Heat pipes operate almost isothermally with a small temperature difference between the lower and the upper ends. A heat exchanger using heat pipes has several advantages when used for waste-heat recovery:

- Good heat transfer characteristics including high thermal conductance
- Compact design
- No large pressure drop
- No moving parts
- No cross-contamination between the fluid streams

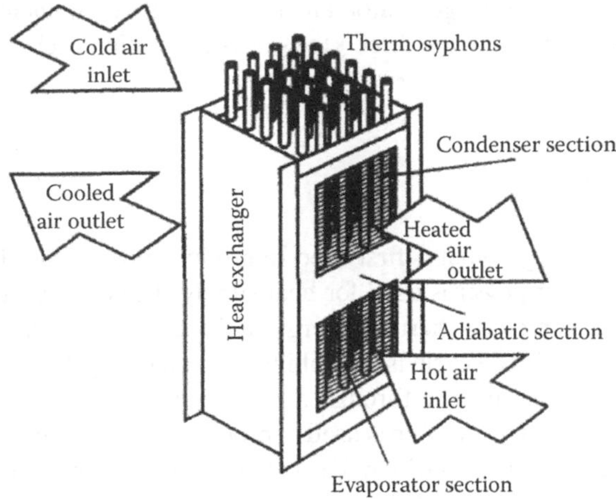

FIGURE 9.13 Heat pipe heat exchanger. (Reprinted from Lukitobudi, A.R., Akbarzadeh, A., Johnson, P.W., and Hendy, P., *Heat Recovery Sys. CHP*, 15, 481, 1995. With permission.)

Heat exchangers with heat pipes are suitable for energy recovery in air conditioning systems (El-Baky and Mohamed, 2007). The incoming fresh air at a high ambient temperature can be pre-cooled by the cold exhaust air stream before it enters the refrigeration equipment. Lukitobudi et al. (1995) designed an air-to-air heat exchanger using 10 heat pipes with water as the working fluid for the recovery of waste heat at a temperature lower than 300°C, in bakeries as shown in Figure 9.13. The lengths of both the evaporator and condenser sections of heat pipes were 300 mm and the central adiabatic section was 150 mm. The outside diameter was 26.7 mm and the wall thickness was 7.65 mm. About 60% of the evaporation section length was filled with the water working fluid. Heat pipes can transfer heat from one stream to another effectively. The effectiveness of the heat exchanger was around 65% for recovery of waste heat in the flue gas from the bakery oven. When thermosyphons are applied to an air-to-air heat exchanger, they can increase the effectiveness of the heat exchanger by reducing the temperature difference needed to transfer the heat between two air streams.

9.5 HEAT AND POWER COGENERATION CYCLES

9.5.1 Cogeneration Cycles and Working Principles

Cogeneration cycles simultaneously produce electricity or power and useful thermal energy from a single energy source such as coal, natural gas, and biomass. Cogeneration cycles can operate at a greater efficiency than those cycles where heat and power are produced in separate processes. Cogeneration can produce a given amount of electricity and processing heat with 30% less fuel than separate production of electricity and processing heat (Hordeski, 2003). The key components of a cogeneration cycle may include a prime mover such as a steam turbine, gas turbine, and internal combustion engine, electrical generator, and heat recovery equipment to recover heat from the prime mover. There are three kinds of cogeneration cycles:

- Top cycle
- Bottom cycle
- Combined cycle

In a top cycle, the energy input is first used to produce power, and the rejected heat from the production of power is used for heat supply. The energy input to a bottom cycle is first used to provide thermal energy and the rejected heat is then used to produce power. A combined cycle is a combination of a top cycle and a bottom cycle. Power is produced in a top cycle through a primary mover, typically a gas turbine. The heat exhausted from the turbine is used to produce steam, which is then expanded in a steam turbine to generate additional electricity or power. In a cogeneration cycle, a prime mover is used to convert thermal or chemical energy in the energy source into power. Internal combustion engine, steam turbine, and gas turbine are three commercially available prime movers.

A cogeneration cycle is used to provide mechanical power and heat. An electrical generator is required to further generate electricity from the mechanical power. Synchronous and induction generators are two basic types of generators. The efficiency of a typical electrical generator is between 95% and 100% at its full load.

9.5.2 Cogeneration Cycles with Internal Combustion Engines as the Mover

Internal combustion engines are commonly used as prime movers in a small cogeneration cycle, i.e., under 15 MW. Internal combustion engines use liquid and gaseous fuels such as natural gas, biogas, gasoline, and diesel to generate shaft power. There are two types of internal combustion engines: spark ignition or Otto cycle engines and diesel engines. In a cogeneration cycle, heat is recovered from the internal combustion engines as they produce shaft power. An internal combustion engine rejects heat from the engine's hot surfaces, exhaust gases, lubricating oil, and cooling jacket. Mull (2001) gives an approximate distribution of energy from the combustion of fuel in an internal combustion engine depending on engine type and speed, which includes

- Shaft power: 33%
- Heat convection and radiation on the surface: 7%
- Heat rejected in cooling jacket: 30%
- Heat rejected in exhaust gas: 30%

The recoverable heat may be the heat that is rejected in the exhaust gases, cooling jacket, and lubricating oil. The temperature of exhaust gases can be as high as 550°C. However, it is necessary to maintain the stack temperature above 150°C to prevent the condensation of moisture in the exhaust gas. Heat rejected in the cooling jacket is another waste-heat source. The temperature of the cooling jacket depends on the cooling fluid, which circulates in the cooling jacket. If water is circulated in the cooling jacket at an elevated pressure, the water temperature can be as high as 125°C. Sometimes, a phase change coolant is used in the cooling jacket. The temperature of lubricating oil is usually up to 70°C.

9.5.3 Cogeneration Cycles with Gas Turbines as the Mover

Gas turbines can produce up to 1000 MW of output power. Small gas turbines are becoming an important power generator. Compared to internal combustion engines, gas turbines have a small size and a high ratio of power to weight. Gas turbines can burn a wide range of liquid or gaseous fuels. They can generate power instantly without the need for a warm-up period and a cooling jacket.

Several cogeneration cycles with gas turbines as the mover are shown in Figure 9.14. As shown in Figure 9.14a, a gas turbine consists of an air compressor to increase the incoming air pressure, a combustor to burn fuels, and an expansion power turbine to extract energy from high-pressure and hot combustion gas flowing through the turbine. Part of the power generated in the power turbine is used to drive the air compressor. The temperature of the hot combustion gas into some turbine units can be as high as 1450°C.

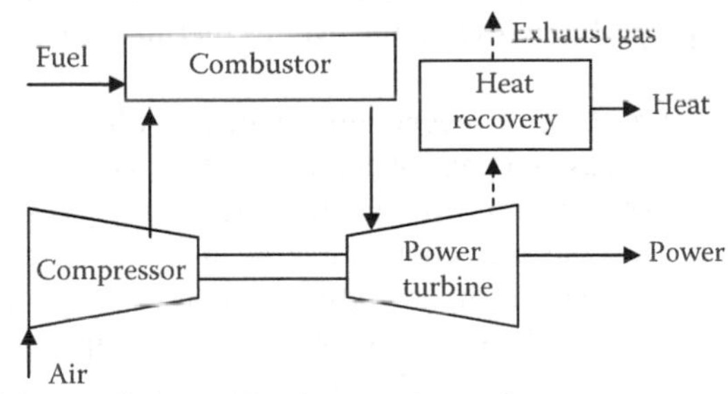

(a) A simple gas turbine cogeneration cycle

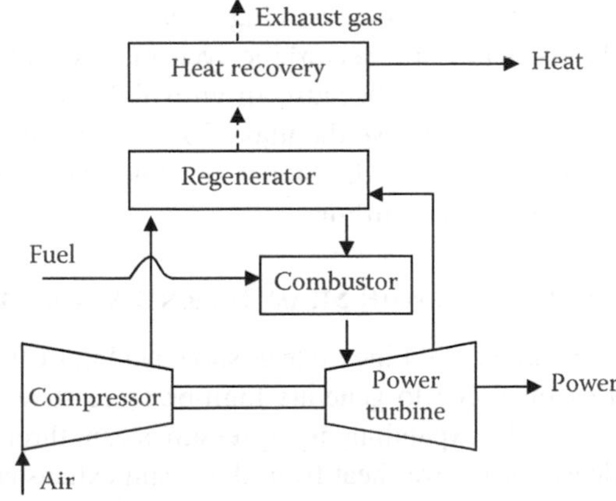

(b) A gas turbine cogeneration cycle with a regenerator to heat compressed air

FIGURE 9.14 Several gas turbine cycles.

(continued)

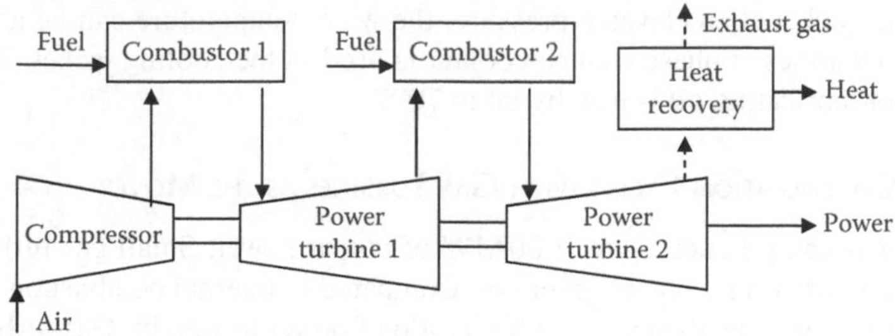

(c) A gas turbine cogeneration cycle with the second combustor and gas turbines

FIGURE 9.14 (continued)

The fuel to electricity efficiency of a gas turbine is typically in the range from 12% to 35%. The efficiency of combustion turbines increases with an increase in turbine inlet temperature and a decrease in exhaust temperature. The efficiency also increases with an increase in the turbine size. Like an internal gas engine, the exhaust gas temperature of a gas turbine is up to 550°C and the minimum temperature should be maintained above 150°C to prevent condensation in the stack. If the heat in exhaust gas can be properly recovered, the overall thermal efficiency of gas turbines can be over 60%.

As shown in Figure 9.14b, a regenerator, which is a heat exchanger to recover waste heat from exhaust gas and preheat compressed air, is usually used to lower the exhaust temperature and thus improve the efficiency of the gas turbine cycle. Alternatively, since the exhaust gas may contain a large amount of excess air, injection of additional fuel to the exhaust gas in an additional burner can increase the temperature of the exhaust gas up to 870°C–980°C, which can be supplied to the second gas turbine for more power generation. In this case, a second low-pressure gas turbine is added to produce additional power with reheated exhaust gas from the first turbine for increased energy efficiency, as shown in Figure 9.14c.

Furthermore, the efficiency of a gas turbine cycle may be increased by increasing the humidity of the compressed air because air with higher humidity may lower the combustion temperature and increase the mass flow rate of flue gas in the turbine. If a large amount of steam is required, a steam generator can be used to recover the heat from the exhaust gas of a gas turbine.

9.5.4 Cogeneration Cycles with Steam Turbines as the Mover

Steam turbines operate on the Rankine cycle as shown in Figure 9.15. A steam turbine cycle consists of a steam boiler to generate high-pressure steam, a steam powered turbine to generate power by expanding high-pressure steam through its blades, and a heat sink or a condenser to recover heat from the steam exhausted from the turbine. The maximum efficiency of a steam turbine cycle is limited to the Carnot efficiency, which is a function of the temperatures of the heat source and sink as given by

$$\eta = \frac{T_H - T_L}{T_H} \tag{9.17}$$

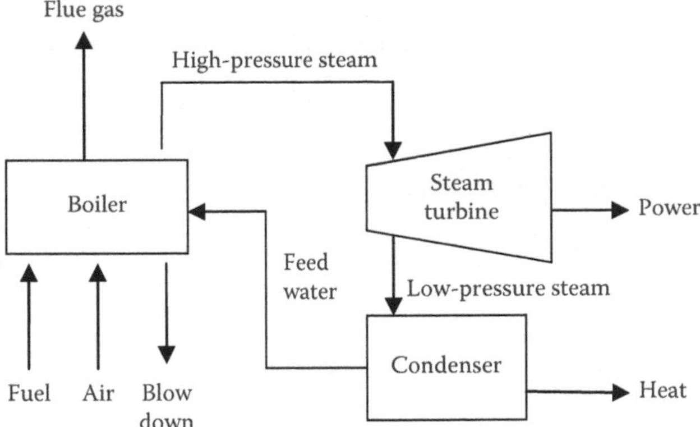

FIGURE 9.15 Schematic of a steam turbine cycle.

where T_H and T_L are the temperatures of the heat source and heat sink in Kelvin, respectively.

The temperature of the heat source, steam, is typically in the range of 190°C–540°C while the condenser temperature could be below 100°C. If the temperatures of the heat source and sink are 400°C and 100°C, respectively, the Carnot efficiency calculated using Equation 9.17 is 44.6%. According to the definition of Carnot efficiency, increasing the steam temperature and decreasing the condensation temperature can increase the Carnot cycle efficiency. In a real steam turbine, the Carnot efficiency is further reduced by the inefficiency of mechanical components and imperfection in the steam flow path. The mechanical efficiency of a steam turbine is typically in the range of 50%–80% when the turbine operates at about 95% of the rated load (Mull, 2001). If the steam turbine discharges condensate at the atmospheric pressure, the condensate temperature is 100°C.

Steam turbine cogeneration systems below 1 MW are generally not cost efficient. The cost range for a 1 MW steam turbine including the turbine, electrical generator and base plates is about 130–330 \$/kW while it drops to less than 100 \$/kW for a 25 MW turbine (Hordeski, 2003).

9.5.5 Cogeneration Cycles with Both Gas Turbine and Steam Turbines as the Mover

Figure 9.16 shows a combined heat and power (CHP) system with both a gas turbine and steam turbines as the mover (Rosen et al., 2005; Balli et al., 2008). The system mainly consists of a combustion chamber, a gas turbine, a high-pressure steam turbine, a low-pressure steam turbine, and electrical generators. The heat recovery from the gas turbine is used to generate steam, which is used in steam turbines for generation of additional electricity. Combustion chambers, gas turbines, and heat recovery steam generators are the main sources of irreversibilities, which account for more than 85% of the overall exergy losses. Therefore, the operating parameters of these components of a CHP system should be optimized to increase the overall exergy efficiency of the CHP system (Cihan et al., 2006).

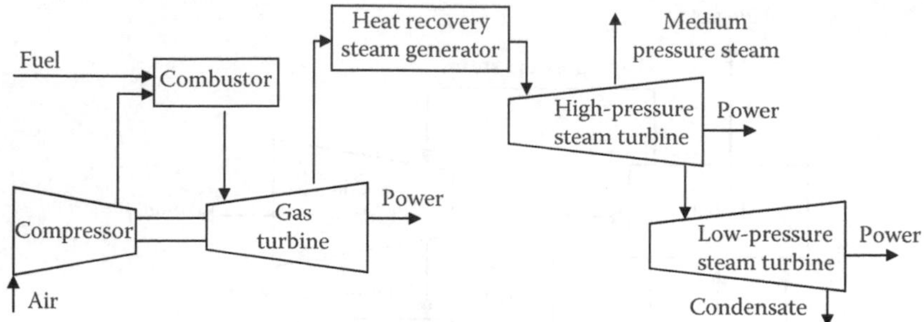

FIGURE 9.16 Schematic of a gas-steam turbine cycle.

9.5.6 Polices and Applications of Combined Heat and Power Generation

The United States has a number of laws to encourage facilities to use energy efficient and economic cogeneration. According to these laws, electric utilities must purchase excess power from cogeneration producers and provide supplemental power for cogeneration producers. However, a cogeneration cycle should meet ownership and efficiency requirements, which include

- No more than 50% of a cogeneration facility is owned by electric utilities or electric utility holding companies, either singly or in partnership.
- Small power plant must be smaller than 80 MW unless powered by waste or renewable resources.
- Minimum of 50% of the useful output of a cogeneration cycle should be in the form of thermal energy.
- Retail power and steam rates for cogeneration producers may be regulated by state bodies under a broad authority and subject to local sales or gross receipt taxes.
- Federal Energy Regulatory Commission (FERC) efficiency is defined as

$$\text{FERC efficiency} = \frac{\text{Power output} + 0.5 \text{ thermal output}}{\text{Lower heating value of fuel input}}$$

The efficiency of a cogeneration cycle should be at least 45% for top cycles. If the amount of useful thermal energy produced by the cogeneration cycle exceeds 15%, the efficiency threshold decreases to 42.5%. There is no efficiency standard for a bottom cycle or a cycle powered with renewable resources.

A number of factors should be analyzed to evaluate the feasibility of a power and heat cogeneration project. A simple estimate is given here. It is supposed that natural gas is used as the energy source of a cogeneration cycle. Natural gas is usually sold in dollars per million Btu ($/MMBtu) while electricity is sold in dollars per kilowatt hour ($/kWh). One MM Btu of natural gas has an equivalent energy content of 293 kWh of electricity. As discussed before, the fuel to electricity efficiency of a gas turbine is usually less than 35% and the overall thermal efficiency of a well-developed cogeneration cycle can be over 60%. In this case, one MM Btu of natural gas can only generate about 102 kWh of electricity and another 0.25 MM Btu of thermal energy. If the electricity price is 0.05 $/kWh and the natural gas price is higher than 6.8 $/MM Btu, the cogeneration system is absolutely unfavorable. However, many other factors such as the demand charge of commercial electricity, electrical and thermal loads in the facility, the compatibility of cogeneration with the existing heat and power systems, cogeneration unit size, and capital costs of the system should be considered for a detailed engineering analysis of the feasibility of a cogeneration system.

9.5.7 Applications of Combined Heat and Power Generation in the Food Industry

Since both processing heat and electricity are required in food processing facilities, a combined heat and power system can be used to efficiently and economically provide electricity or mechanical power and useful heat from the same primary energy source. For example, in a pasta manufacturing factory, the process consumed, on average, about 1.3 MJ of thermal energy and 1.28 MJ of electricity to produce each 1 kg of pasta. A combined heat and power generation plant with a gas turbine as its mover was used in the factory (Panno et al., 2007). A typical gas turbine works above 800°C and the temperature of flue gas is 430°C–540°C. The temperature of the flue gas from a recovery heat exchanger is between 130°C and 160°C, which can be used to produce high-temperature subcooled water for pasta drying at a temperature of about 140°C. The overall efficiency of electricity generation was estimated at 22%–26% while the overall CHP system efficiency was about 70%–80%. The CHP system reduced the primary energy demand by up to 9% and CO_2 emission by up to 9% in the pasta plant (Panno et al., 2007).

9.6 SUMMARY

Novel thermodynamic cycles such as refrigeration cycles, heat pumps, and heat pipes can be powered by low-grade energy sources such as waste heat, solar energy, and geothermal heat. Cogeneration cycles can efficiently produce electricity and heat in food processing facilities at the same time. The adsorption, ejector, and adsorption refrigeration cycles can convert waste heat to useful refrigeration. Although the coefficients of performance of these novel refrigeration systems are significantly lower than those for vapor-compression refrigeration systems, they may be cheaper to operate than conventional vapor-compression cycles powered by electricity. A heat pump can transfer heat from a low-temperature source to that of a higher temperature. Heat pumps have been used in pasteurization, sterilization, evaporation, and drying for energy saving because heat pumps can couple both the cooling generated by their evaporator and the heating generated by their condenser required by these unit operations. Since a heat pipe can enhance heat conduction through a large food item, a large food item with an inserted heat pipe can be cooked and heated more quickly. The rapid cooking and cooling process can save processing time, improve the quality and safety of processed foods, and reduce the energy consumption of the processing facilities. Heat pipes have also been used to recover waste heat at a small temperature difference. Cogeneration produces electricity and thermal energy simultaneously. The overall efficiency of heat and power production can be increased from current levels of 35%–50% in conventional power plants to over 80% in CHP plants.

REFERENCES

Adapa, P.K. and G.J. Schoenau. 2005. Re-circulating heat pump assisted continuous bed drying and energy analysis. *International Journal of Energy Research* 29: 961–972.

Akyurt, M., N.J. Lamfon, Y.S.H. Najjar, M.H. Habeebullah, and T.Y. Alp. 1995. Modeling of waste heat recovery by looped water-in-steel heat pipes. *International Journal of Heat and Fluid Flow* 16: 263–271.

Anon. 1998. New breakthrough in refrigeration. *Food Industry News* 14.

Balli, O., H. Aras, and A. Hepbasli. 2008. Exergoeconomic analysis of a combined heat and power (CHP) system. *International Journal of Energy Research* 32: 273–289.

Berlitz, T., P. Satzger, F. Summerer, F. Ziegler, and G. Alefeld. 1999. Contribution to the evaluation of the economical perspectives of absorption chillers. *International Journal of Refrigeration* 22: 67–76.

Cihan, A., O. Hacihafizoglu, and K. Kahveci. 2006. Energy-exergy analysis and modernization suggestions for a combined-cycle power plant. *International Journal of Energy Research* 30: 115–126.

Critoph, R.E. 1986. Possible adsorption pairs for use in solar cooling. *International Journal of Ambient Energy* 17: 183–190.

El-Baky, M.A.A. and M.M. Mohamed. 2007. Heat pipe heat exchanger for heat recovery in air conditioning. *Applied Thermal Engineering* 27: 795–801.

Hordeski, M.F. 2003. *New Technologies for Energy Efficiency*. New York: Marcel Dekker, Inc.

James, S.J. and D.N. Rhodes. 1978. Cooking beef joints from the frozen or thawed state. *Journal of the Science of Food and Agriculture* 29: 187–192.

James, C., M. Araujo, A. Carvalho, and J. James. 2005. The heat pipe and its potential for enhancing the cooking and cooling of meat joints. *Internal Journal of Food Science and Technology* 40: 419–423.

Jia, X., P. Jolly, and S. Clements. 1990. Heat pump assisted continuous drying. Part 2. simulation results. *International Journal of Energy Research* 14: 771–782.

Jolly, P., X. Jia, and S. Clements. 1990. Heat pump assisted continuous drying. Part 1. simulation model. *International Journal of Energy Research* 14: 757–770.

Ketteringham, L. and S. James. 2000. The use of high thermal conductivity inserts to improve the cooling of cooked foods. *Journal of Food Engineering* 45: 49–53.

Kiatsiriroat, T. and W. Tachajapong. 2002. Analysis of a heat pump with solid desiccant tube bank. *International Journal of Energy Research* 26: 527–542.

Kuzgunkaya, E.H. and A. Hepbasli. 2007. Exergetic performance assessment of a ground-source heat pump drying system. *International Journal of Energy Research* 31: 760–777.

Lukitobudi, A.R., A. Akbarzadeh, P.W. Johnson, and P. Hendy. 1995. Design, construction and testing of a thermosyphon heat exchanger for medium temperature heat recovery in bakeries. *Heat Recovery System & CHP* 15: 481–491.

Mull, T.E. 2001. *Practical Guide to Energy Management for Facilities Engineers and Plant Managers*. New York: ASME Press.

Ozyurt, O., O. Comakli, M. Yilmaz, and S. Karsli. 2004. Heat pump use in milk pasteurization: An energy analysis. *International Journal of Energy Research* 28: 833–846.

Panno, D., A. Messineo, and A. Dispenza. 2007. Cogeneration plant in a pasta factory: Energy saving and environmental benefit. *Energy* 32: 746–754.

Perera, C.O. and M.S. Rahman. 1997. Heat pump dehumidifier drying of food. *Trends in Food Science and Technology* 8: 75–79.

Rosen, A.M., N.M. Le, and I. Dincer. 2005. Efficiency analysis of a cogeneration and district energy system. *Applied Thermal Engineering* 25: 147–159.

Sobhan, C.B., R.L. Rag, and G.P. Peterson. 2007. A review and comparative study of the investigations on micro heat pipes. *International Journal of Energy Research* 31: 664–688.

Soylemez, M.S. 2006. Optimum heat pump in drying systems with waste heat recovery. *Journal of Food Engineering* 74: 292–298.

Stoecker, W.S. and J.W. Jones. 1982. *Refrigeration and Air Conditioning* (2nd ed.), Singapore: McGraw-Hill, Inc.

Strommen, I. and K. Kramer. 1994. New applications of heat pump drying process. *Drying technology* 12: 889–901.

Summerer, F. 1996. Evaluation of absorption cycles with respect to COP and economics. *International Journal of Refrigeration* 19: 19–24.

Sun, D.W. 1996a. Evaluation of a novel combined ejector-absorption refrigeration cycle – I: computer simulation. *International Journal of Refrigeration* 19: 172–180.

Sun, D.W. 1996b. Variable geometry ejectors and their applications in ejector refrigeration systems. *Energy* 21: 919–929.

Sun, D.W. 1997a. Solar powered combined ejector-vapor-compression cycle for air conditioning and refrigeration. *Energy Conversion & Management* 38: 479–491.

Sun, D.W. 1997b. Thermodynamic design data and optimum design maps for absorption refrigeration systems. *Applied Thermal Engineering* 17: 211–221.

Sun, D.W. 1998a. Comparison of the performances of NH_3- H_2O, NH_3-$LiNO_3$ and NH_3-NaSCN absorption refrigeration systems. *Energy Conversation Management* 39: 357–368.

Sun, D.W. 1998b. Evaluation of a combined ejector-vapor-compression refrigeration system. *International Journal Energy Research* 22: 333–342.

Sun, D.W. and I.W. Eames. 1995. Recent developments in the design theories and applications of ejectors-a review. *Journal of the Institute of Energy Sources* 68: 65–79.

Sun, D.W. and I.W. Eames. 1996. Performance characteristics of HCFC-123 ejector refrigeration cycles. *International Journal of Energy Research* 20: 871–885.

Sun, D.W. and I.W. Eames. 1997. Optimisation of ejector geometry and its application in ejector air-conditioning and refrigeration cycles. *Emirates Journal for Engineering Research* 2: 16–21.

Sun, D.W. and L.J. Wang. 2001. Novel refrigeration cycles. In *Advances in Food Refrigeration*, Sun, D.W. (Ed.), pp. 1–69. Leatherhead Publishing. UK: Leatherhead.

Wang, L.J., D.S. Zhu, and Y.K. Tan. 1999. Heat transfer enhancement on the adsorber of adsorption heat pump. *Adsorption* 5: 279–286.

Wang, L.J., L. Lin, D.S. Zhu, L.G. Fang, and Y.K. Tan. 1997. Applications of Solid Adsorption in Refrigeration and Heating Systems. *The Proceedings of the Fourth China-Japan-USA Symposium on Advanced Adsorption Separation Science and Technology*, Guangzhou, China.

Zhu, D.S., X.Y. Kang, L.J. Wang, Y.K. Tan, and E.J. Hu. 1996. Enhancement of thermal conductivity by using polymerzeolite for sorption heat pump. *Application. Refrigeration Science and Technology Proceeding*, International Institute of Refrigeration, Melbourne, Australia.

Part III

Energy Consumption and Saving Opportunities in Existing Food Processing Facilities

10 Energy Consumption in the Food Processing Industry

10.1 INTRODUCTION

The manufacturing industry utilizes almost one-third of the total energy consumed in the United States. Over the past 15 years, the energy intensity of the U.S. economy has been decreased by approximately 50% mainly because of the applications of efficient manufacturing technologies (Fischer et al., 2007). The food industry is recognized as one of eight energy-intensive manufacturing industries in the United States, among which are food, paper, bulk chemicals, petroleum refining, glass, cement, steel, and aluminum (Unruh, 2002). The energy cost in the food industry ranks the third among all input costs behind raw materials and labor (U.S. EPA, 2007).

The North American Industry Classification System (NAICS, 2007) categorized the food manufacturing industry (NAICS code 311) into nine manufacturing sectors, which include

- 3111 Animal food manufacturing
- 3112 Grain and oilseed milling
- 3113 Sugar and confectionery product manufacturing
- 3114 Fruit and vegetable preserving and specialty food manufacturing
- 3115 Dairy product manufacturing
- 3116 Animal slaughtering and processing
- 3117 Seafood product preparation and packaging
- 3118 Bakeries and tortilla manufacturing
- 3119 Other food manufacturing

Beverage manufacturers are part of the tobacco product manufacturing sector under the NAICS. The NAICS provides a common basis for economic data collection and analysis in the United States, Canada, and Mexico.

Energy conservation technologies can reduce the total energy consumption of a process and thus reduce the total production costs. Energy efficiency improvement in the food industry should not be considered to provide only economic benefit in an isolative way since it may also provide benefits for environmental protection, social sustainability, energy supply security, and industrial competitiveness. However, little information on the energy use in different food processing sectors is available. An important issue in the analysis of energy consumption in food processing facilities is the choice of an indicator, which can be used to evaluate the energy intensity and monitor the improvement of energy efficiency in the facilities. The most common indicator is the energy use per unit of output (Ramirez et al., 2006a). Since there is a large variation in products and processes in different food processing facilities, economic values such as total value of shipment and value added via processing are commonly used as a measure of output from a food processing facility (U.S. Census Bureau, 2006).

Two main types of energy: fuels such as coal, natural gas and petroleum oil, and electricity are used in food processing facilities. Steam and electricity are two direct energy carriers used in food processing facilities. Since electricity generation efficiency is only 30%–40%, the energy stored in the fuels consumed to generate electricity is three times the energy in the electricity generated. Steam is generated by combustion of fuels at a conversion efficiency of 80%. Different energy sources may be used in different food processes and unit operations. In energy analyses, factors such as 3.4 and 1.25 are usually used to multiply the electricity and steam consumption for comparison, respectively (Tragardh, 1986).

The goal of energy analysis is to find ways of reducing energy consumption or conservation of energy. Simple approaches to reduce energy use include (Poulsen, 1986)

- Development of new food products that do not require energy-intensive preservation methods
- Reduction of heat loads during processing
- Improvement of the performance of equipment
- Utilization of waste heat or renewable energy such as solar energy in processing

However, energy conservation measures will be adopted in the food industry only if they are cost effective. In food processing facilities, various energy-intensive unit operations such as sterilization/pasteurization, chilling/freezing, and evaporation/dehydration are used to manufacture different food products. Special attention should be paid to these energy-intensive operations.

In this chapter, the energy consumption in the food industry is briefly reviewed with respect to energy indicators, energy sources, and energy uses in different processing sectors, production of different products, and different unit operations. The energy conservation technologies applied to several energy-intensive unit operations including thermal processing, refrigeration, and dehydration are discussed. The energy consumption and energy saving opportunities in each food manufacturing sector are discussed in Chapters 11 through 16.

10.2 OVERVIEW OF ENERGY CONSUMPTION IN THE FOOD INDUSTRY

10.2.1 Energy Consumption in the Food Manufacturing Industry

In food processing facilities, edible raw materials are converted into higher-value food products. The conversion processes utilize significant amounts of labor, machinery, and energy. In the United States, the food processing industry is one of the largest manufacturers. The shipment value from the food industry increased from $458 billion in 2002 to $538 billion in 2006. The shipment values from the food industry were about 11.70% and 10.72% of the total shipment value from all manufacturing industries in 2002 and 2006, respectively. The food industry is also a large energy consumer. The total costs for purchasing fuels and electricity in the food industry increased from $3.55 billion in 2002 to $9.92 billion in 2006, which were 9.14% and 9.57% of the total energy costs for all manufacturing industries (U.S. Census Bureau, 2006). In the European Union, the food and tobacco sector, on average, accounted for about 8% of the total energy demand in the entire manufacturing sector in 2001. In the Netherlands, the food and tobacco sector accounted for about 9% of the total industrial energy demand and 23% of the industrial value added (Ramirez et al., 2006a).

There is a large variation in the shares of commercial energy going into the food production system in developing countries. In some developing countries, the energy consumed in the food processing industry could be very high. In Thailand, the food and beverage manufacturing sector accounted for about 30% of the total energy consumption and generated 16% of the total added value in all manufacturing sectors in 2000 (Bhattacharyya and Ussanarassamee, 2004). In Haiti, Burma, and Nepal, more than 25% of the total commercial energy is used in the food production system. In more industrialized countries like Brazil and India, less than 11% of the total commercial energy goes to the food production system. However, it should be noted that the share of energy consumed in the food industry increases somewhat again with increased income because more urbanization leads to more consumption of processed foods (Parikh and Syed, 1988).

10.2.2 Energy Indicators in the Food Manufacturing Industry

The current energy cost is only about 2% of the total production costs in the food industry in the developed countries (Ramirez et al., 2006a). However, the energy indicator based on economic values has increased from $0.78 energy cost/dollar shipment value in 2002 to $1.84 energy cost/dollar shipment value in the United States in 2006. The energy indicator for the food industry has increased 2.36 times from 2002 to 2006 while the energy indicator for all manufacturing sectors has increased 2.1 times during the same period (U.S. Census Bureau, 2006). Ramirez et al. (2006b) found that the energy indicator based on energy consumption in MJ per ton of finished products in the meat industry of four European countries including France, Germany, Netherlands, and United Kingdom has increased between 14% and 48% in the 1990s. The possible reason for this increase was the stronger hygiene regulations and increased processed products such as frozen and cut meat products to meet the changes in consumer preferences. Generally, an increase in the energy indicator for the food industry could be partially caused by an increase in the prices of major energy sources and tougher government regulations. The changes in food products and processes might also contribute to the increase in the energy indicator.

10.2.3 Energy Sources in the Food Manufacturing Industry

The food processing sector consumed 1085 GJ (or 1029 trillion Btu) delivered energy in the United States in 1998, which was 4.1% of delivered energy consumed within

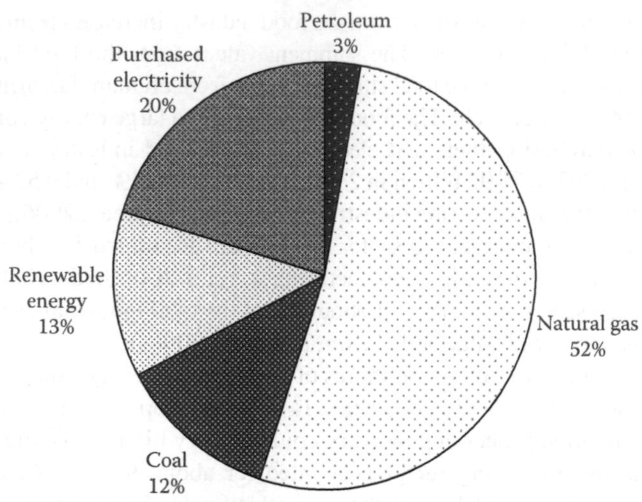

FIGURE 10.1 Delivered energy consumption by the type of fuels in the food manufacturing industry in the United States in 1998.

all industrial sectors. The delivered energy consumption per dollar of output for the whole food industry in the United States was 2.5 MJ/US$in 1998 (Unruh, 2002). In the developing countries, a large share of foods is consumed locally by the people who produce the foods. The foods undergo much less processing and packaging compared to those in the developed countries (Parikh and Syed, 1988).

The main energy sources used in the food manufacturing industry include petroleum, natural gas, coal, renewable energy, and electricity. Figure 10.1 shows the delivered energy consumption in terms of the types of fuels in the food processing industry of the United States in 1998 (data from Unruh, 2002). Natural gas and electricity are the two main energy sources used in the food industry of the United States, which were 52% and 20% of the total energy consumed, respectively, in 1998. While natural gas remains the largest energy source, consumption of renewable energy sources is projected to grow more rapidly as the food industry becomes more proficient in recovering and utilizing process and agricultural wastes. It should be noted that the food manufacturing industry produces about 9% of the electricity with its onsite power systems, 95% of which are heat and power cogeneration systems (U.S. EPA, 2007).

In the developing countries, the postharvest food system requires 2 to 4 times more energy than farming. The share of commercial energy that is used for food processing, such as milling, crushing, and food transport and cooking, ranges between 22% in Africa and 80% in the Near East. Biomass such as wood is a main noncommercial energy source used in the developing regions. The levels of energy consumption in the postharvest system depend on the level of income and the extent of urbanization (Parikh and Syed, 1988). In the developing regions, a large amount of noncommercial energy sources is used for cooking. In Africa, the relative shares of energy used for processing, transport, and cooking in the whole food system are

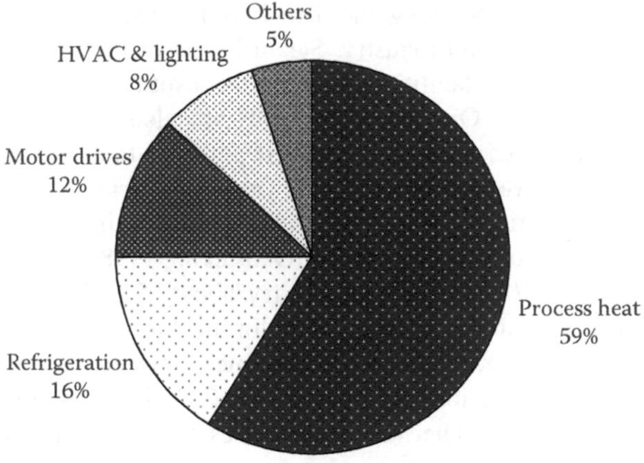

FIGURE 10.2 Energy consumption by the end users.

25:14:61 if only commercial energy is considered and 10:3:81 if both commercial and noncommercial energy are considered. In Latin America, the relative shares are 21:14:65 if only commercial energy is considered and 31:8:61 for the total energy. In the Near East, the relative shares are 12:7:81 and 12:6:82 for commercial energy and total energy, respectively. In the Far East, the shares are 23:11:66 and 18:4:78, respectively (Parikh and Syed, 1988).

The end users of energy in the food industry are process heating, process cooling and refrigeration, machine drive, and miscellaneous users. About half of all energy input is used to process raw materials into products. Fuels are mainly used for process heat and space heating while electricity is used for refrigeration, motor drives, and automation. Figures 10.2 and 10.3 give the energy consumption by the end users.

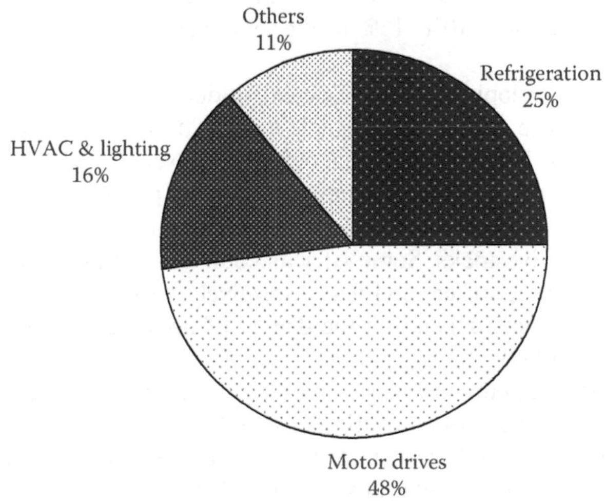

FIGURE 10.3 Electricity consumption by the end users.

Process heat for thermal processing and dehydration consumes approximately 59% of the total energy in the food industry. Steam is one of the important processing mediums in food processing facilities. Boiler fuel constitutes nearly one-third of the total energy consumption. Ovens and furnaces are also widely used in thermal processes. Refrigeration uses approximately 16% of the total energy in the whole food industry. Motor drives represent 12% of the total energy use. The non-process uses including space heating, venting, air conditioning, lighting, and onsite transport consume only about 8% of the total energy (Okos et al., 1998).

As shown in Figure 10.3, in the food industry, about 25% of the electricity is used for process cooling and refrigeration and 48% for machine drive (Okos et al., 1998). However, in the meat sector, refrigeration consumes between 40% and 90% of total electricity use during production time and almost 100% during non-production periods (Ramiez et al., 2006a). The non-process uses for space heating, venting, air conditioning, lighting, and onsite transport consume about 16% of electricity (Okos et al., 1998).

10.2.4 Energy Use in Different Food Manufacturing Sectors

In the United States, the meat manufacturing sector is the largest individual sector in the food industry in terms of shipment value followed by grain and oilseed milling, fruit and vegetable, and bakeries and tortilla. Grain and oilseed milling and meat manufacturing are two large energy consumers in the food industry as shown in Table 10.1, which consume 22.3% and 20.7% of the total energy input into the food industry. However, in other developed countries, the energy demand distribution in different food sectors may be different. Figure 10.4 shows the primary energy demand distribution in different food sectors in the Netherlands in 2001 (Ramirez et al., 2006a). (1) The processing and preserving of fruits and vegetables, (2) production, processing, and preserving of meat products, (3) manufacturing of vegetable oil and animal fat, (4) manufacturing of dairy products, (5) manufacturing of prepared animal feeds, and (6) manufacturing of grain mill, starches, and starch products consumed about 10%–15% of the total energy input into the whole food industry.

In most of the developing countries, cereal products have the largest share among all food products since grains are the staple foods. The shares of cereal products are 65% for Africa, 35% for Latin America, 69% for the Near East, and 77% for the Far East. Some developing countries may have special food product structures. The share of sugar is 90% in Mauritius and 78% in Cuba. Columbia, Mexico, and Argentina have high shares in livestocks. Some African countries, Brazil, and Sri Lanka have high shares in cassava, coffee, and tea, respectively. Generally, the shares of livestock and milk products depend on the degree of urbanization and income levels (Parikh and Syed, 1988).

Different food manufacturing sectors have different energy indicators. In the United States, grain and oilseed milling has the highest energy indicator while the meat manufacturing sector has the lowest energy indicator as shown in Table 10.1. It required \$3.81 to generate 1 dollar of shipment value in the grain and oilseed milling sector, compared to \$1.37 of energy cost per dollar shipment value in the meat manufacturing sector in the United States in 2006.

TABLE 10.1
Energy Use and Indicators in Different Food Manufacturing Sectors in the United States in 2006

Manufacturing Sector	Shipment Value (Million $)	Total Energy Cost (Million $)	Energy Indicator (Cents Energy Cost/$ Shipment Value)	Percent of Electricity Cost Among the Total Energy Cost (%)	Percent of Total Energy Cost in the Whole Industry (%)
3111 Animal food manufacturing	33,988	522	1.54	47.1	5.3
3112 Grain and oilseed milling	57,667	2198	3.81	37.0	22.3
3113 Sugar and confectionery product manufacturing	28,225	577	2.04	36.6	5.8
3114 Fruit and vegetable preserving and specialty food manufacturing	56,279	1302	2.31	44.3	13.1
3115 Dairy product manufacturing	75,428	1184	1.57	52.4	11.9
3116 Animal slaughtering and processing	149,577	2055	1.37	53.4	20.7
3117 Seafood product preparation and packaging	10,849	246	2.27	45.5	2.5
3118 Bakeries and tortilla manufacturing	54,173	911	1.68	54.7	9.2
3119 Other food manufacturing	71,602	926	1.29	52.9	9.3

Source: From U.S. Census Bureau. 2006. *2006 Annual Survey of Manufactures*. Available at http://factfinder.census.gov.

In the United States, although the purchased electricity is only about 20% of the total consumed energy, the food industry spent nearly 47% of its energy expense on average to purchase electricity in 2006 because the electricity price per unit of energy content is higher than that of other energy sources. Bakeries and tortilla manufacturing, meat manufacturing, and dairy product manufacturing sectors spent a little more than 50% of their energy expenses on purchased electricity. The costs of purchased electricity in the grain and oilseed milling and sugar and confectionery sectors were only 37% of their total energy expenses in 2006. The grain and oilseed milling and sugar and confectionery

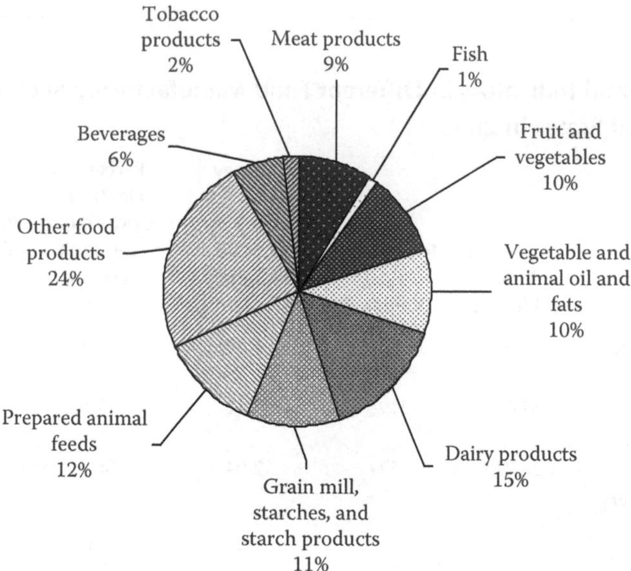

FIGURE 10.4 The primary energy demand distribution in different food sectors in the Netherlands in 2001. (Adapted from Ramirez, C.A., Blok, K., Neelis, M., and Patel, M., *Energy Policy* 34, 1720, 2006a. With permission.)

sectors require high boiler usage. Within the food industry, the dairy product manufacturing and the meat manufacturing sectors particularly have a high electricity demand while the sugar sector has a high fuel demand (Fritzson and Berntsson, 2006).

10.2.5 Energy Use for Production of Different Food Products

The amount of energy used for production of a given amount of products highly depends on the type of the products. Table 10.2 gives the energy use for production of different food products in the Netherlands in 2001 (Ramirez et al., 2006a). The ranking of food commodities in energy consumption does not follow the same pattern as the volumes processed. Production of milk powder and whey powder consumes 9385 and 9870 MJ heat per ton of products, respectively, while production of pasta consumes only 2 MJ of heat per ton of product. Production of wheat starch consumes 2,960 MJ of electricity per ton of product while production of beet pulp consumes only 5 MJ electricity per ton of product.

10.2.6 Energy Use in Different Unit Operations

Table 10.3 gives the energy consumption by different operations in the food industry. The energy consumption for the same operations may differ because of the type of equipment used, operating practice, ambient temperature, local infrastructure, and the skills of staffs. Drying and freezing consume large amounts of energy. On average, a drying process consumes 6 MJ energy to remove 1 kg of water from products and a freezing process consumes 1 MJ (or 0.3 kWh of electricity) to process 1 kg of food products at −20°C.

TABLE 10.2
Energy Use for Production of Different Food Products in the Netherlands in 2001

Product	Specific Electricity Consumption	Specific Fuels and Heat Consumption	Unit
Meat Sector			
Beef and sheep	341	537	MJ/ton dress carcass weight
Pig	465	932	MJ/ton dress carcass weight
Poultry	1008	576	MJ/ton dress carcass weight
Processed meat	750	3950	MJ/ton product
Rendering	234	1042	MJ/ton raw material
Fish Sector			
Fresh fillets	129	6	MJ/ton product
Frozen fish	608	6	MJ/ton product
Prepared and preserved fish	482	1062	MJ/ton product
Smoked and dried fish	12	2077	MJ/ton product
Fish meal	684	6200	MJ/ton product
Fruits and Vegetables			
Potatoes product		5722	MJ/ton product
Unconcentrated juice	250	900	MJ/ton product
Tomato juice	125	4789	MJ/ton product
Frozen vegetables and fruits	738	1800	MJ/ton product
Preserved mushrooms		2898	MJ/ton product
Vegetables preserved with vinegar		2178	MJ/ton product
Tomato ketchup	380	1700	MJ/ton product
Jams and marmalade	490	1500	MJ/ton product
Dried vegetables and fruits	1500	4500	MJ/ton product
Crude and refined oil		672	MJ/ton product
Dairy Products			
Milk and fermented products	241	524	MJ/ton product
Butter	457	1285	MJ/ton product
Milk powder	1051	9385	MJ/ton product
Condensed milk	295	1936	MJ/ton product
Cheese	1206	2113	MJ/ton product
Casein and lactose	918	4120	MJ/ton product
Whey powder	1138	9870	MJ/ton product
Starch and Starch Products			
Wheat starch	2960	8800	MJ/ton product
Maize starch	1000	2331	MJ/ton product
Potato starch	1425	3564	MJ/ton product
Prepared Animal Feeds			
For farm animals		475	MJ/ton product
For pets		2306	MJ/ton product

(continued)

TABLE 10.2 (continued)
Energy Use for Production of Different Food Products in the Netherlands in 2001

Product	Specific Electricity Consumption	Specific Fuels and Heat Consumption	Unit
Sugar			
Refined sugar	555	5320	MJ/ton product
Beet pulp	5	1820	MJ/ton product
Other Products			
Sweet biscuits		4581	MJ/ton product
Waffles and wafers		3195	MJ/ton product
Soups and broths		7659	MJ/ton product
Pasta	648	2	MJ/ton product
Flour	420	30	MJ/ton product
Cacao beans		6384	MJ/ton product
Non-roasted coffee	141	1597	MJ/ton product
Roasted coffee	518	1997	MJ/ton product
Extracts of coffee in solid form		15675	MJ/ton product
Beer	19.5	153	MJ/hl product
Mineral water and soft drinks	133	199	MJ/1000 L product
Unsweetened water and soft drinks	120	360	MJ/1000 L product

Source: Adapted from Ramirez, C.A., Blok, K., Neelis, M., and Patel, M., *Energy Policy* 34, 1720, 2006a. With permission.

10.3 ENERGY EFFICIENCY AND CONSERVATION IN THE FOOD INDUSTRY

10.3.1 Energy and Exergy Efficiencies in Food Processing Facilities

Energy and exergy efficiencies in food processing facilities vary with end users and production lines. Ozdogan and Arikol (1995) investigated the energy and exergy efficiencies of different energy users in the Turkish food industry and their results are given in Table 10.4. In the Turkish food industry, heat at a temperature below 200°C accounts for 97.5% of the total thermal energy demand. The energy and exergy efficiencies vary from 18.5% to 90% and from 13% to 90%, respectively. Motor power had the highest energy and exergy efficiencies while illumination had the lowest energy efficiency and water heating had the lowest exergy efficiency.

Ho et al. (1986) and Ho and Chandratilleke (1987) audited and analyzed the energy consumption of production lines in several multiproduct food processing plants in Singapore. The energy efficiency and exergy efficiencies varied from 23.6% to 81.5% and 7.9% to 26.1%, respectively, for different food processing plants as shown in Table 10.5. The instant noodles plant has the highest energy and exergy efficiencies while the malt beverage plant has the lowest energy and exergy efficiencies.

TABLE 10.3
Energy Consumption by Different Operations in the Food Industry

	Energy Consumption	Unit
Transport		
Truck for short distance	2.7	MJ/km ton
Truck for long distance	1.4	MJ/km ton
Truck for local use	2.0	MJ/km ton
Train	0.22	MJ/km ton
Air cargo	12	MJ/km ton
Ship through coast traffic	0.7	MJ/km ton
Ship through canal	0.15	MJ/km ton
Cold transport	Addition of 20%–30%	—
Preservation		
Heat sterilization of retort	Heat: 0.29 and electricity: 0.02	MJ/kg
Blanching of fish and meat products	0.25	MJ/kg
Freezing to −20°C	0.99	MJ/kg
Freezing to −30°C	1.2	MJ/kg
Heating of spices	0.36	MJ/kg
Drying	6	MJ/kg removed water
Cold Storage (Electricity to refrigeration system)		
Temperature difference at 5°C	1.25	kW/kg
Temperature difference at 10°C	3.75	kW/kg
Temperature difference at 15°C	5.0	kW/kg
Temperature difference at 20°C	6.25	kW/kg
Temperature difference at 35°C	13	kW/kg
Temperature difference at 50°C	20	kW/kg
Packaging		
Form fill and seal of retort pouch	Electricity: 0.066 and heat: 0.11	MJ/kg

Source: Adapted from Singh, R.P., *Energy in Food Processing*. Elsevier Science Publishing Company Inc., New York, 1986. With permission.

Fischer et al. (2007) reported that around 57% of the primary energy inputs into the industry are lost before reaching the intended process activities. Estimates from several studies indicate that on average, savings of 20% to 30% energy can be achieved without capital investment, using only procedural and behavioral changes. Industrial energy consumption can be further cost-effectively reduced by 10% to 20% through well-structured energy management programs that combine energy conservation technologies, operation practices, and management practices.

10.3.2 Energy Conservation in Food Processing Facilities

Prospective energy efficiency improvement measures to be used in the food industry should be technically feasible and economically practical. The development of energy

TABLE 10.4
Energy and Exergy Efficiencies for Different End-Uses in the Turkish Food Industry

End-Use	Energy Efficiency (First Law), %	Exergy Efficiency (Second Law), %
Processing heat	72	18
Space heating	75	13
Water heating	60	10
Illumination	18.5	17
Motive power	90	90
Weighted average	75.1	23.5

Source: Adapted from Ozdogan, S. and Arikol, M., *Energy*, 20, 73, 1995. With permission.

efficiency improvement for food processing facilities requires the evaluation of numerous prospective energy conservation measures for each food manufacturing sector.

Most food processing facilities have large steam or hot water demands. Boilers are the largest fuel user to provide steam as process heat for different unit operations such as sterilization, pasteurization, evaporation, and dehydration in the food industry. Boilers consume about one-third of the total energy or more than half (50%) of the fuel. Appreciable energy is lost from stack flue gas, blowdown water, steam leaks, and poor surface insulation during steam generation and distribution. The energy conservation technologies for boilers and steam distribution systems are discussed in Chapter 4.

Compressed air is another important processing media for conveying foods and for process control in food processing facilities. The production of compressed air can be one of the most expensive processes in manufacturing facilities. These energy conservation technologies include the use of high-efficiency and variable speed motors, reduction of inlet air temperature, use of a cooling or waste-heat

TABLE 10.5
Energy and Exergy Efficiencies for Different End-Uses in the Turkish Food Industry

Production Line	Energy Efficiency (First Law), %	Exergy Efficiency (Second Law), %
Condensed milk plant	59.6	19.6
Bakery plant	43.0	15.5
Instant noodles plant	81.5	26.1
Malt beverage plant	23.6	7.9

Sources: Data from Ho, J.C., Wijeysundera, N.E., and Chou, S.K., *Energy*, 11, 887, 1986; Ho, J.C. and Chandratilleke, T.T., *Appl. Energy*, 28, 35, 1987.

recovery unit for compressors, reduction of air leaks along the air distribution line, reduction of air pressure, and use of a localized air delivery system. These energy conservation technologies for compressed air generation and distribution are discussed in Chapter 5.

Motor drives and refrigerators are two large electricity users in the food industry, which consume about 48% and 25% of the total electricity, respectively. Furthermore, most motors operate in a fashion that requires both real power due to the presence of resistance and reactive power due to the presence of inductance in the motors. Increase in power factors should be considered for improving electrical efficiency and reducing the energy costs of motors. Motors are designed to operate most efficiently under their rated loads. Therefore, an effective way to conserve energy is to match the required loads with the rated loads of motors. The energy conservation technologies for motors are discussed in Chapter 6.

Heating and cooling of foods is achieved in heat exchanger equipment. Heat exchangers also play a key role in waste-heat recovery. Several energy conservation technologies including heat transfer enhancement, fouling removal, optimization of heat exchanger design, and optimization of heat exchanger network have been used to improve the energy efficiency of heat exchangers. These energy conservation technologies for heat exchangers are discussed in Chapter 7.

Any processing air, vapor, and water effluent streams above the ambient temperature may be an energy source. Boiler flue gas, boiler blowdown water, steam condensate, exhaust gas from dryers and ovens, cooling air and water from air compressor and large motors, and vapor from cookers are examples of waste heat sources. By recirculation and recovery of waste heat, the energy consumption of food processing facilities could be cut by 40%. Waste-heat recovery and thermal storage are discussed in Chapter 8.

Table 10.6 is an example to show the potential energy savings in the British industry, which used 2410 PJ of energy in 1986 (Anonymous, 1987). As seen from Table 10.6, the potential for industrial energy savings was between 25% and 34% of the total industrial fuel used. Waste-heat recovery was the biggest contribution to the total energy savings, which could save 8.96%–11.95% of the total industrial fuel use. The use of processing waste as fuels could save another 4.48%–7.47% of the total fuel consumption. Table 10.7 is another example to show the potential energy savings identified for a Nestle factory in Switzerland (Muller et al., 2007). In the Nestle facility, insulating of high-temperature condensate return pipes and using a low-pressure air blower to replace the use of high-pressure compressed air for the sealing operation of process units can save 1127 GJ/year (or 338 MWh/year) fuels and 166 MWh/year electricity, respectively.

10.3.3 Renewable Energy in Food Processing Facilities

Renewable energy technologies, in general, are sometimes considered as direct substitutes for existing technologies. For example, solar and other renewable energy technologies can provide small incremental capacity additions to existing energy systems for short peak demands. Such power generation units usually provide more flexibility in incremental supply than large, long lead-time units such as nuclear power stations. The development of renewable energy technologies except hydroelectric power generation has in part been limited by the existing focus on fossil and nuclear

TABLE 10.6
Potential Energy Savings in British Industry

Type of Energy Saving	Amount (PJ)	Percent of Total Energy Consumption (%)
Waste-heat recovery	216–288	8.96–11.95
Waste as fuel	108–180	4.48–7.47
Improved instrumentation and control	72–108	2.99–4.48
Heat pumps	36	1.49
Process insulation	36	1.49
Improved drying and evaporation practice	36	1.49
Industrial combined heat and power plants	36	1.49
Improved methods of driving machinery	72–108	2.99–4.48
Total savings	612–828	25.39–34.36

Sources: Data taken from Anonymous, *Our Common Future, World Commission on Environment and Development*, Oxford University Press, Oxford, 1987, and Dincer, I. and Rosen, M.A., *Appl. Energy*, 64, 427, 1999.

energy. Food processing facilities usually generate large amounts of organic processing wastes. The processing wastes can also be used as energy sources in the facilities. Bagasse is the main noncommercial energy source used in sugarcane processing facilities. Rice husk and other by-products are also burned for heat and power supply in rice milling facilities. Conversion of food processing wastes into energy products is discussed in Chapters 21 through 27.

TABLE 10.7
Summary of Some Energy Savings Identified in a Nestle Factory

Measure	Energy Type	Energy Saving (MWh/Year)	Estimated Payback (Year)
Replacing compressed air usage by dedicated blower	Electricity	166	2
Regulation of HVAC	Electricity	80	Negligible
Removing standby of air compressors with a VSD unit	Electricity	69	23
Fixed compressed air leakage	Electricity	50	Negligible
Insulating pipes of high-temperature condensate return	Fuels	338	1.5
Vacuum production in dryer	Fuels	150	1
Regulation of steam use	Fuels	50	Negligible

Source: Adapted from Muller, D.C.A., Marechal, F.M.A., Wolewinski, T., and Roux, P.J., *Appl. Thermal Eng.*, 27, 2677, 2007. With permission.

10.4 ENERGY CONSERVATION IN UNIT OPERATIONS

Thermal processes such as pasteurization and sterilization, chilling and freezing, and evaporation and drying are energy-intensive unit operations used in the food industry for food preservation and safety. It is a real need for food manufacturers to combine food safety and quality with energy conservation. The energy saving opportunities for each unit operations include three aspects:

1. Improvement of energy efficiency in existing units
2. Replacement of energy-intensive units with novel units
3. Use of renewable energy sources, particularly food processing wastes

Improvement of the energy efficiency of existing units is still the main consideration for decreasing energy costs. Meanwhile, novel energy conservation technologies such as heat pumps, supercritical-fluid processing, nonthermal sterilization and pasteurization processes, and thermal energy-powered refrigeration cycles have been introduced in food processing facilities. Capital investments should incorporate cutting-edge technologies and processes in anticipation of a future when energy is likely to be an increasing cost of doing business. The energy saving opportunities and technologies for pasteurization and sterilization, evaporation and drying, and chilling/freezing operations are discussed in the following sections.

10.4.1 Pasteurization and Sterilization

10.4.1.1 Maintenance and Optimization of Existing Systems

Thermal pasteurization of liquid foods such as milk and fruit juices is a well-established and effective means of terminal decontamination and disinfection of these products. Thermal processing is also an important method of food preservation in the manufacturing of canned foods and retortable pouches. The basic function of a thermal process is to inactivate pathogens and food spoilage causing microorganisms in foods. Simpson et al. (2006) developed a mathematical model to estimate total and transient energy consumption during heating of retortable shelf-stable foods. According to the results from Simpson et al. (2006), retort insulation can reduce 15%–25% of current energy consumption depending on selected conditions. Furthermore, in batch retort operations, maximum energy demand occurs at the venting step, which lasts only for the first few minutes of the process cycle while very little energy is needed thereafter to maintain the process temperature. An increase in the initial temperature of food products can reduce the peak energy demand in the order of 25%–35%. In addition, it is customary to operate the retorts in a staggered schedule so that no more than one retort is vented at any one time. Thus, operating practice can also reduce the peak energy demand during retorting.

10.4.1.2 Application of Heat Pumps

Pasteurization and sterilization require a heating step and a cooling step. A heat pump can be used to couple the energy flow between the heating unit and the cooling unit.

Ozyurt et al. (2004) designed a liquid-to-liquid heat pump as discussed in Chapter 9 for the pasteurization of milk. The hot pasteurized milk is cooled by the evaporator of the heat pump while the cold raw milk is heated by the condenser of the heat pump. For the pasteurization temperature at 72°C and coagulation temperature at 32°C, the measured COP of the heat pump ranged from 2.3 to 3.1. The heat pump system can save 66% of the primary energy compared to traditional plate and double jacket milk pasteurization systems (Ozyurt et al., 2004).

10.4.1.3 Applications of Nonthermal Processes

Thermal processes are usually considered to be energy intensive. In addition, the slow heat transfer through food products is usually a limiting factor for thermal treatment of food products. Nonthermal pasteurization techniques including food irradiation (Chapter 18), pulsed electric field treatment (Chapter 19) and high-pressure processing (Chapter 20), as well as microwave sterilization (Chapter 21) have also been developed to replace the conventional thermal sterilization and pasteurization processes for saving energy and improving product quality and safety. These nonthermal processing times are usually short. For example, high-pressure processing is exposed to pressure up to 1000 MPa for a few minutes. Pulsed electric field treatment is based on the delivery of pulses at a high electric field intensity of 5–55 kV/cm for a few milliseconds. Food irradiation lasts for several seconds to several minutes. Most alternative preservation processes can achieve the equivalent effect of thermal pasteurization but not sterilization as shown in Table 10.8 (Lado and Yousef, 2002).

Food irradiation is a cold process, which can damage the DNA of living cells effectively so that the living cells become inactivated. Compared with thermal pasteurization, food irradiation is a more efficient pasteurization method for solid foods without causing significant changes in taste and quality of the products. The energy used for food irradiation is very small. The dose of food irradiation is usually less than 10 kGy (energy equivalent to 10 kJ/kg of food). The increase in food temperature due to irradiation is less than 3°C (Loaharanu, 1996).

TABLE 10.8
Inactivation of *Escherichia coli* in Milk by Heat and Alternative Preservation Technologies

Preservation Process	Product	Treatment Conditions	Targeted Bacteria	Log Count Decreased
Heat	Milk	63°C, 16.2 s	*E. coli*	5.9
Gamma irradiation	Milk	10 kGy	*E. coli*	7.0
High-pressure processing	Milk	500 MPa, 25°C, 5 min	*E. coli*	5.9
Pulsed electric field	Milk	22.4 kV/cm, 330 μs	*E. coli*	4.7

Source: Adapted from Lado, B.H. and Yousef, A.E., *Microbes Infection*, 4, 433, 2002. With permission.

Microwave heating has been used for high temperature-short time (HTST) sterilization/pasteurization of foods, particularly thick food items. Because of low thermal conductivity, it is impossible to achieve HTST treatment for food items with a several-centimeter thickness by conventional heating methods. Microwave energy can penetrate into the food items and cause a rapid temperature rise to pasteurization temperatures. Huang and Sites (2007) used a microwave heating system for in-package pasteurization of ready-to-eat meats. They observed that the inactivation rate of *Listeria monocytogens* during microwave pasteurization was 0.41, 0.65, and 0.94 log (CFU/pk)/min at the surface temperature of 65°C, 75°C, and 85°C, respectively. The overall rate of bacterial inactivation for the water immersion pasteurization at the same surface temperatures was only 30%–75% higher than that of microwave in-package pasteurization. However, microwave pasteurization is much faster than water immersion pasteurization.

High-quality fruit juices with sufficient product safety cannot be achieved with conventional thermal sterilization or pasteurization. Also, the need to reduce energy costs stimulates the search for nonthermal techniques such as pulsed electric field treatment and high-pressure processing (Mertens and Knorr, 1992; Toepfl et al., 2006). Application of an external electrical field to a biological cell induces an electrical potential across the cell membrane. If the electrical potential exceeds a critical level, local structural changes of the cell membranes will occur. Consequently, a drastic increase in membrane permeability occurs, which impairs on the irreversible loss of physiological control systems and therefore causes the cell death. Pulsed electric fields treatment has been used to pasteurize liquid foods (Heinz et al., 2003). Application of pulsed electric fields treatment at low temperatures has a potential to provide food products with a fresh-like character and high nutritional value. The application of pulsed electric fields for treatment of liquid foods at 30°C requires a specific energy input of 100 kJ/kg or more (Heinz et al., 2003). Pulsed electric fields treatment is usually considered to have a higher energy input than a thermal process with heat recovery capacity. When operating at elevated treatment temperature and making use of synergetic heat effects, the pulsed electric field energy input might be reduced close to the amount of 20 kJ/kg required for a conventional thermal pasteurization process with 95% of heat recovery (Toepfl et al., 2006). Due to high production costs, commercial applications of pulsed electric fields treatment as an alternative to traditional thermal process have not yet been accomplished (Heinz et al., 2003). The investment costs of a pulsed electric fields unit with a production capacity of 5 t/h are estimated to be in the range of 2–3 million US$ (Toepfl et al., 2006).

A high-pressure process inactivates microbes by targeting the membranes of biological cells. During pressurization, water and acid molecules show increased ionization. This ionization change in living cells causes the major killing effect on living cells during pressurization. Although atomic bonds are barely affected by a high pressure, alternation of proteins or lipids can be observed when exposed to a high pressure. Lethal damage occurs when alternation of proteins or lipids occurs in the membranes of biological cells (Manas and Pagan, 2005; Toepfl et al., 2006). Theoretically, the compression work and energy required for the resulting temperature increase due to pressurization is about 52 kJ/kg and 70 kJ/kg upon compression of pure water up to 600 MPa, respectively.

10.4.1.4 Concentration, Dehydration, and Drying

Concentration, dehydration, and drying are a common unit operation in food processing facilities to lower the moisture content of foods in order to reduce water activity and prevent spoilage or reduce the weight and the volume of food products for transport and storage. Dehydration and drying are an energy-intensive unit operation because of the high latent heat of water evaporation and relatively low energy efficiency of industrial dryers. Industrial dryers consume about 12% of the total energy used in all manufacturing sectors (Bahu, 1991). The typical temperature during air-drying is between 65°C and 85°C. Loss of moisture from food products and high-temperature processing during air drying may cause undesirable effects on the textural properties and nutritional values of the products. Other common drying methods include microwave drying, freeze-drying, and vacuum drying.

A drying process is a simultaneous heat and mass transfer operation. The energy required for evaporation of water, which is dependent of temperature and pressure, is in the range of 2.5–2.7 MJ/kg. However, the total energy input into a conventional dryer is in the range of 4–6 MJ/kg of removed water depending on the thermal efficiency of the drying systems. Several studies have been conducted on exergy analyses of food drying (Midilli and Kucuk, 2003; Dincer and Sahin, 2004; Akpinar, 2004; Akpinar et al., 2005, 2006; Ozgener and Ozgener, 2006; Colak and Hepbasli, 2007; Corzo et al., 2008). The energy and exergy efficiencies of a pasta drying process were found to be 75.5%–77.09% and 72.98%–82.15%, respectively (Ozgeber and Ozgener, 2006). In order to increase the energy efficiency, the air leaving the dryer can be re-circled back to the dryer. The exergy efficiency decreases with the increase in air temperature and velocity. Corzo et al. (2008) conducted exergy analyses of thin layer drying of coroba slices. At drying temperatures from 71°C to 93°C and drying air velocities from 0.82 to 1.18 m/s, the exergy efficiency of the thin layer drying of coroba slices was in the range from 97% and 80%. The exergetic efficiency of drying red pepper slices in a convective-type dryer varied from 97.92% to 67.28% at the inlet temperature from 55°C to 70°C and a drying air velocity of 1.5 m/s (Akpinar, 2004).

Several methods have been used to improve the energy and exergy efficiencies of the evaporation, dehydrating, and drying process. Mechanical processes such as filtration and centrifugation can be used to remove as much water as possible before evaporation and drying. Evaporation is an energy-intensive unit operation. Mechanical recompression evaporation is most commonly used for concentration of dilute solutions in food processing. Membrane technology has a potential to reduce the overall energy consumption in combination with the evaporation technology (Kumar et al., 1999). A multiple-effect evaporator system is a simple series arrangement of several evaporators, which use steam to remove product moisture by evaporation. The evaporated water vapor from food products is collected and used as steam for the next evaporator in the series. This collection and reuse of vapor result in smaller energy requirements to remove product moisture. The greater the number of effects in the series, the smaller the energy consumption. It was found that changing from four-effect evaporators to five evaporators could save 20% energy. However, the number of evaporators in series should also be determined by the economics of the process (Casper, 1977).

Heat pumps have been used to increase the drying efficiency of convectional air dryers (Perera and Rahman, 1997). As discussed in Chapter 9, a heat pump dehumidifier dryer functions in a manner similar to a refrigerator. It consists of a condenser for high-temperature heat exchange, a compressor, an evaporator for low-temperature heat exchange, and an expansion valve to decrease the pressure of the working liquid. A fan is usually used to provide air movement in the drying chamber. In a heat pump dehumidifier, the evaporator is used to remove the moisture in the moist air exiting the drying chamber, and the condenser is used to increase the temperature of the dry air from the evaporator. The hot dry air is then sent back to the drying chamber (Perera and Rahman, 1997; Kiatsiriroat and Tachajapong, 2002; Adapa and Schoenau, 2005). A heat pump can also be used to extract heat from a low-temperature energy source such as geothermal energy through its evaporator and upgrade the extracted heat to a high-temperature heat source at its condenser for drying (Kuzgunkaya and Hepbasli, 2007).

Supercritical fluids such as supercritical carbon dioxide can be used to remove moisture from foods (Brown et al., 2007). Supercritical CO_2 has a low critical temperature and pressure (31.1°C and 7.3 MPa). Therefore, drying with supercritical CO_2 can be operated at a much lower temperature than conventional air-drying. However, since supercritical CO_2 is a nonpolar solvent, the water solubility in supercritical CO_2 is 4 mg/g at 50°C and 20 MPa and 2.5 mg/g at 40°C and 20 MPa (King et al., 1992; Sabirzyanov et al., 2002). Therefore, small quantities of polar co-solvents such as ethanol are usually added into the supercritical CO_2 to increase the solubility of polar water in supercritical CO_2. Supercritical CO_2 drying has been found to generate more favorable re-hydrated textural properties than the air-dried equivalents (Brown et al., 2007).

10.4.1.5 Chilling and Freezing

Food processing facilities make heavy use of refrigeration. It is estimated that the refrigeration systems use as much as 15% of the total energy consumed worldwide. In the whole U.S. food industry, about 25% of the electricity is used for process cooling and refrigeration (Okos et al., 1998). The dairy sector and the meat sector are likely to be the highest and second users of refrigeration. Generally, energy conservation for refrigeration unit operations can be achieved by

- Improved insulation
- Best practice
- Use of novel refrigeration cycles powered by waste heat

Air blast chillers or freezers are widely used in the food industry. The fans of air blast chillers or freezers add a heat load to chillers or freezers during operation. Since the heat generated by fans increases with the required air load, it is critical to optimize the air velocity to minimize the heat generation and maximize the refrigeration effect during air blast chilling or freezing of food products. Harrison and Bishop (1985) proposed that the ratio of fan work divided by the useful refrigeration effect was the best indicator of economic energy usage in a freezing tunnel.

During air blast chilling/freezing, the heat transfer from the cold medium of air to the inside of foods must pass two layers of thermal resistance: external resistance to heat convection between the cold air and the food surface and internal resistance to heat conduction in solid foods. Biot number is the ratio of the internal resistance to the external resistance, which is expressed as (Singh and Heldman, 2001)

$$\text{Bio} = \frac{hl}{k} \tag{10.1}$$

where

- l is a characteristic dimension of the food body
- m is the radius of a round shaped body and half of the thickness of a flat shaped body
- h is the surface convective heat transfer coefficient (W/m^2°C)
- k is the thermal conductivity of foods (W/m°C)

There are

- Bio ≤ 0.1, negligible international resistance to heat conduction
- 0.1 < Bio < 40, finite internal resistance and external resistance
- Bio ≥ 40, negligible external resistance to heat convection (Singh and Heldman, 2001)

According to Equation 10.1, an increase in air velocity will increase the value of h and thus the Biot number. As a general rule in cooling and freezing, the Bio value should not exceed 5 (Mattarolo, 1976).

Example 10.1

Beef carcasses with a half thickness of 0.15 m are cooled in an air blaster cooler at 0°C. Given that the thermal conductivity of beef is 0.35 W/m°C, calculate the surface heat transfer coefficient (or air velocity) for effective cooling. (Suppose that the average surface temperature of carcasses during chilling is 20°C and assume that the shape of the carcasses is spherical)

Solution 10.1

1. Determine the surface heat transfer coefficient
   The Bio value of a cooling or freezing process should not exceed 5. That is,

$$\text{Bio} = \frac{hl}{k} = \frac{h \times 0.15}{0.35} < 5$$

   Therefore, $h < 11$ W/m^2°C
2. Identify the fluid and determine its properties:
   The properties of air are evaluated at

$$T_f = \frac{T_s + T_\infty}{2} = \frac{20 + 0}{2} = 10°\text{C}$$

From Table 7.2 in Chapter 7, the properties of air are

$$\rho = 1.206 \text{ kg/m}^3$$

$$c_p = 1.010 \text{ kJ/kg°C}$$

$$k_{air} = 0.0244 \text{ W/m°C}$$

$$\mu = 17.848 \times 10^{-6} \text{ Pa s}$$

3. Calculate the dimensionless numbers

$$\text{Pr} = \frac{c_p \mu}{k_{air}} = \frac{1.206 \times 1000 \times 17.848 \times 10^{-6}}{0.0244} = 0.88$$

$$\text{Nu} = \frac{hl}{k_{air}} = \frac{11 \times 0.15}{0.0244} = 67.62$$

From Table 7.3 in Chapter 7, for the forced convection past in an immersed sphere, we have

$$\text{Nu} = 2 + 0.60\text{Re}^{0.5}\text{Pr}^{0.33}$$

$$\text{Re} = 13{,}014$$

4. Determine the air velocity from the Reynolds number:

$$\text{Re} = \frac{\rho l u}{\mu} = 13{,}014$$

The air velocity is

$$u = 1.28 \text{ m/s}$$

Example 10.2

Peas with a radius of 0.5 cm are cooled in an air blast cooler at an air temperature of 0°C. Given that the thermal conductivity of peas is 0.40 W/m°C, calculate the surface heat transfer coefficient (or air velocity) for effective cooling. (Suppose that the average surface temperature of peas during chilling is 20°C)

Solution 10.2

1. The Bio value of a cooling or freezing process should not exceed 5. That is,

$$\text{Bio} = \frac{hl}{k} = \frac{h \times 0.005}{0.40} < 5$$

Therefore, $h < 400\,\text{W/m}^2\,°\text{C}$

2. Identify the fluid and determine its properties:
   The properties of air are evaluated at

$$T_f = \frac{T_s + T_\infty}{2} = \frac{20 + 0}{2} = 10°\text{C}$$

From Table 7.2 in Chapter 7, the properties of air are

$$\rho = 1.206\ \text{kg/m}^3$$

$$c_p = 1.010\ \text{kJ/kg°C}$$

$$k = 0.0244\ \text{W/m°C}$$

$$\mu = 17.848 \times 10^{-6}\,\text{Pa s}$$

3. Calculate the dimensionless numbers:

$$\text{Pr} = \frac{c_p \mu}{k} = \frac{1.206 \times 1000 \times 17.848 \times 10^{-6}}{0.0244} = 0.88$$

$$\text{Nu} = \frac{hl}{k_{\text{air}}} = \frac{400 \times 0.005}{0.0244} = 81.97$$

From Table 7.3 in Chapter 7, for the forced convection past in an immersed sphere, we have

$$\text{Nu} = 2 + 0.60\,\text{Re}^{0.5}\,\text{Pr}^{0.33}$$

$$\text{Re} = 19{,}328$$

4. Determine the air velocity from the Reynolds number:

$$\text{Re} = \frac{\rho l u}{\mu} = 19{,}328$$

$$u = 57.21\,\text{m/s}$$

From Examples 10.1 and 10.2, we can see that the surface convective heat transfer coefficient, h, increases with the increase in air velocity. For cooling or freezing of big food items such as beef, a small air velocity should be used. Higher velocities can only lead to a small reduction in the cooling time but require a large increase in fan energy and generate more extra heat load. For cooling or freezing of small food items such as peas, a high air velocity should be used. Air impingement, water spray, or water immersion chillers can also be used to achieve a high surface heat transfer coefficient for small food items.

Novel refrigeration cycles based on liquid–liquid absorption, liquid–solid adsorption, and fluid ejection as discussed in Chapter 9 offer potential energy saving opportunities for food refrigeration. Novel refrigeration cycles such as absorption and adsorption refrigeration cycles can be powered by low-grade waste heat or other renewable energy sources such as geothermal energy and solar energy at a low temperature (e.g., 70°C) (Sun and Wang, 2001).

10.5 SUMMARY

The food processing industry is the fifth biggest consumer of energy in the United States. Because of the increasing energy prices and efforts for the reduction of CO_2 emission, improving the energy efficiency, replacing the existing energy-intensive unit operations with new energy efficient processes, and increasing the use of renewable energy in the food industry have become significant activities. Energy efficiency improvement and waste-heat recovery in the food industry have been a focus in the past decades. Replacement of conventional energy-intensive food processes with novel technologies such as nonthermal processes may provide another potential to reduce energy consumption, reduce production costs, and improve the sustainability of food production. Some novel food processing technologies have been developed to replace traditional energy-intensive unit operations for pasteurization and sterilization, evaporation and dehydration, and chilling and freezing in the food industry. Most energy conservation technologies can be readily transferred from other manufacturing sectors to the food processing sector.

REFERENCES

Adapa, P.K. and G.J. Schoenau. 2005. Re-circulating heat pump assisted continuous bed drying and energy analysis. *International Journal of Energy Research* 29: 961–972.

Akpinar, E.K. 2004. Energy and exergy analyses of drying of red pepper slices in a convective type dryer. *International Communications in Heat and Mass Transfer* 31: 1165–1176.

Akpinar, E.K., A. Midilli, and Y. Bicer. 2005. Energy and exergy of potato drying process via cyclone type dryer. *Energy Conversion and Management* 46: 2530–2552.

Akpinar, E.K., A. Midilli, and Y. Bicer. 2006. The first and second law analyses of thermodynamic of pumpkin drying process. *Journal of Food Engineering* 72: 320–331.

Anonymous. 1987. *Our Common Future, World Commission on Environment and Development*, Oxford: Oxford University Press.

Bahu, R.E. 1991. Energy consumption in dryer design. In *Drying' 91*, Mujumdar, A.S., and I. Filkova (Eds.), pp. 553–557, Amsterdam: Elsevier.

Bhattacharyya, S.C. and A. Ussanarassamee. 2004. Decomposition of energy and CO_2 intensities of Thai industry between 1981 and 2000. *Energy Economics* 26: 765–781.

Brown, Z.K., P.J. Fryer, I.T. Norton, S. Bakalis, and R.H. Bridson. 2008. Drying of foods using supercritical carbon dioxide–investigations with carrot. *Innovative Food and Emerging Technologies* 9: 280–289.

Casper, M.E. 1977. *Energy-Saving Techniques for the Food Industry*. New Jersey: Noyes Data Corporation.

Colak, N. and A. Hepbasli. 2007. Performance analysis of drying of green olive in a tray dryer. *Journal of Food Engineering* 80: 1188–1193.

Corzo, O., N. Bracho, A. Vasquez, and A. Pereira. 2008. Energy and exergy analyses of thin layer drying of coroba slices. *Journal of Food Engineering* 86: 151–161.

Dincer, I. and M.A. Rosen. 1999. Energy, environment and sustainable development. *Applied Energy* 64: 427–440.

Dincer, I. and A.Z. Sahin. 2004. A new model for thermodynamic analysis of a drying process. *International Journal of Heat and Mass Transfer* 47: 645–652.

Fischer, J.R., J.E. Blackman, and J.A. Finnell. 2007. Industry and energy: Challenges and opportunities. *Resource: Engineering & Technology for a Sustainable World* 4: 8–9.

Fritzson, A. and T. Berntsson. 2006. Efficient energy use in a slaughter and meat processing plant-opportunities for process integration. *Journal of Food Engineering* 76: 594–604.

Harrison, M.A. and P.J. Bishop. 1985. Parametric study of economical energy usage in freezing tunnels. *International Journal of Refrigeration* 8: 29–36.

Heinz, V., S. Toepfl, and D. Knorr. 2003. Impact of temperature on lethality and energy efficiency of apple juice pasteurization by pulsed electric fields treatment. *Innovative Food Science and Emerging Technologies* 4: 167–175.

Ho, J.C. and T.T. Chandratilleke. 1987. Thermodynamic analysis applied to a food-processing plant. *Applied Energy* 28: 35–46.

Ho, J.C., N.E. Wijeysundera, and S.K. Chou. 1986. Energy analysis applied to food processing. *Energy* 11: 887–892.

Huang, L. and J. Sites. 2007. Automatic control of a microwave heating process for in-package pasteurization of beef frankfurters. *Journal of Food Engineering* 80: 226–233.

Kiatsiriroat, T. and W. Tachajapong. 2002. Analysis of a heat pump with solid desiccant tube bank. *International Journal of Energy Research* 26: 527–542.

King, M.B., A. Mubarak, J.D. Kim, and T.R. Bott. 1992. The mutual solubilities of water with supercritical and liquid carbon dioxide. *Journal of Supercritical Fluids* 5: 296–302.

Kumar, A., S. Croteau, and O. Kutowy. 1999. Use of membranes for energy efficient concentration of dilute steams. *Applied Energy* 64: 107–115.

Kuzgunkaya, E.H. and A. Hepbasli. 2007. Exergetic performance assessment of a ground-source heat pump drying system. *International Journal of Energy Research* 31: 760–777.

Lado, B.H. and A.E. Yousef. 2002. Alternative food-preservation technologies: Efficacy and mechanisms. *Microbes and Infection* 4: 433–440.

Loaharanu, P. 1996. Irradiation as a cold pasteurization process of food. *Veterinary Parasitology* 64: 71–82.

Manas, P. and R. Pagan. 2005. Microbial inactivation by new technologies of food preservation. *Journal of Applied Microbiology* 98: 1387–1399.

Mattarolo, L. 1976. Technical and engineering aspects of refrigerated food preservation. *Annex 1976–1*, p. 49–69, International Institute of Refrigeration, Paris, France.

Mertens, B. and D. Knorr. 1992. Developments of nonthermal processes for food preservation. *Food Technology* 46: 124–133.

Midilli, A. and H. Kucuk. 2003. Energy and exergy analyses of solar drying process of pistachio. *Energy* 28: 539–556.

Muller, D.C.A., F.M.A. Marechal, T. Wolewinski, and P.J. Roux. 2007. An energy management method for the food industry. *Applied Thermal Engineering* 27: 2677–2686.

North American Industry Classification System (NAICS) Association. 2007. *NAICS Manual and SIC Manual*. Available at http://www.census.gov/naics/2007/NAICOD07.HTM.

Okos, M., N. Rao, S. Drecher, M. Rode, and J. Kozak. 1998. *Energy Usage in the Food Industry.* American Council for an Energy-Efficient Economy. Available at http://www.aceee.org/pubs/ie981.htm.

Ozdogan, S. and M. Arikol. 1995. Energy and exergy analyses of selected Turkish Industries. *Energy* 20: 73–80.

Ozgener, L. and O. Ozgener. 2006. Exergy analysis of industrial pasta drying process. *International Journal of Energy Research* 30: 1323–1335.

Ozyurt, O., O. Comakli, M. Yilmaz, and S. Karsli. 2004. Heat pump use in milk pasteurization: An energy analysis. *International Journal of Energy Research* 28: 833–846.

Parikh, J.K. and S. Syed. 1988. Energy use in the post-harvest food (PHF) system of developing countries. *Energy in Agriculture* 6: 325–351.

Perera, C.O. and M.S. Rahman. 1997. Heat pump dehumidifier drying of food. *Trends in Food Science and Technology* 8: 75–79.

Poulsen, K.P. 1986. Energy use in food freezing industry, Chapter 12. In *Energy in Food Processing*, Singh, R.P. (Ed.), pp. 155–178, New York: Elsevier Science Publishing Company Inc.

Ramirez, C.A., M. Patel, and K. Blok. 2006b. How much energy to process one pound of meat? A comparison of energy use and specify energy consumption in the meat industry of four European countries. *Energy* 31: 2047–2063.

Ramirez, C.A., K. Blok, M. Neelis, and M. Patel. 2006a. Adding apples and oranges: The monitoring of energy efficiency in the Dutch food industry. *Energy Policy* 34: 1720–1735.

Sabirzyanov, A.N., A.P. Il'in, A.R. Akhunov, and F.M. Gumerov. 2002. Solubility of water in supercritical carbon dioxide. *High Temperature* 40: 203–206.

Simpson, R., C. Cortes, and A. Teixeira. 2006. Energy consumption in batch thermal processing: Model development and validation. *Journal of Food Engineering* 73: 217–224.

Singh, R.P. 1986. *Energy in Food Processing.* New York: Elsevier Science Publishing Company Inc.

Singh, R.P. and D.R. Heldman. 2001. *Introduction to Food Engineering* (3rd ed.), San Diego: Academic Press.

Sun, D.W. and L.J. Wang. 2001. Novel refrigeration cycles, Chapter 1, In *Advances in Food Refrigeration.* Sun, D.W., (Ed.), pp.1–69, Leatherhead, United Kingdom: Leatherhead Publishing.

Toepfl, S., A. Mathys, V. Heinz, and D. Knorr. 2006. Review: Potential of high hydrostatic pressure and pulsed electric fields for energy efficiency and environmentally friendly food processing. *Food Reviews International* 22: 405–423.

Tragardh, C. 1986. Energy requirements in food irradiation. In *Energy in Food Processing*, Chapter 12, Singh, R.P. (Ed.), pp. 203–225, New York: Elsevier Science Publishing Company Inc.

U.S. Census Bureau. 2006. *2006 Annual Survey of Manufactures.* Available at http://factfinder.census.gov.

U.S. Environmental Protection Agency (U.S. EPA). 2007. *Energy Trends in Selected Manufacturing Sectors: Opportunities and Challenges for Environmentally Preferable Energy Outcomes.* Available at http://www.epa.gov/ispd/energy/index.html.

Unruh, B. 2002. *Delivered Energy Consumption Projections by Industry in the Annual Energy Outlook.* Available at www.eia.doe.gov/oiaf/analysispaper/industry/pdf/consumption.pdf.

11 Energy Conservation in Grains and Oilseeds Milling Facilities

11.1 INTRODUCTION

The grain and oilseed milling sector (NAICS code 3112) can be divided into three main categories: (1) flour milling and malt manufacturing, (2) starch and vegetable fats and oils manufacturing, and (3) breakfast cereal manufacturing. Flour milling and malt manufacturing is engaged in one or more of the following manufacturing activities: (1) milling flour or meal from grains or vegetables, (2) preparing flour mixes or dough from milled flour, (3) milling, cleaning, and polishing rice, and (4) preparing malt from barley, rye, or other grains. Starch and vegetable fats and oils manufacturing is primarily engaged in one or more of the following manufacturing activities: (1) wet milling corn and vegetables, (2) crushing oilseeds and tree nuts, (3) refining and blending vegetable oils, (3) shortening and margarine, and (4) blending purchased animal fats with vegetable fats. The cereal breakfast foods industry comprises plants primarily engaged in manufacturing three main types of cereal products: ready to eat, hot cooked, and natural breakfast cereals (U.S. Census Bureau, 2006).

In the grain and oilseed milling sector, a large portion of fuels is used for steam generation and a large portion of purchased electricity is used for motors and air compressors. Energy conservation in steam generation and distribution is discussed in Chapter 4. Energy conservation technologies for air compressors and motors are discussed in Chapters 5 and 6, respectively. Heat exchangers are widely used in this sector, and energy conservation technologies for heat exchangers are discussed in Chapter 7. Boilers, heat exchangers, motors, and air compressors may generate waste heat, and waste-heat recovery technologies are discussed in Chapter 8. Since both processing heat and electricity are required by the grain and oilseed processing sector, a combined heat and power system can be used to efficiently and economically provide electricity or mechanical power and useful heat from the same primary energy source. Combined heat and power systems are discussed in Chapter 9.

In this chapter, the main products and their processes in the grain and oilseed milling sector are reviewed. The energy uses and conservation in main unit operations in this sector are discussed. Finally, energy utilization in processing wastes from grain and oilseed processing facilities is discussed.

11.2 ENERGY CONSUMPTION IN THE GRAIN AND OILSEED MILLING SECTOR

The total shipment value from the grain and oilseed milling sector was 57.7 billion in the United States in 2006. The total costs for purchasing fuels and electricity were \$2.2 billion in 2006. It thus required 3.81 ¢ of energy cost per dollar shipment value. Grain and oilseed milling, which is the largest energy consumer in the U.S. food industry, consumed 22.3% of total energy input into the whole food industry and generated 10.7% of the total shipment value in 2006. The cost of purchased electricity in the grain and oilseed milling sector was 37% of the total energy expenses (U.S. Census Bureau, 2006).

Starch and vegetable fats and oils manufacturing industry is the dominant part in the grain and oilseed milling sector in terms of both shipment value and energy input. The total shipment value and energy input costs from the starch and vegetable fats and oils manufacturing industry was 65% and 78% of total shipment value and energy costs in the grain and oilseed milling sector in the United States in 2006. Soybean and other oilseeds processing (excluding the fats and oils refining and blending) contributed to 32% of the total shipment value and 20% of the total energy costs in the grain and oilseed milling sector. Wet corn milling (excluding wet corn milling for fuel ethanol production) contributed to 17% of the total shipment value and consumed 51% of the total energy inputs (U.S. Census Bureau, 2006).

Energy and foods are also the major concerns of developing countries. Cassava is a major source of foods in terms of calories for about 40% of the African population. Gari, starch, and flour are three traditional products from Cassava. Jekayinfa and Olajide (2007) analyzed that the energy consumption including electrical energy, thermal energy, and labor energy for production of gari, starch, and flour from cassava in 18 cassava processing mills in Nigeria were 0.327, 0.357, and 0.345 MJ/kg, respectively. Through their optimization by carefully deciding on the number of persons involved in operations and using efficient and high-capacity processing machines, they found that the minimum energy inputs required for the production of gari, starch, and flour were 0.291, 0.305, and 0.316 MJ/kg, respectively.

11.3 CORN WET MILLING PROCESS

11.3.1 Wet Milling Process and Energy Consumption

Corn wet milling is a process to separate corn into its four main components of starch, germ, fiber, and protein or gluten and further convert the separated components into marketable products including starch for sweeteners and ethanol, and cornstarch, germ for corn oil, and fiber for protein. The materials and energy flow of the wet corn milling process is given in Figure 11.1. Wet corn milling produces corn sweeteners such as glucose, dextrose, and fructose, corn oil, and starches. In recent years, fuel ethanol has become an important product from corn wet milling. Two major products of corn wet milling are corn sweeteners (or corn syrup) and ethanol made from the starch. Cornstarch and oil are two other high-value refined products from the corn wet milling process.

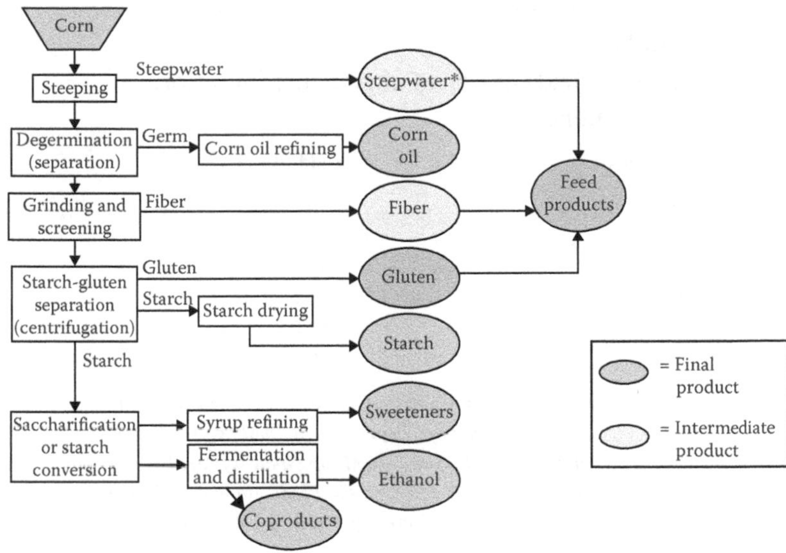

FIGURE 11.1 A typical wet corn milling process. (Reprinted from Galitsky, C., Worrell, E., and Ruth, M., *Energy Efficiency Improvement and Cost Saving Opportunities for Corn Wet Milling Industry-An ENERGY STAR Guide for Energy and Plant Managers*. Ernest Orlando Lawrence Berkeley National Laboratory, 2003. With permission.)

The wet milling process starts with steeping corn kernels in a dilute sulfuric acid solution, where the kernels are soaked in a large stainless steel tank filled with mildly acidic water at about 50°C for 20–36h. During steeping, because the kernels absorb water, the gluten bonds in the corn kernels loosen and starch is readily released from the kernels. The corn is then coarsely ground to release the germ from the kernels for the production of corn oil. The steeping water contains a significant percentage of proteins and sugars at a solid content in the range of 5%–10%. The steeping water is usually evaporated to 45%–50% solids used as commercial products such as animal feeds. The slurry generated by the coarse grinding undergoes degermination using hydrocyclone separators to separate the germ from other components.

The germ is dewatered to a water content of 50%–60% using a screw press. The concentrated germ is then dried typically to a moisture content of 2%–4% using a rotary steam tube dryer. Steam consumption in the dryer is about 120% of the mass of water removed. It is economical feasible to process the corn germ on a large scale. Corn oil is then extracted from the dry germ through a combination of chemical and mechanical processes. The extracted oil is finally refined to oil products.

The slurry after removing germ undergoes fine grinding and screening to release all starch and gluten from the fiber. Fiber is dewatered to a moisture content of 65%–75% using a screen centrifuge and further to 10% using a screw press. This concentrated fiber and the solids from steeping water are mixed and dried to produce corn gluten feed. The starch–gluten mixture undergoes centrifugation to spin out the gluten from the solution. The moist gluten is dewatered to a solid content of 30%–40% using a filter

TABLE 11.1
Typical Yields of Corn Components from Corn Wet Milling

Product	Yield (%)
Steep liquor	6.5
Germ	7.5
Bran/fiber	12.0
Gluten	5.6
Starch	68.0
Losses	0.4

Source: Reprinted from Galitsky, C., Worrell, E., and Ruth, M., *Energy Efficiency Improvement and Cost Saving Opportunities for Corn Wet Milling Industry-An ENERGY STAR Guide for Energy and Plant Managers*. Ernest Orlando Lawrence Berkeley National Laboratory, 2003. With permission.

or centrifuge. The concentrated gluten is then dried to produce corn gluten meal. The moist starch at a solid content of 33%–40% can be dried to produce starch product or modified to produce a broad range of products. The moist starch can also be saccharified into syrup and fermented to ethanol. Evaporation, which is a critical unit operation at the syrup refining step, consumes a large amount of energy. Typical yields of corn components from a wet corn mill are given in Table 11.1 (Galitsky et al., 2003).

Corn wet milling is the most energy-intensive industry. For a typical plant processing 1 million tons of corn per year, the energy costs are approximately $15–$25 million per year while the capital costs are $250–$300 million. In the wet milling process, heat is used for evaporation, drying, and oil extraction. Power is used to convey, grind, and separate large quantities of corn.

Feed drying accounts for the highest amount of fuel consumed in direct use. About 33% of the steam is used in operating the evaporators. About 95% of electricity is used to operate motors and to supply other mechanical power. The remaining 5% of electricity is used in lighting.

11.3.2 Energy Conservation in Steeping

Corn is soaked in the steeping tanks at a temperature of 50°C for up to 50h. Reducing the steep time can reduce the energy required to heat and maintain the corn and steepwater at 50°C. Enzymes are being developed to reduce the steeping time (Johnston and Singh, 2001). Preheating of incoming corn up to the steeping temperature by waste heat is another energy conservation measure in steeping.

11.3.3 Energy Conservation in Dewatering

The energy requirements for removing moisture from germ, fiber, and gluten by mechanic dewatering technologies such as pressing, filtrating, and sedimentation are generally insignificant compared to evaporation. Therefore, as much moisture as possible should be removed mechanically before materials are sent into an evaporator and dryer (Galitsky et al., 2003). Germ is easy to be dewatered from 80% to 90% to less than 50% by a screw press screen. Single screw presses are used almost exclusively to mechanically dewater fiber from 80% to 90% to about 60%. A centrifuge or hydrocyclone system is usually used to dewater starch. Gluten, which is in the form of fine particles, is generally dewatered by vacuum filtration.

11.3.4 Applications of Membrane Separation

Membrane separation processes could be a suitable energy-efficient alternative to evaporation and distillation to remove water in wet corn milling. During membrane separation, there is no heat requirement and no phase change. It was reported that 90% energy savings were observed by replacing evaporation with membrane separation (Rausch, 2002). However, the total energy savings should be carefully justified since high pressures and high recirculation rates during membrane separation require a significant amount of energy (usually electricity). Membranes have been used in the corn wet milling process, such as steepwater concentration by reverse osmosis, concentration of corn syrups, reducing chemical oxygen demand in evaporator overhead, solvent recovery, and oil purification at the corn oil refining stage, concentrating starch, and recovering fresh water (Singh and Cheryan, 1997). Reverse osmosis can effectively increase the concentration of steepwater to 14%–18% total solids content or up to a pressure difference of 1.38 MPa (Singh and Cheryan, 1997). Membrane technologies are discussed in Chapter 17.

11.4 OILSEED MILLING

11.4.1 Oilseed Process and Energy Consumption

Vegetable oils are produced from many oilseed crops. Soybean is the dominant oilseed worldwide, which is about one-half of world oilseed production. Soybeans on a dry basis typically have about 12%–25% oil and 35%–50% protein. Canola is another major oilseed crop grown in many parts of the world. Canola contains over 40% oil and 20% protein. Palm-kernel oil is mainly produced in south Asia and Africa. The top three annual production capacities were 70 million tons in Malaysia, 60 million tons in Indonesia, and 8.7 million tons in Nigeria in 2004. About 90% of palm oil is used as food products worldwide and the other 10% is used as a basic raw material for soap production (Mahlia et al., 2001). On a dry basis, more than 70% of the pulp and 40% of the kernel of palm consists of oil.

There are several major unit operations in oilseed processing, which include extraction, refining, bleaching, and deodorization. For specific oilseeds and oil products, additional unit operations include winterization, hydrogenation, texturing, and fractional crystallization. Vegetable oils are mainly used as food products. Some of these oils such as soybean oil and palm oil have already been evaluated and used as renewable energy sources for biodiesel production (Altin et al., 2001; Sumathi et al., 2008). The production of biodiesel from vegetable oils is discussed in Chapter 26.

Oilseed processing requires a high energy demand. Electricity and fuels such as natural gas and fuel oil are two major types of energy sources used in oilseed processing facilities. The energy consumption for preprocessing of soybeans before extraction is 2.38 MJ/kg of crude soybean oil produced. Among the total energy consumption, 1.19 MJ/kg is used for drying raw soybeans prior to mechanical cleaning, 0.60 MJ/kg is used for conditioning the soybeans to raise the moisture content and the temperature of the soybeans prior to flaking, and 0.51 MJ/kg is used in the expander to compress the soybean flakes into pellets for a more complete oil extraction. The total energy consumption for conventional extraction and separation with hexane is 3.19 MJ/kg of crude soybean oil. Most of the energy (1.38 MJ/kg) is used for hexane distillation from the extracted soybean oil. With potential energy recovery, the net energy consumption for extraction is 2.04 MJ/kg. The hexane loss is about 50 kg of hexane per 1000 kg of crude soybean oil produced (Li et al., 2006). For supercritical CO_2 extraction of soybean oil with energy recovery, the net energy consumption is only 2.31 MJ/kg of crude soybean oil extracted at 60 MPa and 80°C and separated at 5.5 MPa and 20°C. Without accounting for potential energy recovery, supercritical CO_2 extraction and separation uses a total of 7.55 MJ of energy per 1 kg of crude soybean oil produced. Most of the energy (i.e., 6.80 MJ/kg) is used to compress CO_2 during extraction. About 292 kg of CO_2 is lost per 1000 kg of crude soybean oil produced (Li et al., 2006). Supercritical fluid processing is discussed in Chapter 22.

Palm-kernel oil production requires seven unit operations: palm-nut drying, palm-nut cracking, palm-kernel roasting, palm-kernel crushing, palm-kernel oil expression, palm-kernel oil sifting, and palm-kernel oil bottling/pumping. The most common method of extracting oil from palm fruits is the screw press. The energy input to produce palm oil is between 0.1 and 0.16 MJ/L oil depending on the production scale. Approximately 352, 232, and 177 MJ energy is consumed to process one ton of palm nut in small, medium, and large-scale mills, respectively. The electrical energy required to process 1 t of fresh fruit bunches is 20 kWh (or 72 MJ) for a palm oil processing facility with a capacity of 30 t/h (Mahlia et al., 2001). Palm-nut cracking and palm-kernel oil expression each consume about 35%–40% of the total energy inputs (Jekayinfa and Bamghoye, 2004, 2007).

11.4.2 Energy Conservation in Mechanical Oil Extraction Process

There are three basic methods to extract oil from nuts and seeds, which include using a hydraulic press, an expeller press, and solvent extraction. Mechanical pressing is the most common method and it produces a solvent-free and protein-rich meal as a by-product. It is simple in construction and easy in operation and maintenance. The extraction efficiency of the oil extraction machine has been improved from 50% to 80% through the optimization of processing conditions, such as processing pressure, temperature, and moisture conditions of feedstock (Ohlson, 1992); physical pretreatments such as size reduction, cracking, and de-hulling; and thermal treatment such as preheating, dry extrusion, steaming, hot water soaking, blanching, and flaking (Bredesson, 1993; Bargale et al., 1999). The screw press, which uses a screw or continuous press, with a constantly rotating shaft mounted inside a barrel to squeeze oil out under pressure and discharge the cake at the end of the barrel, can achieve 93% efficiency for the extraction of oil from untreated and dehulled sunflower seeds (Isobe et al., 1992). The viscous friction of oilseeds in the screw press generates heat. Viscous dissipation of the mechanical energy into heat inside the material is the major source of energy transformation inside the screw. Specific mechanical energy is the net mechanical energy input divided by the mass flow rate of oilseeds. The specific mechanical energy can be measured indirectly through measuring the electrical energy input to the motor and the efficiency of the motor (Singh and Bargale, 2000) or measuring the thermal energy balance (Zheng et al., 2005). The specific mechanical energy was found to be increased from 81.1 to 104.7 kJ/kg of oilseed when the moisture content of the flaxseed was decreased from 12.6% to 6.3% (Zheng et al., 2005). The electricity consumption for palm oil expression was found to be 107 kJ/kg in a small palm oil plant with annual oil output less than 300 m^3 (Jekayinfa and Bamghoye, 2004). Industrial oil presses driven by an electric motor alone consume large amounts of electricity. The hydraulic press applies the force at one point and transmits the force to anther point using a fluid. The efficiency of a hydraulic pressing system is very high because almost all the input energy can be recovered and reused during the next stroke (Okoye et al., 2008).

11.4.3 Energy Conservation in Oil Extraction Process with Solvents

Mechanical extraction has a low capital cost and no solvent requirement. However, the extraction efficiency of mechanical oil extraction is much less than that of solvent extraction methods (Li et al., 2006). Soybean oil is typically extracted from soybean flakes with hexane in an edible oil process, except some small-scale plants that use mechanical extraction methods. Like other oilseeds, soybeans need to be pretreated before extraction in order to break the cells and make the oil available. Conventional pretreatment includes cracking, hulling, conditioning, flaking, cooking, and steaming. Enzymatic hydrolysis pretreatment has also been investigated to enhance oil availability and extractability. The oil availability and extractability can be increased by 1%–2% and 1.4%–4% using the enzymatic hydrolysis pretreatment (Kashyap et al., 2006). A typical soybean oil production process is shown in Figure 11.2.

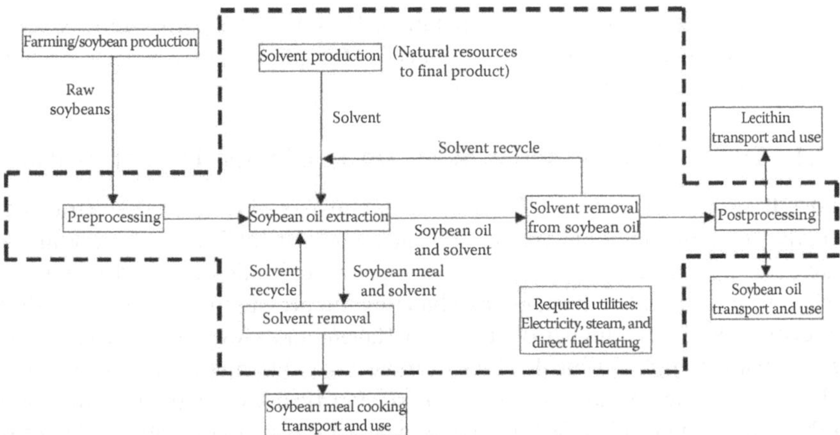

FIGURE 11.2 Overview of a soybean oil production process. (Reproduced from Li Y., Griffing, E., Higgins, M., and Overcash, M., *J. Food Process Eng.*, 29, 429, 2006. Copyright Wiley-Blackwell. With permission.)

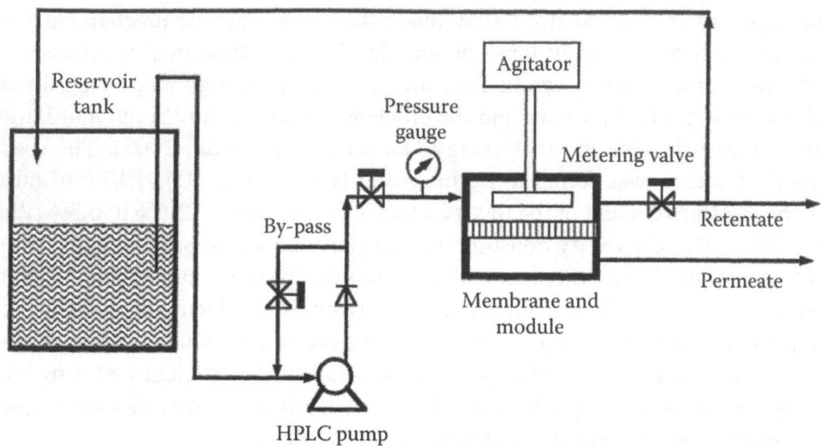

FIGURE 11.3 Schematic diagram of a soybean oil extract separation system. (Reprinted from Wu J.C.S. and Lee, E.H., *J. Membrane Sci.*, 154, 251, 1999. With permission.)

The crude extract normally contains 25%–30% of soybean oil by weight and the remaining 70%–75% is hexane (Wu and Lee, 1999). Removal of more than 70% hexane in the extract by distillation consumes most of the energy input in a typical soybean oil plant. Ultrafiltration or reverse osmosis with membranes can replace the conventional distillation used in processing soybean oil to minimize thermal damage to the oil products and reduce energy consumption for hexane evaporation (Cuperus and Nijhuis, 1993). Wand and Lee (1999) used a cross-flow ultrafiltration ceramic membrane with a pore diameter of 0.02 μm and thickness of ~1 μm as shown in Figure 11.3 to separate hexane from soybean oil. At 392 kPa (or 4 kgf/cm^2) trans-membrane pressure and 120 rpm agitation speed, the concentration of soybean oil was decreased from 33% of feed to 27% in the permeate at 20% rejection. Smaller pore size is essential for efficient separation. However, the permeate flux usually becomes lower when the pore size is reduced. Multiple stages with recycles can be applied to increase the overall rejection.

11.4.4 Energy Conservation in Supercritical CO_2 Oil Extraction Process

Besides hexane, supercritical CO_2 is also a promising solvent for the extraction of seed oils. Compared to liquid organic solvents, supercritical fluids have several major advantages: (1) the dissolving power of a supercritical fluid solvent depends on its density, which is highly adjustable by changing the pressure or temperature; (2) the supercritical fluid has a higher diffusion coefficient and lower viscosity and surface tension than a liquid solvent, leading to more favorable mass transfer (Wang and Weller, 2006). The special advantages of supercritical CO_2 extraction include the fact that there is no residual left in oil and meal; CO_2 is nonflammable and nontoxic; and separation can be easily accomplished by depressurization. However, supercritical CO_2 extraction process consumes electricity for the compression of CO_2. In addition, the solubility of soybean oil in supercritical CO_2 is low (e.g., 57 g/kg CO_2 at 60 MPa and 80°C) (Reverchon et al., 2000). In terms of energy consumption, supercritical CO_2 extraction is not better than hexane extraction. Some ideas have been proposed to improve supercritical CO_2 extraction, which include improving the mass transfer coefficient, adjusting pressure, and using a co-solvent to increase solubility.

Membranes were used to separate the small CO_2 molecule from the large molecules of soybean oil (Wang and Shen, 2005; Jakubowska et al., 2005). Supercritical CO_2 with membrane separation consumes only about 50% of the total energy used in hexane extraction because the membrane filtration avoids reducing the pressure of CO_2 for separation in traditional supercritical CO_2 extraction and thus reduces the energy consumption for recompression of CO_2 after separation. The energy consumption decreases with increasing membrane efficiency (Li et al., 2006).

11.5 DRYING OF GRAINS AND OILSEEDS

Grains and oilseeds are extremely susceptible to spoilage if not properly dried before or during storage. The moisture content of oilseeds also influences the extraction pressing performance. Low moisture content typically produces better oil yield (Singh et al., 2002). For marketing, moisture contents of oilseeds and grains are usually below 15% on a wet basis.

11.5.1 Energy Conservation in In-Bin Drying of Grains and Oilseeds

Different drying methods can be used to remove water from grains and oilseeds. The most common method of drying agricultural products is forced air circulation through the products held in a suitable container such as bins, silos, or chambers. The drying rate depends on the initial moisture content and temperature of the product, air circulation rate, entering conditions of the circulated air, and the design of the containers. The drying of grains and oilseeds must be completed fast enough to prevent rapid product deterioration by microorganisms and enzymes. For in-bin drying of canola seeds, it takes about 15 days or less for a fan to circulate ambient air at a flow rate of 1.5–2 m^3/min to dry one metric ton of canola seeds from the initial moisture of 19% to 10% and 8% average-dry and through-dry moisture contents, in North America in August. The energy consumption by the fan is 9–26 MJ/t seed (Schoenau et al., 1995).

Supplemental heat can raise the air temperature, which is necessary in cold weather, and reduce the air flow rate during in-bin drying. Conventional energy sources such as natural gas, fuel oils, and electricity and renewable energy sources such as solar energy and biomass have been used for heating the drying air. Since the fans used to circulate the drying air through bins consume electricity, it may be less costly to dry grains and oilseeds with the air heated by supplemental heat at a low volumetric flow rate of 1 m^3/min than unheated air at an increased volumetric airflow rate of 2 m^3/min per ton of grains and oilseeds. Solar drying has been found to be more cost effective than other supplemental heating systems with conventional fuels such as natural gas and fuel oils, provided a well-designed solar collector for air heating is available for use in locations with good solar energy availability. A solar collector area of 1–1.2 m^2 for 1 m^3/min of air flow is adequate for most drying situations (Schoenau et al., 1995).

11.5.2 Use of More Efficient Dryers

Some dryers are more efficient than others. Direct dryers are typically more efficient than indirect dryers. The typical efficiency of direct heating is 95%–98% while the energy efficiencies of indirect steam-to-air dryers and air-to-air dryers are 85% and 70%–80%, respectively. Natural gas is a preferred fuel for directly drying corn products (Best Practice Programme, 1997).

11.5.3 Energy Conservation in Fluidized Bed Drying of Grains and Oilseeds

Fluidized bed dryers are mainly suitable for drying granular solids and are operated best on solids with a narrow particle size distribution. Hot flue gases from a combined heat and power generation unit can be used to fluidize a grain or oilseed particle bed. For fluidized bed drying of soybeans, the optimum conditions for minimum energy consumption and maximum capacity include temperature of 140°C, bed depth of 18 cm, air velocity of 2.9 m/s, and fraction of air recirculated of 0.9. Under these conditions, energy consumption was found to be 6.8 MJ/kg water removed during fluidized bed drying of soybeans, compared to the theoretical value of 2.5–2.7 MJ/kg required for evaporation of pure water. However, since hot flue gas at a low or no cost can be used as a drying media, the dryer can also be considered as a waste-heat recovery unit (Soponronnarit et al., 2001).

11.5.4 Application of Heat Pumps in Drying of Grains and Oilseeds

It is necessary to examine different methods to improve the energy efficiency of the drying process. Heat pumps have been used to increase the efficiency of convection drying (Perera and Rahman, 1997). As discussed in Chapter 9, a heat pump dehumidifier dryer functions in a manner similar to a refrigerator. It consists of a condenser for high-temperature heat exchange, an evaporator for low-temperature heat exchange, a compressor to increase the pressure of the working liquid, and an expansion valve to decrease the pressure of the working liquid. A fan is usually used to provide air movement in the drying chamber. The evaporator of the heat pump is used to remove the moisture from moist air exiting from the drying chamber and the condenser of the heat pump is used to increase the temperature of the drying air. A heat pump can also be used to extract and upgrade low-temperature renewable energy such as geothermal energy to a high-temperature heat source for drying (Kuzgunkaya and Hepbasli, 2007).

11.5.5 Novel Drying Processes for Grains and Oilseeds

Supercritical fluids such as supercritical CO_2 can be used to remove moisture from foods (Brown et al., 2008). Supercritical CO_2 has a low critical temperature and pressure (31.1°C and 7.3 MPa). Therefore, supercritical CO_2 drying can be operated at a much lower temperature than conventional air drying. However, since supercritical CO_2 is a nonpolar solvent, the water solubility in supercritical CO_2 is only 4 mg/g at 50°C and 20 MPa and 2.5 mg/g at 40°C and 20 MPa (King et al., 1992; Sabirzyanov et al., 2003). Therefore, small quantities of polar co-solvents such as ethanol are usually added in the supercritical CO_2 to increase the solubility of polar substances such as water in supercritical CO_2 (Brown et al., 2007).

Energy consumption and drying time affect the efficiency of drying grains and oilseeds. The low thermal conductivity of grains and oilseeds leads to a slow air-drying process. Far-infrared radiation (Afzal et al., 1999) and microwave heating (Zhang et al., 2006) have been used to reduce the drying time and increase the energy efficiency of a drying process. Afzal et al. (1999) found that the use of far-infrared radiation significantly increased the drying rate in thin layer drying of barley and the energy consumption was reduced considerably. The optimum far-infrared intensity was found to be 0.333 W/cm^2.

11.6 ENERGY UTILIZATION OF BY-PRODUCTS IN GRAIN AND OILSEED PROCESSING

11.6.1 By-Products in Grain and Oilseed Processing Facilities

Starch processing plants produce large amounts of solid residues such as bran and diluted wastewater. The need for fresh water in a corn wet milling plant is as high as 1.5 m^3/ton of corn (Kollacks and Rekers, 1988). The flour is washed with water to produce starch and the starch processing residue is a commercially accepted feedstock for ethanol production (Nguyen, 2003). The starch processing wastewater contains some solids, which can be recovered as a potential renewable source (Nguyen, 2003; Verma et al., 2007). Nguyen (2003) reported that the distillery effluent from a starch-based ethanol plant had 3.3% total solids.

Oilseed processing facilities generate large amounts of solid and liquid residues. A palm oil mill produces a large amount of residues including empty fruit bunches, fibers, shells, and liquid palm oil mill effluent, which is equivalent to almost three times the amount of oil produced by biomass (Husain et al., 2003). For processing of 1 ton of fresh fruit bunches, the mill generates 140 kg of fiber and 60 kg of shell (Mahlia et al., 2001). Olive oil production facilities produce by-products including solid residue of olive husk and liquid residue of olive mill wastewater. The average amount of olive husk and olive mill wastewater produced are both 40% of treated olive mass for a batch-pressure process, and 50% and 95% for a continuous-centrifugation process. The heating value of olive husk is in the range of 14–18 MJ/kg and the biological oxygen demand (BOD) of olive mill wastewater is 23–100 g/l (Caputo et al., 2003). These by-products, which have a low content of sulfur, have a huge potential to be used in cogeneration power plants for heat and power supply in oilseed processing facilities. The by-products from oil processing facilities can also be converted to favorable energy products.

Waste treatment technologies for energy recovery represent an interesting alternative for a sustainable disposal of residues from grain and oilseed processing facilities because these technologies are able to reduce the environmental impact and generate electric energy to be sold or to meet the energy needs of mills. Biological conversion technologies including anaerobic digestion (Chapter 24) and fermentation (Chapter 25) and thermochemical conversion technologies including combustion, gasification, pyrolysis, and hydrothermal liquefaction (Chapter 27) can be used to convert solid and liquid residues from grain and oilseed processing facilities into energy products. The selection of the conversion technologies depend on the characteristics of available residues and the market of the final products. The remaining residues after conversion can be used as a fertilizer for grain and oilseed plantation.

11.6.2 Biological Conversion of By-Products

Biological conversion technologies including both aerobic and anaerobic processes can reduce BOD and COD concentrations and enable removal of organic and inorganic suspended solids. A biological conversion process is usually used to treat liquid residue or very wet solid residue from a grain and oil mill. Compared to an aerobic or composting process, an anaerobic process can produce biogas, which is a valuable by-product to be used as an energy source in the mill. An anaerobic digestion process usually has an aerobic follow-up treatment to enhance the treatment. However, due to high investment costs and complex process management, the biological conversion technology is suited for industrial-scale mills or as a centralized treatment facility to serve several mills (Caputo et al., 2003).

Grain residuals are rich in cellulose and hemicellulose, which can be a renewable source for enzymatic production of soluble sugars as feedstocks for ethanol fermentation (Hang, 2004). Wheat bran consists of three main components of residual starch, cellulose, and hemicellulose. Choteborska et al. (2004) produced sugar solution from wheat bran for ethanol fermentation. They first treated the wheat bran with starch degrading enzymes to remove the starch from the bran. The maximum yield of sugars was 52.1 g/100 g of starch-free wheat bran obtained using 1% of sulfuric acid at 130°C for 40 min. As the wastewater from starch processing facilities contains some residual starch, it can also be converted to ethanol by fermentation. The solids in the wastewater can be economically recovered by ultrafiltration (Nguyen, 2003).

Palm oil mill effluent has been converted into biogas using an anaerobic digestion process. The energy yield was 20.5 MJ/kg COD at a hydraulic retention time of 31 days for a two-phase mesophilic anaerobic digestion process (Ng et al., 1987). The liquid residue can also be treated by other methods such as membrane filtration.

11.6.3 Thermochemical Conversion of By-Products

Thermochemical conversion processes such as combustion, gasification, and hydrolysis usually prefer to have a dried feedstock. Thermochemical conversion presents several environmental advantages such as the reduction of mass and volume of disposed solids, the reduction of pollutants, and the potential for energy recovery. Different types of dryers including drum dryers, belt dryers, and fluidized bed dryers have been used to dry the solid residues from grain and oil mills. The main disadvantage is the high-energy demand associated with drying. The dried solid residues can be combusted or co-fired with other fuels. The shell and fiber alone in a palm oil processing facility can generate more than enough energy to meet the energy demand of the palm oil processing facility (Mahlia et al., 2001). The analysis of mass and energy balances showed that it is possible to have surplus power ranging between 1 and 7 MW beyond its energy demand for a facility with a fruit bunch processing

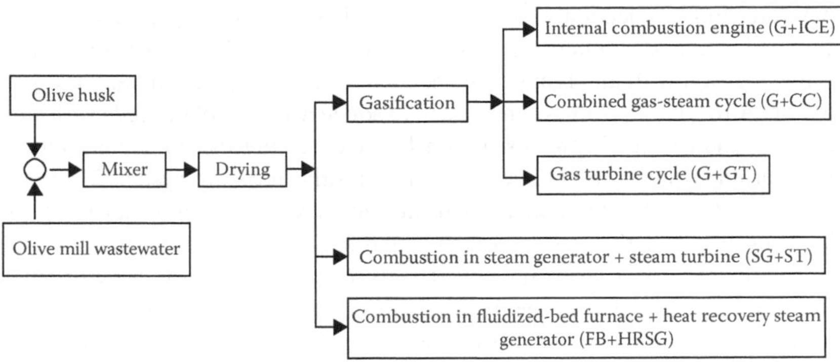

FIGURE 11.4 A centralized-combined waste-to-energy system for conversion of olive processing wastes. (Reprinted from Caputo, A.C., Scacchia, F. and Pelagagge, P.M., *Appl. Thermal Eng.*, 23, 197, 2003. With permission.)

capacity between 18 and 60 ton/h (Arrieta et al., 2007). Sookkumnerd et al. (2005) reported that rice mills with a processing capacity from 45 to 120 tons per day are financially favorable to invest in steam engines for converting rice husk into power for the rice mills.

Several types of combustors have been developed to use low-calorie fuels such as grain and oilseed processing by-products. Traditional combustors such as fixed bed combustors are in general technically inefficient and economically unacceptable for the combustion of grain and oilseed processing residues. Fluidized bed combustors are suitable for a wide range of solid wastes. The combustion efficiency was found to be from 86% to 95% for the combustion of olive oil mill waste in a 13.2 cm internal diameter fluidized bed reactor at a combustion intensity of 812 kg/m^{2}h and bed height of 0.1–0.15 m (Abu-qudais, 1996).

A centralized, combined gasification and combustion system, as shown in Figure 11.4, was proposed and analyzed by Caputo et al. (2003) to treat the by-products from olive mills and convert them into power and heat. In this system, the solid residue of olive husk and the liquid residue of olive mill wastewater are mixed and then dried to reduce the moisture content from 69% to 15% in a rotary dryer using the hot gas produced by the combustion of part of the dried residue. The remaining dried residue is fed to a gasifier to produce syngas with a low heating value of 5.86 MJ/kg at a cold gas efficiency of 70%. The cleaned syngas is fed to a combined gas-steam cycle for heat and power generation. Although the thermal efficiency of the system for treating mixed residue is low, it has been proved to be economically profitable when savings from avoided external disposal costs are accounted for (Caputo et al., 2003).

11.7 SUMMARY

Grain oilseed milling is energy intensive. Corn wet milling undergoes a steeping operation followed by a series of separation operations to convert the starch, germ, fiber, and protein in the corn kernels into an array of marketable products. Reduction of steeping time by pretreating corn with enzymes and preheating corn to the steeping temperature and application of mechanical and membrane separation to dewater dilute slurries can save energy use in a wet corn mill. Mechanical pressing and solvent extraction with an organic solvent such as hexane are commonly used in oil extraction. A mechanical press can be used to squeeze oil out of oilseeds with a high-oil content. Industrial oil presses driven by electric motors alone consume large amounts of electricity. The efficiency of a hydraulic pressing system is very high because almost all the input energy can be recovered and reused during the next stroke. For oilseeds with a low oil content, solid-solvent extraction method is usually used. The removal of a large amount of solvent in the crude exact by distillation consumes a large amount of energy. Ultrafiltration or reverse osmosis with membranes can replace conventional distillation to minimize thermal damage to products and reduce energy consumption for solvent recovery in an oil mill. Wet grains and oilseeds are extremely susceptible to spoilage. Energy conservation technologies for drying of grains and oilseeds should be an integrated part of the energy management project in the grain and oilseed milling sector. The solid and liquid residues from grain and oilseed processing facilities can be converted into energy products via either a biological or a thermochemical conversion process. This conversion can not only provide enough energy to meet the energy demand of the facilities but also help to dispose of the residues for reduced environmental pollution. The ash from the conversion processes can be used as fertilizer for grain and oilseed plantation.

REFERENCES

Abu-qudais, M. 1996. Fluidized-bed combustion for energy production from olive cake. *Energy* 21: 173–178.

Afzal, T.M., T. Abe, and Y. Hikida. 1999. Energy and quality aspects during combined FIR-convection drying of barley. *Journal of Food Engineering* 42: 177–182.

Altin, R., S. Cetinkaya, and H.S. Yucesu. 2001. The potential of using vegetable oil fuels as fuel for diesel engines. *Energy Conversion and Management* 42: 529–538.

Arrieta, F.R.P., F.N. Teixeira, E. Yáñez, E. Lora, and E. Castillo. 2007. Cogeneration potential in the Columbian palm oil industry: Three case studies. *Biomass and Bioenergy* 31: 503–511.

Bargale, P.C., R.J. Ford, F.W. Sosulski, D. Wulfsohn, and J.I. Irudayaraj. 1999. Mechanical oil expression from extruded soybean. *Journal of American Chemists' Society* 76: 223–229.

Best Practice Programme. 1997. *Good Practice Guide 149: Rotary Drying in the Food and Drink Industry*. Available at http://www.energy-efficiency.gov.uk/indes.cfm.

Bredesson, D.K. 1993. Mechanical oil extraction. *Journal of the American Oil Chemists' Society* 60: 211–213.

Brown, Z.K., P.J. Fryer, I.T. Norton, S. Bakalis, and R.H. Bridson. 2008. Drying of foods using supercritical carbon dioxide–investigations with carrot. *Innovative Food and Emerging Technologies* 9: 280–289.

Caputo, A.C., F. Scacchia, and P.M. Pelagagge. 2003. Disposal of by-products in olive oil industry: waste-to-energy solutions. *Applied Thermal Engineering* 23: 197–214.

Choteborska, P., B. Palmarola-Adrados, M. Galbe, G. Zacchi, K. Melzoch, and M. Rychtera. 2004. Processing of wheat bran to sugar solution. *Journal of Food Engineering* 61: 561–565.

Cuperus, F.P. and H.H. Nijhuis. 1993. Applications of membrane technology to food processing. *Trends in Food Science and Technology* 4: 277–282.

Galitsky, C., E. Worrell, and M. Ruth. 2003. *Energy Efficiency Improvement and Cost Saving Opportunities for Corn Wet Milling Industry-An ENERGY STAR Guide for Energy and Plant Managers.* Ernest Orlando Lawrence Berkeley National Laboratory.

Hang, Y.D. 2004. Management and utilization of food processing wastes. *Journal of Food Science* 69: 104–107.

Husain, Z., Z.A. Zainal, and M.Z. Abdullah. 2003. Analysis of biomass-residue-based cogeneration system in palm oil mills. *Biomass and Bioenergy* 24: 117–124.

Isobe, S., F. Zuber, K. Uemura, and A. Noguchi. 1992. A twin-screw press design for oil extraction of dehulled sunflower seeds. *Journal of American Oil Chemists' Society* 69: 884–889.

Jakubowska, N., Z. Polkowska, J. Namiesnik, and A. Przyjazny. 2005. Analytical applications of membrane extraction for biomedical and environmental liquid sample preparation. *Critical Review of Analytical Chemistry* 35: 217–235.

Jekayinfa, S.O. and A.I. Bamghoye. 2004. Energy requirements for palm-kernel oil processing operations. *Nutrition and Food Science* 34: 166–173.

Jekayinfa, S.O. and A.I. Bamghoye. 2007. Development of equation for estimating energy requirements in palm-kernel oil processing operations. *Journal of Food Engineering* 79: 322–329.

Jekayinfa, S.O. and A.I. Bamgboye. 2008. Energy use analysis of selected palm-kernel oil mills in south western Nigeria. *Energy* 33: 81–90.

Jekayinfa, S.O. and J.O. Olajide. 2007. Analysis of energy usage in the production of three selected cassava-based foods in Nigeria. *Journal of Food Engineering* 82: 217–226.

Johnston, D.B. and V. Singh. 2001. Use of proteases to reduce steep time and SO_2 requirements in a corn wet-milling process. *Cereal Chemistry* 78: 405–411.

Kashyap, M.C., Y.C. Agrawal, P.K. Ghosh, D.S. Jayas, B.C. Sarkar, and B.P.N. Singh. 2006. Enzymatic hydrolysis pretreatment to solvent extraction of soybrokens for enhanced oil availability and extractability. *Journal of Food Process Engineering* 29: 664–674.

King, M.B., A. Mubarak, J.D. Kim, and T.R. Bott. 1992. The mutual solubilities of water with supercritical and liquid carbon dioxides. *The Journal of Supercritical Fluids* 5: 296–302.

Kollacks, W.A. and C.J.N. Rekers. 1988. Five years of experience with the application of reverse osmosis on light middlings in a corn wet milling plant. *Starch* 40: 88–94.

Kuzgunkaya, E.H. and A. Hepbasli. 2007. Exergetic performance assessment of a ground-source heat pump drying system. *International Journal of Energy Research* 31: 760–777.

Li, Y., E. Griffing, M. Higgins, and M. Overcash. 2006. Life cycle assessment of soybean oil production. *Journal of Food Process Engineering* 29: 429–445.

Mahlia, T.M.I., M.Z. Abdulmuin, T.M.I. Alamsyah, and D. Mukhlishien. 2001. An alternative energy source from palm wastes industry for Malaysia and Indonesia. *Energy Conversion and Management* 42: 2109–2118.

Ng, W.J., K.K. Chin, and K.K. Wong. 1987. Energy yields from anaerobic digestion of palm oil mill effluent. *Biological Wastes* 19: 257–266.

Nguyen, M.H. 2003. Alternatives to spray irrigation of starch waste based distillery effluent. *Journal of Food Engineering* 60: 367–374.

Ohlson, I.S.R. 1992. Modern processing of rapeseed. *Journal of American Oil Chemists' Society* 69: 195–198.

Okoye, C.N., J. Jiang, and Y.H. Liu. 2008. Design and development of secondary controlled industrial palm kernel nut vegetable oil expeller plant for energy saving and recuperation. *Journal of Food Engineering* 87: 578–590.

Perera, C.O. and M.S. Rahman. 1997. Heat pump dehumidifier drying of food. *Trends in Food Science and Technology* 8: 75–79.

Rausch, K.D. 2002. Front end to backpipe: Membrane technology in the starch processing industry. *Starch/Starke* 54: 273–284.

Reverchon, E., M. Poletto, L.S. Osseo, and M. Somma. 2000. Hexane elimination from soybean oil by continuous packed tower processing with supercritical CO_2. *Journal of American Oil Chemists' Society* 77: 9–14.

Sabirzyanov, A.N., R.A. Shagiakhmetov, F.R. Gabitov, A.A. Tarzimanov, and F.M. Gumerov. 2003. Water solubility of carbon dioxide under supercritical and subcritical conditions. *Theoretical Foundations of Chemical Engineering* 37: 51–53.

Schoenau, G.J., E.A. Arinze, and S. Sokhansanj. 1995. Simulation and optimization of energy systems for in-bin drying of canola grain (rapeseed). *Energy Conversion and Management* 36: 41–59.

Singh, J. and P.C. Bargale. 2000. Development of a small capacity double stage compression screw press for oil expression. *Journal of Food Engineering* 43: 75–82.

Singh, N. and M. Cheryan. 1997. Membrane technology in corn wet milling. *American Association of Cereal Chemists, Inc* 42: 520–525.

Singh, K.K., D.P. Wiesenborn, K. Tostenson, and N. Kangas. 2002. Influence of moisture content and cooking on screw pressing of Crambe Seed. *Journal of American Oil Chemists' Society* 79: 165–170.

Sookkumnerd, C., N. Ito, and K. Kito. 2005. Financial viabilities of husk-fueled steam engines as an energy-saving technology in Thai rice mills. *Applied Energy* 82: 64–80.

Soponronnarit, S., T. Swasdisevi, S. Wetchacama, and W. Wutiwiwatchai. 2001. Fluidized bed drying of soybeans. *Journal of Stored Products Research* 37: 133–151.

Sumathi, S., S.P. Chai, and A.R. Mohamed. 2008. Utilization of oil palm as a source of renewable energy in Malaysia. *Renewable and Sustainable Energy Reviews*, in press.

U.S. Census Bureau. 2006. *2006 Annual Survey of Manufactures*. Available at http://factfinder.census.gov.

Verma, M., S.K. Brar, R.D. Tyagi, R.Y. Surampalli, and J.R. Valero. 2007. Starch industry wastewater as a substrate for antagonist *Trichoderma viride* production. *Bioresource Technology* 98: 2154–2162.

Wang, L.J. and W. Shen. 2005. Chemical and morphological stability of Aliquat 336/PVC membranes in membrane extraction: A preliminary study. *Separation and Purification Technology* 46: 51–62.

Wang, L.J. and C.L. Weller. 2006. Recent advances in extraction of natural products from plants. *Trends in Food Science and Technology* 17: 300–312.

Wu, J.C.S. and E.H. Lee. 1999. Ultrafiltration of soybean oil/hexane extract by porous ceramic membranes. *Journal of Membrane Science* 154: 251–259.

Zhang, M., J. Tang, A.S. Mujumdar, and S. Wang. 2006. Trends in microwave-related drying of fruits and vegetables. *Trends in Food Science & Technology* 17: 524–534.

Zheng, Y.L., D.P. Wiesenborn, K. Tostenson, and N. Kangas. 2005. Energy analysis in the screw pressing of whole and dehulled flaxseed. *Journal of Food Engineering* 66: 193–202.

12 Energy Conservation in Sugar and Confectionary Processing Facilities

12.1 INTRODUCTION

The sugar and confectionery product manufacturing sector (NAICS code 3113) is engaged in one or more of the following manufacturing activities: (1) processing sugarcane and sugar beet for a new product such as sugar or chocolate and (2) further processing sugar and chocolate for other products. Sugar is mainly produced from cane and beet. The sugar content of sugar beet is about 25% higher than that of sugarcane. About one-fourth of the world's sugar production comes from sugar beet and the production capacity was 40 million tons in 1999 (Erdal et al., 2007). About 70% of the sugar consumed in the United States is produced from sugarcane sugar while the remaining 30% is beet sugar. In the sugar industry, the most energy-intensive unit operations are extraction, juice purification, evaporation, and crystallization (Urbaniec, 2004).

Almost all steps in the refining process consume power for pumping and centrifugation operations and heat for evaporation and drying. The energy consumption of a sugar manufacturing facility depends on the energy demand of the individual unit operations and the efficiency of energy transformation in the heat exchanger network. The heat exchanger network can be optimized for maximum energy recovery. In the sugar and confectionery product manufacturing sector, a large portion of fuels is used for steam generation and a large portion of purchased electricity is used to run motors. Energy conservation technologies for steam generation and distribution are discussed in Chapter 4. Energy conservation technologies for motors are discussed in Chapter 6. Heat exchangers are widely used in this sector, and energy conservation technologies for heat exchangers are discussed in Chapter 7. Boilers and motors may generate waste heat, and waste-heat recovery technologies are discussed in Chapter 8. Since both processing heat and electricity are required by the sugar processing sector, a combined heat and power system can be used to efficiently and economically provide electricity or mechanical power and useful heat from the same primary energy source. The combined heat and power systems are discussed in Chapter 9.

In this chapter, the main products and their processes in the sugar and confectionery product manufacturing sector are reviewed. The energy uses and conservation of main unit operations in this sector are discussed. Finally, energy utilization of the processing wastes from sugar and confectionery product manufacturing facilities is discussed.

12.2 OVERVIEW OF MAIN PROCESSES

The basic sugar processes consist of slicing, diffusion, juice purification, evaporation, crystallization, and recovery of sugar. As shown in Figure 12.1, the mass flows from the top to the bottom while the energy flows from the left to the right. Sugar beet is first cleaned and washed to remove soil, stones, and organic matter from the beet. Cleaned and sliced beet is delivered to the extraction unit where raw juice is extracted. Extraction of sugar from sliced beet requires heating to 75°C–80°C with steam (e.g., 110°C or higher).

The resulting pulp is dewatered by mechanical pressing, followed by drying to produce dried pulp. The dried pulp is pelletized for storage and transportation. The concentration of raw sugar juice is about 15%. The extracted raw juice is purified using lime–milk in order to remove impurities. The carbonation slurry is concentrated in filters and part of it is recycled and the rest is further concentrated to carbonation lime. The juice is then delivered to multiple-effect evaporators. The vapor generated in the evaporators is used for extraction, juice heating, crystallization, and other processing heat. The energy needed to heat the first effect evaporator is supplied from a boiler. Concentrated juice at a concentration of 70% or higher from the last evaporator is delivered to multi-stage crystallization. Evaporating crystallization requires heating with steam (e.g., 120°C or higher). The crystallized sugar is recovered from the syrup by centrifugation. The syrup from the last crystallization stage is called molasses (Urbaniec et al., 2000; Grabowski et al., 2001; Krajnc et al., 2007).

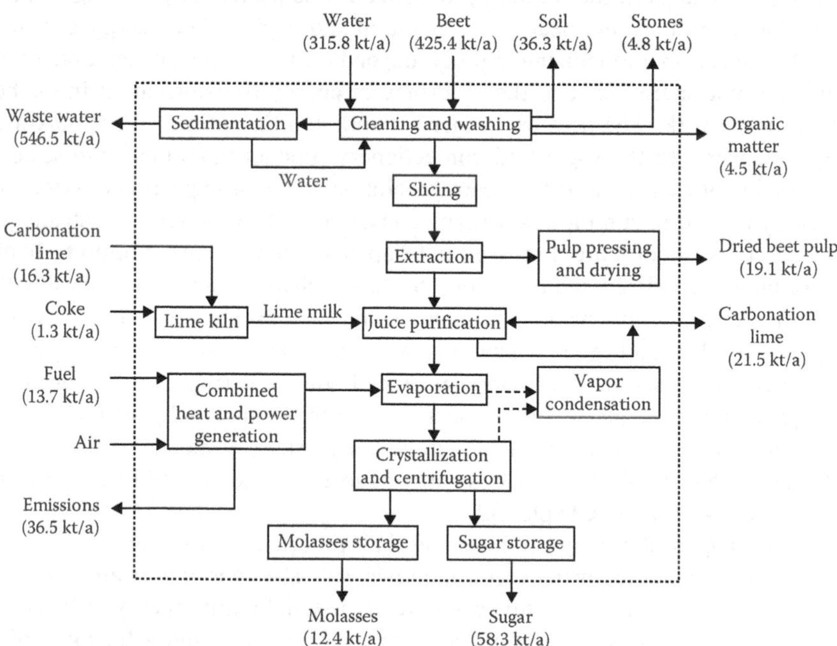

FIGURE 12.1 A flow sheet of sugar processing. (Reprinted from Krajnc, D., Mele, M., and Glavič, P. *J. Cleaner Prod.*, 15, 1240, 2007. With permission.)

12.3 ENERGY CONSUMPTION IN THE SUGAR AND CONFECTIONERY PRODUCT MANUFACTURING SECTOR

The total shipment value from the sugar and confectionery product manufacturing sector was $28.23 dollars in the United States in 2006. The total costs for purchasing fuels and electricity were $0.58 billion in 2006. About 2.04 ¢ of energy cost is required to produce each dollar shipment value of sugar and confectionery products in the United States. The sugar and confectionery product manufacturing sector consumed 5.8% of the total energy input into the food industry and generated 4.80% of the total shipment value in 2006. The cost of purchased electricity in sugar and confectionery product manufacturing sector was 37% of the total energy expenses. Sugar manufacturing contributed to 54% of the total energy costs in this sector (U.S. Census Bureau, 2006).

The steady engineering improvement of the equipment necessary to make various unit operations function efficiently has gradually reduced the energy requirements for sugar processing. Energy consumption in the sugar facilities can be lowered by the simultaneous optimization of the evaporation and crystallization processes (Urbaniec, 2004). Use of the processing wastes for energy supply creates possibilities for zero-waste emissions and self-sufficiency in heat and power supply in sugar production facilities (Grabowski et al., 2001; Krajnc et al., 2007).

12.4 ENERGY CONSERVATION IN MAIN UNIT OPERATIONS

12.4.1 Energy Conservation in Dewater and Drying

It was reported that industrial dryers consume on average about 12% of the total energy used in all manufacturing sectors (Bahu, 1991). Dehydration or drying processes are simultaneous heat and mass transfer operations. It is necessary to examine different methods to improve the energy efficiency of dryers. Mechanical processes such as filtration and centrifugation can be used to remove as much water as possible before drying. The British Sugar Beet Factory at Wissington operates six presses and three rotary dryers to dry its pulp. It was reported that water was expelled from the wet beet pulp by a mechanical screw press at a rate of 8.69 kg/s, consuming energy at a rate of 23 kJ/kg of water removed. The dryers removed water at a rate of 6.88 kg/s, consuming energy at a rate of 2.907 MJ/kg of water removed. Using mechanical dewatering, therefore, saved 55.8% in primary energy use (Best Practice Programme, 1997).

12.4.2 Energy Conservation in Evaporator

Multi-stage evaporation is widely used in the sugar industry. A multiple-effect evaporator system is a simple, series arrangement of several evaporators, which use steam to remove product moisture by evaporation. The evaporated water vapor from food products is collected and used as the steam for the next evaporator in the series. This collection and reuse of vapor result in smaller energy requirements to remove product moisture. The greater the number of effects in the series, the smaller the energy consumption. It was found that changing from four-effect evaporators to five-effect evaporators could save 20% energy. However, the number of evaporators in series should also be determined by the economics of the process (Casper, 1977). The total heating surface installed in a multi-stage evaporator station of a large sugar factory may be of the order of 40,000 m^2 with a unit up to 8,000 m^2 (Urbaniec, 2004). Evaporation is a link between the purification of juice from low concentration and the crystallization of sugar from high concentration in the evaporator. The initial and final values of juice concentration in the evaporator station determine the flow of vapors that can be used for process heating. The final concentration of juice in the evaporator could be as high as 72%–75% (Urbaniec, 2004). To avoid the deterioration of juice quality at a high temperature and uncontrolled crystal formation at a low temperature, the juice temperature during evaporation is usually controlled between 90°C and 128°C. There, multi-stage evaporation (e.g., six stages) is used, and the average temperature interval between two stages is 6°C–7°C. A thin film evaporator with plate or tubular heat transfer surfaces is usually used for intensive heat transfer. The temperature difference between the juice and processing heat stream is 3°C–4°C.

In order to save energy, multi-stage evaporation combined with heat recovery and the retrofit of evaporator station should be designed jointly with the retrofit of the heat exchanger network. Plate heat exchangers may effectively replace shell and tube heat exchangers where no extreme temperatures or pressures are required. Plate heat exchangers have higher heat transfer efficiencies, low operational, installation, and investment costs, and high flexibility. Furthermore, plate heat exchangers can maintain high product quality because of shorter residence time. Plate evaporators are expected to replace standard falling film evaporators.

Traditionally, evaporation is one of the most energy-intensive unit operations used for concentrating thin sugar juice. Application or combination of low-energy consuming membrane separation can save a considerable amount of energy. Compared to evaporation, since membrane separation is not involved in phase change, the energy consumption of membrane separation is thus very low. A two-stage reverse osmosis system has been studied for pre-concentrating sugar syrup (Madaeni and Zereshki, 2008).

12.4.3 Energy Conservation in Crystallization

Krajnc et al. (2007) reported that simultaneous optimization of the evaporation and evaporating crystallization processes in a sugar processing facility could reduce process steam consumption by 6% and increase sugar production by 1%. Urbaniec et al. (2000) introduced a new approach for the retrofit design of energy systems in sugar processing facilities. The energy saving was estimated at 29% and the payback period was 4 years. The environmental impact of the process optimization is significant due to the decreased pollutant emissions from the combustion of the fuel used for steam generation. The energy consumption during crystallization is a function of juice concentration as shown in Figure 12.2 (Urbaniec, 2004). Grabowski et al. (2001) introduced a novel sugar manufacturing process, which replaced the evaporating crystallization of sugar from concentrated juice with cooling crystallization. The energy saving from the novel process is promising.

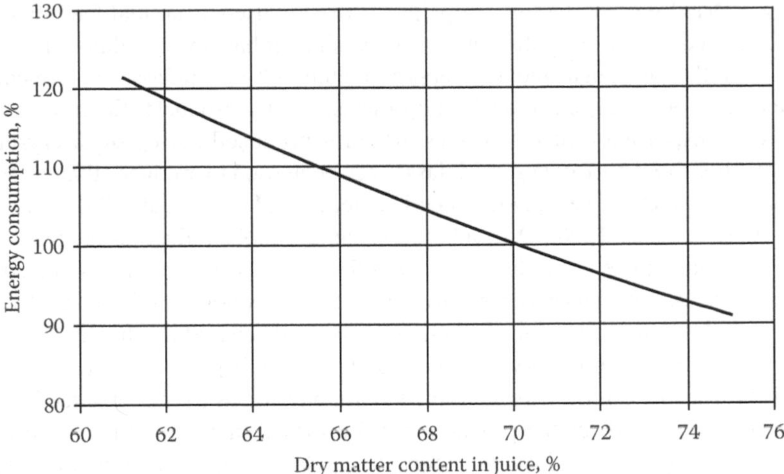

FIGURE 12.2 Relative energy consumption in three-stage sugar crystallization versus concentration of juice at evaporator outlet. (Reprinted from Urbaniec, K., *J. Food Eng.*, 61, 505, 2004. With permission.)

12.5 COGENERATION OF HEAT AND POWER

In cane sugar refineries, there is an economical dual use of steam to considerably increase the efficiency of boilers. The high-pressure steam is first used to operate a turbine for electricity generation, and then its exhaust steam is utilized as processing steam (Urbaniec et al., 2000). The cogeneration of steam and electricity has become common in the sugarcane industry worldwide (Mbohwa, 2003; Mbohwa and Fukuda, 2003). Cane sugar refineries generate most of their own electricity. Because of the dual use of steam, the overall energy efficiency of the combined steam and electricity system in sugar refineries is 75%–80% compared to the 35% of electricity generation alone.

In a typical sugar plant, thermal energy is supplied by steam at a relatively high pressure of 0.35 MPa from a combined heat and power plant. Energy in steam is transformed in three steps: (1) generating medium pressure vapor at a pressure of 0.08–0.25 MPa and hot condensates, (2) supplying vapor and condensates for processing heating and generating low-pressure vapor at 0.02 MPa in crystallization, and (3) condensing surplus vapor and dissipating waste heat to the environment (Urbaniec, 2004).

12.6 USING PROCESSING WASTE FOR PRODUCTION OF RENEWABLE ENERGY

The beet sugar industry produces large masses of beet pulp, molasses, and carbonation lime, which consume large amounts of energy and water. Carbonation lime is used for soil pH correction. Beet sugar processing facilities generate beet pulp and molasses. Dried pulp in the form of pellets is mainly used as animal feed. Beet pulp cannot be used as a direct substitute for wood since it has lower cellulose and lignin contents. In the beet sugar industry, approximately 90% of molasses is fermented to produce alcohol, yeast, citric acid, and other specialty products. Ethanol has been used as a transportation fuel. Molasses, which is produced during sugar crystallization, contains about 50% of sugar. Molasses has been used to produce ethanol (Moriya et al., 1989; Patil et al., 1989; Doelle and Doelle, 1990; Doelle et al., 1991; Cachot and Pons, 1991; Roukas, 1996). A combined sugar–ethanol production process, which uses the primary juice to produce sugar and the secondary juice to produce ethanol, is a promising option in terms of the environment and economics (Krajnc et al., 2007). Ethanol production from sugar processing wastes is discussed in Chapter 25.

A cane-based sugar factory produces nearly 30% of bagasse out of its total crushing. Many research efforts have attempted to use bagasse as a renewable feedstock for power generation and for the production of bio-based materials. In a modern cane sugar processing facility, all plant energy demands are supplied from cane bagasse. More energy effective cogeneration system can reduce the bagasse consumed for the energy supply in processing cane (Lobo et al., 2007). Sugarcane bagasse can be co-fired with fossil fuels or other biomass for heat and power supply (Kuprianov et al., 2006; Turn et al., 2006). Kuprianov et al. (2006) report that the combustion efficiency is 95%–96% for co-firing of sugarcane bagasse and rice husk in a conical fluidized bed combustor at rice husk energy fractions greater than 0.6. Sugarcane bagasse can also be thermochemically converted into gaseous or liquid fuels. A cyclone gasifier was used to gasify bagasse powder at 39–53 kg/h and 820°C–850°C. The heating value of the syngas produced at an oxygen equivalence ratio from 0.18 to 0.25 was in the range of 3.5–4.5 MJ/Nm^3 dry gas (Gabra et al., 2001a,b). Sugarcane bagasse without densification is very bulky and inhomogeneous. Currently, most bagasse is used in inefficient combustion devices connected to a steam cycle. Biomass pellets have become a popular form of feedstock for power generation and residential heating due to their easier handling, transportation, storage, and conversion. Pelletization of bagasse would improve the energy efficiency of conversion processes and reduce emission (Erlich et al., 2006). The thermochemical conversion of solid sugar processing wastes into energy products is discussed in Chapter 27.

12.7 SUMMARY

A massive amount of energy is needed to produce sugar from beet and cane. Minimization of energy consumption and utilization of renewable energy sources in sugar processing facilities can improve their sustainability. The energy consumption in a sugar processing facility is determined by extraction, juice purification, evaporation, and crystallization. The reduction of energy consumption in sugar production usually includes improvements in its energy systems comprising power plants, multi-effect evaporators, and process heating equipment. Simultaneous optimization of evaporation and crystallization processes in sugar processing facilities can significantly reduce energy consumption. Membrane technologies, which are not involved in phase change, can be used to replace or combine with evaporation units to save energy for concentrating sugar juice. Combined sugar–ethanol production is an environmentally and economically favorable process.

REFERENCES

Bahu, R.E. 1991. Energy consumption in dryer design. In *Drying' 91*, Mujumdar, A.S. and I. Filkova (Eds.), pp. 553–557. Amsterdam: Elsevier.

Best Practice Programme. 1997. *Good Practice Guide 149: Rotary Drying in the Food and Drink Industry.* Available at http://www.energy-efficiency.gov.uk/indes.cfm.

Cachot, T. and M. Pons. 1991. Improvement of alcoholic fermentation on cane and beet molasses by supplementation. *Journal Fermentation Bioengineering* 71: 24–27.

Casper, M.E. 1977. *Energy-Saving Techniques for the Food Industry.* New Jersey: Noyes Data Corporation.

Doelle, H.W., L.D. Kennedy, and M.B. Doelle. 1991. Scale-up of ethanol production from sugar cane using Zymomonas mobilis. *Biotechnology Letter* 13: 131–136.

Doelle, M.B. and H.W. Doelle. 1990. Sugar-cane molasses fermentation by Zymomonas mobilis. *Applied Microbiology and Biotechnology* 33: 31–35.

Erdal, G., K. Esengün, H. Erdal, and O. Gündüz. 2007. Energy use and economical analysis of sugar beet production in Tokat province of Turkey. *Energy* 32: 35–41.

Erlich, C., E. Bjornbom, D. Bolado, M. Giner, and T.H. Fransson. 2006. Pyrolysis and gasification of pellets from sugar cane bagasse and wood. *Fuel* 85: 1535–1540.

Gabra, M., E. Pettersson, R. Backman, and B. Kjellstrom. 2001a. Evaluation of cyclone gasifier performance for gasification of sugar cane residue-Part 1: Gasification of bagasse. *Biomass and Bioenergy* 21: 351–369.

Gabra, M., E. Pettersson, R. Backman, and B. Kjellstrom. 2001b. Evaluation of cyclone gasifier performance for gasification of sugar cane residue-Part 2: Gasification of cane trash. *Biomass and Bioenergy* 21: 371–380.

Grabowski, M., J. Kleme, K. Urbaniec, G. Vaccari, and X.X. Zhu. 2001. Minimum energy consumption in sugar production by cooling crystallisation of concentrated raw juice. *Applied Thermal Engineering* 21: 1319–1329.

Krajnc, D., M. Mele, and P. Glavič. 2007. Improving the economic and environmental performances of the beet sugar industry in Slovenia: Increasing fuel efficiency and using by-products for ethanol. *Journal of Cleaner Production* 15: 1240–1252.

Kuprianov, V.I., K. Janvijitsakul, and W. Permchart. 2006. Co-firing of sugar cane bagasse with rice husk in a conical fluidized-bed combustor. *Fuel* 85: 434–442.

Lobo, P.C., E.F. Jaguaribe, J. Rodrigues, and F.A.A. da Rocha. 2007. Economics of alternative sugar cane milling options. *Applied Thermal Engineering* 27: 1405–1413.

Madaeni, S.S. and S. Zereshki. 2008. Reverse osmosis alternative: Energy implication for sugar industry. *Chemical Engineering and Processing: Process Intensification* 47: 1075–1080.

Mbohwa, C. 2003. Bagasse energy cogeneration potential in the Zimbabwean sugar industry. *Renewable Energy* 28: 191–204.

Mbohwa, C. and S. Fukuda. 2003. Electricity from bagasse in Zimbabwe. *Biomass and Bioenergy* 25: 197–207.

Moriya, K., H. Shimoi, S. Sato, K. Saito, and M. Tadenuma. 1989. Ethanol fermentation of beet molasses by a yeast resistant to distillery waste water and 2-deoxy-glucose. *Journal of Fermentation Bioengineering* 67: 321–323.

Patil, S.G., D.V. Gonhale, and B.G. Patil. 1989. Novel supplements enhance the ethanol production in cane molasses fermentation by recycling yeast cell. *Biotechnology Letter* 11: 213–216.

Roukas, T. 1996. Ethanol production from non-sterilized beet molasses by free and immobilized *Saccharomyces cerevisiae* cells using fed-batch culture. *Journal of Food Engineering* 27: 87–96.

Turn, S.Q., B.M. Jenkins, L.A. Jakeway, L.G. Blevins, R.B. Williams, G. Rubenstein, and C.M. Kinoshita. 2006. Test results from sugar cane bagasse and high fiber cane co-fired with fossil fuels. *Biomass and Bioenergy* 30: 565–574.

U.S. Census Bureau. 2006. *2006 Annual Survey of Manufactures*. Available at http://factfinder.census.gov.

Urbaniec, K. 2004. The evolution of evaporator stations in the beet-sugar industry. *Journal of Food Engineering* 61: 505–508.

Urbaniec, K., P. Zalewski, and X.X. Zhu. 2000. A decomposition approach for retrofit design of energy systems in the sugar industry. *Applied Thermal Engineering* 20: 1431–1442.

13 Energy Conservation in Fruit and Vegetable Processing Facilities

13.1 INTRODUCTION

The fruit and vegetable processing sector (NAICS code 3114) is engaged in one or more of the following manufacturing activities: (1) frozen fruit, frozen juices, frozen vegetables, and frozen specialty foods and (2) canned, pickled, and dried fruits, vegetables, and specialty foods (U.S. Census Bureau, 2006). The representative products include canned juice, canned specialty foods, frozen fruits and vegetables, and dried fruits and vegetables.

In the fruit and vegetable processing sector, a large portion of fuels is used for steam generation, and a large portion of purchased electricity is used for motors and air compressors. Energy conservation of steam generation and distribution is discussed in Chapter 4. Energy conservation technologies for air compressors and motors are discussed in Chapters 5 and 6, respectively. Heat exchangers are widely used in the fruit and vegetable processing sector; energy conservation technologies for heat exchangers are discussed in Chapter 7. Boilers, heat exchangers, motors, and air compressors may generate waste heat; waste-heat recovery technologies are discussed in Chapter 8. Since both processing heat and electricity are required by the fruit and vegetable processing sector, a combined heat and power system can be used to efficiently and economically provide electricity or mechanical power and useful heat from the same primary energy source. Combined heat and power systems are discussed in Chapter 9.

In this chapter, the main fruit and vegetable products and their processes are reviewed. The energy uses and conservation in main unit operations in the fruit and vegetable processing sector are discussed. Finally, energy utilization of fruit and vegetable processing wastes is discussed.

13.2 MAIN PRODUCTS AND PROCESSES

13.2.1 Material and Energy Flow in Canned Fruit and Vegetable Processing Facilities

In a canned fruit and vegetable processing facility, raw fruits and vegetables are processed into canned fruits, vegetables, juices, tomato sauces, jellies, and preserves. Within the fruit and vegetable processing industry, there is wide variation in plant size, location, products, and patterns of energy consumption. The typical material

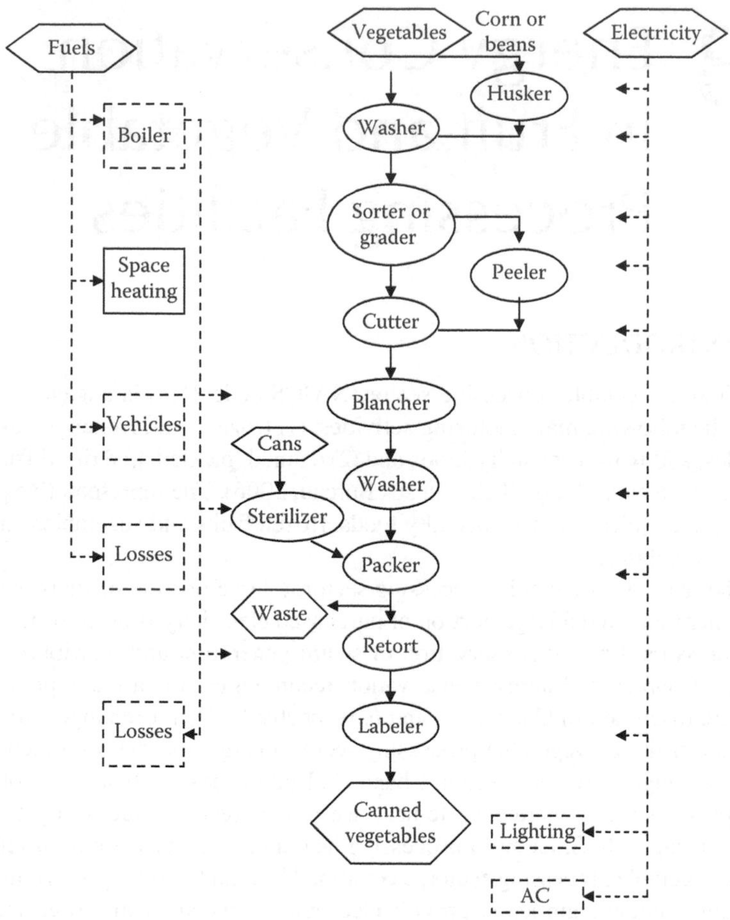

FIGURE 13.1 Typical material and energy flow for processing of canned vegetables.

and energy flows for processing canned vegetables and fruits are given in Figures 13.1 and 13.2. Major energy consuming unit operations for canned vegetables include cutting, peeling or blanching, packing, and retorting. The major energy consuming unit operations for canned fruits and fruit juices include trimming and dicing, peeling, pressing, pasteurization of juices, and retorting.

For the production of canned fruits and vegetables, most fuels (e.g., 80%) are used in boilers to produce steam for such operations as blanching, brine or syrup making, syruping, container sterilization, filling, exhausting, retorting, and closing. The purchased electricity is used for lighting, refrigeration, and power.

13.2.2 Freezing and Dehydrating Fruits and Vegetables

Freezing is used to produce frozen fruit, frozen juices, frozen vegetables, and frozen specialty foods. Freezing and drying of fruits and vegetables are much simpler than canning processes. However, the preparation of fruits and vegetables for further freezing, drying, and canning is the same. The difference among these processes is the

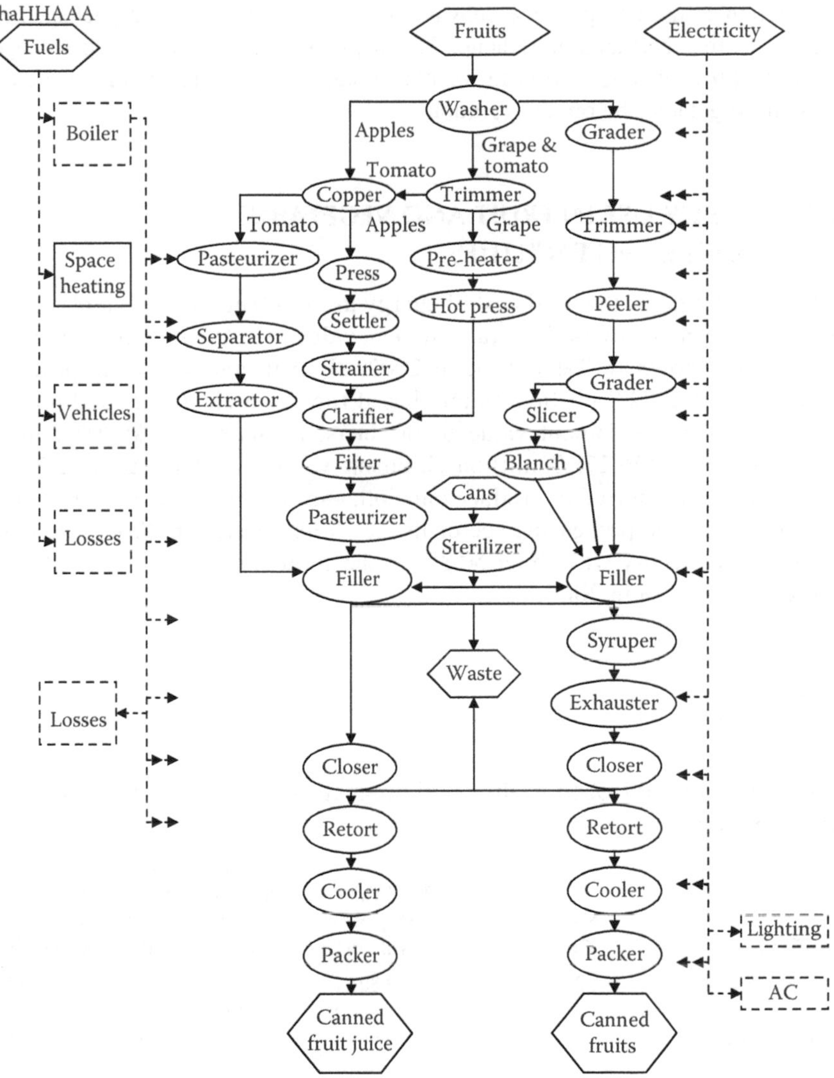

FIGURE 13.2 Typical material and energy flow for processing of canned fruit juices and fruits.

preservation and packaging operations. Major energy consuming steps in the frozen fruit processes are trimming, peeling, slicing, and freezing. The major energy consuming operations for processing frozen vegetables are husking or vining, cutting, blanching, and freezing. Energy consuming operations for processing frozen juice concentrate are extraction, concentration, blending, and freezing. Freezers can be divided into

- Direct contact plate, band, and drum freezers
- Static air and air blast freezer
- Immersion freezers
- Liquid nitrogen, fluorocarbon, and solid carbon dioxide freezers

Dehydration is used to produce dehydrated fruits, vegetables, and packaged soup mixtures from dehydrated ingredients. The major energy consuming step is dehydration although energy is also used for blanching, peeling, and dicing or slicing of fruits and vegetables before dehydration.

13.3 ENERGY USE IN FRUIT AND VEGETABLE PROCESSING FACILITIES

In the United States, the total cost of energy used in fruit and vegetable processing facilities was 1.81% of the total shipment value in 2002. It increased to 2.31% in 2006. As shown in Table 13.1, in 2002, all fruit and vegetable processing facilities spent $972.2 million on purchased fuels and electricity and produced $53,804.7 million shipment value of products, compared to $1,302.4 million energy cost and $56,278.7 million shipment value in 2006. About 2.31 ¢ of energy cost was required to produce one dollar shipment value of products in the fruit and vegetable processing sector in the United States in 2006. The cost of purchased electricity was about 50.3% of the total energy cost in 2002, which decreased to 44.3% in 2006.

TABLE 13.1
Fuels and Electricity Used in the Food Processing Facilities in United States (2002–2006)

Classification	NAICS Code	Year	Total Value of Shipments (Million $)	Cost of Purchased Fuels (Million $)	Cost of Purchased Electricity (Million $)	Total Cost of Energy (Million $)
Fruit and vegetable preserving and specialty food manufacturing	3114	2002	53,804.7	482.9	489.3	972.2
		2003	53,068.2	517.5	496.7	1,014.2
		2004	53,322.6	574.3	498.5	1,072.8
		2005	54,950.7	723.3	537.0	1,260.3
		2006	56,278.7	725.6	576.8	1,302.4
Frozen fruits and vegetables	31141	2002	21,907.2	169.1	262.8	431.9
		2003	22,050.6	197.8	265.0	462.7
		2004	22,042.0	204.6	262.6	467.2
		2005	23,664.8	243.6	293.8	537.4
		2006	24,340.4	252.9	314.4	567.3
Canned and dehydrated fruits and vegetables	31142	2002	31,897.5	313.8	226.5	540.3
		2003	31,017.6	319.7	231.8	551.5
		2004	31,280.6	369.7	235.9	605.6
		2005	31,285.9	479.8	243.2	723.0
		2006	31,938.3	472.7	262.4	735.1

Source: Data from U.S. Census Bureau, *2006 Annual Survey of Manufactures*, 2006.

13.4 ENERGY CONSERVATION IN FRUIT AND VEGETABLE PROCESSING

13.4.1 Energy Conservation in Blanching

Blanching is a thermal treatment applied to raw vegetables before preservation processes like canning, freezing, or drying. Its main objective is the inactivation of the enzymes in the vegetables that are responsible for deterioration reactions, such as undesirable color, flavor, or texture changes. Blanching may be accomplished with hot water or steam. Some products are blanched in hot water to preserve color and texture. Steam blanching is quicker than hot water blanching and there is less leaching of nutrients from products during steam blanching. All blanchers are energy intensive. Steam blanchers require 75%–90% less energy than hot water blanchers. Normally 2–8 kg of steam is required to process 1 kg of products (Gould, 1996).

Most blanched products are immediately cooled following blanching. The most efficient blanching system is the integrated blancher/cooler developed by the Danish company Cabinplant International S.A. in 1981 (Arrogui et al., 2003). The integrated blancher/cooler has three sections: preheating, blanching, and precooling. A perforated belt carries the vegetables through all blancher sections. Vegetables are heated by hot water in the preheating section to a temperature that will assure a correct blanching treatment. During preheating, the hot water, usually recovered from the cooling section, is sprayed over the vegetables, collected under the belt, and recirculated in a counterflow in the preheating section. In the blanching section, the vegetables are heated quickly using hot water at a temperature of 85°C–95°C. In the precooling section, the products are similarly cooled in a counterflow with a tap water. In the integrated blancher/cooler, heat recovered from the cooling section is used to preheat the incoming raw materials. In conventional screw conveyor blanchers, the specific energy consumption is typically 1 MJ/kg of vegetables while the specific energy consumption of an integrated blancher/cooler is only 0.16 MJ/kg of vegetables (Togeby et al., 1986). Changing from steam to hot water blanching has technologically feasible potential savings.

13.4.2 Energy Conservation in Pasteurization and Sterilization

Thermal processing is an important method for the preservation of canned fruits and vegetables. Waheed et al. (2008) reports that the types of energy used in the manufacturing of orange juice in Nigeria were 18.51% electrical, 80.91% steam, and 0.58% manual. The average energy intensity was found to be 1.12 MJ/kg of canned orange juice manufactured. The most energy-intensive unit operation was pasteurization at 0.932 MJ/kg, followed by packaging at 0.119 MJ/kg. The amount of energy consumed in pasteurization depends on the process with or without a heat recovery unit. Fruit and vegetable juices are traditionally pasteurized by a thermal process in a batch or continuous operation. High-temperature and short-time (HTST) pasteurization has become a common method for thermal treatment of juices. Thermal sterilization of canned foods is a mature technology. Recent developments have been directed toward better energy utilization, more efficient production, automation, lighter, more convenient, more appealing packaging, and better organoleptic quality (Durance, 1997).

Good control, variable retort temperature processing, and good retort insulation can reduce the energy consumption of a retort. Proper control of a thermal process is essential for food safety and energy consumption (Kumar et al., 2001). Variable retort temperature processing can provide a means of increasing product quality and the energy efficiency of the thermal process by improving the uniformity of heat penetration within the product particles in a can (Durance, 1997). Simpson et al. (2006) developed a mathematical model to calculate the energy consumption in a batch retort. They found that retort insulation accounted for 15%–25% of energy loss depending on operating conditions. In batch retort operations, maximum energy demand occurs only at the venting step during the first few minutes of the process cycle while very little energy is needed thereafter in maintaining process temperature. The increase in initial product temperature could reduce the peak energy demand by 25%–35%. It can also reduce the peak energy demand by operating the retorts in a staggered schedule so that no more than one retort is venting at any time.

Since the pasteurization process of fruit juice involves both heating usually with steam or hot water and cooling usually with chilled water or chilled brine solution, a heat pump discussed in Chapter 9 can be used to couple the heating and cooling operations during pasteurization. The hot juice after pasteurization can be cooled by the evaporator of the heat pump while the cold raw juice can be heated by the condenser of the heat pump. A heat pump system was reported to save 66% of the primary energy compared to traditional plate and double jacket pasteurization systems (Ozyurt et al., 2004).

13.4.3 Applications of Nonthermal Pasteurization

Although pasteurization is an efficient method in inactivation of microorganisms in canned fruits, vegetables, and juices, the high-temperature heat used during pasteurization may affect the overall quality of final products (Charles-Rodriguez et al., 2007). Nonthermal pasteurization is a promising alternative to thermal pasteurization. Among the nonthermal pasteurization technologies, pulsed electric field treatment and high-pressure processing have been intensively investigated. Other novel processes such as irradiation (Tran and Farid, 2004; Song et al., 2006; Keyser et al., 2008), ultrafiltration (Ortega-Rivas et al., 1998), and microwave heating (Canumir et al., 2002) have also been studied for pasteurizing fruit juices. Microwave heating is a thermal process but a microwave can penetrate into foods and achieve uniform heating through the foods.

Pulsed electric field treatment, which is discussed in Chapter 19, can be used to inactivate enzyme activity and microorganisms in liquid foods such as fruit and vegetable juices and soups at a low temperature (Qui et al., 1998; Hodgins et al., 2002; Heinz et al., 2003; Elez-Martinez and Martin-Belloso, 2007; Nguyen and Mittal, 2007; Aguiló-Aguayo et al., 2008). Pulsed electric field treatment can retain more high-value thermally sensitive nutrients such as vitamin C and antioxidant in the juices and soups than a thermal pasteurization method (Elez-Martinez and Martin-Belloso, 2007). Pulsed electric field treatment is also used to increase the yield and quality of fruit juices during juice extraction. Schilling et al. (2007) found that the juice yield was increased with increasing the electric field intensities during pulsed electric field treatment of apple mash and the quality of the juice was not significantly affected by the treatment. Pulsed electrical field treatment usually consumes relatively higher energy than traditional thermal treatment with heat recovery ability. The electrical energy consumption can be directly calculated from measured voltage and current. Geveke et al. (2007) found that the electric energy consumption determined by the measured voltage and current was 180 kJ/L for a 3.3 log reduction in *E. coli* at an outlet temperature of 65°C during pulsed electric field treatment of orange juice. The energy consumption of a pulsed electric field process can be reduced by optimizing the design of pulse generation circuits, optimizing the operation of pulsed electric field treatment, and using a regenerative heat exchanger to minimize the cooling requirement. A combined treatment of pulsed electric field and mild-temperature heat for gentle microbial inactivation can also save energy (Heinz et al., 2003).

High-pressure treatment at 40°C–60°C and 200–600 MPa, which is discussed in Chapter 20, has been applied to juice, dairy products, fruits and vegetables, meat products, and fish products (Basak and Ramaswamy, 1996; Parish, 1998; Norton and Sun, 2008). Houska et al. (2006) reported that a high-pressure pasteurization process at a pressure of 500 MPa for 10 min was able to inactivate the viable microorganisms present in raw broccoli juice by more than 5 logs. The high-pressure treated broccoli juices were found to be comparable in sulforaphane content and anti-mutagenic activity with frozen juice. A high-pressure process is environmentally friendly and can retain the fresh-like characteristics of foods better than thermal treatment. High-pressure treatment can be applied to packaged foods. A high-pressure process also provides a potential to reduce the energy requirement for food pasteurization and may contribute to improved energy efficiency in the food industry. However, due to high costs of capital investment, maintenance, and operation for high-pressure equipment, high-pressure pasteurization units with a maximum pressure up to 600 MPa are more suitable for the treatment of highly valuable foods or highly heat-sensitive foods such as flavors, vitamins, and functional biopolymers. It is unlikely that high-pressure processing will replace food canning or freezing in terms of production economics.

13.4.4 Energy Conservation in Freezing of Fruits and Vegetables

Freezing of water consumes a large amount of energy. When water is cooled in a freezer, above the freezing point, the specific heat of liquid water is 4.18 kJ/kg°C. During the change of its phase from liquid to solid at 0°C, a latent heat of 335 kJ/kg must be removed. Below the freezing point, the specific heat of ice is about 2 kJ/kg°C. Refrigeration energy can be conserved by proper maintenance and waste-heat recovery. Installation, maintenance, and operation of doors on cold storage and freezing facilities are always important to save energy. A large portion of the energy losses is caused by the inefficient heat exchange in evaporators and condensers. It is technologically feasible and economically practicable to install heat exchangers to recover heat from refrigeration condenser coils, water towers, and compressor rooms. The optimization of the defrost cycle is desirable and practicable.

The selection of a freezer can have an important effect on energy consumption. In a blast freezer, additional energy is required to operate fans and the resulting heat from the fans increases the refrigeration load. The total power input per unit of frozen products might be two to three times higher in a blast freezer than in a plate freezer for the same freezing rate. However, not all products can be frozen by a plate freezer. The plate freezer requires the products to have a regular shape so that they can be closely held between two plates to achieve a good contact between the product surface and plates.

Since an air blast freezing process requires energy to operate its circulating fans and adds heat to the freezing load, the energy savings by shortening the freezing time and optimizing the air velocity are significant. Harrison and Bishop (1985) proposed that the ratio of the required power by fans divided by the useful refrigeration effect was the best indicator of economical energy usage in a freezing tunnel.

During air blast chilling/freezing, the heat transfer from the cold air to the inside of foods must pass two layers of thermal resistance: the external resistance between cold air and the surface of the foods and the internal resistance inside the solid foods. Biot number is the ratio of the internal resistance to the external resistance. As a general rule for food cooling and freezing, the Biot value should not exceed 5 (Mattarolo, 1976). A high air velocity can be used for cooling or freezing small food items such as peas while a low air velocity (i.e., around 1 m/s) should be used for cooling or freezing large food items such as a beef carcass. Air impingement, water spray, or water immersion chiller can also be used to achieve a Biot value of around 5 for small food items such as peas, and diced fruits and vegetables.

Rapid freezing can maintain the quality and nutrition of foods. Huan et al. (2003) investigated the effect of airflow blockage and guide in spiral quick freezers on energy saving by measuring the velocity distribution in the freezers and analyzing the airflow fields for different designs. Their study shows that the airflow pattern inside freezers plays a key role in energy efficiency, freezing time, and production rate. Through the optimization of the airflow blocking boards and the guide boards, the average air velocity in the freezer would be increased to 2.5–2.7 times compared with the original design. For freezing of bean curd, the freezing time would thus be reduced by 78%–85% and the energy efficiency and production rate would be increased by approximately 18%–28%.

Mechanical compression freezers consume electricity to achieve refrigeration effect. Novel refrigeration cycles based on liquid–liquid absorption, liquid–solid adsorption, and fluid ejection as discussed in Chapter 9, which can be powered by low-grade thermal energy, offer potential energy saving opportunities. Novel refrigeration cycles such as absorption and adsorption refrigeration cycles can be powered by waste heat or other abundant renewable energy sources such as geothermal energy and solar energy at a low temperature (e.g., 70°C) (Wang and Sun, 2001).

13.4.5 Energy Conservation in Concentration and Drying

Concentration and drying are energy-intensive unit operations because of the high latent heat of water evaporation and relatively low energy efficiency of industrial evaporators and dryers. Concentration can be achieved by evaporation or a membrane process. Fruit juices have been traditionally concentrated by vacuum evaporation. In order to decrease energy demand during evaporation, multiple stage evaporators are usually used. It was found that changing from four-effect evaporators to five-effect evaporators could save 20% energy. However, the number of evaporators in series should also be determined by the economics of the process affected by the capital investment (Casper, 1977). Thermal vapor recompression or mechanical vapor recompression systems may be used to recover the heat from the vapor from the evaporators in series. Membrane processes have been developed to produce concentrated fruit juice for improved product quality and reduced energy consumption (Molinari et al., 1995; Jiao et al., 2004). Membrane filtration consumes about 0.014–0.036 MJ/kg water removed, which is significantly smaller than the energy consumption during evaporation. The restriction imposed by the membrane filtration is that the membrane concentration can only reach a maximum dry weight of 12%–20%.

Drying is one the oldest methods used for the preservation of vegetables and fruits and production of food ingredients. The common drying methods used in the fruit and vegetable processing sector include cabinet drying, tunnel drying, drum drying, spray drying, thin layer drying (Corzo et al., 2008), vacuum drying (Durance and Wang, 2002; Cui et al., 2004; Regier et al., 2005), and freeze drying (Sun et al., 2007). It is necessary to conduct thermodynamic analyses for the optimization of the design and operation of a drying system in terms of energy and exergy efficiencies (Corzo et al., 2008). Hot air is widely used as a drying media. Air velocity, temperature, relative humidity, and the fraction of air recycled are important parameters in the design and operation of an air-drying system. The specific energy consumption by an air-drying system varies significantly with the operating parameters. Koyuncu et al. (2004) found that the energy requirement was 6.47 kWh/kg (or 23.3 MJ/kg) and 25.25 kWh/kg (or 90.9 MJ/kg) for drying of chestnuts at 50°C and 0.5 m/s, and 40°C and 1.0 m/s in a parallel air flow dryer, respectively. Sarsavadia (2007) reported that the specific energy consumption during drying of onion was between 23.5 and 62.1 MJ/kg for a hot air drying process with an airflow rate of 2.43–8.09 kg/min and air temperature of 55°C–75°C. The maximum saving in total energy up to 70.7% could be achieved by recycling the exhaust air up to 90% in the hot air dryer. The energy consumption for spray drying can also be reduced by re-circulating air. Velic et al. (2003) found that the maximum ratio of recirculation air was 60% and the fuel saving was approximately 14% for a spray drying process.

A heat pump, which is discussed in Chapter 9, has been used to increase the drying efficiency of a convectional drying process (Perera and Rahman, 1997; Braun et al., 2002). A heat pump dehumidifier is used to dry and heat air. Like a refrigerator, a heat pump dehumidifier consists of a condenser for high-temperature heat exchange, an evaporator for low-temperature heat exchange, a compressor to increase the pressure of its working fluid, and an expansion valve to decrease the pressure of the working liquid. The evaporator of a heat pump is used to remove the moisture from moist air exiting from a drying chamber, and the condenser of the heat pump is used to increase the temperature of the dry air from the evaporator. The hot and dry air can then be re-used in the drying chamber. A fan is usually used to provide air movement in the drying chamber (Perera and Rahman, 1997). Heat pump dehumidifier dryers are more energy efficient than conventional hot-air dryers. Perera and

Rahman (1997) compared the performance of heat pump dehumidifier dryers with hot-air and vacuum dryers. The energy efficiency of a heat pump dehumidifier dryer can be as high as 95%. For a heat pump dehumidifier drying system, electrical energy is required to drive the compressor and air fan. Furthermore, additional energy is required to heat products and the drying chamber and to replace any energy loss through the system. Since the operations of compressors and fans generate heat, the compressors and fans can be placed in the drying chamber so that the heat produced by them can be used for the drying process.

An air cycle heat pump dryer is another design. Compared to conventional vapor compression heat pumps, the air cycle heat pump, which is based on Brayton cycle, uses air as its working fluid, removing moisture and recovering heat from the air that leaves the dryer and adding heat to the air that enters the dryer. The cost of an air cycle heat pump dryer would be less than that of a vapor compression heat pump dryer. An air cycle heat pump dryer can improve the energy efficiency by 40% over an electric dryer (Braun et al., 2002).

Microwaves can provide an efficient means to transfer energy into foods so that the moisture inside the foods can be quickly removed. Microwave assisted air drying is a rapid dehydration technique that can be applied to specific foods, particularly fruits and vegetables (Zhang et al., 2006; Orsat et al., 2007; Ozbek and Dadali, 2007; Gowen et al., 2008). Microwaves can be applied at the final stage of hot-air drying to reduce the drying time while it may be too costly to remove water using electrical energy at the constant-rate drying period (Funebo and Ohlsson, 1998). Microwave assisted hot air drying even at a very low microwave power input can reduce the drying time to half or a third of the hot-air drying time for drying different vegetables such as potatoes, mushrooms and carrots (Riva et al., 1991; Bouraoui et al., 1994; Prabhanjan et al., 1995). Sharma and Prasad (2006) reported that a microwave power of 40 W, air temperature of 70°C, and air velocity of 1 m/s have the lowest specific energy consumption for combined microwave and hot air drying of garlic cloves from an initial moisture content of 185% to a final moisture content of 6% on a dry basis. Microwave has also been used as an energy source in vacuum drying to reduce drying time and to improve energy efficiency and product quality (Durance and Wang, 2002; Cui et al., 2004; Regier et al., 2005). However, currently the price of a microwave-assisted vacuum dryer is $5–$12/watt of microwave heating capacity, compared to $0.5–$1/watt of hot air capacity (Durance and Wang, 2002). Microwave has also been used as an energy source in other drying methods such as freeze-drying (Sun et al., 2007).

Solar energy, geothermal, and waste heat can be used as energy in drying (Ivanova et al., 2003). Sarsilmaz et al. (2000) investigated a rotary column cylindrical dryer equipped with a specifically designed solar collector for drying of apricots. Since the availability of solar energy depends on the time of the day and varies in different seasons, it is necessary to store excess solar energy and release it when the energy availability is inadequate or when energy is not available (Devahastin and Pitaksuriyarat, 2006). The temperature of waste-heat sources and renewable energy sources such as geothermal sources may be too low to be used in a drying process. In this case, a heat pump can be used to extract heat from low-temperature energy sources and upgrade it to a higher temperature for drying (Kuzgunkaya and Hepbasli, 2007).

13.5 ENERGY UTILIZATION OF VEGETABLE AND FRUIT PROCESSING WASTES

13.5.1 Vegetable and Fruits Processing Wastes

In the United States, the fruit and vegetable processing industry generates approximately 11 million tons of by-product wastes along with over 4.3×10^{11} L of effluent wastewaters. Discrete solids include culls, leaves, trimmings, stems, peels, pods, husks, cobs, silk, and defective parts of processed vegetables and fruits. Fruit and vegetable processing wastes have not been used as a feedstock for ethanol production despite many favorable features. The discrete solids are usually dried, pelletized, and sold as low-value animal feeds. The production of animal feeds from discrete solid wastes consumes a large amount of energy. The prices of animal feeds often are not high enough to cover the production costs (Wilkins et al., 2007).

Vegetable and fruit processing wastes contain mainly starch, cellulose, hemicellulose, soluble sugars, and organic acids. Research has already been done to develop high-value energy products such as ethanol and biogas from fruit and vegetable processing wastes. Fruit and vegetable processing wastes are characterized by high percentages of moisture (e.g., >80%) and volatile solids (e.g., >95%) and a very high biodegradability. High-moisture vegetable and fruit processing wastes are difficult to be converted into heat and power, or fuels in a thermochemical conversion process without an auxiliary drying process, as discussed in Chapter 27. The wet fruit and vegetable processing waste can be converted into biogas through an anaerobic digestion process or ethanol through a fermentation process. The details of anaerobic digestion and fermentation of food processing wastes are discussed in Chapters 24 and 25, respectively.

13.5.2 Anaerobic Digestion of Vegetable and Fruit Wastes

Vegetable and fruit processing wastes can be treated either aerobically or anaerobically to reduce the pollutant and pathogen risk and recover energy from the wastes. An aerobic process is to convert the wastes to carbon dioxide and water by aerobic respiration. Aerobically treated organic wastes could be used as a fertilizer (Van Heerden et al., 2002). Anaerobic digestion is a biological process, in which organic matter is degraded to a gaseous mixture of biogas in the absence of oxygen. Biogas, which mainly consists of methane and carbon dioxide, can be used for energy to replace fossil natural gas. Methane could be as high as 60% by volume (Kalia et al., 2000). If biogas produced by anaerobic digestion of biomass is used for electricity generation, the overall conversion efficiency from biomass to electricity is about 10%–16% (McKendry, 2002).

One great advantage of anaerobic digestion is that it can be used to treat very wet and pasty organic wastes or liquid wastes. Anaerobic digestion is a commercially proven technology and is widely used for treating high moisture content organic wastes (i.e., >80%–90% moisture) (McKendry, 2002). For treatment of food processing wastes, anaerobic digestion can not only produce methane for energy but also destroy pathogenic bacteria present in the wastes and reduce pollutant emission from the wastes. Recent advances in anaerobic digestion technologies have made them possible to compete with other methods for the treatment of diverse food processing wastes.

The methane yield of anaerobic digestion of fruit and vegetable processing wastes is very high. The CH_4 yield in anaerobic digestion of fruit and vegetable solid wastes was reported to be as high as 0.53 m^3/kg VS with 100% VS conversion for digestion of damaged banana (Gunaseelan, 1997). Linke (2006) found that both biogas yield and CH_4 in the biogas for anaerobic digestion of solid wastes from potato processing in a completely stirred tank reactor decreased with the increase in organic loading rate. For the organic loading rate in the range from 0.8 to 3.4 kg/m^3 day, biogas yield and CH_4 obtained were from 0.85 to 0.65 m^3/kg and from 58% to 50%, respectively. Knol et al. (1978) found that the maximum loading rate for stable digestion of a variety of fruit and vegetable solid wastes ranges from 0.8 to 1.6 kg VS/m^3 day.

13.5.3 Production of Fermentable Sugars from Vegetable and Fruit Processing Wastes

Ethanol has been a key industrial chemical for many years. In the recent years, fuel ethanol fuel has been considered to be more environmentally friendly than fossil fuels. Fuel ethanol has been seen as a replacement for gasoline. Food processing wastes can be explored as the potentially low-cost and abundant feedstock for production of ethanol. Most of the carbohydrate present in fruit and vegetable wastes is composed of soluble sugars and easily hydrolysable polysaccharides. There are two waste streams of solid wastes and effluent wastewater from fruit and vegetable processing facilities, which can be used to produce fermentable sugars.

Apple pomace is the main by-product from apple cider and juice processing facilities and accounts for about 25% of the original fruit mass. Apple pomace typically contains between 66.4%–78.2% moisture and 9.5%–22.0% carbohydrates. Fermentable sugars in apple pomace such as glucose, fructose, and sucrose can be converted to ethanol (Hang et al., 1981; Hang, 1987; Ngadi and Correia, 1992). It was observed that ethanol yields on glucose from the apple pomace and yeast cells were from 0.33 to 0.37 g/g and from 2.6×10^{-8} g/colony forming unit (CFU) to 3.14×10^{-8} g/CFU, respectively. As ethanol concentration increased from 1% to 18% g/g (by dry weight), the specific growth rate decreased almost to zero because ethanol inhibited the performance of yeast (Ngadi and Correia, 1992).

During grapefruit processing, about half of the fruit is expressed as juice and the remaining residue is the peel waste consisting of peels, seeds, and segment membranes (Braddock, 1999). Cellulose, pectin, and hemicellulose in grapefruit peel waste can be hydrolyzed by pectinase and cellulase enzymes to monomer sugars, which can be used by microorganisms to produce ethanol and other fermentation products. Wilkins et al. (2007) found that 5 mg pecinase/g peel dry matter and 2 mg cellulase/g peel dry matter supplemented with 2.1 mg β-glucosidase/g peel dry matter at an optimum pH value of 4.8 and a temperature of 45°C were the lowest loadings to yield the most glucose from grapefruit peel waste.

Reverse osmosis method provides a promising potential to recover 1.42 million tons of fermentable sugars at a 20% sugar concentrate from the effluent wastewater in the fruit and vegetable processing sector in the United States. The recovered sugar concentrate can then be used for production of liquid fuels. Because reverse osmosis processing is relatively expensive, excessively dilute wastewater that contains less than 0.2% sugars is too costly to be recovered. The potential for the production of ethanol from the recovered sugars is between 750 and 900 million liters per year in the United States. Economic analysis showed that the overall ethanol production costs should, on average, be only 40% of current ethanol prices in the United States. Furthermore, the recovery of sugars from the wastewater can reduce the biological oxygen demand of the wastewater and thus reduce the disposal cost (Blondin et al., 1983).

Recently developed solid-state fermentation technology can directly use water-insoluble materials for microbial growth and metabolism. It is usually carried out in solid or semi-solid systems in the near absence of free water or reduced water content compared with traditional fermentation. Solid fermentation has already been investigated to produce ethanol from fruit and vegetable residues (Laufenberg et al., 2003).

SUMMARY

Large amounts of fruits and vegetables are processed into canned, frozen, or dried products. The preparation of vegetables and fruits for freezing, drying, and canning is the same. The difference among these processes is the preservation and packaging methods. Energy efficiencies of different unit operations, particularly blanching, pasteurization, sterilization, freezing, concentration, and drying in the fruit and vegetable processing sector have been improved. Many novel energy saving technologies such as the integrated blancher and cooler, nonthermal pasteurization, novel refrigeration cycles, membrane concentration, heat pumps, and microwaves have been introduced into the fruit and vegetable processing sector. Fruit and vegetable processing facilities generate large amounts of processing wastes. These processing wastes usually have high moisture and carbohydrate contents. These processing wastes can be converted into biogas through an anaerobic digestion process for heat and power generation or ethanol through a fermentation process.

REFERENCES

Aguiló-Aguayo, I., I. Odriozola-Serrano, L.J. Quintão-Teixeira, and O. Martín-Belloso. 2008. Inactivation of tomato juice peroxidase by high-intensity pulsed electric fields as affected by process conditions. *Food Chemistry* 107: 949–955.

Arrogui, C., A. Lopez, A. Esnoz, and P. Virseda. 2003. Mathematic model of an integrated blancher/cooler. *Journal of Food Engineering* 59: 297–307.

Basak, S. and H.S. Ramaswamy. 1996. Ultra high pressure treatment of orange juice: A kinetic study on inactivation of pectin methyl esterase. *Food Research International* 29: 601–607.

Blondin, G.A., S.J. Comiskey, and J.M. Harkin. 1983. Recovery of fermentable sugars from process vegetable wastewaters. *Energy Agriculture* 2: 21–36.

Bouraoui, M., P. Richard, and T. Durance. 1994. Microwave and convective drying of potato slices. *Journal of Food Process Engineering* 17: 353–363.

Braddock, R.J. 1999. *Handbook of Citrus By-Products Processing Technology*. New York: John Wiley and Sons.

Braun, J.E., P.K. Bansal, and E.A. Groll. 2002. Energy efficiency analysis of air cycle heat pump dryers. *International Journal of Refrigeration* 25: 954–965.

Canumir, J.A., J.E. Celis, J. de Bruijn, and L.V. Vidal. 2002. Pasteurization of apple juice by using microwave. *Lebensmittel-Wissenschaft und-Technologie* 35: 389–392.

Casper, M.E. 1977. *Energy-Saving Techniques for the Food Industry.* New Jersey: Noyes Data Corporation.

Charles-Rodriguez, A.V., G.V. Nevarez-Moorillon, Q.H. Zhang, and E. Ortega-Rivas. 2007. Comparison of thermal processing and pulsed electric fields treatment in pasteurization of apple juice. *Food and Bioproducts Processing* 85: 93–97.

Corzo, O., N. Bracho, A. Vasquez, and A. Pereira. 2008. Energy and exergy analyses of thin layer drying of coroba slices. *Journal of Food Engineering* 86: 151–161.

Cui, Z.W., S.Y. Xu, and D.W. Sun. 2004. Microwave-vacuum drying kinetics of carrot slices. *Journal of Food Engineering* 65: 157–164.

Devahastin, S. and S. Pitaksuriyarat. 2006. Use of latent heat storage to conserve energy during drying and its effect on drying kinetics of a food product. *Applied Thermal Engineering* 26: 1705–1713.

Durance, T.D. 1997. Improving canned food quality with variable retort temperature processes. *Trends in Food Science & Technology* 8: 113–118.

Durance, T.D. and J.H. Wang. 2002. Energy consumption, density, and rehydration rate of vacuum microwave-hot-air convection-dehydrated tomatoes. *Journal of Food Science* 67: 2212–2216.

Elez-Martinez, P. and O. Martin-Belloso. 2007. Effects of high intensity pulsed electric field processing conditions on vitamin C and antioxidant capacity of orange juice and gazpacho, a cold vegetable soup. *Food Chemistry* 102: 201–209.

Funebo, T. and T. Ohlsson. 1998. Microwave-assisted air dehydration of apple and mushroom. *Journal of Food Engineering* 38: 353–367.

Geveke, D.J., C. Brunkhorst, and X. Fan. 2007. Radio frequency electric fields processing of orange juice. *Innovative Food Science & Emerging Technologies* 8: 549–554.

Gould, W.A. 1996. *Unit Operations for the Food Industries.* Timonium, Maryland: CTI Publications, Inc.

Gowen, A.A., N. Abu-Ghannam, J. Frias, and J. Oliveira. 2008. Modeling dehydration and rehydration of cooked soybeans subjected to combined microwave-hot-air drying. *Innovative Food Science & Emerging Technologies* 9: 129–137.

Gunaseelan, V.N. 1997. Anaerobic digestion of biomass for methane production: A review. *Biomass and Bioenergy* 13: 83–114.

Hang, Y.D. 1987. Production of fuels and chemicals from apple pomace. *Food Technology* 41: 115–117.

Hang, Y.D., C.Y. Lee, E.E. Woodams, and H.J. Cooley. 1981. Production of alcohol from apple pomace. *Applied Environmental Microbiology* 42: 1128–1129.

Harrison, M.A. and P.J. Bishop. 1985. Parametric study of economical energy usage in freezing tunnels. *International Journal of Refrigeration* 8: 29–36.

Heinz, V., S. Toepfl, and D. Knorr. 2003. Impact of temperature on lethality and energy efficiency of apple juice pasteurization by pulsed electric fields treatment. *Innovative Food Science and Emerging Technologies* 4: 167–175.

Hodgins, A.M., G.S. Mittal, and M.W. Griffiths. 2002. Pasteurization of fresh orange juice using low-energy pulsed electrical field. *Journal of Food Science* 67: 2294–2299.

Houska, M., J. Strohalm, K. Kocurova, J. Totusek, D. Lefnerova, J. Triska, N. Vrchotova, V. Fiedlerova, M. Holasova, D. Gabrovska, and I. Paulickova. 2006. High pressure and foods-fruit/vegetable juices. *Journal of Food Engineering* 77: 386–398.

Huan, Z., Y. Ma, and S. He. 2003. Airflow blockage and guide technology on energy saving for spiral quick-freezer. *International Journal of Refrigeration* 26: 644–651.

Ivanova, D., K. Enimanev, and K. Andonov. 2003. Energy and economic effectiveness of a fruit and vegetable dryer. *Energy Conservation and Management* 44: 763–769.

Jiao, B., A. Cassano, and E. Drioli. 2004. Recent advances on membrane processes for the concentration of fruit juices: a review. *Journal of Food Engineering* 63: 303–324.

Kalia, V.C., V. Sonakya, and N. Raizada. 2000. Anaerobic digestion of banana stem waste. *Bioresource Technology* 73: 191–193.

Keyser, M., I. Muller, F.P. Cilliers, W. Nel, and P.A. Gouws. 2008. Ultraviolet radiation as a nonthermal treatment for the inactivation of microorganisms in fruit juice. *Innovative Food Science & Emerging Technologies* 9: 348–354.

Knol, W., M.M. van der Most, and J. de Waart. 1978. Biogas production by anaerobic digestion of fruit and vegetable waste. A preliminary study. *Journal of Science of Food and Agriculture* 29: 822–830.

Koyuncu, T., U. Serdar, and I. Tosun. 2004. Drying characteristics and energy requirement for dehydration of chestnuts (*Castanea sativa* Mill). *Journal of Food Engineering* 62: 165–168.

Kumar, M.A., M.N. Ramesh, and S.N. Rao. 2001. Retrofitting of a vertical retort for on-line control of the sterilization process. *Journal of Food Engineering* 47: 89–96.

Kuzgunkaya, E.H. and A. Hepbasli. 2007. Exergetic performance assessment of a ground-source heat pump drying system. *International Journal of Energy Research* 31: 760–777.

Laufenberg, G., B. Kunz, and M. Nystroem. 2003. Transformation of vegetable waste into value added products: (A) the upgrading concept; (B) practical implementations. *Bioresource Technology* 87: 167–198.

Linke, B. 2006. Kinetic study of thermophilic anaerobic digestion of solid wastes from potato processing. *Biomass and Bioenergy* 30: 892–896.

Mattarolo, L. 1976. Technical and engineering aspects of refrigerated food preservation. Annex 1976-1, pp. 49–69, International Institute of Refrigeration, Paris, France.

McKendry, P. 2002. Energy production from biomass (part 2): Conversion technologies. *Bioresource Technology* 83: 47–54.

Molinari, R., R. Gagliardi, and E. Drioli. 1995. Methodology for estimating saving of primary energy with membrane operations in industrial processes. *Desalination* 100: 125–137.

Ngadi, M.O. and L.R. Correia. 1992. Kinetics of solid-state ethanol fermentation from apple pomace. *Journal of Food Engineering* 17: 97–116.

Nguyen, P. and G.S. Mittal. 2007. Inactivation of naturally occurring microorganisms in tomato juice using pulsed electric field with and without antimicrobials. *Chemical Engineering and Processing* 46: 360–365.

Norton, T. and D.W. Sun. 2008. Recent advances in the use of high pressure as an effective processing technique in the food industry. *Food and Bioprocess Technology* 1: 2–34.

Orsat, V., W. Yang, V. Changrue, and G.S.V. Raghavan. 2007. Microwave-assisted drying of biomaterials. *Food and Bioproducts Processing* 85: 255–263.

Ortega-Rivas, E., E. Zarate-Rodriguez, and G.V. Barbosa-Canovas. 1998. Apple juice pasteurization using ultrafiltration and pulsed electric fields. *Food and Bioproducts Processing* 76: 193–198.

Ozbek, B. and G. Dadali. 2007. Thin-layer drying characteristics and modeling of mint leaves undergoing microwave treatment. *Journal of Food Engineering* 83: 541–549.

Ozyurt, O., O. Comakli, M. Yilmaz, and S. Karsli. 2004. Heat pump use in milk pasteurization: An energy analysis. *International Journal of Energy Research* 28: 833–846.

Parish, M.E. 1998. Orange juice quality after treatment by thermal pasteurization or isostatic high pressure. *Lebensmittel-Wissenschaft und-Technologie* 31: 439–442.

Perera, C.O. and M.S. Rahman. 1997. Heat pump dehumidifier drying of food. *Trends in Food Science and Technology* 8: 75–79.

Prabhanjan, D.G., H.S. Ramaswamy, and G.S.V. Raghavan. 1995. Microwave-assisted convective air drying of thin layer carrots. *Journal of Food Engineering* 25: 283–293.

Qui, X., S. Sharma, L. Tuhela, M. Jia, and Q.H. Zhang. 1998. An integrated PEF pilot platn for continuous nonthermal pasteurization of fresh orange juice. *Transactions of the ASAE* 41: 1069–1074.

Regier, M., E. Mayer-Miebach, D. Behsnilian, E. Neff, and A. Schuchmann. 2005. Influences of drying and storage of lycopene-rich carrots on the carotenoid content. *Drying Technology* 23: 989–998.

Riva, M., A. Schiraldi, and L. Di Cesare. 1991. Drying of *Agaricus biosporus* mushroom by microwave-hot air combination. *Lebensmittel Wissenschaft und Technologie* 24: 479–482.

Sarsavadia, P.N. 2007. Development of a solar-assisted dryer and evaluation of energy requirement for the drying of onion. *Renewable Energy* 32: 2529–2547.

Sarsilmaz, C., C. Yildiz, and D. Pehlivan. 2000. Drying of apricots in a rotary column cylindrical dryer supported with solar energy. *Renewable Energy* 21: 117–127.

Schilling, S., T. Alber, S. Toepfl, S. Neidhart, D. Knorr, A. Schieber, and R. Carle. 2007. Effects of pulsed electric field treatment of apple mash on juice yield and quality attributes of apple juices. *Innovative Food Science & Emerging Technologies* 8: 127–134.

Sharma, G.P. and S. Prasad. 2006. Optimization of process parameters for microwave drying of garlic cloves. *Journal of Food Engineering* 75: 441–446.

Simpson, R., C. Cortes, and A. Teixeira. 2006. Energy consumption in batch thermal processing: Model development and validation. *Journal of Food Engineering* 73: 217–224.

Song, H.P., D.H. Kim, C. Jo, C.H. Lee, K.S. Kim, and M.W. Byun. 2006. Effect of gamma irradiation on the microbiological quality and antioxidant activity of fresh vegetable juice. *Food Microbiology* 23: 372–378.

Sun, D.W. and L.J. Wang. 2001. Novel refrigeration cycles, in: Sun, D.W. (Ed.), *Advances in Food Refrigeration*, Chapter 1, pp. 1–69, Leatherhead Publishing: Leatherhead, U.K.

Sun, H., H. Zhu, H. Feng, and L. Xu. 2007. Thermoelectromagnetic coupling in microwave freeze-drying. *Journal of Food Process Engineering* 30: 131–149.

Togeby, M., N. Hansen, E. Mosekilde, and K.P. Poulsen. 1986. Modeling energy consumption, loss of firmness and enzyme inactivation in an industrial blanching process. *Journal of Food Engineering* 5: 251–267.

Tran, M.T.T. and M. Farid. 2004. Ultraviolet treatment of orange juice. *Innovative Food Science & Emerging Technologies* 5: 495–502.

U.S. Census Bureau. 2006. *2006 Annual Survey of Manufactures.* Available at: http://factfinder.census.gov.

Van Heerden, I., C. Cronje, S.H. Swart, and J.M. Kotze. 2002. Microbial, chemical and physical aspects of citrus waste composting. *Bioresource Technology* 81: 71–76.

Velic, D., M. Bilic, S. Tomas, and M. Planinic. 2003. Simulation, calculation and possibilities of energy saving in spray drying process. *Applied Thermal Engineering* 23: 2119–2131.

Waheed, M.A., S.O. Jekayinfa, J.O. Ojediran, and O.E. Imeokparia. 2008. Energetic analysis of fruit juice processing operations in Nigeria. *Energy* 33: 35–45.

Wilkins, M.R., W.W. Widmer, K. Grohmann, and R.G. Cameron. 2007. Hydrolysis of grapefruit peel waste with cellulase and pectinase enzymes. *Bioresource Technology* 98: 1596–1601.

Zhang, M., J. Tang, A.S. Mujumdar, and S. Wang. 2006. Trends in microwave-related drying of fruits and vegetables. *Trends in Food Science & Technology* 17: 524–534.

14 Energy Conservation in Dairy Processing Facilities

14.1 INTRODUCTION

The dairy processing sector (NAICS code 3115) comprises establishments that manufacture dairy products from raw milk, processed milk, and dairy substitutes. Dairy processing facilities range in size from very small family-owned, single plant operations manufacturing one or two products to large multimillion dollar corporations operating several plants and processing a wide range of products. The main dairy products include fluid milk, cheese, and butter cheese from raw milk and processed milk, ice cream, frozen yogurts, and dry, condensed, concentrated, and evaporated dairy and dairy substitute products.

In the dairy processing sector, large amounts of fuels are used for steam generation, and large portion of purchased electricity is used for motors and air compressors. Energy conservation in steam generation and distribution is discussed in Chapter 4. Energy conservation technologies for air compressors and motors are discussed in Chapters 5 and 6, respectively. Heat exchangers are widely used in the dairy processing sector. Energy conservation technologies for heat exchangers are discussed in Chapter 7. Boilers, heat exchangers, motors, and air compressors may generate waste heat. Waste-heat recovery technologies are discussed in Chapter 8. Since both processing heat and electricity are required by the dairy processing sector, a combined heat and power generation system can be used to efficiently and economically provide electricity or mechanical power and useful heat from the same primary energy source. Combined heat and power systems are discussed in Chapter 9.

In this chapter, the main dairy products and their processes are reviewed. The energy uses and conservation in the main unit operations in the dairy processing sector are discussed. Finally, conversion of dairy processing wastes into energy is discussed.

14.2 MAIN PRODUCTS AND UNIT OPERATIONS IN THE DAIRY PROCESSING SECTOR

The main unit operations and products in the dairy sector are given in Figure 14.1 (Ramirez et al., 2006b). Fluid milk, cheese, butter, ice cream, and dried milk powders are the main dairy products. Whey is a main by-product from the cheese production process in the dairy sector. Heating, cooling, concentration, drying, and cleaning are

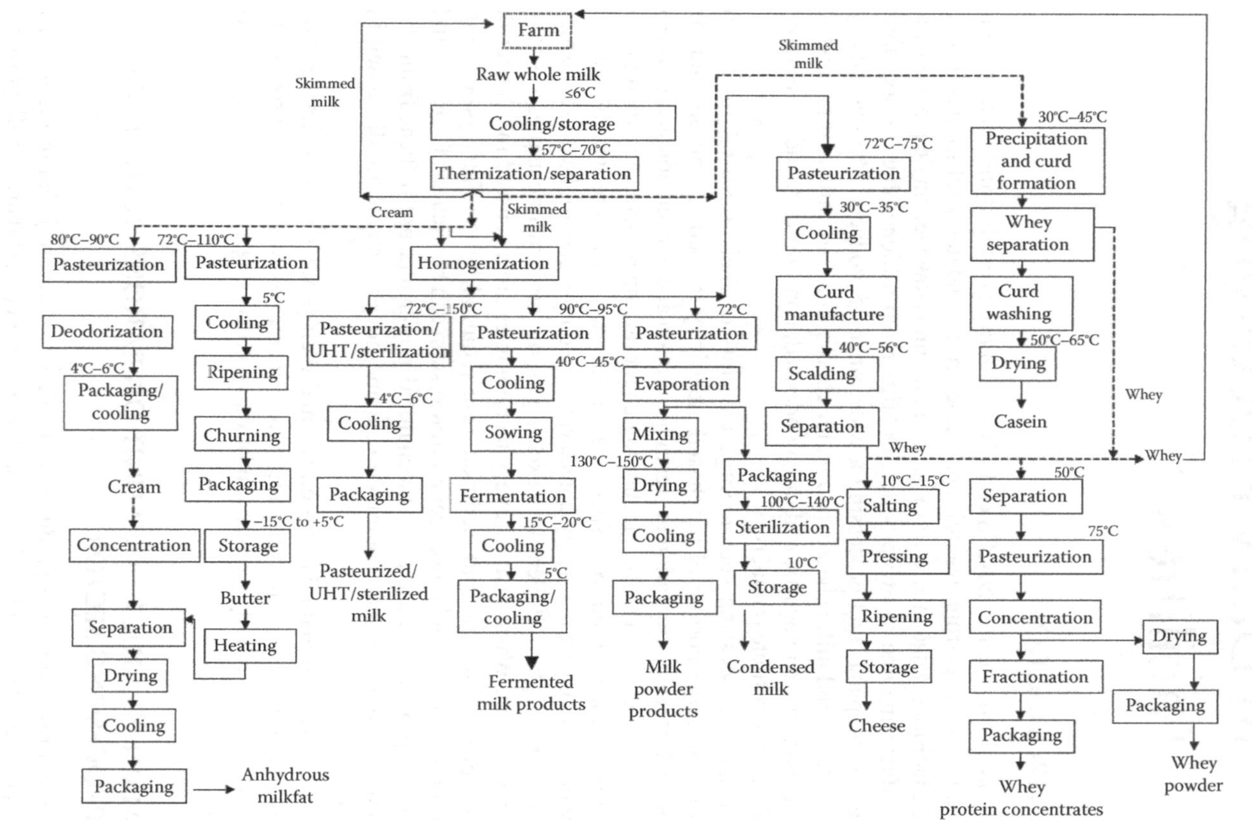

FIGURE 14.1 A flow sheet of dairy products and processes in the dairy industry. (Reprinted from Ramirez, C.A., Patel, M., and Blok, K., *Energy*, 31, 1984, 2006b. With permission.)

five main operations in the dairy sector. These five main operations account for about 50% and 96% of energy consumption in production of fluid milk and dry milk products, respectively (Ramirez et al., 2006b).

14.2.1 Fluid Milk and Its Process

Fluid milk is the dominant product from the dairy sector. Liquid milk is one of the important dairy products. The fat content of raw milk is 4% or higher. Low-fat or fat free liquid milk products are produced by partially skimming or completely skimming the raw milk. Liquid milk can be pasteurized at 72°C for 15 s, sterilized at 115°C for 20 min, or treated at 138°C–150°C for 1–4 s. Condensed milk is also called as concentrated milk. Sugar may be added to condensed milk. Vitamins such as vitamin A and D may also be added in fluid milk.

14.2.2 Ice Cream, Butter, and Cheese and Their Processes

The fat-rich stream from the skimming process, which is used to produce low-fat or fat free milk products, is cream. The fat content of the cream is usually from 35% to 45%. The cream is further used to produce ice cream and butter. Cream must be pasteurized to ensure that the products are free of pathogenic bacteria. Low-fat cream can be homogenized but heavy cream is usually not homogenized due to high fat contents and thus high viscosity. For production of butter, the cream is ripened. Sometimes, cultures are added into the cream to ferment sugars in the cream into lactic acid and desirable flavor and aroma for production of cultured butter. The ripened cream is further held at a cool temperature to crystallize the fat globules for around 15 h. Cream is then agitated and butter granules are formed in a churning process. At the end of churning, there are two phases left: semisolid butter and liquid buttermilk. The liquid buttermilk is drained out of the butter. Salt may be added to the butter to improve its flavor and extend its shelf life.

Cheese is a milk concentrate. The main solids in cheese are protein and fat. Cheese can be categorized according to the moisture content, the fat content, or the curing characteristics. Rennet or natural cheese is manufactured straight from milk by using bacteria culture and proteolytic enzymes. Lactic bacteria convert milk sugars, mainly lactose, into lactic acid. The proteolytic enzymes or rennet cause the coagulation of the proteins in milk, mainly casein proteins into curd. The curd is then heated to remove part of its moisture. The liquid is then drained out of the curd and the liquid residue is whey. Fresh cheese thus has a high degree of acidity and is not subjected to a proteolytic ripening process. Processed cheese is made from rennet cheese and subjected to further thermal treatment to increase its shelf stability (Ramirez et al., 2006b).

14.2.3 Dry Milk Products and Processes

Dry milk products include whole milk powder, which typically contains 2%–5% water content. Fat free milk powder contains 2% or less moisture and 1.5% or less milk fat. Whey is the liquid residue of cheese and casein production. Milk powder is an important dairy product, which can be used as a compensation for low fresh milk supplies. It can also be used to develop a number of modern food products such as whey powders, baby-food products, coffee whiteners, dried caseins, and ice-cream powders. Milk powder is produced in two steps: vacuum evaporation and spray drying to remove most of the water from liquid milk. Vacuum evaporation can improve the final product quality and reduce energy consumption. The energy consumption in modern multiply-effect evaporators with stream recompression is about 10 times lower than in spray drying (Caric, 1994). Milk powder manufacturing process has the highest energy demand per ton of finished products in the whole dairy sector (Knipschildt and Andersen, 1994).

14.3 ENERGY USE IN THE DAIRY SECTOR

In the United States, the dairy product manufacturing sector had a total shipment value of 75.43 billion dollars while the total energy cost was 1.18 billion dollars in 2006. It required 1.57 ¢ energy to generate one dollar shipment value in the dairy processing sector. The energy cost in the dairy sector was about 11.9% of the total energy cost in the whole food industry while the total shipment value was 14.0% of the total shipment value from the whole food industry in the United States. Electricity cost was 52.4% of total energy costs (U.S. Census Bureau, 2006).

In the Netherlands, the energy demand for the dairy sector was 15% of the total demand in the food industry in 2001 (Ramirez et al., 2006a). Energy use for production of different dairy products in the Netherlands in 2001 is given in Table 14.1 (Ramirez et al., 2006a). The large use of fuels is primarily for drying milk at a specific fuel consumption rate of 9.385 MJ/kg of product and whey drying at a specific fuel consumption rate of 9.87 MJ/kg of product. The major uses of electricity are for the processing of products, refrigeration, and air compressors. Among the many unit operations used in the dairy sector, pasteurization, cooling, and drying are the major energy users. The main distribution of energy by unit operations is given in Table 14.2 (Ramirez et al., 2006b).

TABLE 14.1
Energy Use for Production of Different Dairy Products in the Netherlands in 2001

Product	Specific Electricity Consumption	Specific Fuels and Heat Consumption	Unit
Milk and fermented products	241	524	MJ/t product
Butter	457	1285	MJ/t product
Milk powder	1051	9385	MJ/t product
Condensed milk	295	1936	MJ/t product
Cheese	1206	2113	MJ/t product
Casein and lactose	918	4120	MJ/t product
Whey powder	1138	9870	MJ/t product

Source: Data taken from Ramirez, C.A., Blok, K., Neelis, M., and Patel, M., *Energy Policy,* 34, 1720, 2006a. With permission.

TABLE 14.2
Primary Energy Demand in Dutch Dairies in 2000

Product	Operations	Energy Consumption (%)
Fluid milk	Reception, thermization	2
	Storage	7
	Centrifugation/homogenization/pasteurization	38
	Packing	9
	Cooling	19
	Pressurized air	0.5
	Cleaning in place	9.5
	Water provision	6
	Building (lighting and space heating)	9
Cheese	Reception, thermization	19
	Cheese processing	14
	Cheese treatment/storage	24
	Cooling	19
	Pressurized air	5
	Cleaning in place	19
Butter	Cooling	66
	Pressurized air	8
	Cleaning in place	26
Milk powder	Thermization/pasteurization/centrifugation	2.5
	Thermal concentration/evaporation	45
	Drying	51
	Packing	1.5

Source: Adapted from Ramirez, C.A., Patel, M., and Blok, K., *Energy*, 31, 1984–2004, 2006. With permission.

14.4 POTENTIAL ENERGY CONSERVATION MEASURES

14.4.1 Energy Conservation in Pasteurization and Cooling

Various methods have been used to preserve milk products. Thermal treatment is a traditional method used to prevent microbial spoilage. Thermal treatment requires a large thermal demand as shown in Table 14.3. The most common thermal process of milk is pasteurization at a temperature below 100°C. Other processes are sterilization at 115°C–120°C for 20–45 min, ultra high-temperature treatment at 140°C–165°C for a few seconds, radiation with UV, high-pressure treatment at 40°C–60°C and 200–600 MPa, pulsed electric field treatment, and microwave treatment. Among these measures, only sterilization and ultra high-temperature treatment have been widely used. Pasteurization consumes a small amount of energy since 90%–94% of the heat input of an energy-efficient pasteurization process can be recovered. Sterilization and ultra high-temperature treatment, which can be used to preserve dairy products

TABLE 14.3
Effects of Different Preservation Methods for Milk

Method	Shelf Life	Energy Consumption in the Netherlands (10^6 kWh/year)
None	1 day	0
Pasteurization	6 days	13
Microfiltration	>21 days	18
UHT	3 months	73
Sterilization	6 months	166
PEF	Unknown	7

Source: Reprinted from Juriaanse, A.C., *Trends Food Sci. Technol.*, 10, 303, 1999. With permission.

for a period longer than 5 months, have more energy demand than pasteurization (Ramirez et al., 2006b).

Since a pasteurization process of milk involves both heating and cooling stages, Ozyurt et al. (2004) designed a liquid-to-liquid heat pump for the pasteurization of milk as discussed in Chapter 9. The heat source for the evaporator of the heat pump is hot pasteurized milk while the heat sink for the condenser of the heat pump is cold raw milk. The raw milk is pre-heated in an economizer by the hot milk from the condenser of the heat pump, and then sent to the condenser for heating. When the milk reaches the set pasteurization temperature in the condenser, it is sent to the evaporator of the heat pump and cooled to the coagulation temperature. Finally, the milk is sent to a pasteurized milk tank. For the pasteurization temperature at 72°C and coagulation temperature at 32°C, the measured COP of the heat pump ranged from 2.3 to 3.1. The heat pump system saved 66% of the primary energy compared to traditional plate and double jacket milk pasteurization systems (Ozyurt et al., 2004).

Nonthermal treatment or less intensive heat treatment has been developed. Bacteria and spores can be removed from milk at a low temperature by microfiltration. Pulsed electric field (PEF) can also be used to inactivate bacteria and spores at a low temperature (Reina et al., 1998; Bendicho et al., 2002; Craven et al., 2008). At added *Pseudomonas* levels of 10^3 and 10^5 CFU/mL, the microbial shelf life of PEF-treated milk was extended by at least 8 days at 4°C compared with untreated milk. The total microbial shelf life of the PEF-treated milk was 13 and 11 days for inoculation levels of 10^3 and 10^5 CFU/mL, respectively. The results indicate that PEF treatment is useful for the reduction of *Pseudomonads*, the major spoilage bacteria of milk (Craven et al., 2008). The pulsed electric field treatment is discussed in Chapter 19. These nonthermal treatment technologies can conserve the properties of the product and reduce the energy consumption, as shown in Table 14.3 (Juriaanse, 1999).

14.4.2 Energy Conservation in Concentration and Drying

Milk and whey drying processes are large fuel users. Concentration and drying are energy-intensive unit operations because of the high latent heat of water evaporation and relatively low energy efficiency of the industrial evaporators and dryers. Concentration can be achieved by evaporation or membrane filtration. In order to decrease energy demand during evaporation, multiple-stage evaporators are usually used. A multiple-effect evaporator system is a simple, series arrangement of several evaporators, which use steam to remove product moisture by evaporation. The evaporated water vapor from food products are collected and used as the steam for the next evaporator in the series. This collection and reuse of vapor result in smaller energy requirements to remove product moisture (Riberiro and Andrade, 2003). The greater the number of effects in the series, the smaller the energy consumption. It was found that changing from four-effect evaporators to five evaporators can save 20% energy. However, the number of evaporators in series should also be determined by the production economics (Casper, 1977). Thermal vapor recompression or mechanical vapor recompression systems may be used to recover the heat from the vapor from the evaporators in series. Membrane filtration consumes about 0.014–0.036 MJ/kg water removed, which is a significantly smaller amount of energy than evaporation. The restriction imposed by membrane filtration is that the membrane concentration can only reach a maximum dry weight of 12%–20%. Combined membrane filtration and evaporation is thus increasingly used in the dairy processing sector (Ramirez et al., 2006b).

Spray drying is the most popular technique in the dairy sector. In practice, a spray dryer consists of one to three stages. The specific energy demands for 1, 2, and 3 stage spray driers are 4.9, 4.3, and 3.4 MJ/kg water removed, respectively (Ramirez et al., 2006b). Compared to evaporation with waste heat recovery, the energy consumption per kilogram of water removed in spray drying is 10–20 times higher than that in evaporation. Therefore, it is important to concentrate the liquid milk before drying. Heat recovery from the spray drying of milk and whey also offers potential for energy savings. Waste-heat recovery is discussed in Chapter 8.

A heat pump, as discussed in Chapter 9, has been used to increase the drying efficiency of convectional drying. Like a mechanic compression refrigerator, a heat pump dehumidifier consists of a condenser, a compressor, an evaporator, and an expansion valve. A fan is usually used to circulate air through the external heat exchange surface of the evaporator of a heat pump to remove its moisture at a low temperature and then the external heat exchanger surface of the condenser to increase the temperature of dry air and into a drying chamber (Perera and Rahman, 1997; Kiatsiriroat and Tachajapong, 2002; Adapa and Schoenau, 2005). A heat pump can also be used to upgrade a low-temperature energy sources such as waste heat and renewable energy such as geothermal energy for drying (Kuzgunkaya and Hepbasli, 2007).

14.4.3 Energy Conservation through Cleaning of Fouling Layers

Fouling and cleaning of heat exchangers is a serious industrial problem in the dairy sector. It was reported that fouling caused an increase of up to 8% in the energy consumption in fluid milk plants and another 21% of the total energy consumption was used to clean milk pasteurization plants (Ramirez et al., 2006b). Although the chemistry of the fouling process is understood, the interaction between the chemistry and the fluid mechanics in heat exchangers makes it difficult to deal with the fouling problem (Fryer and Belmar-Beiny, 1991; Visser and Jeurnink, 1997). The dairy industry has been confronted with fouling on metal surfaces of plate heat exchangers. The major components in fouling deposits are heat-sensitive whey proteins and heat-induced precipitation of calcium phosphate salts. Fouling may cause an increased pressure drop, heat transfer resistance, and microbial growth at the fouling places (Changani et al., 1997; Visser and Jeurnink, 1997).

14.5 ENERGY UTILIZATION OF DAIRY BY-PRODUCTS

Cheese whey is an abundant source of biomass that is suitable for production of alcohol. Whey, which is the liquid residue of cheese and casein production, is one of the troublesome by-products of the dairy industry. Fluid whey contains about 6%–6.5% solids. Approximately 13 million tons of fluid whey is produced annually in the United States. The main components of dried whey powder include 42%–45% (disaccharides), 20%–23% of protein, 10% of Na, K, and Ca, and 25% of ash (Fischer and Bipp, 2005). The yeast of *Saccharomyces cerevisiae,* which is the microorganism used in the existing ethanol industry, lacks the ability to ferment lactose. Co-immobilized energy of β-galactosidase and yeast of *S. cerevisiae* have been used to produce ethanol from whey (Staniszewski et al., 2007).

14.6 SUMMARY

Thermal processing and cooling are the two main unit operations in pasteurization of milk products in the dairy processing sector. A heat pump can be used to couple the heating and cooling operations during pasteurization of milk. Nonthermal processes such as ultra-filtration, high-pressure, and pulsed electric field treatments are also promising alternatives to traditional thermal pasteurization to reduce the energy demand in the dairy sector. Concentration and drying are two energy-intensive operations used in the dairy industry to produce concentrated milk, milk powder, and whey powders. Multiple-stage evaporators and spray driers can reduce the specific energy consumption in concentration and drying. Novel measures such as membrane filtration can also reduce the energy demand to concentrate milk products. Vapor compressors and heat pumps can be used to recover the waste heat from evaporators and driers. Fouling could increase the energy demand in a liquid milk plant by 8%. Reduction of fouling in heat exchangers can significantly save energy in the dairy sector. Whey, which is rich in lactose and protein, is a by-product from the cheese production process in the dairy industry. It can be used as an energy source to produce fuel ethanol.

REFERENCES

Adapa, P.K. and G.J. Schoenau. 2005. Re-circulating heat pump assisted continuous bed drying and energy analysis. *International Journal of Energy Research* 29: 961–972.

Bendicho, S., G.V. Barbosa-Canovas, and O. Martin. 2002. Milk processing by high intensity pulsed electric fields. *Trends in Food Science and Technology* 13: 195–204.

Caric, M. 1994. Milk powders: General production. In *Concentrated and Dried Dairy Products*. New York: VCH Publishers, Inc.

Casper, M.E. 1977. *Energy-Saving Techniques for the Food Industry*. New Jersey: Noyes Data Corporation.

Changani, S.D., M.T. Belmar-Beiny, and P.J. Fryer. 1997. Engineering and chemical factors associated with fouling and cleaning in milk processing. *Experimental Thermal and Fluid Science* 14: 392–406.

Craven, H.M., P. Swiergon, S. Ng, J. Midgely, C. Versteeg, M.J. Coventry, and J. Wan. 2008. Evaluation of pulsed electric field and minimal heat treatments for inactivation of pseudomonads and enhancement of milk shelf life. *Innovative Food Science and Emerging Technologies* 9: 211–216.

Fischer, K. and H.P. Bipp. 2005. Generation of organic acids and monosaccharides by hydrolytic and oxidative transformation of food processing residues. *Bioresource Technology* 96: 831–842.

Fryer, P.J. and M.T. Belmar-Beiny. 1991. Fouling of heat exchanger in the food industry: A chemical engineering perspective. *Trends in Food Science & Technology* 2: 33–37.

Juriaanse, A.C. 1999. Changing pace in food science and technology: Examples from dairy science show how descriptive knowledge can be transferred into predictive knowledge. *Trends in Food Science & Technology* 10: 303–306.

Kiatsiriroat, T. and W. Tachajapong. 2002. Analysis of a heat pump with solid desiccant tube bank. *International Journal of Energy Research* 26: 527–542.

Knipschildt, M.E. and G.G. Andersen. 1994. Drying of milk and milk products. In *Modern Dairy Technology vol.1*, Robinson, R.K. (Ed.), London: Chapman & Hall.

Kuzgunkaya, E.H. and A. Hepbasli. 2007. Exergetic performance assessment of a ground-source heat pump drying system. *International Journal of Energy Research* 31: 760–777.

Ozyurt, O., O. Comakli, M. Yilmaz, and S. Karsli. 2004. Heat pump use in milk pasteurization: An energy analysis. *International Journal of Energy Research* 28: 833–846.

Perera, C.O. and M.S. Rahman. 1997. Heat pump dehumidifier drying of food. *Trends in Food Science and Technology* 8: 75–79.

Ramirez, C.A., K. Blok, M. Neelis, and M. Patel. 2006a. Adding apples and oranges: The monitoring of energy efficiency in the Dutch food industry. *Energy Policy* 34: 1720–1735.

Ramirez, C.A., M. Patel, and K. Blok. 2006b. From fluid milk to milk power: Energy use and energy efficiency in the European dairy industry. *Energy* 31: 1984–2004.

Reina, L.D., Z.T. Jin, Q.H. Zhang, and A.E. Yousef. 1998. Inactivation of *Listeria monocytogenes* in milk by pulsed electric field. *Journal of Food Protection* 61: 1203–1206.

Riberiro, Jr., C.P. and M.H.C. Andrade. 2003. Performance analysis of the milk concentrating system from a Brazilian milk powder plant. *Journal of Food Process Engineering* 26: 181–205.

Staniszewski, M., W. Kujawski, and M. Lewandowska. 2007. Ethanol production from whey in bioreactor with co-immobilized enzyme and yeast cells followed by pervaporative recovery of product-kinetic model predictions. *Journal of Food Engineering* 82: 618–625.

Sun, D.W. and L.J. Wang. 2001. Novel refrigeration cycles, in: Sun, D. W. (Ed.), *Advances in Food Refrigeration*, Chapter 1, pp. 1–69, Leatherhead Publishing: Leatherhead, U.K.

U.S. Census Bureau. 2006. *2006 Annual Survey of Manufactures*. Available at http://factfinder.census.gov.

Visser, J. and T.J.M. Jeurnink. 1997. Fouling of heat exchangers in the dairy industry. *Experimental Thermal and Fluid Science* 14: 407–424.

15 Energy Conservation in Meat Processing Facilities

15.1 INTRODUCTION

The meat manufacturing sector (NAICS code 3115) is engaged in one or more of the following manufacturing activities: (1) slaughtering animals, (2) preparing processed meats and meat by-products, and (3) rendering and refining animal fat, bones, and meat scraps. The meat sector comprises the production and preservation of different meats including beef, pork, sheep, goat, and poultry and the further processing of all these types of meats. The industry also includes cutting and packing of meats from purchased carcasses (U.S. Census Bureau, 2006).

In meat processing facilities, fuels such as coals and natural gas are used to provide process heat and space heating while electricity is used for refrigeration and motor drives. An analysis conducted by Ramirez et al. (2006a) showed the energy consumption in the meat industry increased between 14% and 32% in four European countries including France, Germany, the Netherlands, and United Kingdom because of stronger hygiene regulations and increased freezing and cutting of meat products. Sausages and prepared meats such as cured meats, smoked meats, canned meats, and frozen meats have increased significantly in recent years. The facilities for prepared meats are significantly more energy intensive than the slaughtering facilities. In the meat manufacturing sector, a large portion of fuels is used for steam generation, and a large portion of purchased electricity is used for motors and air compressors. Energy conservation in steam generation and distribution is discussed in Chapter 4. The energy conservation technologies for air compressors and motors are discussed in Chapters 5 and 6, respectively. Heat exchangers are widely used in this sector, and the energy conservation technologies for heat exchangers are discussed in Chapter 7. Boilers, heat exchangers, motors, and air compressors may generate waste heat, and waste-heat recovery technologies are discussed in Chapter 8. Since both processing heat and electricity are required by the meat processing sector, a combined heat and power system can be used to efficiently and economically provide electricity or mechanical power and useful heat from the same primary energy source. Combined heat and power systems are discussed in Chapter 9.

In this chapter, the main products and their processes in the meat manufacturing sector are reviewed. Energy uses and conservation in the main unit operations in this sector are discussed. Finally, energy utilization of processing wastes from the meat manufacturing facilities is discussed.

15.2 OVERVIEW OF MAIN PROCESSES

The process flow of the meat sector is given in Figure 15.1. Meat processing plants vary considerably by animal species processed and, in turn, by the degree of processing, i.e., primarily slaughter, or slaughter and cutting, or slaughter, cutting, and extensive by-product rendering. Major unit operations of a meat process include slaughter, hide removal, eviscerating, trimming, rendering, and cooling. The primary product may leave the plant as carcass meats, primal cuts, or boxed meat. Further processing is also carried out in some plants to produce sausage and other prepared meats.

15.3 ENERGY USE IN THE MEAT MANUFACTURING SECTOR

The total shipment value from the meat manufacturing sector was $149.6 billion in the United States in 2006. The total costs for purchasing fuels and electricity were $2.1 billion in 2006. About 1.37 cents of energy cost per dollar shipment value was required from the meat manufacturing sector in the United States in 2006. The meat manufacturing sector is the second largest energy consumer behind the grains and oilseed milling sector in the U.S. food industry. In 2006, the meat processing sector consumed 20.7% of total energy input into the whole food industry while it generated 27.8% of the total shipment value. The cost of purchased electricity in the meat manufacturing sector was 53.4% of the total energy expenses (U.S. Census Bureau, 2006).

The amount of energy used for the production of a given amount of products is highly dependent on the type of products. Table 15.1 gives the energy use for the production of different meat products in the Netherlands in 2001 (Ramirez et al., 2006b). In general, livestock slaughtering is not energy intensive. However, meat processing is much more energy intensive than slaughtering but the energy insensitivity varies considerably by products.

Table 15.2 shows the distribution of delivery energy in the meat processing sector in terms of energy function. Fossil fuels are mainly used for processing heat while the main use of electricity is cooling (Ramirez et al., 2006a). The energy and physical production data from the meat industry in four European countries including France, Germany, the Netherlands, and United Kingdom show that freezing of poultry, pig, and cattle products may need 120–260 kWh/ton (or 0.432–0.936 MJ/kg), 115 kWh/ton (or 0.414 MJ/kg), and 80 kWh/ton (or 0.288 MJ/kg), respectively. Cutting and deboning consume 60 kWh of electricity and 215 MJ of fuels per ton of finished product (Ramirez et al., 2006a). In the past 15 years, the energy use per ton of meat products in these four European countries increased by 14%–48%. The strong hygiene regulations caused between one- and two-thirds of the increase in energy consumption. The increases in frozen, cut, and deboned meat products also increased the energy consumption per unit of marketed products (Ramirez et al., 2006a).

15.4 POTENTIAL ENERGY CONSERVATION MEASURES

15.4.1 Optimization of Heat Exchange Network

Optimization of a heat exchanger network can decrease the external energy demand in meat processing facilities. Heat pinch analysis is based on thermodynamic principles to

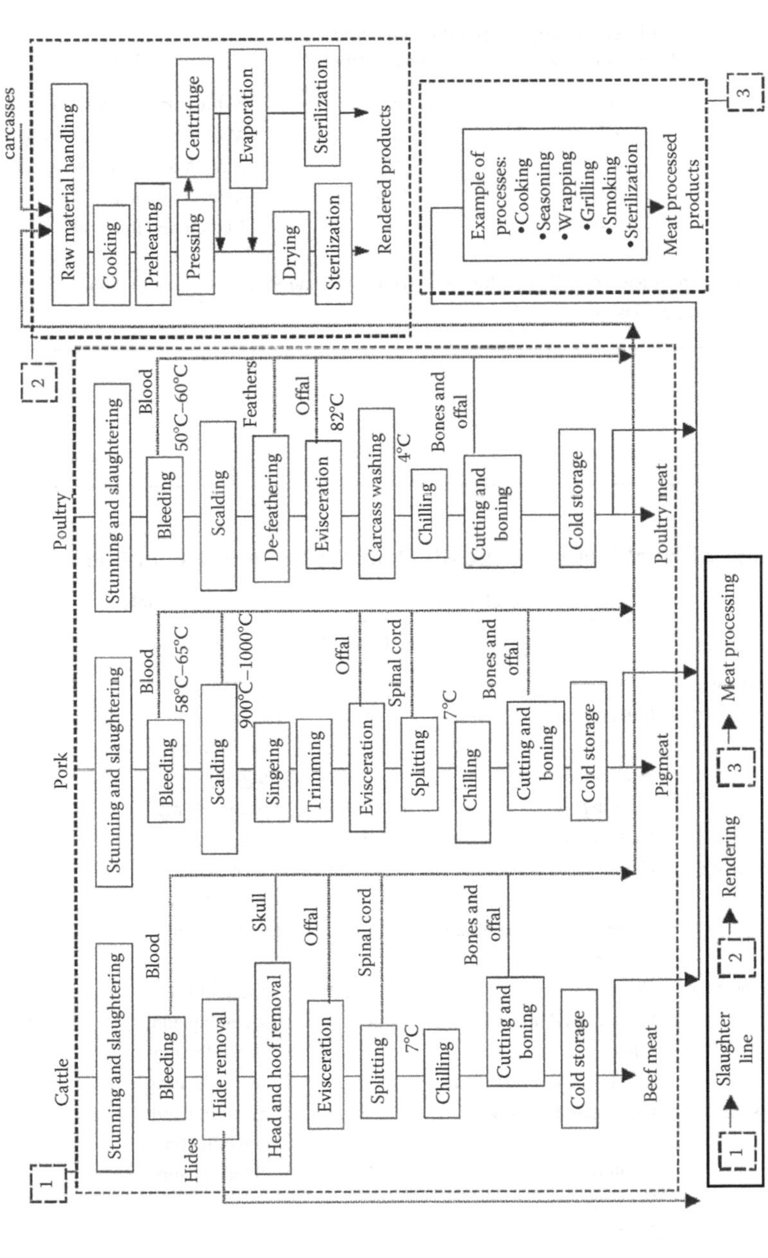

FIGURE 15.1 Process flow diagram of the meat sector. (Reprinted from Ramirez, C.A., Patel, M., and Blok, K. *Energy*, 31, 2047, 2006a. With permission.)

TABLE 15.1
Energy Use for Production of Different Meat Products in the Netherlands in 2001

Product	Specific Electricity Consumption	Specific Fuels and Heat Consumption	Unit
Beef and sheep	341	537	MJ/ton dress carcass weight
Pig	465	932	MJ/ton dress carcass weight
Poultry	1008	576	MJ/ton dress carcass weight
Processed meat	750	3950	MJ/ton product
Rendering	234	1042	MJ/ton raw material

Source: Data taken from Ramirez, C.A., Blok, K., Neelis, M., and Patel, M., *Energy Policy*, 34, 1720, 2006b. With permission.

TABLE 15.2
Distribution of Final Energy in the Meat Sector by Energy Function

	Electricity		Fuels	
Sector	Activity	Percent	Activity	Percent
Pork slaughtering	Cooling	49–70	Gas oven	60–65
	Slaughtering	5–30	Cleaning and disinfecting	18–20
	Water cleaning	5–7	Singeing	15
	Lighting	2–8	Space heating	7
	Evisceration	3		
Cattle slaughtering	Cooling	45–70	Cleaning and disinfecting	80–90
	Compressed air, lighting, and machines	30	Space heating	10–20
Poultry slaughtering	Cooling	52–60	Singeing	60
	Compressed air and machines	30	Cleaning and disinfecting	30
	Lighting and ventilation	4	Space heating	10
Meat processing	Cutting and mixing	40	Cleaning and disinfecting	25
	Cooling	40	Space heating	15
	Packing	10		
	Lighting	10		

TABLE 15.2 (continued)
Distribution of Final Energy in the Meat Sector by Energy Function

	Electricity		Fuels	
Sector	**Activity**	**Percent**	**Activity**	**Percent**
Rendering	Drying	23	Drying	61
	Grinding and pressing	17	Grinding and pressing	17
	Compressed air, lighting, and machines	12	Meal sterilization	8
	Milling plant	8	Fat treatment	3
	Vacuum evaporation	6	Vacuum evaporation	2
	Meal sterilization	2	Space heating	1

Source: Reprinted from Ramirez, C.A., Patel, M., and Blok, K. *Energy*, 31, 2047, 2006b. With permission.

identify appropriate changes that can have a positive impact on energy savings. The heat pinch analysis is to answer the following questions: what are the minimal external heating and cooling demands for a particular process? What is the maximum heat that can be recovered through internal heat exchangers? How should a heat exchanger network be designed so that internal heat exchanging is optimized? A heat pinch analysis starts with the analysis of heat and mass balances of a process. In the analysis, the streams that need to be heated or cooled are identified. The energy demands at various process temperatures, the net external heating and cooling demands, and the pinch temperature are determined (Fritzson and Berntsson, 2006a). Heat exchanger design and network optimization are discussed in Chapter 7. A heat pump, as discussed in Chapter 9, can be used to recover the net surplus of heat below the pinch temperature through its evaporator and supply a higher-temperature heat through its condenser.

15.4.2 Energy Conservation in Direct Fuel Use

Application of heat recovery units can also decrease the external energy demand in meat processing facilities. The fuel input in a meat processing plant can be broken down into direct use and boiler use. The direct use of fuel inputs is primarily accounted for by four end-use activities: (1) smokehouses, cook ovens, cooking tanks, and smoke generators; (2) hog singers; (3) space heating; and (4) afterburners in the plants. Afterburners are usually used to control air pollution by completely burning the flue gas. Space heating in meat processing plants has the highest technologically and economically feasible potential for saving direct energy use. Energy saving in space heating can be achieved by lowering temperature levels, scheduling of space heater use, and proper maintaining of doors, windows, and vents. Energy conservation in hog singers, smokehouses, and cook ovens usually requires capital investment for better maintenance activities and replacement of old units with more efficient ones. The economic feasibility of these energy conservation technologies should be carefully justified. Other efficiency improvement measures include optimization in operations, production, and maintenance.

15.4.3 Energy Conservation in Refrigeration

15.4.3.1 Energy Conservation in Refrigeration through Maintenance and Optimization

On average, the refrigeration system consumes as much as 15% of the total energy. The meat manufacturing sector is the second largest user of refrigeration behind the dairy sector. In the meat sector, refrigeration constitutes between 40% and 90% of total electricity use during production time and almost 100% during nonproduction periods (Ramiez et al., 2006a). Gigiel and Collett (1989) pointed out that the factors in descending order of importance in a chiller operation were effective cooling, low weight loss, and low energy consumption. The primary goal of a chilling process is to achieve effective cooling for the microbial safety of meat products. Since meat products are sold by weight, weight loss directly decreases the economic value of the products. Energy consumption is related to ambient temperature, and the energy consumed and heat generated by the fan associated with the evaporator of a refrigerator.

Fritzson and Berntsson (2006a,b) that found more than 10% of the shaft work or electricity used in a refrigeration plant can be saved in a modern slaughtering and meat processing plant and 15% can be saved in a less modern plant. This saving can be achieved by decreasing the temperature difference between refrigeration media and the needed temperature in the refrigeration and freezing rooms. The method with the largest potential for energy saving is to install a heat pump for the recovery and upgrading of available waste heat in a refrigeration plant. The recovered heat can meet all of the external fuel demands except processing steam.

Gigiel and Collett (1989) concluded that a target energy saving of 40% was possible with existing technologies, which would increase the profits in the meat industry by 26%. To reduce energy consumption, the infiltration and base loads should be kept as small as possible. Infiltration load can be reduced by keeping the chiller doors closed at all times except when meats are being loaded or unloaded. Base load can be reduced by reducing the evaporator fan energy consumption once the peak product load has been removed, maintaining effective chiller insulation, and ensuring that the room is efficiently sealed.

15.4.3.2 Energy Conservation in Refrigeration through Reduced Weight Loss

During meat chilling, energy is required by the chiller to remove the sensible heat for the reduction of meat temperature and the latent heat of water evaporation due to weight loss. After slaughtering carcasses are usually cooled down to a temperature of 7°C or less in chillers. It was found that the weight loss during chilling of beef carcasses was between 1.1% and 2% after 24 h and between 1.5% and 2.3% after 48 h (Gigiel and Collett, 1989). Energy consumption in chillers must be assessed against their primary purpose, which is to reduce the temperature of foods. However, when weight loss occurs during chilling, the weight loss not only increases the energy consumption for the removal of latent heat of evaporation but also causes the loss of the economic value of products.

Example 15.1

Suppose the initial temperature of carcasses after slaughtering is 45°C, the specific heat of the carcasses is 3.5 kJ/kg, the weight loss during chilling is 2%, and the latent heat of water evaporation is 2500 kJ/kg. What is the theoretical specific energy requirement to reduce the temperature of carcasses from 45°C to 7°C?

Solution 15.1

The sensible heat for temperature reduction is

$$Q_s = c_p \Delta T = 3.5\,[\text{kJ/kg°C}] \times (45-7)\,[\text{°C}] = 133\,\text{kJ/kg}$$

The latent heat for water evaporation is

$$Q_l = \lambda w = 2500\,[\text{kJ/kg}] \times 2\% = 50\,\text{kJ/kg}$$

The total theoretical energy requirement is thus 183 kJ/kg. A significant percentage of energy consumption (around 25% of the total energy use) is for the removal of the latent heat of water evaporation due to weight loss during chilling.

However, suppose that the energy efficiency of a chiller is as low as 50% and a mechanic compression chiller consumes electricity at a COP of 3. In this case, the specific electricity consumption for chilling of carcasses is 122 kJ/kg. Gigiel and Collett (1989) found that the specific energy consumption for chilling of beef carcasses in 14 beef chillers in United Kingdom was 115 kJ/kg.

15.4.3.3 Energy Conservation in Refrigeration through Reduced Cooling Load

A modified hot processing strategy can be used to remove the low-value cuts from the dressed carcass along with associated bone and fat before cooling. The remaining high-value meat is then chilled in the usual manner. The low-value cuts are immediately processed. The modified hot processing strategy could reduce the refrigeration load for beef carcass chilling by 51%. It can also reduce the chilling time for the carcasses. However, the amount of energy to be saved depends on the methods used to further process the low-value cuts, which is more than 7% of the carcass mass (McGinnis et al., 1994).

15.4.3.4 Applications of a Novel Waste-Heat-Powered Refrigeration System

Novel refrigeration cycles based on liquid–liquid absorption, liquid–solid adsorption, and fluid ejection as discussed in Chapter 9, which can be powered by low-grade thermal energy, offer potential energy saving opportunities in meat processing facilities. Novel refrigeration cycles such as absorption and adsorption refrigeration cycles can be powered by waste heat or other renewable energy sources such as geothermal energy and solar energy at a low temperature (e.g., 70°C) (Sun and Wang, 2001).

15.4.4 Nonthermal Processing of Meat Products

Thermal processes are usually considered to be energy intensive. In addition, the slow heat transfer through food products due to the low thermal conductivity of foods is usually a limiting factor for thermal treatment of food products. Nonthermal pasteurization techniques including food irradiation (Chapter 18), pulsed electric field treatment (Chapter 19), and high-pressure processing (Chapter 20), as well as microwave sterilization (Chapter 21) have been developed to replace or combine with conventional thermal sterilization and pasteurization processes for saving energy and improving product quality and safety. These nonthermal processing times are usually short. For example, during high-pressure processing, foods are exposed to pressure up to 600 MPa for a few minutes. Pulsed electric field treatment is based on the delivery of pulses at a high electric field intensity of 5–55 kV/cm for a few milliseconds. Food irradiation occurs for several seconds to several minutes. Most alternative preservation processes can achieve the equivalent of pasteurization but not sterilization (Amymerich et al., 2008).

15.5 ENERGY UTILIZATION OF MEAT PROCESSING WASTES

15.5.1 Meat Processing Wastes

In the meat industry, slaughterhouses are producing increasing amounts of organic solid by-products and wastes. The organic solid wastes generated in a poultry slaughterhouse include blood, feather, offal, feet, head, trimmings, and bones. A broiler is about 1.8–1.9 kg before it is slaughtered. During slaughtering and processing, about 40% of the broiler mass enters into the waste stream (Salminen and Rintala, 2002). Solid organic wastes from slaughterhouses may contain several species of microorganisms including potential pathogens (Salminen and Rintala, 2002). In addition, animals may accumulate various metals, drugs, and other chemicals added in their feed for nutritional and pharmaceutical purposes (Haapapuro et al., 1997). The disposal and processing of animal wastes should destroy potential pathogens present in the wastes for the prevention of animal and public health.

Meat processing wastes can be converted into energy products, which can reduce the costs for both energy supply and waste disposal. Using the slaughter waste as a fuel in a boiler in a slaughter and meat processing plant usually has a lower efficiency and a less favorable economy than using it in a large combined heat and power plant (Fritzson and Berntsson, 2006b). Meat wastes can also be converted to liquid and gaseous fuels by a biological conversion, thermochemical conversion, or chemical conversion process. Anaerobic digestion, which is a biological process, has been widely used to treat slaughterhouse wastewaters. The details of anaerobic digestion of slaughterhouse wastes are given in Chapter 24. The fats generated during meat rendering can be used as a feedstock for biodiesel production, which is discussed in Chapter 26.

15.5.2 Anaerobic Digestion of Slaughterhouse Wastes

Anaerobic digestion has been used to treat solid slaughterhouse wastes such as cattle paunch wastes, blood, and settlement tank solids (Banks, 1994; Banks and Wang,

1999). Banks (1994) treated cattle and lamb paunch contents, blood, and process wastewaters with an organic loading of 0.36 kg COD/m^3 d in a 105 m^3 continuously stirred tank digester. The yield of biogas at 0.18 m^3/kg COD was observed at a hydraulic retention time of 43 days. Bank and Wang (1999) further investigated the treatment of cattle blood and rumen paunch contents in a two-phase process with uncoupled solids and liquid retention times. The two-phase process allowed a higher overall loading at 3.6 kg TS/m^3 d. Salminen et al. (2000) treated solid poultry slaughterhouse waste at a loading rate of 0.8 kg VS/m^3 d and a hydraulic retention time of 50 days in a 2 L continuously stirred tank digester. A methane yield of 0.55 m^3/kg VS was observed.

15.5.3 Biodiesel Production from Animal Fats

Animal fats contain a large amount of saturated fatty acids, which are almost 50% of the total fatty acids. Due to the high content of saturated fatty acids, animal fats have the unique properties of high melting point and high viscosity. They are solid at room temperature. The most commonly used method to produce biodiesel from crude vegetable oils and animal fats is to transesterify vegetable oils and animal fats with light alcohols such as methanol and ethanol to an alkyl ester, which has lower and higher volatility than the feedstocks of vegetable oils and animal fats (Ma and Hanna, 1999; Tashtoush et al., 2004).

Ma et al. (1998a,b) investigated the effects of free fatty acid and water contents in beef tallow on its transesterification reaction with methanol. Their results showed that the water content of beef tallow should be kept below 0.06% w/w and free fatty acid content should be kept below 0.5%, w/w (acid value less than 1) in order to achieve the best conversion. Water content is a more critical variable in the transesterification process than free fatty acids. The requirements for fats with very low free fatty acids and water contents restrict the use of the fats as a low-cost feedstock in a traditional biodiesel production process with an alkali catalyst. Because the solubility of methanol in beef tallow is only 19% w/w at 100°C, mixing is essential to disperse the methanol in the beef tallow to start the reaction effectively. Ma et al. (1999) found that once the two phases were mixed and the reaction was started, stirring was no longer needed. It was observed that after the reaction was finished, there was 60% w/w of unreacted methanol in the beef tallow ester phase and 40% w/w in the glycerol phase. The optimum operation sequence was to recover the unreacted methanol using vacuum distillation after transesterification, separate ester and glycerol phases, and then purify the beef tallow methyl esters (Ma et al., 1998b).

Since animal fats, particularly waste fats, may contain 10%–15% of free fatty acids, basic catalysts such as sodium hydroxide used in traditional biodiesel production can easily react with free fatty acids in the animal fats to form soaps. An acidic catalyst such as 2.25 M H_2SO_4 was used for the transesterification of animal fats into biodiesel (Tashtoush et al., 2004). Ethanol was found to be better than methanol for conversion of waste animal fat into biodiesel in terms of the high conversion ratio and lower viscosity of the biodiesel. There is an interaction between reaction time and temperature. The optimum temperature and time for the transesterification of waste animal fat into esters with ethanol were found to be 50°C and 2 h, respectively (Tashtoush et al., 2004).

15.6 SUMMARY

The meat manufacturing sector has a high demand in fuels for process heat supply and electricity due to the heavy use of refrigeration and motors. Different energy conservation technologies can be used to save energy in the meat manufacturing sector. Optimization of heat exchange network and operations, energy conservation technologies for direct fuel use in smokehouses and cook ovens, energy conservation technologies for steam generation and distribution, and energy conservation technologies for refrigeration systems are particularly important to the meat manufacturing sector. Meat processing facilities generate large amounts of processing wastes such as slaughterhouse wastes and animal fats. The slaughterhouse wastes can be converted to biogas via anaerobic digestion. Animal fats can be used to produce biodiesel via a transesterification process.

REFERENCES

Amymerich, T., P.A. Picouet, and J.M. Monfort. 2008. Decontamination technologies for meat products. *Meat Science* 78: 114–129.

Banks, C.J. 1994. Anaerobic digestion of solid and high nitrogen content fractions of slaughterhouse wastes. *Environmentally Responsible Food Processing, AlChE Symp. Ser.* 90: 48–55.

Banks, C.J. and Z. Wang. 1999. Development of a two phase anaerobic digester for the treatment of mixed abattoir wastes. *Water Science and Technology* 40: 67–76.

Fritzson, A. and T. Berntsson. 2006a. Efficient energy use in a slaughter and meat processing plant-opportunities for process integration. *Journal of Food Engineering* 76: 594–604.

Fritzson, A. and T. Berntsson. 2006b. Energy efficiency in the slaughter and meat processing industry-opportunities for improvements in future energy markets. *Journal of Food Engineering* 77: 792–802.

Gigiel, A. and P. Collett. 1989. Energy consumption, rate of cooling and weight loss in beef chilling in UK slaughter houses. *Journal of Food Engineering* 10: 255–272.

Haapapuro, E.R., N.D. Barnard, and M. Simon. 1997. Review-animal waste used as livestock feed: danger to human health. *Preventive Medicine* 26: 599–602.

Ma, F. and M.A. Hanna. 1999. Biodiesel production: A review. *Bioresource Technology* 70: 1–15.

Ma, F., L.D. Clements, and M.A. Hanna. 1998a. The effects of catalyst, free fatty acids and water on transesterification of beef tallow. *Transactions of the ASAE* 41: 1261–1264.

Ma, F., L.D. Clements, and M.A. Hanna. 1998b. Biodiesel fuel from animal fat. Ancillary studies on transesterification of beef tallow. *Industrial and Engineering Chemistry Research* 37: 3768–3771.

Ma, F., L.D. Clements, and M.A. Hanna. 1999. The effect of mixing on transesterification of beef tallow. *Bioresource Technology* 69: 289–293.

McGinnis, D.S., J.L. Aalhus, B. Chabot, C. Gariepy, and S.D.M. Jones. 1994. A modified hot processing strategy for beef: Reduced electrical energy consumption in carcass chilling. *Food Research International* 27: 527–535.

Ramirez, C.A., M. Patel, and K. Blok. 2006a. How much energy to process one pound of meat? A comparison of energy use and specify energy consumption in the meat industry of four European countries. *Energy* 31: 2047–2063.

Ramirez, C.A., K. Blok, M. Neelis, and M. Patel. 2006b. Adding apples and oranges: the monitoring of energy efficiency in the Dutch food industry. *Energy Policy* 34: 1720–1735.

Salminen, E. and J. Rintala. 2002. Anaerobic digestion of organic solid poultry slaughterhouse waste—a review. *Bioresource Technology* 83: 13–26.

Salminen, E., J. Rintala, L.Y. Lokshina, and V.A. Vavilin. 2000. Anaerobic batch degradation of solid poultry slaughterhouse waste. *Water Science and Technology* 41: 33–41.

Sun, D.W. and L.J. Wang. 2001. Novel refrigeration cycles, in: Sun, D.W. (Ed.), *Advances in Food Refrigeration*, Chapter 1, pp. 1–69, Leatherhead Publishing: Leatherhead, U.K.

Tashtoush, G.M., M.I. Al-Widyan, and M.M. Al-Jarrah. 2004. Experimental study on evaluation and optimization of conversion of waste animal fat into biodiesel. *Energy Conversion and Management* 45: 2697–2711.

U.S. Census Bureau. 2006. *2006 Annual Survey of Manufactures*. Available at http://factfinder.census.gov.

16 Energy Conservation in Bakery Processing Facilities

16.1 INTRODUCTION

The bakery processing sector (NAICS code 3116) is engaged in one or more of the following manufacturing activities: (1) bread and bakery products; (2) cookies, crackers, and pastas; and (3) tortillas (U.S. Census Bureau, 2006).

In a bakery facility, thermal energy is used for bake ovens and boilers while electricity is used for machinery, electric bake ovens, refrigeration systems, and lighting. Energy conservation in steam generation and distribution is discussed in Chapter 4. Energy conservation technologies for air compressors and motors are discussed in Chapters 5 and 6, respectively. Heat exchangers are widely used in the dairy sector, and the energy conservation technologies for heat exchangers are discussed in Chapter 7. Boilers, heat exchangers, motors, and air compressors may generate waste heat, and waste-heat recovery technologies are discussed in Chapter 8. Since both processing heat and electricity are required by the bakery processing sector, a combined heat and power system can be used to efficiently and economically provide electricity or mechanical power and useful heat from the same primary energy source. Combined heat and power systems are discussed in Chapter 9.

In this chapter, the main bakery products and their processes are reviewed. Energy uses and conservation in main unit operations in the dairy sector are discussed. Finally, energy utilization of bakery processing wastes is discussed.

16.2 OVERVIEW OF MAIN PRODUCTS AND PROCESSES

The baking industry makes bread, buns, rolls, dough, cookies, crackers, pastas, tortillas, and other products that are either baked or frozen. A typical bakery production process is shown in Figure 16.1 (Kannan and Boie, 2003). In this bakery facility, wheat is milled into flour. Flour stored in a silo is transported to the kneading machine pneumatically. Flour, fermented broth, and other ingredients are mixed in the kneading machine to make dough for bread/bun/rolls, while the mixer is used to make cake batter. Dough is molded into bread/bun/rolls by machines. The molded breads are fermented in a fermentor and baked at 180°C–300°C, depending on the type of breads. A portion of the bread dough is baked partially and frozen by blowing cold air at −38°C (shock freezer) and stored in the freezer at −23°C. The bread moulds and baking trays are washed in hot water at 40°C. Molded bun dough is loaded in the auto-fermentor, which is a storage room cum fermentor where the temperature is

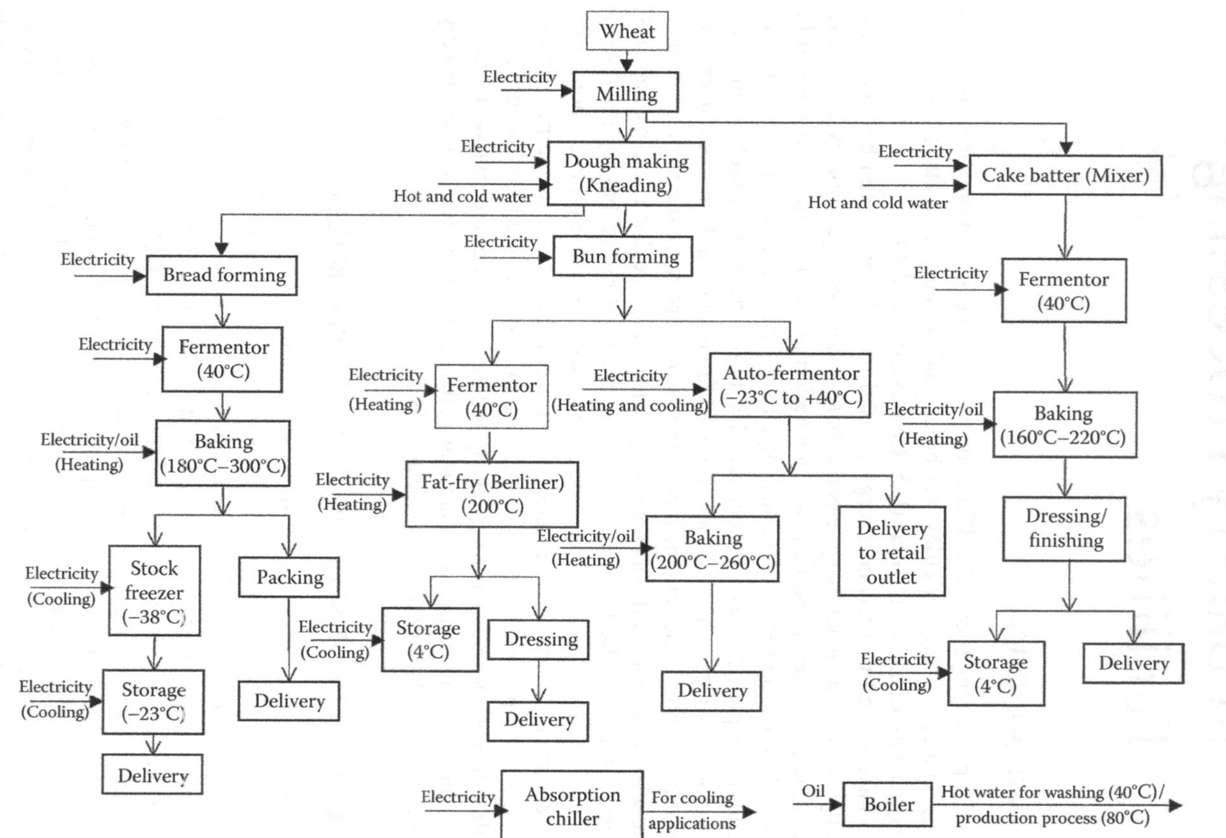

FIGURE 16.1 Production process diagram of a bakery. (Reprinted from Kannan, R. and Boie, W., *Energy Conv. Manag.*, 44, 945, 2003. With permission.)

automatically controlled between −23°C and 40°C. While loading dough in the auto-fermentor, the temperature is −23°C and is gradually increased to 40°C over 16 h during which period fermentation takes place. Then the auto-fermentor is cooled to −23°C for the next loading of bun dough. The fermented bun dough is baked at 200°C–260°C. A small portion of the bun dough is delivered to their retail outlets, where they are baked. Besides, some bun dough is baked in fat at 200°C (Berliner). Part of the baked goods are stored in the cold storage room at 4°C or −23°C depending on the storage period.

The major machineries are the miller, mixer/kneading machine, bun and bread former, fermentor, bake ovens, cold storage unit, and boilers. The bake ovens can be electrical ovens or fuel (oil or gas) fired ovens. Hot water at 40°C–50°C is mainly used for washing baking plates, molds, and trays. Frozen dough and partially baked bakery products continue to grow. Refrigeration is a common unit operation used by many food processing sectors. The energy conservation technologies are addressed in Chapter 10.

16.3 ENERGY CONSUMPTION IN THE BAKERY PROCESSING SECTOR

The total shipment value from the bakery processing sector was 54.2 billion in the United States in 2006. The total costs for purchasing fuels and electricity were $0.9 billion in 2006. In general, the baking industry is a moderate user of energy per unit of output. About 1.68 cents of energy cost per dollar shipment value was required in the bakery manufacturing sector in the United States in 2006. The bakery manufacturing sector consumed 9.2% of the total energy input into the whole food industry and generated 10.1% of the total shipment value in 2006. The cost of purchased electricity in the bakery manufacturing sector was 54.7% of the total energy expenses (U.S. Census Bureau, 2006).

Bakery processing facilities use large amounts of electricity and fuels to operate ovens, refrigeration systems, and compressed air system. Most bakeries are small-scale industries. In German bakeries, about 72% of the total energy requirement is thermal energy at 180°C–360°C and 28% of electricity. A 950 m^2 modern bakery production unit processes 2 tons of flour per day for bread, buns, rolls, cakes, and pastries. The bakery consumed annually about 225 MWh electricity at a load of 125 kW and 57,700 L of furnace oil. The estimated specific energy consumption could be in the range of 4.57–6.80 MJ/kg of processed flour depending on the type of fuels used in baking ovens and the type of bread baked, size of the bakery, and the number of shifts. Baking ovens consumed the largest portion of energy. About 73% of the total energy was consumed in baking, and 6% each for fermentation and cleaning (Kannan and Boie, 2003).

Ho et al. (1986) analyzed the first law and second law efficiencies of thermodynamics in a bakery plant. They found that the first and second law efficiencies of the bakery plant were 43% and 15.5%, respectively. They recommended that improvements in both first and second law efficiencies for the bakery plant could be achieved by reducing the mass of dough moulds and recovery of energy in the bake oven exhaust to heat the air in the proofing oven.

16.4 POTENTIAL ENERGY CONSERVATION MEASURES

16.4.1 Operation of Bake Ovens on Their Full Load

Baking ovens offer the greatest potential for energy savings. The energy consumption of a baking oven is significantly affected by the load. The specific energy consumption will increase by 5% at a 75% load while it will increase by 16% at a 50% load (Kannan and Boie, 2003). Therefore, it is important to operate a baking oven on its full load. Dough requiring high temperature should be baked first, followed by low-temperature dough baking. It should also eliminate long initial heating periods without any product being baked.

16.4.2 Recovery of Waste Heat from Bake Ovens

The flue gas from a bake oven can be used to generate hot water or preheat air supplied into oil-fired bake ovens. The flue gas from a 10 m^3 bake oven could be used to heat 300 L water to 70°C daily, and this could improve the energy consumption by 10%–15%. The hot flue gas can also be used to preheat the ambient air into an oil fired bake oven. This can save energy and reduce the maintenance cost of the oven by preventing tar build-up due to the high moisture of ambient air (Kannan and Boie, 2003).

Heat pipes can dissipate substantial quantities of heat with a minimal temperature gradient. This is achieved as a result of good heat transfer through the mass transfer of the working fluid inside the pipe from one end of the heat pipe to the other end with phase changes. A heat pipe can transfer heat several hundred times better than a solid copper rod of the same diameter. A heat exchanger using thermosyphon heat pipes was designed and used to recover heat from flue gas at a temperature of around 300°C from a bake oven in the bakery industry (Lukitobudi et al., 1995). The heat pipe technology is discussed in Chapter 9.

16.4.3 Applications of Microwave Baking Technology

Halogen lamp–microwave combination baking is a new technology in the bakery industry. Microwaves can penetrate into the products and thus rapidly heat the products from the inside. Microwave baking cannot induce browning and crust on the surface of breads and cakes (Zhang and Datta, 2006). Halogen lamp heating provides near-infrared radiation light with high frequency and low penetration depth, which is focused on the surface of food products and increases the surface temperature for browning. A halogen lamp–microwave combination baking takes the time-saving advantage of microwave heating and the browning and crisping advantages of halogen lamp heating (Keskin et al., 2004; Sumnu et al., 2005).

Halogen lamp–microwave combination baking can increase the quality of bakery products compared with other baking methods such as conventional, microwave, and infrared baking. The combined baking process can reduce the conventional baking time by about 75%. Bakery products prepared by the combined baking unit has a specific volume and crust color similar to the conventionally baked products but the weight loss and firmness of breads are still higher than those of conventionally baked ones (Keskin et al., 2004; Sumnu et al., 2005).

16.4.4 Energy Conservation in Drying of Pasta

A modern pasta drying process at high temperatures has four steps: shaking pre-dryer to enhance the mechanical strength of pasta, pre-dryer at 90°C to reduce water content of pasta from 35% to 16%, dryer at 75°C–80°C to reduce the water content to 12.5%, and product cooling to a temperature of about 35°C (Panno et al., 2007). The process consumes, on average, about 1.3 MJ of thermal energy and 1.28 MJ of electricity to produce each one kilogram of pasta. The energy cost in a pasta plant is about 6% of the total production cost as shown in Table 16.1 (Panno et al., 2007). Ozgener and Ozgener (2006) found that the energy efficiency of a pasta drying process widely varied from 7.55% to 77.09%. The exergy efficiency of a pasta drying process was 72.98%–82.15%. The highest energy losses occurred in the drying chamber walls due to air leakages. The energy efficiency could be thus improved by isolation of the drying chamber.

16.4.5 Application of Combined Heat and Power Generation

Since both processing heat and electricity are required in bakery facilities, a combined heat and power (CHP) system can be used to efficiently and economically provide electricity or mechanical power and useful heat from the same primary energy source. A combined heat and power generation plant with a gas turbine as its mover has been used in a pasta factory (Panno et al., 2007). A typical gas turbine works above 800°C and the temperature of flue gas is 430°C–540°C. The temperature of flue gas from a recovery heat exchanger is between 130°C and 160°C, which can be used to produce subcooled water for pasta drying at a temperature of about 140°C. The overall electricity generating efficiency was estimated at 22%–26% while the overall CHP system efficiency was about 70%–80%. The CHP system can reduce the primary energy demand by up to 9% and CO_2 emission by up to 9% in the pasta plant (Panno et al., 2007).

TABLE 16.1
Cost Distribution in the Pasta Production Process

Cost Item	Cost Percentage (%)
Raw material	77
Labor	14
Electric energy	3
Thermal energy	3
Packaging	3

Source: Reprinted from Panno, D., Messineo, A., and Dispenza, A., *Energy*, 32, 746, 2007. With permission.

16.5 SUMMARY

The bakery industry is a moderate user of energy per unit of product output. However, the total energy demand of the bakery sector is high because of its large production. Bake ovens consume a large portion of total energy input. Ovens in bakeries offer the greatest potential for technologically feasible savings of energy. The energy saving opportunities for ovens include improved operating procedures to operate the oven on full load, waste-heat recovery from flue gas of ovens and use of novel baking technologies such as microwave baking technology. Since both processing heat and electricity are required in bakery processing facilities, a combined heat and power system can be used to efficiently and economically provide electricity or mechanical power and useful heat from the same primary energy source.

REFERENCES

Ho, J.C., N.E. Wijeysundera, and S.K. Chou. 1986. Energy analysis applied to food processing. *Energy* 11: 887–892.

Kannan, R. and W. Boie. 2003. Energy management practices in SME-case study of a bakery in Germany. *Energy Conversion and Management* 44: 945–959.

Keskin, S.O., G. Sumnu, and S. Sahin. 2004. Bread baking in halogen lamp-microwave combination oven. *Food Research International* 37: 489–495.

Lukitobudi, A.R., A. Akbarzadeh, P.W. Johnson, and P. Hendy. 1995. Design, construction and testing of a thermosyphon heat exchanger for medium temperature heat recovery in bakeries. *Heat Recovery Systems & CHP* 15: 481–491.

Ozgener, L. and O. Ozgener. 2006. Exergy analysis of industrial pasta drying process. *International Journal of Energy Research* 30: 1323–1335.

Panno, D., A. Messineo, and A. Dispenza. 2007. Cogeneration plant in a pasta factory: Energy saving and environmental benefits. *Energy* 32: 746–754.

Sumnu, G., S. Sahin, and M. Sevimli. 2005. Microwave, infrared and infrared-microwave combination baking of cakes. *Journal of Food Engineering* 71: 150–155.

Zhang, J. and A.K. Datta. 2006. Mathematical modeling of bread baking process. *Journal of Food Engineering* 75: 78–89.

U.S. Census Bureau. 2006. *2006 Annual Survey of Manufactures*. Available at: http://factfinder.census.gov.

Part IV

Energy Efficiency and Conservation in Emerging Food Processing Systems

17 Membrane Processing of Foods

17.1 INTRODUCTION

Separation processes play an important role in food processing facilities for concentration of liquid foods, recovery of by-products from processing wastewater, removal of contaminants from liquid foods, and treatment of wastewater. Evaporation is widely used to concentrate dilute streams such as milk, fruit juices, and sugar juices in the food industry. Evaporation is an energy-intensive unit operation. Mechanical vapor recompression is the most common technology used to improve the energy efficiency of an evaporation process. Membrane filtration, which is driven by pressure difference, is an alternative method to energy-intensive separation processes with a phase change such as distillation and evaporation for concentrating dilute streams with less energy consumption. Membrane filtration has a potential to save 30%–50% of energy used by distillation and evaporation (Kumar et al., 1999). Membrane filters can also be used to remove suspended solids, dissolved solids, and microbes from a liquid solution. In the food processing industry, membrane technologies can be used to

- Concentrate dilute liquid foods such as milk and juice
- Remove microbes in liquid foods for cold pasteurization
- Recover dissolved solids such as sugars from processing wastewater
- Treat processing wastewater

In this chapter, membrane technology is reviewed. The energy consumption of membrane processing is discussed. Finally, applications of membranes in different food processing sectors are reviewed.

17.2 OVERVIEW OF A MEMBRANE PROCESS

During membrane separation, fluid–solid mixtures such as sugar juice, milk, and fruit juices are separated by fluid passage through a porous barrier, which retains most of the solids as shown in Figure 17.1. Through the membrane, the feed stream is separated into two streams: permeate or filtrate stream passing through the membrane and retentate or concentrate rejected by the membrane. The separation is driven by the pressure difference across the membrane, which is called transmembrane pressure. The membrane is usually made of ceramic or polymeric materials. Ceramic membranes can easily meet extreme pH, temperature, and solids loading requirements compared with polymeric membrane filters.

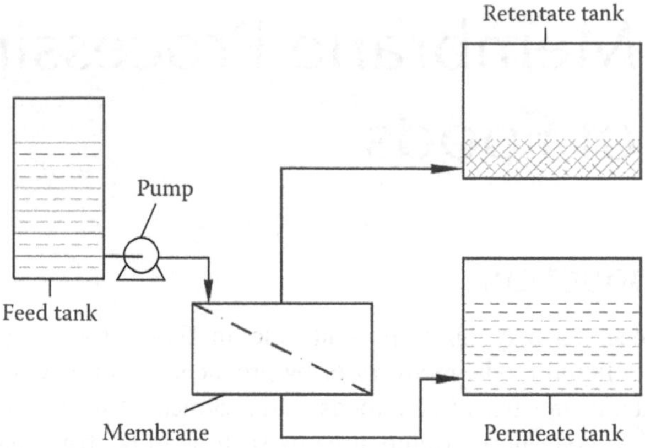

FIGURE 17.1 A typical membrane separation process.

Hollow fiber, tubular, and spiral configurations are three common designs of membrane filtration systems. The spiral design is inexpensive. The cost of the current spiral membrane system is about 300 $/m^2$. However, the spiral design cannot accommodate suspended solids. Hollow fiber design can accommodate suspended solids but cannot accommodate high pressures. The cost of a hollow fiber membrane system is 900 $/m^2$. Tubular design can accommodate suspended solids and high pressure. However, the tubular design is expensive and the cost of the system is 1300 $/m^2$.

The size ranges for different membrane processes are given in Table 17.1. Microfiltration can separate suspended particles typically such as bacteria from 0.1 to 10 μm. Ultrafiltration can separate suspended particles such as proteins from 0.001 to 0.1 μm. Reverse osmosis separates small dissolved molecules such as salts, sugars, and organic acids from 0.5 to 1 nm. The cost of filtration of dilute water solution containing both suspended and dissolved solids is in the range of 0.1–0.25 ¢/L of water filtrate. The filtrate flux, $\dot{V}$, of a given membrane filtration system can be determined by

$$\dot{V} = \frac{A\Delta P}{\mu\left(R_{\mathrm{f}} + R_{\mathrm{c}}\right)} \tag{17.1}$$

TABLE 17.1
Size Range for Different Membrane Processes

Type of Membrane	Pore Size (μ)	Example Retains
Convectional filtration	Several hundred	—
Microfiltration	0.1–10	Bacteria, cell debris, fat globules
Ultrafiltration	0.001–0.1	Macrosolutes (viruses, pyrogens, proteins, peptides)
Reverse osmosis	0.005–0.001	Ions, sugars

where

ΔP is the pressure drop across the membrane
μ is the viscosity of the liquid
A is the cross-sectional area
R_c is the resistance of the cake accumulated on the membrane surface (or fouling layer)
R_f is the resistance of the filter (membrane) and a constant depending on the type of membrane

17.3 ENERGY CONSUMPTION IN MEMBRANE PROCESSING

Energy is required for the feed pump, recirculation pumps, pretreatment, heat treatment, cleaning and sterilization, and control equipment. The energy required by recirculation pumps is related to the recirculation flow rate and the pressure drop of the system. Membrane filtration consumes about 14–36 kJ/kg of water removed, which is a significantly smaller amount of energy than that consumed by evaporation. The energy consumption for evaporation with mechanical vapor-recompression is 350 kJ/kg of removed water. Additional energy savings could be achieved by processing hot feed and recovering heat in the hot permeate (Kumar et al., 1999).

Criscuoli et al. (2008) found that the vacuum membrane system performed better than the direct contact membrane system. The pure water flux through a vacuum membrane system with a 0.2 μm polypropylene membrane at 40 cm^2 was 56.2 kg/m^{2}h at a feed flow rate of 235 L/h, a feed temperature of 59.2°C, and a permeate pressure of 1 kPa. The low permeate fluxes make microfiltration too expensive for many food processing applications. Another restriction imposed by membrane filtration is that the membrane concentration can only reach a maximum dry weight of 12%–20%. Combined membrane filtration and evaporation is thus increasingly used in the dairy processing sector (Ramirez et al., 2006). The use of a membrane to pre-concentrate, prior to the final concentration with mechanical vapor recompression, has a potential to save significant amounts of energy (Kumar et al., 1999).

Process streams in the food industry contain many complex compounds including proteins, fats, acids, and sugars. These compounds are known to cause the fouling of the membranes. Fouling is a challenging issue in membrane applications. The fouling layer increases the energy consumption. In this case, membranes are regularly cleaned with caustic solutions. A cleaning protocol should be developed to prevent the membrane from heavily fouling and restore the performance of the membrane system. Air bubbling techniques have been used to control fouling in submerged membrane systems (Fane et al., 2005). Vibrating technology has also been used to remove suspended solids (Kumar et al., 1999).

17.4 APPLICATION OF MEMBRANE TECHNOLOGY IN FOOD PROCESSING FACILITIES

17.4.1 Applications of Membrane Separation in Wet Grain Milling

Membrane separation processes are a suitable energy-efficient alternative to evaporation and distillation to remove water in wet corn milling. During membrane separation, there is no heat requirement and no phase change. It has been reported that 90% energy savings were observed (Rausch, 2002). However, the total energy savings should be carefully justified since high transmembrane pressure drop and recirculation rates during membrane separation require a significant amount of energy, usually electricity. Membranes have been used in corn wet milling processes, such as steepwater concentration by reverse osmosis, concentration of corn syrups, reducing chemical oxygen demand in evaporator overhead, solvent recovery in corn oil extraction with solvents, oil purification in the corn oil refining stage, concentrating starch, and recovering fresh water (Singh and Cheryan, 1997). Reverse osmosis can effectively increase the concentration of steepwater to 14%–18% total solid content or up to a pressure difference of 1.38 MPa (Singh and Cheryan, 1997).

17.4.2 Applications of Membrane Separation in Vegetable Oil Extraction with Solvents

Extraction of soybean oil with hexane is a common industrial practice. The raw extract normally contains 25%–30% of soybean oil by weight and the remaining 70%–75% is hexane (Wu and Lee, 1999). Removal of more than 70% hexane in the extract by distillation consumes most of the energy cost in a typical soybean oil plant. Ultrafiltration or reverse osmosis with membranes can replace conventional distillation used in the processing of soybean oil to minimize thermal damage to products and reduce energy consumption for hexane evaporation (Cuperus and Nijhuis, 1993). Wang and Lee (1999) used a cross-flow ultrafiltration ceramic membrane with a pore diameter of 0.02 μm and thickness of ~1 μm to separate hexane from soybean oil. At a 0.392 MPa transmembrane pressure and 120 rpm agitation speed, the concentration of soybean oil was decreased from 33% of the feed to 27% in the permeate at 20% rejection. Smaller pore size is essential for effective separation. However, the permeate flux usually becomes low when the pore size is reduced. Multiple stages with recycles can be applied to increase the overall rejection.

Supercritical CO_2 is a promising alternative solvent to hexane for the extraction of vegetable oil. Supercritical fluid processing of foods is addressed in Chapter 22. Membranes were also used to separate the small CO_2 molecule from the large molecules of soybean oil (Wang and Shen, 2005; Jakubowska et al., 2005). Supercritical CO_2 with membrane separation consumes only about 50% of the total energy used in hexane extraction because the membrane filtration avoids reducing the pressure of CO_2 for separation in traditional supercritical CO_2 extraction. The energy consumption decreases with increasing membrane efficiency (Li et al., 2006).

17.4.3 Applications of Membrane Separation in Concentration of Liquid Foods

Traditionally, evaporation is one of the most energy-intensive unit operations used for concentrating thin sugar juice and fruit juices. Application or combination of low-energy consuming membranes can significantly save energy. Compared to evaporation, membrane separation is not involved in phase change and the energy consumption of membrane separation is thus low. A two-stage reverse osmosis system has been studied for pre-concentrating sugar syrup (Madaeni and Zereshki, 2008).

17.4.4 Applications of Membrane Separation in Recovery of Sugars from Fruit and Vegetable Processing Wastewater

Reverse osmosis method can recover 1.42 million tons of fermentable sugars at a 20% sugar concentrate from the effluent wastewater in the fruit and vegetable processing sector in the United States. The recovered sugar solution can be used for the production of liquid fuels. Because reverse osmosis processing is relatively expensive, excessively dilute wastewater, which contains less than 0.2% sugars is too costly to recover the dissolved sugars using the reverse osmosis method. The potential for the production of ethanol from the recovered sugars is between 750 and 900 million liters per year in the United States. Furthermore, the recovery of sugars from wastewater could reduce the biological oxygen demand of the wastewater and thus reduce the disposal cost (Blondin et al., 1983).

17.4.5 Applications of Membrane Separation in Pasteurization of Liquid Foods

Microfiltration and ultrafiltration have been used to remove bacteria and spores from milk and juice at a low temperature (Ortega-Rivas et al., 1998; Juriaanse, 1999; Carneiro et al., 2002). The shelf life of milk treated with microfiltration was found to be more than 21 days, compared to 6 days of thermally pasteurized milk (Juriaanse, 1999). Membrane treatment is an efficient technique to inactivate microbes in apple juice and pineapple juice (Ortega-Rivas et al., 1998; Carneiro et al., 2002). Carneiro et al. (2002) found that the permeate flux was 100 L/m^{2}h for cold sterilization of pineapple juice with a tubular polyethersulfone membrane with a 0.3 μm pore size operating at 25°C and 100 kPa. The microfiltered pineapple juice, which was kept at 8°C for 28 days, was still safe to drink.

17.5 SUMMARY

Energy-efficient membrane technologies have been used to replace or combine with energy-intensive unit operations of distillation and evaporation to concentrate dilute solutions in different food processing facilities such as wet corn milling, vegetable oil extraction, and sugar manufacturing. Membrane treatment has also been used to recover solid by-products from wastewater, for example, sugars from fruit and vegetable processing wastewater. Microfiltration and ultrafiltration have been used to remove bacteria and spores from liquid foods such as milk and juice at a low temperature to achieve a cold pasteurization effect on these liquid foods.

REFERENCES

Blondin, G.A., S.J. Comiskey, and J.M. Harkin. 1983. Recovery of fermentable sugars from process vegetable wastewaters. *Energy in Agriculture* 2: 21–36.

Carneiro, L., I. dos Santos Sa, F. dos Santos Gomes, V.M. Matta, and L.M.C. Cabral. 2002. Cold sterilization and clarification of pineapple juice by tangential microfiltration. *Desalination* 148: 93–98.

Criscuoli, A., M.C. Carnevale, and E. Drioli. 2008. Evaluation of energy requirements in membrane distillation. *Chemical Engineering and Processing* 47: 1098–1105.

Cuperus, F.P. and H.H. Nijhuis. 1993. Applications of membrane technology to food processing. *Trends in Food Science and Technology* 4: 277–282.

Fane, A.G., A. Yeo, A. Law, K. Parameshwaran, F. Wicaksana, and V. Chen. 2005. Low pressure membrane processes-doing more with less energy. *Desalination* 185: 159–165.

Jakubowska, N., Z. Polkowska, J. Namiesnik, and A. Przyjazny. 2005. Analytical applications of membrane extraction for biomedical and environmental liquid sample preparation. *Critical Reviews in Analytical Chemistry* 35: 217–235.

Juriaanse, A.C. 1999. Changing pace in food science and technology: Examples from dairy science show how descriptive knowledge can be transferred into predictive knowledge. *Trends in Food Science & Technology* 10: 303–306.

Kumar, A., S. Croteau, and O. Kutowy. 1999. Use of membranes for energy efficient concentration of dilute steams. *Applied Energy* 64: 107–115.

Li, Y., E. Griffing, M. Higgins, and M. Overcash. 2006. Life cycle assessment of soybean oil production. *Journal of Food Process Engineering* 29: 429–445.

Madaeni, S.S. and S. Zereshki. 2008. Reverse osmosis alternative: Energy implication for sugar industry. *Chemical Engineering and Processing: Process Intensification* 47: 1075–1080.

Ortega-Rivas, E., E. Zarate-Rodriguez, and G.V. Barbosa-Canovas. 1998. Apple juice pasteurization using ultrafiltration and pulsed electric fields. *Food and Bioproducts Processing* 76: 193–198.

Ramirez, C.A., M. Patel, and K. Blok. 2006. From fluid milk to milk power: Energy use and energy efficiency in the European dairy industry. *Energy* 31: 1984–2004.

Rausch, K.D. 2002. Front end to backpipe: Membrane technology in the starch processing industry. *Starch/Starke* 54: 273–284.

Singh, N. and M. Cheryan. 1997. Membrane technology in corn wet milling. *American Association of Cereal Chemists, Inc.* 42: 520–525.

Wang, L.J. and W. Shen. 2005. Chemical and morphological stability of Aliquat 336/PVC membranes in membrane extraction: A preliminary study. *Separation and Purification Technology* 46: 51–62.

Wu, J.C.S. and E.H. Lee. 1999. Ultrafiltration of soybean oil/hexane extract by porous ceramic membranes. *Journal of Membrane Science* 154: 251–259.

18 Energy Efficiency and Conservation in Food Irradiation

18.1 INTRODUCTION

Irradiation pasteurizes foods by exposing them to high-energy electrons ($\lambda = 10^{-11}$ to 10^{-13} m) such as x-rays, electron beams, and gamma rays, which are similar to light waves ($\lambda = 4 \times 10^{-7}$ to 7×10^{-7} m), ultraviolet waves ($\lambda = 10^{-7}$ m), or microwaves ($\lambda = 10^{-3}$ to 10^{0} m) but have shorter wave length (Tragardh, 1986; Farkas, 2006; Nayak et al., 2007; Thomas et al., 2008). Electron-beam technology is currently being developed as a safer alternative to gamma radiation generated by radioactive isotopes (Lado and Yousef, 2002). Irradiation has been claimed as a new preservation method with a lower energy demand than conventional ones. The radiation energy, usually at a dose below 10 kGy (or 10 kJ/kg product), causes changes in molecules by breaking chemical bonds. Benefits of food irradiation include reduced storage losses, extended shelf life, and improved microbial safety of foods. Including irradiation processes in the food supply chain also affects other operations such as packaging, transport, and storage (Tragardh, 1986).

The U.S. Food and Drug Administration regulated that the kinetic energy of an electron beam for direct irradiation is limited to 10 MeV and the kinetic energy limit for x-ray irradiation is 7.5 MeV to avoid the possibility of food activation via photonuclear reactions (Miller et al., 2003; Farkas, 2006). In the United States, the first commercial food irradiation plant was opened in Tampa, Florida, in 1992 to treat fruits and vegetables. Irradiated vegetables and fruits have been sold in small quantities since 1992 and irradiated poultry was introduced in the marketplace in 1993. Irradiation has also been used to treat red meats, seafood, poultry, and shell eggs (Derr et al., 1995; Farkas, 1998). It was reported that there was significant saving by reducing spoilage losses from 10% of nonirradiated strawberries to about 2% of irradiated ones (Marcotte, 1992). In this chapter, the food irradiation process and its applications are briefly reviewed. The energy consumption of an irradiation process is discussed.

18.2 OVERVIEW OF MAIN FOOD IRRADIATION PROCESSES

18.2.1 Irradiation Mechanism

Ionizing radiation damages the DNA effectively so that living cells become inactivated. Therefore, microorganisms, insect gametes, and plant meristems are prevented from reproduction. The radiation energy causes changes in molecules by breaking chemical bonds. At small doses, irradiation inhibits or modifies food spoilage microorganisms. Medium doses will kill or genetically alter microorganisms so that they cannot reproduce to cause spoilage. High doses will sterilize foods. The effect of radiation is sensitive to the chemical and physical structure of cells of microorganisms and the ability of cells to recover from radiation injury. Cells that are unable to repair their radiation-damaged DNA will die. Sublethally injured cells are often subject to mutations. Ionizing radiations generate hydroxyl radicals from water, which remove hydrogen atoms from the sugar and the bases of the DNA strands (Lado and Yousef, 2002).

18.2.2 Irradiation Process

Three sources of ionizing radiation are used in commercial radiation processing: gamma rays from radioactive sources, electron beams, and x-rays. Gamma rays are produced by a radioactive source. The latter types of radiation are produced by electron accelerators powered by electricity (Leboutet and Aucouturier, 1977; Morrison, 1990). Since the supply of radioactive sources is limited, the use of accelerators is therefore critical to the development of food irradiation technologies (Sivinski and Sloan, 1985).

Gamma radiation uses a radioactive source, usually cobalt-60, generating gamma radiation with energies of 1.17 and 1.33 MeV. This isotope is produced on irradiation of nonradioactive cobalt-59. Cobalt-60 is encapsulated in several layers of stainless steel and cannot contaminate any materials including foods in the irradiation chamber. As an alternative to cobalt-60, the U.S. Department of Energy introduced cesium-137, a gamma emitter obtained from reprocessed nuclear waste. The radioactive source is usually stored in a pool of water when not in use and is raised into an irradiation chamber for irradiation treatment. Gamma radiation has a high penetrating power, which is suitable for treatment of large items such as chickens and drums of foods (Kilcast, 1995). In the industry, the majority of facilities use cobalt-60 because it has a stronger gamma ray and it is insoluble in water.

The second source of ionizing radiation is high-energy electrons from machines. These machines have the advantage that they can be switched off when not in use, leaving no radiation hazard. However, a major limitation of electron beams for food use is their limited penetration depth, up to a maximum of about 8 cm in food for the maximum permitted energy of 10 MeV, while the maximum penetration depth for 10 MeV electrons in high moisture foods is only 3.9 cm. They can be used for products such as grain on a conveyor or low-density foods such as ground spices. They can also be used to remove surface contamination on prepared meals (Kilcast, 1995). There are three basic industrial electron accelerators: (1) direct current (DC) machines, (2) Rhodotrons, and (3) pulsed and continuous wave linear accelerators (Alimov et al., 2000). The relationships between beam energy and power for these three types of accelerators are given in Figure 18.1 (Alimov et al., 2000). Low-energy applications with electron beams up to 1 MeV are adequately created by DC machines. Pulsed and continuous wave linear accelerators can generate higher beam power. The cost of electron accelerators varies from 3 $/W for a low-cost accelerator to 20 $/W for a high-cost accelerator (Alimov et al., 2000).

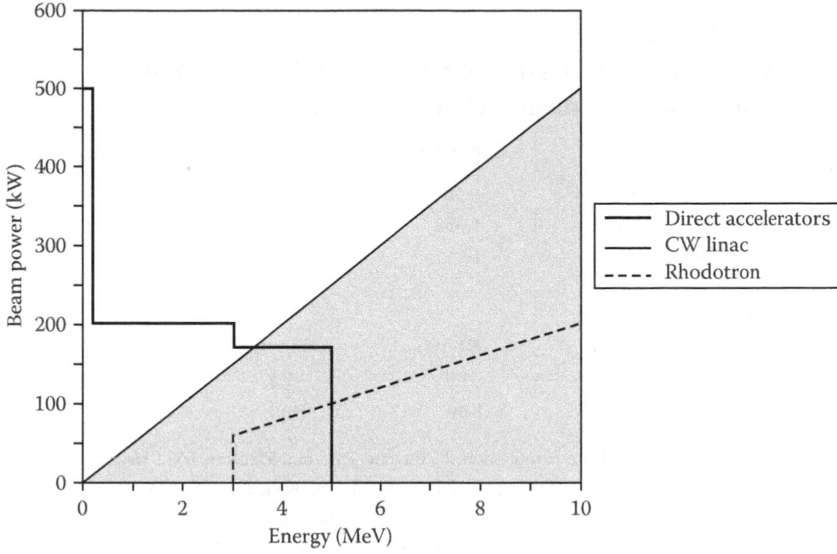

FIGURE 18.1 Relationship between beam energy and power for three types of electron accelerators. (Reprinted from Alimov, A.S., Knapp, E.A., Shvedunov, V.I., and Trower, W.P., *Appl. Radiation Isotopes*, 53, 815, 2000. With permission.)

The third radiation source is x-rays, which are generated from bombardment of a metal target by electrons. They are less well developed and are not used for food treatment because of the low conversion efficiency of electrons to x-rays. However, their advantages include high penetration power and switch-off capability (Kilcast, 1995). The specifications of a 7.5 MeV x-ray linear accelerator is given in Table 18.1 (Eichenberger et al., 2005).

TABLE 18.1
Systems Specifications of a 7.5 MeV Linear Accelerator X-Ray System

Parameter	Value
Kinetic energy (MeV)	7.5
Average power (kW)	100
Average beam current (Amps)	0.013
Duty cycle	~0.04
Pulse duration (μs)	125
Repetition rate (Hz)	325
Peak beam current (Amps)	0.33
Peak beam power (MW)	2.5
Radio frequency power efficiency (%)	60

Source: Reprinted from Eichenberger et al., *Proceedings of Pulsed Power Conference, 2005 IEEE*, 1274–1277, Monterey, California. With permission.

TABLE 18.2
Summary of the Depth and Efficiency of Three Ionized-Irradiation Technologies Used in the Food Industry

	Gamma Ray	X-Ray	E-Beam
Power source (kW)	~ 50	25	35
Source energy (MeV)	1.33	5	5–10
Processing capacity to deliver a dose rate of 4 kGy (tons/h)	12	10	5–10
Penetration depth (cm)	80–100	80–100	8–10
Dose uniformity ratio	~1.7	~1.5	Moderate
Dose rate (kGy/h)	Low	High	High

Sources: Reprinted from Amymerich, T., Picouet, P.A., and Monfort, J.M., *Meat Sci.*, 78, 815, 2008. With permission; Adapted from Koutchma, T. 2006. Emerging technologies in food processing and packaging: Irradiation. In: *Proceedings of the Fourth International Feeding Congress*. With permission.

The depth and efficiency of three ionized-irradiation technologies used in the food industry are given in Table 18.2 (Koutchma, 2006). The main characteristics of three radiation sources are summarized below (Kilcast, 1995):

1. Gamma rays with cobalt-60
   - High penetrating power
   - High efficiency
   - Permanent radioactive source
   - Source replenishment needed
   - Low throughput
2. Electron beams
   - Low penetrating power
   - High efficiency
   - Switch-on and -off capacity
   - High throughput
   - Power and cooling needed
   - Technically complex
3. X-rays
   - High penetrating power
   - Switch-on and -off capacity
   - Low efficiency
   - High throughput
   - Power and cooling needed
   - Technically complex

Irradiation treatment costs range from 1 to 15 ¢/kg of foods, and large volumes of foods can lower treatment costs. Average treatment costs per kilogram of irradiated foods are similar for electron accelerators and cobalt-60 irradiators. Cobalt-60 is less expensive than electrons when annual volumes are below 23 million kilograms. Irradiation of fruits with x-rays is more expensive than with cobalt-60 (Morrison, 1990).

Radiation treatment does not cause significant temperature rise in products. Irradiation can be applied through packaging materials. However, because of potential changes in the materials caused by irradiation, the safety of packaging materials must be tested before use. When commercial packaging films including polyethylene (LDPE), amide 6-amide 6.6 copolymer (PA6-PA6.6), and poly ethylene terephthalate (PET) are exposed to gamma irradiation doses up to 100 kGy, there are no observable changes in optical properties of PET while there are changes in optical properties of LDPE and PA6-PA6.6 depending on the used dose, which might be caused by cross-linking and degradation due to irradiation (Moura et al., 2004). Pentimalli et al. (2000) investigated the effects of gamma irradiation at a dose from 1 to 100 kGy on polystyrene, poly-butadiene, styrene-acrylonitrile, high-impact polystyrene, and acrylonitrile-butadiene-styrene used in food packaging with NMR. They found that the effect of gamma irradiation on polystyrene is negligible even at high doses and polystyrene was thus confirmed as an ideal polymer for food packaging use during irradiation. Styrene-acrylonitrile copolymers are not as resistant to radiation as polystyrene itself but are still fairly stable. Poly-butadiene very easily undergoes both cross-linking and degradation.

During food irradiation, prepackaged foods are usually moved on a conveyor into a thick-walled room and exposed to the electron beams for several seconds to several minutes or gamma rays for 15–45 min (de Nruyn, 2000).

18.2.3 Applications of Food Irradiation

Irradiation has been used in more than 50 countries to treat one or more food products. Irradiated foods worldwide increased from about 200,000 tons in 1997 to 257,000 tons in 1999. Ionizing radiation is a safe and effective method for eliminating bacterial pathogens from food products and disinfesting fruits and vegetables. Specifically, irradiation has been used for (Kilcast, 1995):

- Elimination or reduction of pathogenic organisms
- Retardation of the decay process and destruction of spoilage organisms
- Reduction of wastes due to premature ripening, germination, or sprouting
- Disinfestations of plants and plant products

Low irradiation doses, less than 1 kGy, can be used to inhibit sprouting in produces such as potatoes and onions, prevent insect infestation of grains and fruits, and delay ripening of certain fruits. However, the extension of shelf life is the main goal of radiation processing of foods at slightly higher doses, e.g., 1–3 kGy. Although thermal pasteurization is a well-established and effective method to decontaminate and disinfect liquid foods such as milk and fruit juices, it is not suitable for solid foods. Irradiation, which is a cold process, is more suitable for pasteurization of solid foods such as nuts, cereals, spices, and tea without causing significant changes in taste and quality of the products (Loaharanu, 1996; Golge and Ova, 2008; Thomas et al., 2008).

Irradiation was used to treat approximately 90,000 ton of spices, herbs, and dry vegetable seasonings in 2002 (Rubio, 2003; Farkas, 2006). Radiation decontamination of dry ingredients, herbs, and enzyme preparations is a technically feasible, economically viable, and safe physical process. The procedure is direct, simple, and highly efficient, and requires no additives. The radiation dose requirement is moderate in the range of 3–10 kGy. The flavor, texture, and other properties of the ingredients are not affected by irradiation at a moderate dose. Irradiation can be carried out in commercial containers and result in considerable savings of energy, compared to other preservation technologies (Farkas, 1984).

Irradiation has also been used to treat vegetables, fruits, grains, and tea (Farkas, 2006; Cabeza et al., 2007). In the United States, the first commercial food irradiation plant was opened in Tampa, Florida, in 1992 to treat fruits and vegetables.The first commercial x-ray irradiator for irradiation of fruits started operating in Hawaii in July 2000. Irradiation has been used to treat refrigerated fresh ground beef and frozen beef patties to eliminate or significantly reduce *Escherichia coli* O157:H7 and other hazardous bacteria since May 2000. Irradiation can also be used to eliminate pathogens such as *E. coli* O157:H7, *Salmonella* spp., and *Listeria monocytogenes* from ready-to-eat (RTE) food products including hot dogs, bologna, lettuce, cilantro, sprouts and seeds, and frozen vegetables (Sommers et al., 2004; Golge and Ova, 2008).

Irradiation doses at 0.0, 0.5, 1.0, 3.0, and 5.0 kGy applied to pine nut kernels have been found to induce no significant change in fatty acid composition, color, and texture properties of the kernels. Furthermore, the effect of storage on the properties was also insignificant (Golge and Ova, 2008). However, the exposure to irradiation results in an increase in cell wall permeabilization, which may lead to softening of tissue, thus affecting the textural properties of some food products. Nayak et al. (2007) found that textural properties such as hardness, cohesiveness, springiness, gumminess, and chewiness decreased with an increase in irradiation doses up to 12 kGy. They investigated a calcium pretreatment method to reduce the extent of damage to food texture during irradiation. Irradiation can also be combined with other preservation methods such as heating, cooling, modified atmosphere, or vacuum packaging to minimize the development of undesirable sensory and chemical changes in some foods (Thakur and Singh, 1995). The ionizing radiation dose required to eliminate pathogens from foods depends on products, formulation, and temperature. The need to eliminate bacterial pathogens from food products must always be balanced with the maintenance of product quality. In addition to determining the effective ionizing radiation doses required for pathogen elimination, the effects of irradiation on product chemistry, nutritional value, and organoleptic quality should also be determined.

18.3 ENERGY USE, EFFICIENCY, AND CONSERVATION

18.3.1 Efficiency of Food Irradiation

Food irradiation with electron beams and x-ray consumes high-quality electricity. Gamma rays use radioactive materials such as cobalt-60. The energy efficiency is affected by photon utilization, conversion efficiency, and self-absorption. An overall energy efficiency of 4% can be achieved from a 5 MeV x-ray machine and 8% can be achieved from a 7.5 MeV machine. A 7.5 MeV x-ray machine source with a 100 kW electron beam power is comparable in efficiency to that of a gamma source. A 5 MeV x-ray processing facility using a 100 kW accelerator at the x-ray converter can process 2.5 tons of frozen seafood per hour (Eichenberger, 2005). Electronvolt (symbol eV) is a unit of energy. It is the amount of energy equivalent to that gained by a single unbound electron when it is accelerated through an electrostatic potential difference of one volt.

18.3.2 Irradiation Dose and Energy Consumption

The absorbed dose is considered the most important parameter in an irradiation process but the effectiveness of the treatment also depends on the sensitivity of microorganisms, the extrinsic characteristic of environments such as temperature and pH value, and the intrinsic characteristics of foods such as composition and additives (Amymerich et al., 2008). The dose of radiation is measured in the SI unit known as Gray (Gy). One Gray (Gy) dose of radiation is equal to 1 Joule of energy absorbed per kilogram of food material. In radiation processing of foods, the doses are generally measured in kGy (=1000 Gy). In 1997, an FAO/IAEA/WHO study group on high-dose irradiation carried out a safety study on food irradiated with a dose above 10 kGy (10 kJ/kg). They found that irradiation of any food products up to an overall dose of 10 kGy caused no toxicological hazards. Few foods even tolerated a dose above 10 kGy. Irradiation of foods up to a dose of 10 kGy introduced no special microbiological or nutritional problems (Molins et al., 2001). Long-term animal feeding studies with foods irradiated with a dose as high as 70 kGy have not shown any treatment-related adverse health effect (Diehl, 2002).

The amount of radiation energy required to control microorganisms in foods varies with the resistance of a specific microorganism to radiation (Farkas, 2006).The energy used for irradiation of foods is extremely small and the maximum dose of food irradiation is usually less than 10 kGy (equivalent to 10 kJ/kg). Tables 18.3 and 18.4 give the guidelines of dose requirements for various applications of food irradiation,

TABLE 18.3
Food Permitted to Be Irradiated Under U.S. FDA's Regulations

Food	Dose (kGy)	Electricity Consumption at 5% Efficiency (kWh/ton)
Fresh pork	0.3–1	1.67–5.56
Fresh foods	1	5.56
Foods	1	5.56
Dry enzyme preparations	10	55.56
Dry spices/seasonings	30	166.67
Poultry	3	16.67
Frozen meats (NASA)	44	244.44
Refrigerated meat	4.5	25
Frozen meat	7	38.89
Shell eggs	3	16.67
Seeds for sprouting	8	44.44

TABLE 18.4
Food Permitted to Be Irradiated under European Commission's Regulations

Food	Dose (kGy)	Electricity Consumption at 5% Efficiency (kWh/ton)
Blood products	10	55.56
Cereal flakes	10	55.56
Rice flour	4	22.22
Casin/caseinates	6	33.33
Shrimps	5	27.78
Egg white	3	16.67
Frog legs	5	27.78
Arabic gum	3	16.67

which were issued by U.S. FDA (Morehouse, 2002) and the European Commission (Neyssen, 2000).

18.3.3 Temperature Increase during Food Irradiation

The dose of food irradiation is usually less than 10 kGy, which is equivalent to 10 kJ/kg of foods. Suppose the specific heat of foods is 3.5 kJ/kg°C, the increase in food temperature due to irradiation is only 2.86°C at a dose of 10 kGy. Therefore, food irradiation is a cold pasteurization process and the thermal effect of the process is negligible.

18.4 SUMMARY

Food irradiation, which is a cold process, damages the DNA effectively so that living cells become inactivated. Compared with thermal pasteurization, food irradiation is a more efficient pasteurization method for solid foods, without causing significant changes in taste and quality of the products. There are three main radiation sources: gamma rays, x-rays, and electron beams. Gamma rays and x-rays have high penetrability, which can be used to treat food even in pallet-size containers. X-rays are not well developed for food use because of their low electron conversion efficiency. Unlike gamma rays and x-rays, which can penetrate pallet loads of foods, the maximum penetration depth for 10 MeV electrons in high moisture foods is only 3.9 cm. However, electron beams can be switched off when not in use. The energy level used for irradiation of foods is very small. The dose of irradiation of foods is usually less than 10 kGy (energy equivalent to 10 kJ/kg of foods). The increase in food temperature due to irradiation is less than 3°C.

REFERENCES

Alimov, A.S., E.A. Knapp, V.I. Shvedunov, and W.P. Trower. 2000. High-power CW LINAC for food irradiation. *Applied Radiation and Isotopes* 53: 815–820.

Amymerich, T., P.A. Picouet, and J.M. Monfort. 2008. Decontamination technologies for meat products. *Meat Science* 78: 114–129.

Cabeza, M.C., I. Cambero, L. de la Hoz, and J.A. Ordóñez. 2007. Optimization of E-beam irradiation treatment to eliminate *Listeria monocytogenes* from ready-to-eat (RTE) cooked ham. *Innovative Food Science & Emerging Technologies* 8: 299–305.

de Nruyn, I.N. 2000. The application of high dose food irradiation in South Africa. *Radiation Physics and Chemistry* 57: 223–5.

Derr, D.D., D.L. Engeljohn, and R.L. Griffin. 1995. Progress of food irradiation in the United States. *Radiation Physics and Chemistry* 46: 681–688.

Diehl, J.F. 2002. Food irradiation-past, present and future. *Radiation Physics and Chemistry* 63: 211–215.

Eichenberger, C., D. Palmer, S.L. Wong, G. Robison, B. Miller, and D. Shimer. 2005. 7.5 MeV high average power linear accelerator system for food irradiation applications. In: *Proceedings of Pulsed Power Conference, 2005 IEEE*, 1274–1277, Monterey, California.

Farkas, J. 1984. Radiation decontamination of dry food ingredients and processing aids. *Journal of Food Engineering* 3: 245–264.

Farkas, J. 1998. Irradiation as a method for decontaminating food: A review. *International Journal of Food Microbiology* 44: 189–204.

Farkas, J. 2006. Irradiation for better foods. *Trends in Food Science & Technology* 17: 148–152.

Golge, E. and G. Ova. 2008. The effects of food irradiation on quality of pine nut kernels. *Radiation Physics and Chemistry* 77: 365–369.

Kilcast, D. 1995. Food irradiation: Current problems and future potential. *International Biodeterioration & Biodegradation* 36: 279–296.

Koutchma, T. 2006. Emerging technologies in food processing and packaging: Irradiation. In: *Proceedings of the Fourth International Feeding Congress,* Colombia, Medellin.

Lado, B.H. and A.E. Yousef. 2002. Alternative food-preservation technologies: Efficacy and mechanisms. *Microbes and Infection* 4: 433–440.

Leboutet, H. and J. Aucouturier. 1977. Theoretical evaluation of induced radioactivity in food products by electron-or x-ray beam sterilization. *Radiation Physics and Chemistry* 25: 233–242.

Loaharanu, P. 1996. Irradiation as a cold pasteurization process of food. *Veterinary Parasitology* 64: 71–82.

Marcotte, M. 1992. Irradiated strawberries enter the US market. *Food Technology* 46: 80–86.

Miller, R.B., G. Loda, R.C. Miller, R. Smith, D. Shimer, C. Seidt, M. MacArt, H. Mohr, G. Robison, P. Creely, J. Bautista, T. Oliva, L.M. Young, and D. DuBois. 2003. A high-power electron linear accelerator for food irradiation applications. *Nuclear Instruments and Methods in Physics Research* 211: 562–570.

Molins, R.A., Y. Motarjemi, and F.K. Kaferstein. 2001. Irradiation: A critical control point in ensuring the microbiological safety of raw foods. *Food Control* 12: 347–356.

Morehouse, K.M. 2002. Food irradiation-US regulatory considerations. *Radiation Physics and Chemistry* 63: 281–4.

Morrison, R.M. 1990. Economics of food irradiation: Comparison between electron accelerators and cobalt-60. *International Journal of Radiation Applications and Instrumentation. Part C. Radiation Physics and Chemistry* 35: 673–679.

Moura, E.A.B., A.V. Ortiz, H. Wiebeck, A.B.A. Paula, A.L.A. Silva, and L.G.A. Silva. 2004. Effects of gamma radiation on commercial food packaging films—study of changes in UV/VIS spectra. *Radiation Physics and Chemistry* 71: 201–204.

Nayak, C.A., K. Suguna, K. Narasimhamurthy, and N.K. Rastogi. 2007. Effect of gamma irradiation on histological and textural properties of carrot, potato and beetroot. *Journal of Food Engineering* 79: 765–770.

Neyssen, P.J.G. 2000. Practical implications of developments in legislation on food irradiation in the European Union. *Radiation Physics and Chemistry* 57: 215–217.

Pentimalli, M., D. Capitani, A. Ferrando, D. Ferri, P. Ragni, and A.L. Segre. 2000. Gamma irradiation of food packaging materials: An NMR study. *Polymer* 41: 2871–2881.

Rubio, T. 2003. Legislation and application of food irradiation. Prospects and controversies. *Ernaehrung/Nutrition* 27: 18–22.

Sivinski, J.S. and D.P. Sloan. 1985. The role of linear accelerators in industry. *Nuclear Instruments and Methods in Physics Research Section B: Beam Interactions with Materials and Atoms* 10–11: 981–986.

Sommers, C., X. Fan, B. Niemira, and K. Rajkowski. 2004. Irradiation of ready-to-eat foods at USDA'S Eastern Regional Research Center-2003 update. *Radiation Physics and Chemistry* 71: 511–514.

Thakur, B.R. and R.K. Singh. 1995. Combination processes in food irradiation. *Trends in Food Science & Technology* 6: 7–11.

Thomas, J., R.S. Senthilkumar, R.R. Kumar, A.K.A. Mandal, and N. Muraleedharan. 2008. Induction of γ irradiation for decontamination and to increase the storage stability of black teas. *Food Chemistry* 106: 180–184.

Tragardh, C. 1986. Energy requirements in food irradiation. In *Energy in Food Processing*, Chapter 12, pp. 203–225, R.P. Singh (Ed.), New York: Elsevier Science Publishing Company Inc.

19 Energy Efficiency and Conservation in Pulsed Electric Fields Treatment

19.1 INTRODUCTION

Sterilization and pasteurization are usually achieved by a thermal process. Thermal food processes are facing several challenges. It is difficult to control the process for the set time–temperature requirement. If the temperature is lower than the specified limit, the pasteurization may be inefficient for the safety of products. Overheating may cause thermal damage to products. Furthermore, thermal sterilization and pasteurization inactivate microorganisms present in foods but also degrade the quality and nutritional value of foods.

Application of an external electrical field to a biological cell induces an electrical potential across the cell membrane. If the electrical potential reaches a critical level, a drastic increase in the permeability of the cell membrane occurs, which impairs on the irreversible loss of physiological control systems and therefore causes cell death (Heinz et al., 2003). Pulsed electric field (PEF) technology has been shown to be an attractive alternative to thermal treatments. Since a treatment at an electric field strength of 20–40 kV/cm will destroy the structure of solid foods, PEF treatments seem impossible for pasteurization of solid foods and are limited to liquid foods (Toepfl et al., 2006). PEF technology has been used to pasteurize several liquid foods such as liquid egg products (Gongora-Nieto et al., 2003) and apple juice (Heinz et al., 2003). In this chapter, the PEF process is reviewed. The energy efficiency of PEF is analyzed. Finally, the opportunities for improving the energy efficiency of a PEF process are discussed.

19.2 PROCESS OVERVIEW

19.2.1 Pulsed Electric Fields Treatment Mechanism

When a living cell is suspended in an electrical field, an electric potential is induced across the membrane of the cell. The electric potential causes an electrostatic charge separation in the cell membrane based on the dipole nature of the membrane molecules (Bryant and Wolfe, 1987). When the transmembrane potential exceeds a critical value of approximately 1 V, the repulsion between charge-carrying molecules initiates the formation of pores in weak areas of the membrane. This results in a drastic increase in permeability, which leads to an equilibration of electrochemical and electric potential difference between the cytoplasma and the surrounding media and impairs on the irreversible loss of physiological control systems and causes cell

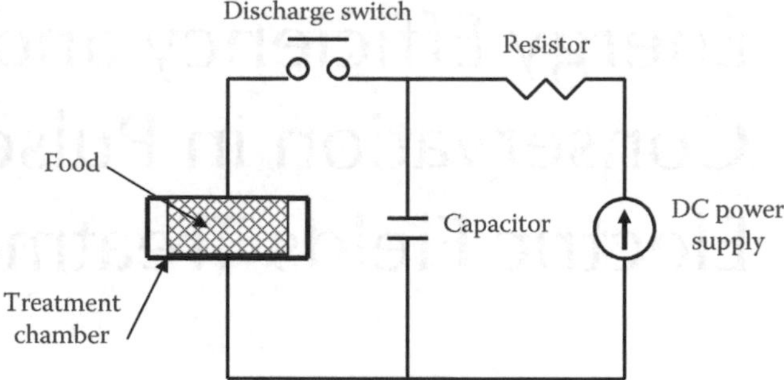

FIGURE 19.1 A simple circuit for a pulsed electric field treatment of foods with exponential decay pulses.

death. A lethal effect on living cells is observed when the transmembrane electric potential exceeds the critical value by a wide margin.

The degree of microbial inactivation by PEF treatment strongly depends on the process parameters including field strength, specific energy input, treatment temperature, and the properties of food products (Gaskova et al., 1996; Heinz et al., 1999; Muraji et al., 1999; Heinz et al., 2003).

19.2.2 Pulsed Electric Fields Treatment Process

A simple circuit for a PEF treatment of foods with exponential decay pulses is given in Figure 19.1. In practice, the time between pulses is much longer than the pulse width. Therefore, the generation of pulses involves slow charging and fast discharging of an electrical energy storage device such as a capacitor. A high-voltage power

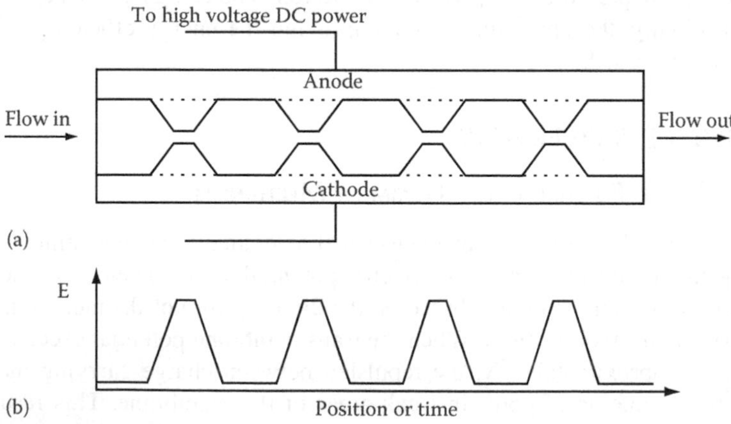

FIGURE 19.2 A continuous flow treatment chamber. (a) Chamber design and (b) electric field as a function of time or location as fluid flows. (Reprinted from Zhang, Q.H., Barbosa-Canovas, G.V., and Swanson, B.G., *J. Food Eng.*, 25, 261, 1995. With permission.)

supply is needed to charge the capacitor. A treatment chamber consists of two electrodes that are held in position by insulation materials. A PEF process can operate in either a batch or continuous mode depending on the treatment chamber design. A continuous treatment chamber design is given in Figure 19.2.

Pulsed shaped voltage is required for the preservation of foods with PEF. PEF treatment can be described by its

- Form of pulse
- Voltage or electric field
- Pulse duration time
- Total number of pulses

The pulse forms generated by a variety of circuits include square pulse, trapezoidal pulse, exponential decay pulse, and oscillatory and over-damped oscillatory pulse (De Haan and Willcock, 2002). The square form is the preferred pulse shape. The pulse is characterized by its certain minimum threshold voltage established during a certain minimum time. For a slow rising pulse, the voltage will be below the threshold level for some time, which contributes to energy dissipation but not to bacterial inactivation. Voltages higher than the threshold level will cause extra energy dissipation, which may not necessarily contribute to better bacterial inactivation.

The ions present in foods are electrical charge carriers and foods, particularly liquid foods, are electrical conductors. To inactive the microorganism in foods, a high-voltage PEF at tens of kilovolts per centimeter (e.g., 30 kV/cm) within the foods is needed. The high-pulsed electric field is generated by high-voltage electrodes as shown in Figure 19.1. Electric field strength is determined by the voltage across the electrodes and the distance between two electrodes as given by

$$E = \frac{V}{d} \tag{19.1}$$

For a certain electric field strength, the required voltage will increase with an increase in the gap between two electrodes.

A large flux of electrical current must flow through a piece of food in a treatment chamber for a very short period (i.e., microseconds) (Zhang et al., 1995). The pulse width is very short at several microseconds. Usually, a product is treated for several pulses to a couple of hundreds of pulses. The total treatment time is thus less than one second (Gongora-Nieto et al., 2003).

19.2.3 Pulsed Electric Fields Treatment Applications

PEF treatment can be used to inactivate enzyme activity and microorganisms in liquid foods such as fruit and vegetable juices (Qui et al., 1998; Hodgins et al., 2002; Heinz et al., 2003; Elez-Martinez and Martin-Belloso, 2007; Aguiló-Aguayo et al., 2008), milk (Reina et al., 1998; Bendicho et al., 2002; Craven et al., 2008), liquid egg (Gongora-Nieta et al., 2003), model beer (Ulmer et al., 2002), and nutrient broth (Selma et al., 2004). The peroxidase activity in tomato juice causes color deterioration, off-flavor formation, and loss of nutrients. The conventional thermal treatment to inactivate the enzyme can damage valuable properties of the juice. High-intensity PEF are effective in inactivating peroxidase in tomato juice (Nguyen and Mittal, 2007; Aguiló-Aguayo et al., 2008). Nguyen and Mittal (2007) found that there was about 4.4 log reduction in microbial counts with PEF treatment of tomato juice at a field strength of 80 kV/cm, 20 pulsed, 50°C, and in the presence of 100 U/mL nision. The treated juice, which was not aseptically filled, was stored at 4°C for 28 days without any significant microbial growth. At added *Pseudomonas* levels of 10^3 and 10^5 CFU/mL, the microbial shelf life of PEF-treated milk was extended by at least 8 days at 4°C compared with untreated milk. The total microbial shelf life of the PEF-treated milk was 13 and 11 days for inoculation levels of 10^3 and 10^5 CFU/mL, respectively. The results indicate that PEF treatment is useful for the reduction of pseudomonads, the major spoilage bacteria of milk (Craven et al., 2008).

PEF treatment can retain more high-value nutrients such as vitamin C and antioxidant in juices and soups than the thermal pasteurization method. Orange juice and gazpacho retained 87.5%–98.2% and 84.3%–97.1% of vitamin C, respectively, after pulsed electric field treatment (Elez-Martinez and Martin-Belloso, 2007).

PEF was also used to improve the extraction of intracellular compounds of plant and animal tissue (Toepfl et al., 2006). Schilling et al. (2007) found that the juice yield increased with increasing electric field intensities during PEF treatment of apple mash, and the quality of the juice was not significantly affected by the treatment. The PEF treatment can increase mass transfer. An almost total permeabilization of apple and potato tissue can be achieved with an electric energy input in the range of 1–5 kJ/kg of product, compared with 20–40 kJ/kg for mechanical, 60–100 kJ/kg for enzymatic, and above 100 kJ/kg for thermal degradation of plant tissue (Toepfl et al., 2006). PEF can increase the yield by 40% at a pressure of 0.2–0.3 MPa and 12% at a pressure of 3 MPa compared to untreated samples (Bazhal and Vorobiev, 2000). Similarly, PEF can enhance the mass transfer during drying (Toepfl et al., 2006).

19.3 ENERGY EFFICIENCY AND CONSUMPTION

19.3.1 Energy Efficiency

A circuit is needed to generate a pulsed electric field. The efficiency of the circuit is defined as the ratio of the minimum energy that is required to establish a pulse with the voltage amplitude and duration time over the actual energy that is needed to generate an actual pulse that is an envelope of the required pulse (De Haan and Willcock, 2002). According to this definition of energy efficiency, De Haan and Willcock (2002) compared the energy performance of pulse generation circuits for pulsed electric fields. As shown in Table 19.1, for highly damped RLC circuits with load resistance R, inductor L, and capacitor C, the maximum theoretical efficiency is only between 37% and 47%. For oscillating damped sinusoids, the maximum efficiency is between 47% and 52%. The energy efficiency of a trapezoid pulse is 100%.

TABLE 19.1
Overview of Efficiencies and Per Unit Charging Voltage

Circuit Type	Pulse Type	Maximum Efficiency (%)
Hard switched	Trapezoid	100
Transmission line	Trapezoid	100
RLC ($\alpha/\omega < 1$)	Oscillatory	47–52
RLC ($\alpha/\omega > 1$)	Damped	37–47
RLC ($\alpha/\omega > 1$)	Decay	37

19.3.2 Energy Consumption

The specific energy input, E_s, can be calculated by (Heinz et al., 2003)

$$E_s = f\frac{1}{m}\int_0^{\infty} k(T)E(t)^2\,dt \tag{19.2}$$

In the above equation, E_s is the specific energy input (kJ/kg), f is the repetition rate, m is the mass flow rate (kg/s), $k(T)$ is the media conductivity (S/m), and $E(t)$ is the electric field strength. The specific heat input varies from 5 to 120 kJ/kg for treatment of apple juice ($k(T) = 0.3$ S/m) at a mass flow rate of 3 kg/h with a pulsed electric field at an electric field strength from 8 to 40 kV/cm and the pulse repletion rate in the range between 2 and 95 Hz. A higher energy input results in higher inactivation rates (Heinz et al., 2003).

It should be noted that the electricity supply can be higher than the specific energy input since the energy efficiency of the pulsed field treatment system is smaller than 100%, usually around 50% depending on the pulse type as given in Table 19.1. Aronsson et al. (2005) reported that energy requirements were 160 kJ/kg and 384 kJ/kg for a 5 log cycle inactivation of *Saccharomyces cerevisiae* and *Escherichia coli*, respectively. The electrical energy consumption can be directly calculated from the measured voltage and current. Geveke et al. (2007) found that the electric energy consumption determined by the measured voltage and current was 180 kJ/L for a 3.3 log reduction in *E. coli* at an outlet temperature of 65°C during PEF treatment of orange juice. PEF treatment of milk provides an opportunity to increase the shelf life of fresh milk for distribution to distant markets. The greatest inactivation of *Pseudomonas* milk (>5 logs) was achieved by PEF processing at 55°C with 31 kV/cm. The specific energy requirement was 139.4 kJ/L. Heat treatment at the same temperature without PEF treatment caused minimal inactivation of *Pseudomonas* (only 0.2 logs). Therefore, the inactivation of the *Pseudomonas* in the milk was caused by the PEF treatment rather than the heat applied to the milk (Craven et al., 2008).

Gongora-Nieto et al. (2003) found that the liquid whole egg pasteurized with PEF required an optimum energy supply of 357 kJ/L to extend the shelf life of the product to be around 26 days. Energy inputs beyond this value negligibly extend the shelf life of the product. For a certain field strength to inactive a microorganism, the energy delivered to foods proportionally increases with the electrical conductivity of the foods, which is given by

$$Q = \frac{V^2}{R} \tag{19.3}$$

where

V is the voltage between two electrodes
R is the electrical resistance of the food

In terms of energy consumption, the PEF treatment is more suitable for treating foods with a low electric conductivity or high electric resistance.

19.3.3 Temperature Increase during PEF Treatment

Not all energy delivered by electric pulses is effectively used to inactive microorganisms in foods. The energy with the pulse generates heat due to ohmic heating, thus increasing the temperature of the foods. The maximum temperature increase, ΔT, due to the ohmic heating can be calculated by

$$\Delta T = \frac{Q}{\rho c_p} \tag{19.4}$$

In the above equation, ρ and c_p are the density and specific heat of the food. Suppose the density and specific heat of the egg are 1000 kg/m^3 and 3.5 kJ/kg, the maximum temperature increase in the liquid eggs during pulsed electric treatment at 357 kJ/L is 102°C.

To prevent the thermal damage to the product during PEF treatment, cooling is needed to remove the generated heat and maintain a low product temperature. The cooling energy is a direct function of the energy delivered by the pulse electric fields to the product. For PEF pasteurization of liquid whole egg, the minimum energy for pulsed electrical field and cooling was thus 714 kJ/L to obtain a microbiologically stable egg product for 26 days under refrigerated storage (Gongora-Nieta et al., 2003). For the treatment of apple juice and skim milk with PEF at a temperature of 30°C, the specific energy requirements were reported to be 705 kJ/kg and 548 kJ/kg to achieve 2.7 and 2 log cycle inactivation, respectively. However, to maintain treatment temperature below 30°C, intermediate cooling is needed and the total energy requirement is above 1400 kJ/kg (Evrendilek and Zhang, 2005).

Pulsed electrical field treatment usually consumes relatively high energy than traditional thermal treatment with heat recovery capability. Whole egg products with additives are currently pasteurized at 61.1°C for 3.5 min in the United States. For a thermal pasteurization process, the theoretical thermal energy required to increase the temperature from initial 25°C to 61.1°C is

$$Q = \rho c_p (T_2 - T_1) = 1\,[\text{kg/L}] \times 3\,[\text{kJ/kg}^\circ\text{C}] \times (61.1 - 25)[^\circ\text{C}] = 108\ \text{kJ/L}$$

However, it should be addressed that for a thermal process, only a small percentage of energy is used to increase the temperature of the product while the remaining large percentage of energy is used to maintain the operation of the unit at its set point. It should also be addressed that relatively high-energy consumption by PEF treatment can be justified by lower thermal damage, superior quality, and extended shelf life under normal refrigerated storage of the PEF-treated products, compared to the product treated by a thermal process.

19.3.4 Energy Conservation

There are four main ways to reduce the energy costs of PEF treatment:

- Optimization of the design of pulse generation circuits
- Optimization of the operation of pulsed electric fields treatment
- Use of a regenerative heat exchanger to minimize the cooling requirement
- Use of synergistic effects of elevated treatment temperature and PEF treatment

A circuit is needed to generate a pulse. If the voltage generated by a circuit is below a certain level, the voltage is ineffective in terms of bacterial inactivation. The associated dissipation in the load of the circuit is the energy loss. However, the excess voltage may not contribute more to the inactivation of microorganisms but to excess energy loss. Therefore, it is critical to design an efficient circuit to generate sufficient pulsed electric fields while reducing the energy loss (De Haan and Willcock, 2002).

During PEF pasteurization, electricity is required to generate pulsed electrical fields while cooling is needed to remove the heat generated in products by pulsed electrical fields and keep a low temperature of the products for preventing any thermal damage to the products. Therefore, optimization of processing conditions can minimize the energy consumption by the PEF and the cooling units.

A regenerative heat exchanger between the low-temperature raw food materials and high-temperature heated foods following PEF treatment may also provide an opportunity to minimize the cooling requirement to maintain the low product temperature during PEF treatment (Gongora-Nieto et al., 2003).

PEF treatment is usually considered to have a higher energy input than a thermal treatment. Increasing treatment temperature can improve treatment efficiency and reduce the energy consumption of PEF treatment. Using synergistic effects on microbial inactivation at an elevated treatment temperature of 35°C–65°C, the energy consumption by the PEF could be reduced from above 100 kJ/kg to less than 40 kJ/kg for a reduction of 6 log cycles (Heinz et al., 2003). When operating at elevated treatment temperature and making use of synergetic heat effects, the energy input of PEF treatment might be reduced close to the amount of 20 kJ/kg required for conventional thermal pasteurization with 95% of heat recovery capability (Toepfl et al., 2006). Furthermore, combined effects of temperature at 50°C and pulsed electric fields at 40 kV/cm for 100 μs treatment can cause significantly higher inactivation in apple juice peroxidase and polyphenoloxidase than conventional pasteurization at 72°C for 26 s (Riener et al., 2008).

19.4 SUMMARY

PEF treatment has been considered a promising alternative to traditional thermal sterilization and pasteurization in the food industry. Compared to a traditional thermal process, pulsed electric fields can lower thermal damage to foods, maintain superior food quality, and extend the shelf life of products. Pulsed electric fields have been used to treat different foods, particularly liquid foods such as liquid egg products, juices, and milk. PEF treatment usually consumes relatively higher energy than an efficient thermal treatment. The energy consumption of a PEF process can be reduced by optimizing the design of pulse generation circuits, optimizing the operation of PEF treatment, and using a regenerative heat exchanger to minimize the cooling requirement. A combined treatment of PEF and mild heat for gentle microbial inactivation can also save energy.

REFERENCES

Aguiló-Aguayo, I., I. Odriozola-Serrano, L.J. Quintão-Teixeira, and O. Martín-Belloso. 2008. Inactivation of tomato juice peroxidase by high-intensity pulsed electric fields as affected by process conditions. *Food Chemistry* 107: 949–955.

Aronsson, K., U. Ronner, and E. Borch. 2005. Inactivation of *Escherichia coli*, *Listeria innocua* and *Saccharomyces cerevisiae* in relation to membrane permeabilization and subsequent leakage of intracellular compounds due to pulsed electric field processing. *International Journal of Food Microbiology* 99: 19–32.

Bazhal, M. and E. Vorobiev. 2000. Electrical treatment of apple cossettes for intensifying juice pressing. *Journal of the Science of Food and Agriculture* 80: 1668–1674.

Bendicho, S., G.V. Barbosa-Canovas, and O. Martin. 2002. Milk processing by high intensity pulsed electric fields. *Trends in Food Science and Technology* 13: 195–204.

Bryant, G. and J. Wolfe. 1987. Electromechanical stress produced in the plasma membranes of suspended cells by applied electrical fields. *Journal of Membrane Biology* 96: 129–139.

Craven, H.M., P. Swiergon, S. Ng, J. Midgely, C. Versteeg, M.J. Coventry, and J. Wan. 2008. Evaluation of pulsed electric field and minimal heat treatments for inactivation of pseudomonads and enhancement of milk shelf life. *Innovative Food Science and Emerging Technologies* 9: 211–216.

De Haan, S.W.H. and P.R. Willcock. 2002. Comparison of energy performance of pulse generation circuit for PEF. *Innovative Food Science and Emerging Technologies* 3: 349–356.

Dunn, J.E. and J.S. Pearlman. 1987. *Methods and Apparatus for Extending the Shelf Life of Fluid Food Products.* U.S. Patent 4,695,472.

Elez-Martinez, P. and O. Martin-Belloso. 2007. Effects of high intensity pulsed electric field processing conditions on vitamin C and antioxidant capacity of orange juice and gazpacho, a cold vegetable soup. *Food Chemistry* 102: 201–209.

Evrendilek, G.A. and Q.H. Zhang. 2005. Effect of pulse polarity and pulse delaying time on pulsed electric fields-induces pasteurization of *E. coli* O157:H7. *Journal of Food Engineering* 68: 271–276.

Gaskova, D., K. Sigler, B. Janderova, and J. Plasek. 1996. Effect of high voltage electric field pulse on yeast cell: Factors influencing the killing efficiency. *Bioelectrochemistry and Bioenergetics* 39: 195–202.

Gongora-Nieto, M.M., P.D. Pedrow, B.G. Swanson, and G.V. Barbosa-Ganovas. 2003. Energy analysis of liquid whole egg pasteurized by pulsed electric fields. *Journal of Food Engineering* 57: 209–6.

Geveke, D.J., C. Brunkhorst, and X. Fan. 2007. Radio frequency electric fields processing of orange juice. *Innovative Food Science & Emerging Technologies* 8: 549–554.

Heinz, V., S.T. Phillips, M. Zenker, and D. Knorr. 1999. Inactivation of *Bacillus subtilis* by high intensity pulsed electric fields under close to isothermal conditions. *Food Biotechnology* 13: 155–168.

Heinz, V., S. Toepfl, and D. Knorr. 2003. Impact of temperature on lethality and energy efficiency of apple juice pasteurization by pulsed electric fields treatment. *Innovative Food Science and Emerging Technologies* 4: 167–175.

Hodgins, A.M., G.S. Mittal, and M.W. Griffiths. 2002. Pasteurization of fresh orange juice using low-energy pulsed electrical field. *Journal of Food Science* 67: 2294–2299.

Muraji, M., H. Taniguchi, W. Tatebe, and H. Berg. 1999. Examination of the relationship between parameters to determine electropermeability of *Saccharomyces cerevisiae. Bioelectrochemistry and Bioenergetics* 48: 485–488.

Nguyen, P. and G.S. Mittal. 2007. Inactivation of naturally occurring microorganisms in tomato juice using pulsed electric field with and without antimicrobials. *Chemical Engineering and Processing* 46: 360–365.

Qui, X., S. Sharma, L. Tuhela, M. Jia, and Q.H. Zhang. 1998. An integrated PEF pilot plant for continuous nonthermal pasteurization of fresh orange juice. *Transactions of the ASAE* 41: 1069–1074.

Reina, L.D., Z.T. Jin, Q.H. Zhang, and A.E. Yousef. 1998. Inactivation of *Listeria monocytogenes* in milk by pulsed electric field. *Journal of Food Protection* 61: 1203–1206.

Riener, J., F. Noci, D.A. Cronin, D.J. Morgan, and J.G. Lyng. 2008. Combined effect of temperature and pulsed electric fields on apple juice peroxidase and polyphenoloxidase inactivation. *Food Chemistry* 109: 402–407.

Schilling, S., T. Alber, S. Toepfl, S. Neidhart, D. Knorr, A. Schieber, and R. Carle. 2007. Effects of pulsed electric field treatment of apple mash on juice yield and quality attributes of apple juices. *Innovative Food Science & Emerging Technologies* 8: 127–134.

Selma, M.V., M.C. Salmeron, M. Valero, and P.S. Fernandez. 2004. Control of *Lactobacillus plantarum* and *Escherichia coli* by pulsed electric fields in MRS Broth, nutrient broth and carrot-apple juice. *Food Microbiology* 21: 193–203.

Toepfl, S., A. Mathys, V. Heinz, and D. Knorr. 2006. Review: Potential of high hydrostatic pressure and pulsed electric fields for energy efficiency and environmentally friendly food processing. *Food Reviews International* 22: 405–423.

Ulmer, H.M., V. Heinz, M.G. Gaenzle, D. Knorr, and R.F. Vogel. 2002. Effects of pulsed electric fields on inactivation and metabolic activity of *Lactobacillus plantarum* in model beer. *Journal of Applied Microbiology* 93: 326–335.

Zhang, Q.H., G.V. Barbosa-Canovas, and B.G. Swanson. 1995. Engineering aspects of pulsed electric field pasteurization. *Journal of Food Engineering* 25: 261–281.

20 Energy Efficiency and Conservation in High-Pressure Food Processing

20.1 INTRODUCTION

High-pressure processing technology has been used to destroy microorganisms and extend the shelf life of foods (Toepfl et al., 2006). High-pressure treatment can be used for the preservation of packaged foods as an alternative to thermal treatment. Unlike thermal treatment, high-pressure treatment, which can treat foods at an ambient temperature, can maintain the quality of fresh foods with little effect on their sensory and nutritional values. Hydrostatic pressure treatment is uniform and instantaneous. During high-pressure treatment, no deformation occurs on packaged foods because the extremely high pressures are isostatic—even on all sides and surfaces of the package. High-pressure treatment is independent of the size, shape, and composition of foods (Norton and Sun, 2008). Another high-pressure application in the food industry is high-pressure freezing and thawing. High-pressure assisted freezing and thawing offer unique opportunities for product development and product quality improvement (Lebail et al., 2002a; Urrutia et al., 2007). In this chapter, the high-pressure processing mechanism and applications are reviewed. The energy consumption and energy conservation opportunities related to high-pressure processes in the food industry are discussed.

20.2 PROCESS OVERVIEW

20.2.1 Process Mechanism of High-Pressure Sterilization/Pasteurization

The effect of high pressure on food chemistry and microbiology is governed by Le Chatelier's principle: when a system at equilibrium is disturbed, the system then responds in a way that tends to minimize the disturbance. That is, under equilibrium conditions, a process associated with a volume reduction is favored by an increase in pressure and vice versa.

High-pressure processing inactivates microbes by targeting the membranes of biological cells. During pressurization, water and acid molecules show increased ionization. These minor changes in the biochemistry of living cells cause the majority of microbial killing effects. Although atomic bonds are barely affected by high pressure, alteration of proteins or lipids can be observed when exposed to high pressure. Lethal damage occurs when alteration of proteins or lipids occurs in membranes of biological cells (Manas and Pagan, 2005; Toepfl, et al., 2006).

Vegetative cells are usually sensitive to pressures over 100 MPa and die rapidly at pressures higher than 500 MPa. However, there are many exceptions to this rule. The chemical and physical environment can affect the effectiveness of the high-pressure inactivation. The amount of free water present in a food is particularly important for an effective high-pressure inactivation process. At least 40% of free water is needed to achieve killing of vegetative microbes. The technique is thus not suitable for the decontamination of dry materials. Bacterial spores are very resistant to pressure and the application of pressure higher than 1000 MPa for several hours will not significantly affect the levels of spores found in some types of foods. Bacterial spores appear to show increased sensitivity to pressure if the temperature is elevated greater than 40°C. Some experiments have shown that bacterial spores and enzymes are resistant to high pressures. Therefore, the use of a combination of heat and pressure for food preservation has been suggested (Wimalaratne and Farid, 2008). Adiabatic heating occurs in most food materials during high-pressure processing and this is proportional to the compressibility of foods: air (e.g., cell vacuoles and entrapped air in food matrices) is very compressible and will give rise to significant temperature elevation. Oils and fats are moderately compressible, and water is less compressible.

20.2.2 Process Mechanism of High-Pressure Freezing/Thawing

Water undergoes different phases when subjected to high pressure at different temperatures, as shown in Figure 20.1 (Lebail et al., 2002a). At atmospheric pressure, water will freeze at 0°C. Higher pressures decrease the freezing point of water to a minimum of –22°C at 207.5 MPa. This phenomenon offers opportunities for subzero food storage without ice crystal formation (by pressurization and then decreasing temperature: ABC

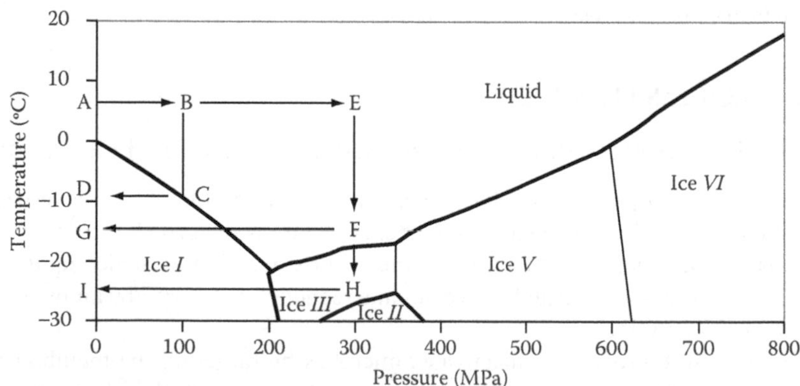

FIGURE 20.1 Phase diagram of water (example pressure-assisted freezing: ABCD, pressure-assisted thawing: DCBA, pressure shift freezing: ABEFG, pressure-induced thawing: GFEBA, freezing to ice III: ABEFHI, thawing to ice III: IHFEBA). (Reprinted from Lebail, A., Chevalier, D., Mussa, D.M., and Ghoul, M., *Int. J. Refrigeration*, 25, 504, 2002a. With permission.)

in Figure 20.1), rapid freezing of foods (by pressurization, decreasing temperature, and then decompression: ABCD in Figure 20.1), and rapid thawing of conventionally frozen foods (by pressurization, increasing temperature, and then decompression: DCBA in Figure 20.1). High-pressure assisted freezing and thawing can improve the quality of frozen and thawed foods by generating small ice crystals under high pressure.

20.2.3 High-Pressure Generation and System Design

A high-pressure food process generally relies on the application of isostatic or hydraulic pressures in the range from 100 to 1000 MPa. The process can be carried out in any type of hydraulic fluids but water is often preferred for ease of operation and its compatibility with food materials. Liquids such as water are relatively incompressible and store much less energy in their compressed state than gases. The risk from explosion is thus greatly reduced by the use of incompressible fluids as pressurizing medium (Earnshaw, 1996).

Liquid or solid foods can be contained in flexible containers immersed in a hydraulic fluid in a high-pressure process. During high-pressure processing, the vessel is closed and sealed. A pressure is applied on the hydraulic fluid to a desired level. Following high-pressure treatment, the vessel is decompressed and the products are unloaded. Typical production cycles for pasteurization are only 3 to 8 min. High-pressure equipment is very specialized and relatively expensive. Pressure vessels are constructed of forged steel or reinforced with tensioned wire windings. Typical vessel volumes are 100 to 500 L. The production costs for high-pressure pasteurization are around 25 cents per batch for a 100 L system and 7 cents per batch for a 500 L unit. For liquid foods that can be pumped, in-line continuous systems are also being investigated. Liquid foods can be compressed directly in a pressure vessel. A high-pressure food processing system includes a pressure vessel, pressurizing unit, and ancillaries. A high pressure can be generated by a direct or indirect method as shown in Figure 20.2a, a pressure intensifier such as a piston, which is located within the pressure vessel, is used to deliver the high pressure to the product. In an indirect pressurization system as shown in Figure 20.2b, a pressure intensifier such as a high-pressure pump, which is separate from the high-pressure vessel, is used to first increase the pressure of a pressuring medium such as water to the desired levels and then convey the high-pressure medium to the pressure vessel.

20.2.4 Applications of High-Pressure Processes

20.2.4.1 High-Pressure Pasteurization and Sterilization

High-pressure and high-temperature (above 0°C) applications include high-pressure sterilization and pasteurization. A high-pressure process has been applied to pasteurize and sterilize juice, dairy products, fruits and vegetables, meat products, and fish products (Basak and Ramaswamy, 1996; Parish, 1998; Polydera et al., 2003; Norton and Sun, 2008). A high-pressure treatment can retain the quality of foods but reduce the microorganisms in the foods. The flavor of high-pressure treated orange juice stored at 4°C for 16 weeks or 8°C for 4 weeks was found to be close to that of fresh or frozen juice. The microflora in juice was reduced to below detectable limits (Parish, 1998). The shelf life

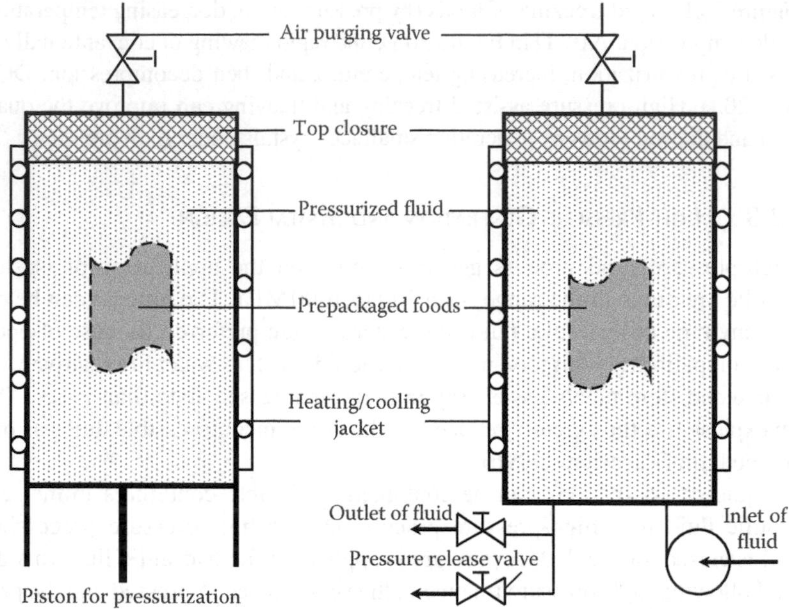

FIGURE 20.2 Schematics of direct and indirect high-pressure food processing systems.

of orange juice processed at a pressure of 500 MPa and temperature of 35°C for several minutes was increased by 11% when stored in a bottle at 15°C and 65% when stored at 0°C, compared to that of orange juice pasteurized at 80°C for 30 s (Polydera et al., 2003). A high-pressure process offers an opportunity to inactivate microorganisms at a low or moderate temperature. The energy required for compression during high-pressure processing is far less than that required by a thermal process for heating and cooling.

A high-pressure process is also a promising nonthermal processing technology to inactivate the anti-nutritional factors of grains while preserving food quality and constituents. During high-pressure processing, allergenic proteins from rice grains are solubilized while no apparent alteration in color, shape, or size of treated kernels occurs at a moderate pressure (Estrada-Giron et al., 2005). Houska et al., (2006) reported that a high-pressure pasteurization process at a pressure of 500 MPa for 10 min was able to inactivate viable microorganisms present in raw broccoli juice by more than 5 logs. The high-pressure treated broccoli juices were found to be comparable in sulforaphane content and anti-mutagenic activity with frozen juice.

For pasteurization purpose, the treatment pressure is usually in the range of 300–600 MPa for a short period of time from seconds to minutes to inactivate the vegetative pathogenic and spoilage microorganisms by more than 4 log units. For sterilization, the pressure should be over 600 MPa and combination with a high temperature is needed because some spores are resistant even to a pressure over than 1000 MPa at a temperature higher than 45°C–75°C (Amymerich et al., 2008; Wimalaratne and Farid, 2008).

20.2.4.2 High-Pressure Freezing and Thawing

High-pressure low-temperature applications include high-pressure assisted freezing and thawing (Yoshioka et al., 1996; Chevalier et al., 1999; Okamoto and Suzuki, 2002; Cheftel et al., 2002; Lebail et al., 2002a,b). The formation of ice within foods during convectional freezing may cause damage to tissues and cell wall of the foods, which leads to the loss of flavor and nutrients. The rate and final temperature determine the size and location of the ice crystals formed in foods (Norton and Sun, 2008). It is well known that a fast freezing rate results in a fine ice structure by forming a lager number of small ice crystals, which causes less damage to the structure of a product. The main benefits of high-pressure and low-temperature processes include (1) freezing point depression (e.g., −21°C at 210 MPa, compared with 0°C at atmospheric pressure), (2) reduced latent heat of fusion (e.g., 190 kJ/kg at 210 MPa, compared with 334 kJ/kg at atmospheric pressure, (3) reduced specific volume, and (4) crystallization of higher level ice polymorphs with greater density than water (Norton and Sun, 2008).

High-pressure assisted freezing can form instantaneous and homogenous ice crystals through the entire volume of a food by depressing the freezing point and generating a high degree of supercooling. The mass ratio of ice to water during adiabatic expansion for pressure from 210 MPa may reach up to 0.36 kg of ice per kilogram of water (Sanz et al., 1997). High-pressure thawing is a reverse process of high-pressure freezing. High-pressure thawing can maintain better quality in foods than traditional thawing with hot water or air (Yoshioka et al., 1996). Furthermore, high-pressure thawing can reduce the drip volume and the microbial load, compared to thawing in water at atmospheric pressure (Chevalier et al., 1999; Okamoto and Suzuki, 2002; Lebail et al., 2002). The drip volume decreases with the increase in holding time at a high pressure. The pressure holding time is the most important parameter in the reduction of drip loss, compared to pressurization rate and thawing rate (Chevalier et al., 1999).

20.3 ENERGY USE, EFFICIENCY, AND CONSERVATION

20.3.1 Energy Use for High-Pressure Generation

Energy is required for compression work and adiabatic heating due to compression during high-pressure processing. Liquids are usually used as high-pressure medium. The relationship between volume and pressure of a liquid can be determined from its compressibility, β, which is a measure of relative volume change of a liquid as response to a pressure change (Fine and Millero, 1973):

$$\beta = -\frac{1}{V}\frac{\partial V}{\partial p} \tag{20.1}$$

where

V is the volume (m^3)
p is the pressure (Pa)

The volume of liquid can thus be expressed as a function of pressure:

$$V = e^{(-\beta p)} \tag{20.2}$$

The compressibility of pure water is $\beta = 4.6 \times 10^{-10}$ m²/N at 25°C. Therefore, when water is pressurized to 600 MPa, its volume will approximately be contracted by 24% under an isothermal condition at 4% per 100 MPa increase in pressure. The specific work of compression can be expressed as

$$W = -\int V \mathrm{d}p \quad (20.3)$$

By integration of the above equation for pure water under an isothermal condition, the compression work is about 52 kJ/kg upon compression up to 600 MPa.

20.3.2 Temperature Increase during High-Pressure Processing

For sterilization with a high-pressure process, adiabatic heat due to compression can provide additional synergetic sterilization effects (Hartmann et al., 2003). It should be noted that the product will return to its initial temperature or even below after the pressure releases (Toepfl et al., 2006). A high pressure may cause changes in physical properties of foods, which should be considered in the calculation of temperature change during high-pressure processing. A high pressure has been found to increase the density of a food by about 3.5% of its original value for each 100 MPa increment in pressure (Otero et al., 2006). The specific heat of water at 1°C was found to decrease linearly from 4216 J/kg K at atmospheric pressure to 3488 J/kg K at 600 MPa (Lemmon et al., 2005). The thermal conductivity of foods does not change significantly under pressure (Hartmann et al., 2003).

All compressible materials change their temperature during compression (Ting et al., 2002). During compression, an adiabatic heating process occurs. The temperature increase during adiabatic and isentropic compression can be calculated by

$$\frac{\mathrm{d}T}{\mathrm{d}p} = \frac{\alpha}{\rho}\frac{T}{c_\mathrm{p}} \quad (20.4)$$

where α, ρ, and c_p are the thermal expansivity, density, and specific heat of the compressed fluid, respectively.

The main ingredient of most foods is water and thus the thermodynamic properties of water can be used to estimate the temperature increase on compression of high moisture foods. The compression heating in fat-containing foods can be up to three times higher than that for water. Suppose the average thermal expansivity, density, and specific heat of pure water are $\alpha = 2.5 \times 10^{-4}$ K^{-1}, $\rho = 900$ kg/m³, and $c_\mathrm{p} = 3.85$ kJ/kg°C at 25°C, respectively, during high-pressure processing. The temperature increase per 100 MPa for pure water is ~3°C. For a high-pressure process of pure water at 600 MPa, the temperature increase due to pressurization is 18°C. The heat for the increase of water temperature is about 70 kJ/kg upon compression up to 600 MPa. Table 20.1 gives the measured adiabatic heat of compression in different food systems (Patazca et al., 2007).

Since different food components may have different compression heating coefficients, the adiabatic heating caused by fluid compression can lead to a significant temperature distribution through the treated food. The temperature increase in solid components such as fat is higher than that of liquid components (Abdul Ghani and Farid, 2007).

TABLE 20.1
Adiabatic Heat of Compression in Different Food Systems

Substance at 25°C	Pressure (MPa)				Temperature Increase per 100 MPa Compression (°C/100 MPa)			
Vegetable oil	156.1	293.8	446.7	586.0	9.2	8.0	7.0	6.6
Mayonnaise dressing	149.3	318.6	468.0	574.6	7.2	5.7	5.0	5.3
Cream cheese	146.4	316.1	447.6	533.3	4.9	4.8	4.6	4.7
Egg yolk	161.4	338.9	472.0	544.8	4.5	4.4	4.3	4.3
Hass avocado	148.9	316.5	480.4	580.0	4.1	4.0	3.6	3.7
Beef ground raw	150.4	324.8	457.1	540.7	3.0	3.3	3.4	3.2
Egg whole	142.1	328.4	470.7	552.3	3.2	3.4	3.5	3.3
Whole milk	145.4	326.6	472.5	556.8	3.1	3.2	3.2	3.2
Gravy beef	153.6	307.7	448.9	534.5	3.1	3.0	2.9	3.0
Chicken breast	146.2	322.6	462.6	542.3	3.0	3.0	3.1	3.0
Egg white	163.0	332.5	465.0	541.2	2.8	2.8	2.8	2.9
Skim milk	148.3	314.6	458.8	544.8	2.9	3.0	2.9	3.0
Honey	157.5	299.1	449.3	581.6	3.5	3.2	3.1	2.9

Source: Reprinted from Patazca, E., Koutchma, T., and Balasubramaniam, V.M., *J. Food Eng.*, 80, 199, 2007. With permission.

20.3.3 Energy Savings with High-Pressure Processes

A high-pressure process requires power to increase pressure, and part of the power consumed is converted to heat for the temperature increase due to pressurization. Theoretically, the compression work and energy required for temperature increase due to pressurization are about 52 kJ/kg and 70 kJ/kg upon compression of pure water up to 600 MPa, respectively. Furthermore, it should be noted that high-pressure sterilization/pasteurization does not require a cooling process and decompression will decrease the temperature of the product. The theoretical total energy input into a high-pressure process at 600 MPa for processing pure water is thus about 122 kJ/kg.

For a thermal sterilization/pasteurization process, the temperature of a product should be increased to a desired value (e.g., 110°C) with a heating unit, maintained at this temperature for a while, and then decreased to a storage temperature with a cooling unit. The energy efficiency of a thermal process is usually below 50%. Suppose the temperature of the product with a specific heat of 4.18 kJ/kg°C is increased from 25°C to 110°C with a heating unit at an energy efficiency of 50%, the total thermal energy input is 711 kJ/kg. More energy is needed to decrease the temperature of the product after sterilization/pasteurization. Suppose a mechanic refrigeration unit with a coefficient of performance of 3 (cool energy generated to electricity consumed) is used to cool the product at an energy efficiency of 50%, the total electricity is 237 kJ/kg (or 0.066 kWh/kg). Therefore, the total energy input for thermal sterilization/pasteurization is much higher than that for a high-pressure process.

Toepfl et al. (2006) compared the energy consumption for sterilization of a 0.1 m diameter can using three methods: (1) thermal treatment with 50% heat recovery, (2) combined thermal with high-pressure treatment with 50% heat recovery, and (3) combined thermal with high-pressure treatment with 50% heat and compression energy recovery. The thermal treatment was recorded to take 130 min to reach an *F*-value of 2.4 min in the center of the can (110°C) in an autoclave, which was heated from 80°C to 140°C. The specific energy input required for the sterilization of cans with thermal treatment with 50% heat recovery was 300 kJ/kg. A combined thermal and high-pressure process was developed to achieve a similar inactivation of spores of *Clostridium*. During the combined process, the temperature of the high-pressure vessel was maintained at 105°C. The core temperature of the can increased to 80°C after 105 min. A pressure of 800 MPa was applied for 10 min and the heat of compression caused a temperature increase to 110°C. After that, decompression caused a temperature drop to 87°C. The specific energy input for the combined process was reduced to 270 kJ/kg due to shorter residence time. If 50% of the compression energy can be recovered using a two-vessel system or pressure storage, the specific energy input is 242 kJ/kg. This means that a combined thermal and high-pressure sterilization process can reduce 20% of the specific energy input.

Although high-pressure equipment is considered to be generally more expensive than conventional processing/packaging systems, significant energy cost savings may be accumulated over time through the use of high pressure rather than high temperature. Furthermore, during high-pressure processing, the products are not sterilized by means of heat pasteurization but pressure pasteurization. Since no heat is applied, the foods do not experience flavor, color, texture, aroma, or nutrient degradation commonly associated with a thermal process. It has been proved that the high-pressure process does not adversely affect food nutrient values.

20.4 SUMMARY

Pasteurization and sterilization at a high temperature may result in thermal damage and reduce the freshness of foods, particularly fruits and vegetables. A high-pressure process is environmentally friendly and can retain the fresh-like characteristics of foods better than thermal treatment. High pressure has also been used to assist freezing and thawing of foods for improved quality and reduced drip loss by generating small size ice crystals. High-pressure treatment can be applied to packaged foods. A high-pressure process provides a potential to reduce the energy requirement for pasteurization and sterilization of foods and contributes to the improvement of energy efficiency in the food industry. However, due to high investment costs and cost-intensive maintenance and service for high-pressure equipment, high-pressure pasteurization units with a maximum pressure up to 600 MPa are more suitable for the treatment of high-value foods or highly heat sensitive foods such as flavors, vitamins, and functional biopolymers. It is unlikely that a high-pressure process will be used to replace food thermal processing or freezing.

REFERENCES

Abdul Ghani, A.G. and M.M. Farid. 2007. Numerical simulation of solid-liquid food mixture in a high pressure processing unit using computational fluid dynamics. *Journal of Food Engineering* 80: 1031–1042.

Amymerich, T., P.A. Picouet, and J.M. Monfort. 2008. Decontamination technologies for meat products. *Meat Science* 78: 114–129.

Basak, S. and H.S. Ramaswamy. 1996. Ultra high pressure treatment of orange juice: A kinetic study on inactivation of pectin methyl esterase. *Food Research International* 29: 601–607.

Cheftel, J.C., M. Thiebaud, and E. Dumay. 2002. Pressure-assisted freezing and thawing of foods: A review of recent studies. *High Pressure Research* 22: 601–611.

Chevalier, D., A. Le Bail, J.M. Chourot, and P. Chantreau. 1999. High pressure thawing of fish (whiting): Influence of the process parameters on drip losses. *Lebensmittel-Wissenschaft und-Technologie* 32: 25–31.

Earnshaw, R. 1996. High pressure food processing. *Nutrition & Food Science* 2: 8–11.

Estrada-Giron, Y., B.G. Swanson, and G.V. Barbosa-Canovas. 2005. Advances in the use of high hydrostatic pressure for processing cereal grains and legumes. *Trends in Food Science & Technology* 16: 194–203.

Fine, R.A. and F.J. Millero. 1973. Compressibility of water as a function of temperature and pressure. *Journal of Chemical Physics* 59: 5529–5536.

Hartmann, C., A. Delgado, and J. Szymczyk. 2003. Convective and diffusive transport effects in a high pressure induced inactivation process of packed food. *Journal of Food Engineering* 59: 33–44.

Houska, M., J. Strohalm, K. Kocurova, J. Totusek, D. Lefnerova, J. Triska, N. Vrchotova, V. Fiedlerova, M. Holasova, D. Gabrovska, and I. Paulickova. 2006. High pressure and foods-fruit/vegetable juices. *Journal of Food Engineering* 77: 386–398.

Lebail, A., D. Chevalier, D.M. Mussa, and M. Ghoul. 2002a. High pressure freezing and thawing of foods: A review. *International Journal of Refrigeration* 25: 504–513.

Lebail, A., D. Mussa, J. Rouille, H.S. Ramaswamy, N. Chapleau, M. Anton, M. Hayert, L. Boillereaux, and D. Chevalier. 2002b. High pressure thawing. Application to selected sea-foods. *Progress in Biotechnology* 19: 563–570.

Lemmon, E.W., M.O. McLinden, and D.G. Friend. 2005. Thermophysical properties of fluid systems. In Linstron, P.J. and Mallard W.G. (Eds.), NIST Chemistry WebBook, NIST standard reference database number 69, June 2005. Available at http://webbook.nist.gov.

Manas, P. and R. Pagan. 2005. Microbial inactivation by new technologies of food preservation. *Journal of Applied Microbiology* 98: 1387–1399.

Norton, T. and D.W. Sun. 2008. Recent advances in the use of high pressure as an effective processing technique in the food industry. *Food Bioprocess Technology* 1: 2–34.

Okamoto, A. and A. Suzuki. 2002. Effects of high hydrostatic pressure-thawing on pork meat. *Progress in Biotechnology* 19: 571–576.

Otero, L., A. Ousegui, B. Guignon, A. LeBail, and P.D. Sanz. 2006. Evaluation of the thermophysical properties of typose gel under pressure in the phase change domain. *Food Hydrocoll* 20: 449–460.

Parish, M.E. 1998. Orange juice quality after treatment by thermal pasteurization or isostatic high pressure. *Lebensmittel-Wissenschaft und-Technologie* 31: 439–442.

Patazca, E., T. Koutchma, and V.M. Balasubramaniam, 2007. Quasi-adiabatic temperature increase during high pressure processing of selected foods. *Journal of Food Engineering* 80: 199–205.

Polydera, A.C., N.G. Stoforos, and P.S. Taoukis. 2003. Comparative shelf life study and vitamin C loss kinetics in pasteurized and high pressure processed reconstituted orange juice. *Journal of Food Engineering* 60: 21–29.

Sanz, P.D., L. Otero, C. de Elvira, and J.A. Carrasco. 1997. Freezing processes in high-pressure domains. *International Journal of Refrigeration* 20: 301–307.

Toepfl, S., A. Mathys, V. Heinz, and D. Knorr. 2006. Review: Potential of high hydrostatic pressure and pulsed electric fields for energy efficiency and environmentally friendly food processing. *Food Reviews International* 22: 405–423.

Urrutia, G., J. Arabas, K. Autio, S. Brul, M. Hendrickx, A. Kakolewski, D. Knorr, A. LeBail, M. Lille, A.D. Molina-Garcia, A. Ousegui, P.D. Sanz, T. Shen, and S.V. Buggenhout. 2007. SAFE ICE: Low-temperature pressure processing of foods: Safety and quality aspects, process parameters and consumer acceptance. *Journal of Food Engineering* 83: 293–315.

Wimalaratne, S.K. and M.M. Farid. 2008. Pressure assisted thermal sterilization. *Food and Bioproducts Processing*, in press.

Yoshioka, K., A. Yamada, and T. Maki. 1996. Application of high pressurization to fish meat: Changes in the physical properties of carp skeletal muscle resulting from high pressure thawing. *Progress in Biotechnology* 13: 369–374.

21 Energy Efficiency and Conservation in Microwave Heating

21.1 INTRODUCTION

Microwaves are high-frequency electromagnetic waves generated by magnetrons and klystrons. The frequency range of microwaves is between 300 MHz and 300 GHz, which corresponds to a wavelength from 1 to 0.01 m. For industrial heating purposes, the available microwave frequencies are 915 and 2450 MHz. In conventional heating, heat is transferred into foods by conduction and convection. In microwave heating, heat is generated directly inside food materials, causing a much faster temperature rise than in conventional heating (Bengtsson and Ohlsson, 1974; Venkatesh and Raghavan, 2004). However, when heating thick foods by microwave or heating foods with large variations in compositions, microwaves may cause uneven heating inside the foods. In this case, heat conduction caused by the temperature gradient inside the foods also plays an important role in heat transfer.

Microwaves provide a rapid and convenient heating method in the food industry. The molecular structure of food materials affects the ability of microwaves to interact with the materials and transfer energy to the materials. Microwaves have been used for thawing, sterilization, dehydration, cooking, and baking in the food industry. In this chapter, the microwave process and applications are reviewed. The energy efficiency and conservation strategies of a microwave heating process are discussed.

21.2 PROCESS OVERVIEW

21.2.1 Microwave Heating Mechanism

In conventional heating, a material is heated by conduction or convection. Microwave heating is based on the material's ability to absorb electromagnetic radiation and convert it to heat. When polar molecules in materials, such as water in foods, are exposed to microwave radiation, the dipolar rotation in the presence of an alternating electric field generates heat due to molecular friction, which rapidly heats the materials. Another mechanism in microwave heating is ionic conduction. Positive and negative ions of dissolved salts in food also interact with the electric field by migrating toward the oppositely charged regions of the electrical field and disrupt hydrogen bonds with water to generate additional heat.

21.2.2 Microwave Heating Systems

Microwaves can provide an energy source during thermal food processing. Microwave-assisted systems can be operated in either batch or continuous modes. Continuous-flow systems are considered to yield more uniform heat treatment than bulk batch-mode heating (LeBail et al., 2000).

Figure 21.1 shows a batch microwave-assisted vacuum dehydration system. Under vacuum, water can rapidly diffuse to the surface of foods and evaporate into the vacuum chamber. Vacuum can reduce the water vapor concentration at the surface of the products. Vacuum lowers the boiling point of water in the interior of the products. Therefore, a large vapor pressure gradient is created between the food interior and surface, causing a rapid drying rate. However, no convection occurs in vacuum. Microwave can be used to provide the thermal energy needed for water evaporation under vacuum.

Microwave energy can replace the steam in a heat exchanger for continuous pasteurization of liquid food products. A continuous-flow microwave heating system shown in Figure 21.2 has been developed for the pasteurization of liquid food products such as milk. A helical glass coil is located in two microwave oven cavities. The heating coils are connected with a circulating pump at the inlet and a water-cooled helical condenser at the exit (LeBail et al., 2000). In a traditional heat exchanger for pasteurization of liquid foods, heat is transferred from the surface of tubes to the bulk of food. As microwave can generate internal heat, a continuous-flow microwave-assisted pasteurization system can keep the tube surface temperature at a relatively lower level and thus reduce the problems associated with surface fouling.

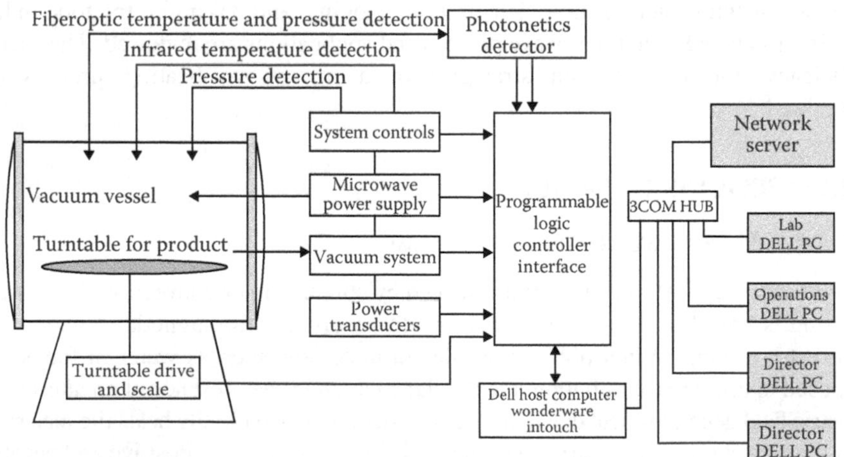

FIGURE 21.1 Microwave-assisted vacuum dehydration system. (Reproduced from Clary, C.D., Wang, S.J., and Petrucci, V.E., *J. Food Sci.*, 70, 344, 2005. Copyright Wiley-Blackwell. With permission.)

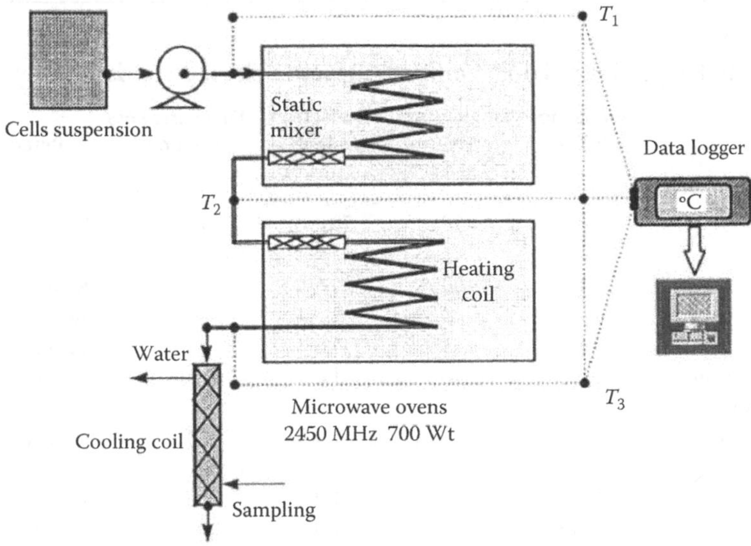

FIGURE 21.2 Continuous flow microwave-assisted pasteurization system. (Reproduced from LeBail, A., Koutchma, T., and Ramaswamy, H.S., *J. Food Process Eng.*, 23, 1, 2000. Copyright Wiley-Blackwell. With permission.)

21.3 FACTORS AFFECTING MICROWAVE HEATING

21.3.1 Dielectric Properties

Penetration of microwave energy inside a material is a function of its dielectric properties. The permittivity of foods describes how the food materials interact with electromagnetic radiation. The permittivity, ε, is a complex quantity with a real component of dielectric constant, ε', and an imaginary component of dielectric loss factor, ε''. The permittivity is usually expressed as

$$\varepsilon = \varepsilon' - j\varepsilon'' \tag{21.1}$$

$$\tan\delta = \frac{\varepsilon''}{\varepsilon'} \tag{21.2}$$

The dielectric constant is related to the capacitance of a substance and the ability of the substance to store electrical energy. The dielectric loss factor is related to energy dissipation (Ryynanen, 1995). The dielectric properties of food components and selected foods are given in Table 21.1.

The dielectric properties of food products are mainly determined by their chemical composition. Water is a strongly polar solvent and it is the major chemical constituent of most food products. The dielectric constant and dielectric loss factor of water at 20°C and 2450 MHz are 78 and 13.4, respectively. Salts depress the dielectric

TABLE 21.1
Dielectric Properties of Food Components and Selected Foods

	Frequency (MHz)	Temperature (°C)	Dielectric Constant, ε'	Dielectric Loss Factor, ε''	References
Food Components					
Water	2450	20	78	13.4	Kudra et al., 1992
Ice	2450	−12	3.2	0.003	Tang et al., 2002
Fat (milk)	2450	20	2.6	0.2	Kudra et al., 1992
Oil (corn)	2450	25	2.5	0.14	Tang et al., 2002
Oil (corn)	915	25	2.6	0.18	Tang et al., 2002
Protein (sodium caseinate)	2450	20	1.6	0	Kudra et al., 1992
Carbohydrates (lactose)	2450	20	1.9	0	Kudra et al., 1992
Air	2450	20	1.0	0	Tang et al., 2002
Air	915	20	1.0	0	Tang et al., 2002
Selected Food Products					
Cooked ham	2450	25	60	42	Tang et al., 2002
Cooked ham	915	25	61	96	Tang et al., 2002
Cooked beef	2450	25	72	23	Tang et al., 2002
Cooked beef	915	25	76	36	Tang et al., 2002
Potato	2450	25	54	16	Tang et al., 2002
Potato	915	25	65	20	Tang et al., 2002
Asparagus	2450	21	71	16	Tang et al., 2002
Asparagus	915	21	74	21	Tang et al., 2002

constant due to the binding of free water molecules by the ions of dissolved salts and elevate the dielectric loss factor due to the addition of conductive charge carriers.

The effects of water and salt contents on the dielectric properties depend on the manner in which they are bound or restricted in their movement by other food components. The permittivity of aqueous solutions is decreased by two reasons: the replacement of water by a substance with a lower permittivity and the binding of water molecules. Organic solids such as fat, protein, and carbohydrate, which are usually dielectrically inert with a dielectric constant smaller than 3 and a dielectric loss factor smaller than 0.2, depress the permittivity of foods. Fat content affects both the dielectric constant and the dielectric loss factor. The dielectric constant and dielectric loss factor of low-fat ground beef are higher than those of high-fat ground beef (Gunnasekaran et al., 2005). The structure of bound water appears to be intermediate between those of ice and free water. Sakai et al. (2005) used 1% agar gel as a model food and 0%–40% sucrose and 0%–2% sodium chloride to adjust the dielectric properties. They found that the addition of sucrose primarily decreased the dielectric constant while the addition of sodium chloride primarily increased the dielectric loss factor and slightly decreased the dielectric constant. Hydrogen bonds restrict the free movement of water, and thus the permittivity.

21.3.2 Temperature and Frequency

The dielectric properties of food materials are also affected by temperature and microwave frequency. Ice has a much lower dielectric constant and dielectric loss factor than liquid water. Below the freezing point, the dielectric constant and dielectric loss factor increase with increasing temperature as liquid water in foods increases. There is a dramatic increase in both dielectric constant and dielectric loss factor during thawing of foods due to the melting of ice. Above the freezing point, both the dielectric constant and the dielectric loss factor decrease slightly with increasing temperature. However for salted foods, the dielectric loss factor significantly increases with increasing temperature (Piyasena et al., 2003). Both dielectric constant and dielectric loss factor decrease with the increase in microwave frequency. Therefore, power absorption and radiation penetration during microwave heating are more effective at lower frequencies than at higher ones (Oliveira and Franca, 2002).

The temperature increase in foods during microwave heating depends on both the dielectric and thermophysical properties of foods. The physical properties of a food material including specific heat, density, and thermal conductivity, which vary with temperature and food composition, will also affect the increase in temperature during microwave heating.

21.3.3 Shape and Size of Food Items

When a microwave propagates into a food material, it decays. Therefore, the region away from the surface is heated slower than the surface as the food item size increases. Microwave absorption is generally calculated by Maxwell's equation (Hill and Marchant, 1994) or Lambert's equation (Romano et al., 2005). Yang and Gunasekaran (2004) compared the temperature distribution in model food cylinders using Maxwell's equation and Lambert's law during pulsed microwave heating. They found that predictions using both methods were statistically accurate compared to the measured temperatures although the power absorption based on Maxwell's equation was more accurate than that based on Lambert's law. Maxwell's equation can accurately describe the propagation and absorption of microwave but it is too complex. Lambert's law considers an exponential decay of microwave energy within foods, which can be expressed as (Campanone and Zaritzky, 2005)

$$P = P_0 e^{(-2\alpha x)} \tag{21.3}$$

where

P is the power at a distance of x from the surface (W/m^2)

P_0 is the surface power (W/m^2)

x is the maximum distance measured from the surface (m)

α is the attenuation factor, which is a function of the dielectric constant and dielectric loss factor:

$$\alpha = \frac{2\pi}{\lambda_0}\sqrt{\frac{\varepsilon'\left(\left(1+\left(\frac{\varepsilon''}{\varepsilon'}\right)^2\right)^{1/2} - 1\right)}{2}} \tag{21.4}$$

where

λ_0 is the free space wavelength

$\lambda_0 = c/f$ (c is the speed of light in a free space, $c = 3 \times 10^8$ m/s)

The penetration depth, d, is defined as the depth where the microwave field is reduced to 1/e (or 36.8%) of its surface value. The penetration depth is thus given by

$$d = \frac{1}{2\alpha} \tag{21.5}$$

Most common food products have a penetration depth in the range from 0.5 to 1.0 cm. The penetration depth decreases with the increase in salt content (Piyasena et al., 2003). The penetration depth of microwave decreases with the increase in microwave frequency and temperature. Microwave at 915 MHz can penetrate to a greater depth than microwave at 2450 MHz. The microwave penetration depth is much smaller than that of radio frequency heating at 10–300 MHz (Zhang et al., 2006). Therefore, microwave heating is significantly dependent on the size and shape of foods. In larger samples, a large temperature gradient occurs from the surface toward the center during microwave heating. In smaller food items, microwave heating is more uniform (Vilayannur et al., 1998; Oliveira and Franca, 2002).

21.4 APPLICATIONS OF MICROWAVE HEATING IN THE FOOD INDUSTRY

21.4.1 Microwave Dehydration

Drying is a complex process involving simultaneous coupled heat and mass transfer. At the beginning of drying, the drying rate increases with the increase in temperature. After the temperature reaches the set value, the drying rate is constant at a high rate because the surface of the material contains free moisture. At the end of the constant rate period, moisture has to be transported from the inside of the material to the surface, and the drying rate decreases. Therefore, there are several stages during drying of food materials: (1) increasing rate at the beginning, (2) constant rate period in the middle, and (3) falling rate period at the end.

Microwave-assisted combination drying is a rapid dehydration technique that can be applied to specific foods, particularly to fruits and vegetables (Zhang et al., 2006; Orsat et al., 2007; Ozbek and Dadali, 2007; Gowen et al., 2008). Funebo and Ohlsson (1998) used a microwave-air assisted process to dehydrate apple and mushroom. They found that it was too costly to remove water using electrical energy during the constant-rate period, compared to thermal energy. The final stage of drying takes much of the total drying time. Microwave can be applied at the final stage of drying to reduce the drying time. Microwave-assisted hot air drying at even a very low microwave power could reduce the drying time to half or a third of the hot-air drying time for drying different vegetables such as potatoes, mushrooms and carrots (Riva et al., 1991; Bouraoui et al., 1994; Prabhanjan et al., 1995).

Microwave has also been as an energy source in vacuum drying to improve energy efficiency and product quality (Cui et al., 2004; Clary et al., 2005). As shown in Figure 21.1, when applying microwave energy, the generation of internal heat increases the vapor pressure inside the foods. The vapor pressures force the moisture to move to the surface where the moisture is readily removed by the ambient environment. Microwave-assisted vacuum drying of lycopene-rich carrots took less than 2h compared with 4.5–8.5h in convection air-drying at 50°C–70°C (Regier et al., 2005). The drying rate for microwave-assisted vacuum drying of tomato was found to be 18 times that of hot air convection drying. The drying time for microwave-assisted vacuum drying was only 0.8h compared to 14.75h for hot air convection drying at 70°C (Durance and Wang, 2002). Microwave has also been used as an energy source in other traditional drying methods such as freeze-drying (Sun et al., 2007). However, the current price of a microwave-assisted vacuum dryer is \$5–\$12 per watt of microwave heating capacity, compared to \$0.5–\$1 per watt of hot air capacity (Durance and Wang, 2002).

At low moisture levels, the permittivity and the loss factor of microwave are low, which means that a small part of the microwave energy is absorbed and converted into heat during drying. However, the specific heat and thermal conductivity of foods with low moisture are also low. Therefore, a low microwave power can be used during the falling-rate period at the final stage of drying to prevent any overheating and scorching. Another challenge related to microwave heating is that the final temperature of products in microwave drying is difficult to be controlled, compared to conventional air-drying in which the product temperature never rises higher than the air temperature.

21.4.2 Microwave Pasteurization

Microwave heating has been used for high-temperature short-time (HTST) sterilization/pasteurization of foods, particularly thick food items. Because of the low thermal conductivity, it is impossible to achieve HTST treatment for food items with a several-centimeter thickness by conventional heating methods. Microwave energy can penetrate into the food items and cause a rapid temperature rise to the desired pasteurization temperature. A continuous-flow microwave-assisted pasteurization system is shown in Figure 21.2 (LeBail et al., 2000).

Huang and Sites (2007) used a microwave heating system for in-package pasteurization of ready-to-eat meats. They observed that the inactivation rate of *Listeria monocytogenes* during microwave pasteurization was 0.41, 0.65, and 0.94 log (CFU/pk)/min at the surface temperature of 65°C, 75°C, and 85°C, respectively. The overall rate of bacterial inactivation for the water immersion pasteurization at the same surface temperatures was 30%–75% higher than that of microwave in-package pasteurization.

21.4.3 Microwave Thawing

It often takes several days to thaw frozen foods at a cold-storage temperature in the food industry. If accelerated conventional heating methods such as warm air or water are used, contamination of foods may occur. Microwave heating can significantly reduce the processing time and weight loss. The main advantages of microwave thawing include a high thawing speed, uniformity of thawing, flexibility, and less floor space requirement.

The microwave absorptivity in frozen materials is much smaller than that of their thawed state. Thermal runaway heating may occur due to temperature dependent dielectric properties (Lee and Marchant, 2004). However, both the dissipation of microwave energy and the energy required to increase the food temperature increase with the increase of food temperature at a comparable rate during microwave heating. This minimizes the effect of runaway heating during microwave thawing. However, microwave thawing of frozen foods is more nonuniform than traditional thawing methods. If a microwave thawing process is improperly controlled and the surface temperature of foods increases to a very high level, microwave thawing even causes a greater moisture loss than conventional thawing (Ni et al., 1999).

21.4.4 Microwave Cooking

Microwave cooking of unfrozen foods can reduce cooking time and achieve a uniform temperature profile inside foods. Since the surface temperature of foods during microwave cooking may easily reach 100°C, a combination of microwave heating with steam may be necessary to keep a low weight loss and prevent the waste of expensive energy used by a microwave heating unit for water evaporation at a high surface temperature.

21.4.5 Microwave Baking

Halogen lamp–microwave combination baking is a new technology in the bakery industry. Microwave can penetrate into products and thus rapidly heat the products from the inside. However, microwave baking cannot induce browning and crust on the surface of breads and cakes (Zhang and Datta, 2006). Halogen lamp heating provides near-infrared radiation light with high frequency and low penetration depth, which is focused on the surface of food products and increases the surface temperature for browning. A halogen lamp–microwave combination baking takes the time-saving advantage of microwave heating and the browning and crisping advantages of halogen lamp heating (Keskin et al., 2004; Sumnu et al., 2005).

Halogen lamp–microwave combination baking can increase the quality of bakery products compared with other baking methods such as conventional, microwave, and infrared baking. The combined baking process can reduce the conventional baking time by about 75%. Bakery products prepared by the combined baking unit have a specific volume and crust color similar to the conventionally baked products but the weight loss and firmness of breads are still higher than conventionally baked ones (Keskin et al., 2004; Sumnu et al., 2005).

21.4.6 Microwave-Assisted Extraction

Microwave-assisted extraction (MAE) offers rapid delivery of energy to a total volume of solvent and solid matrix with subsequent heating of the solvent and solid matrix, efficiently and homogeneously (Wang and Weller, 2006). Because water within the plant matrix absorbs microwave energy, cell disruption is promoted by internal superheating, which facilitates desorption of chemicals from the matrix, improving the recovery of chemicals (Kaufmann et al., 2001). Kratchanova et al. (2004) observed using scanning electron micrographs that microwave pretreatment of fresh orange peels led to destructive changes in the plant tissue. These changes in the plant tissue due to microwave heating gave a considerable increase in the yield of extractable pectin. Furthermore, the migration of dissolved ions increased solvent penetration into the matrix and thus facilitated the release of chemicals. The effect of microwave energy is thus strongly dependent on the dielectric susceptibility of both the solvent and the solid plant matrix.

As MAE depends on the dielectric susceptibility of the solvent and matrix, better recoveries can be obtained by moistening samples with a substance that possesses a relatively high dielectric constant such as water. If a dry biomaterial is rehydrated before extraction, the matrix itself can thus interact with microwaves and hence facilitate the heating process. Microwave heating leads to the expansion and rupture of cell walls and is followed by the liberation of chemicals into the solvent (Eskilsson and Bjorklund, 2000). In this case, the surrounding solvent can have a low dielectric constant and thus remains cold during extraction. This method can be used to extract thermo-sensitive compounds such as essential oils (Brachet et al., 2002). However, it was found that it was impossible to perform a good MAE for completely dry as well as for very wet samples when a nonpolar solvent such as hexane was used as the extraction solvent (Molins et al., 1997). Plant particle size and size distribution usually have a significant influence on the efficiency of MAE. The particle sizes of the extracted materials are usually in the range of 100 μm–2 mm (Eskilsson and Bjorklund, 2000). Fine powder can enhance the extraction because the limiting step of the extraction is often the diffusion of chemicals out of the plant matrix and the larger surface area of a fine powder provides contact between the plant matrix and the solvent. For example, for MAE of cocaine, finely ground coca powder was more easily extracted than large particles (Brachet et al., 2002).

Solvent choice for MAE is dictated by the solubility of the extracts of interest, by the interaction between solvent and plant matrix, and finally by the microwave absorbing properties of the solvent determined by its dielectric constant. Csiktusnadi Kiss et al. (2000) investigated the efficiency and selectivity of MAE for the extraction of color pigments from *paprika* powder using 30 extracting solvent mixtures. Their results showed that both the efficacy and selectivity of MAE depend significantly on the dielectric constant of the extraction solvent mixture. Usually, the chosen solvent should possess a high dielectric constant and strongly absorb microwave energy. Solvents such as ethanol, methanol, and water are sufficiently polar to be heated by microwave energy (Brachet et al., 2002). Nonpolar solvents with low dielectric constants such as hexane and toluene are not potential solvents for MAE. The extracting selectivity and the ability of the solvent to interact with microwaves can be modulated by using mixtures of solvents (Brachet et al., 2002). One of the most commonly used mixtures is hexane–acetone (Eskilsson and Bjorklund, 2000). A small amount of water (e.g., 10%) can also be incorporated in nonpolar solvents such as hexane, xylene, or toluene to improve the heating rate (Eskilsson and Bjorklund, 2000). During extraction, the solvent volume must be sufficient to ensure that the solid matrix is entirely immersed. Generally, a higher ratio of solvent volume to solid matrix mass in conventional extraction techniques can increase the recovery. However, in the MAE, a higher ratio may give lower recoveries. This is probably due to inadequate stirring of the solvent by microwaves (Eskilsson et al., 1999).

Temperature is another important factor contributing to the recovery yield. Generally, elevated temperatures result in improved extraction efficiencies. However, for the extraction of thermolabile compounds, high temperatures may cause the degradation of extracts. In this case, the chosen power during MAE has to be set correctly to avoid excess temperatures, leading to possible solute degradation (Font et al., 1998).

MAE has been considered as a potential alternative to traditional solid–liquid extraction for the extraction of metabolites from plants. It has been used to extract nutraceuticals for several reasons: (1) reduced extraction time, (2) reduced solvent usage, and (3) improved extraction yield. MAE is also comparable to other modern extraction techniques such as supercritical fluid extraction because of its process simplicity and low cost. Considering economic and practical aspects, MAE is a strong novel extraction technique for the extraction of nutraceuticals. However, additional filtration or centrifugation is necessary to remove the solid residue during MAE. Furthermore, the efficiency of microwaves can be very poor when either the target compounds or the solvents are nonpolar or when they are volatile (Wang and Weller, 2006).

MAE can extract nutraceutical products from plant sources in a faster manner than conventional solid–liquid extractions. MAE of the puerarin from the herb *Radix puerariae* could be completed within 1 min (Guo et al., 2001). MAE (80% methanol) could dramatically reduce the extraction time of *ginseng* saponins from 12 h using conventional extraction methods to a few seconds (Kwon et al., 2003). It took only 30 s to extract cocaine from leaves with the assistance of microwave energy quantitatively similar to that obtained by conventional solid–liquid extraction for several hours (Brachet et al., 2002). For extracting an equivalent amount and quality of tanshinones from *Salvia miltiorrhiza bunge*, MAE only needed 2 min, whereas extraction at room temperature, Soxhlet extraction, ultrasonic extraction, and heat reflux extraction needed 24 h, 90 min, 75 min, and 45 min, respectively (Pan et al., 2002). Williams et al. (2004) found that MAE was efficient in recovering approximately 95% of the total capsaicinoid fraction from *capsicum* fruit in 15 min compared with 2 h for the reflux and 24 h for the shaken flask methods.

A higher extraction yield can be achieved in a shorter extraction time using MAE. A 12 min MAE could recover 92.1% of artemisinin from *Artemisia annua L*, while several-hour Soxhlet extraction could achieve only about 60% recovery (Hao et al., 2002). A 4–5 min MAE (ethanol–water) of glycyrrhizic acid from *licorice* root achieved a higher extraction yield than extraction (ethanol–water) at room temperature for 20–24 h (Pan et al., 2000). For the extraction of tea polyphenols and caffeine from green tea leaves, a 4 min MAE achieved a higher extraction yield than an extraction at room temperature for 20 h, ultrasonic extraction for 90 min, and heat reflux extraction for 45 min (Pan et al., 2003). Shu et al. (2003) reported that the extraction yield of ginsenosides from *ginseng* root obtained by a 15 min MAE (ethanol–water) was higher than that obtained by 10 h conventional solvent extraction (ethanol–water).

MAE can also reduce solvent consumption. Focused MAE was applied to the extraction of withanolides from air-dried leaves of *Iochroma gesnerioides* (Kaufmann et al., 2001). The main advantages of MAE over Soxhlet extraction are associated with the drastic reduction in organic solvent consumption (5 mL vs. 100 mL) and extraction time (40 s vs. 6 h). It was also found that the presence of water in the solvent of methanol had a beneficial effect and allowed faster extractions than with organic solvent alone.

21.5 ENERGY EFFICIENCY, CONSUMPTION, AND CONSERVATION DURING MICROWAVE HEATING

21.5.1 Energy Efficiency

During microwave heating, electrical energy is first converted into microwave energy. The microwave then interacts with foods and is converted into heat. Therefore, there are two efficiencies: microwave generation efficiency and microwave absorption efficiency, which are defined as

$$\text{Generation efficiency} = 100 \times \frac{\text{Microwave output power}}{\text{Input electrical power}} \tag{21.6}$$

$$\text{Absorption efficiency} = 100 \times \frac{\text{Thermal energy absorbed by foods}}{\text{Microwave output power} \times \text{heating duration}} \tag{21.7}$$

The microwave generation efficiency is calculated by dividing the rated microwave output power of a microwave oven by the actual measured input electrical power to the oven. To calculate the microwave absorption efficiency, the thermal energy absorbed by foods should be determined. Usually a known quantity of cold water is heated in the microwave oven for a specific period. The thermal energy absorbed by water is determined by the energy balance:

$$Q = mc_p(T_2 - T_1) \tag{21.8}$$

where

Q is the thermal energy absorbed by water (J)
m is the quantity of water (kg)
c_p is the specific heat of water (J/kg°C)
T_2 and T_1 are the final and initial temperatures, respectively, of water (°C)

The absorption efficiency can be calculated by dividing the thermal energy absorbed by the microwave output power and heating duration. If we combine both generation efficiency and absorption efficiency, we can get the total thermal efficiency, which can be expressed as

$$\text{Thermal efficiency} = 100 \times \frac{\text{Thermal energy absorbed by foods}}{\text{Input electrical energy}} \tag{21.9}$$

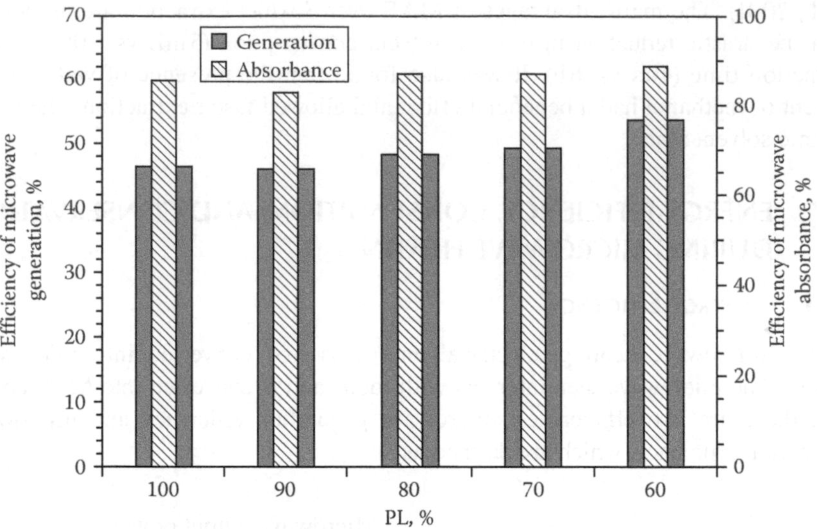

FIGURE 21.3 Efficiency of microwave generation and absorption by water at various power levels. (Reprinted from Lakshmi, S., Chakkaravarthi, A., Subramanian, R., and Singh, V., *J. Food Eng.*, 78, 715, 2007. With permission.)

Lakshmi et al. (2007) found that although the absorption of microwave energy by water is 86%–89%, the conversion efficiency of electrical energy to microwave energy is only about 50% as shown in Figure 21.3. The total thermal efficiency was only around 44%. During microwave heating, moisture loss may occur due to the evaporation of moisture at a high temperature. In this case, part of the thermal energy delivered to the foods releases moisture from evaporation rather than increases the temperature of the foods. Comparison among microwave cooker, electric cooker, and pressure cooker for cooking rice showed that microwave oven had the shortest heating time among these three types of cookers. However, the electric cooker was the most energy efficient (Lakshmi et al., 2007).

21.5.2 Energy Consumption

The power requirement for a given application of microwave heating depends on the enthalpy change involved and the power efficiency obtainable under practical working conditions. Table 21.2 gives approximate values of the heat and electricity required to process 1 kg of food products for different applications of microwave heating.

Sharma and Prasad (2006) compared the specific energy consumption between convective and microwave-convective drying of garlic cloves. The specific energy consumption was defined as the total energy supplied divided by the amount of water removed during drying. A combination of microwave at 40 W and convective air at 70°C and 1 m/s could result in a 70% energy saving as compared to a convective drying process with air at the same conditions. Drying with hot air took a long time, had low energy efficiency, and resulted in substantial degradation of quality attributes such as color, nutrients, and flavor, especially during the falling rate period

TABLE 21.2
Approximate Values of Heat and Electricity Required to Process 1 kg of Food Products during Different Applications of Microwave Heating

Process	Heat Requirement (MJ/kg)	Electricity Requirement at 45% Energy Efficiency (kWh/kg)
Thawing	0.45	0.28
Cooking	0.60	0.37
Blanching	0.33	0.20
Pasteurization	0.36	0.22
Finish drying from 25% to 2%	0.72	0.44
Baking	0.24	0.15
Insect deinfestation	0.08	0.05

(Zhang et al., 2006). The final drying stage takes much of the whole drying time due to the falling drying rate. The addition of microwave, especially at the final drying state, can significantly reduce the drying time, causing a reduction in energy consumption and improvement of thermal energy efficiency (Xu et al., 2004).

Durance and Wang (2002) compared the drying time, energy consumption, and energy costs of (1) microwave-assisted vacuum drying in a pilot-scale batch 20 kW, 2450 MHz vacuum microwave under 6.65 kPa vacuum, (2) convection drying with hot air at 70°C, 12% RH, and 1.5 m^3/s in a pilot-scale conveyer food dehydrator with 0.6 m × 2.0 m belt, and (3) their combination for drying 13.8 kg of tomatoes. The results are given Table 21.3. The drying rate of microwave-assisted vacuum drying was 18 times that of hot air convection drying. The drying time of microwave-assisted vacuum drying was only 0.8 h compared to 14.75 h for hot air convection drying. There was only a slight falling rate effect observed during microwave-assisted vacuum drying. The microwave-assisted vacuum drying process consumed the least energy. However, because a microwave oven uses high-quality electricity, a combined 70% hot air convection drying and 30% microwave-assisted drying could achieve the lowest energy cost. Energy efficiency depends on the production scale and heat recovery system. The reported data were obtained from a single-pass convection air dryer without air recirculation. Energy efficiency was defined as the latent heat for water evaporation at 2.5 MJ/kg water divided by the energy consumed by the dryers for removing 1 kg of water. The energy efficiency of the hot air convection dryer was only 8.4%, compared to 29.1% for the microwave-assisted vacuum dryer (Durance and Wang, 2002).

However, it should be noted that a microwave oven consumes a high-quality energy source of electricity. Energy cost is not only determined by the energy efficiency but also the price of the energy source. Electricity is usually generated by burning fossil fuels at a conversion efficiency of 35%. When transmission costs are included, electricity is often two to four times as expensive as fossil fuels such as natural gas. Suppose the energy efficiency for a microwave oven to convert electricity

TABLE 21.3
Comparison of Energy Efficiency and Cost of Microwave-Assisted Vacuum Drying (MAVD) with Hot Air Convection Drying (HACD)

	100% HACD[a]	95% HACD/ 5% MAVD[a]	85% HACD/ 15% MAVD[a]	70% HACD/ 30% MAVD[a]	100% MAVD[a]
HACD time, h	14.75	12	9	6	0
MAVD time, h	0	0.05	0.12	0.23	0.81
Total time, h	14.75	12.05	9.12	6.23	0.81
HACD energy MJ/kg water	29.9	25.0	20.9	17.3	0
MAVD energy MJ/kg water	0	9.5	9.1	8.9	8.6
Average energy MJ/kg water	29.9	24.2	19.1	14.8	8.6
HACD energy cost \$/kg water[b]	0.27	0.23	0.19	0.17	0
MAVD energy cost \$/kg water[c]	0	0.20	0.19	0.19	0.18
Average energy cost \$/kg water	0.27	0.23	0.19	0.17	0.18

Source: Reproduced from Durance, T.D. and Wang, J.H., *J. Food Sci.*, 67, 2212, 2002. Copyright Wiley-Blackwell. With permission.

[a] Percentages of water removed by HACD and MAVD.

[b] Natural gas cost at \$9/GJ (or \$8.53/MBtu).

[c] Electricity cost at \$21/GJ (\$0.076/kWh).

to heat is 50%, and electricity price is \$0.05/kWh, it requires \$2.78 to provide 100 MJ heat, compared to \$0.63 for a heated hot-air process by natural gas at a price of \$6/MBtu (or \$0.02/kWh) and energy efficiency of 90% for converting natural gas to heat. Therefore, the energy cost of useful heat for a natural-gas heated hot-air drying process is only about 20%–30% of that of a microwave heating process. However, the useful heat received by foods for temperature increase or phase change during a conventional heating process may be as low as 10% of the total energy input. This means 90% of the heat is lost during heating. Since microwave energy can penetrate into foods and cause a temperature rise, heat loss is negligible. In this case, microwave heating is comparable with conventional heating methods in terms of total energy costs.

21.5.3 Potential Energy Conservation Measures

Microwave heating alone is usually considered uneconomic unless very positive effects can be obtained in terms of increased yield and quality. Combination microwaveheating with some conventional sources of heat such as steam and hot air will be necessary. Microwave has been combined with hot air in drying to reduce the drying time and thus the energy consumption (Bouraoui et al., 1994; Riva et al., 1991; Prabhanjan et al., 1995).

Microwave penetrates into products to generate heat inside the products. Penetration of microwave energy inside a product is a function of its dielectric properties. As the product size increases, the regions inside the product away from the surface are not heated satisfactorily due to the decaying of microwave energy as it propagates into the product. Therefore, overheating of some regions may occur. Overheating could increase the energy consumption due to high moisture loss from the overheated region. Pulsed microwave heating is an effective way to reduce the overheating effects (Gunasekaran and Yang, 2007).

High power efficiency is also a desirable characteristic for a microwave oven since the electric energy is expensive. It is possible to achieve high power efficiencies by placing products to be heated at an optimal position within a microwave chamber for which microwave power transferred to the products is maximized. Pedreno-Molina et al. (2007) investigated a procedure for iteratively moving a product within a microwave oven until a maximum average temperature increment is detected across the product.

21.6 SUMMARY

Microwaves can provide a fast heating method in the food industry since they can penetrate into a food material to generate heat directly inside the food material. Microwave heating has been used for dehydration, pasteurization, thawing, cooking, baking, and extraction in the food industry. Under favorable conditions, a uniform temperature distribution can be achieved during microwave heating, which improves the yield and quality of foods. In addition, a microwave heating process is compact and flexible and can be operated in a continuous mode. However, the investment and processing costs associated with microwave heating could be high. Microwave heating is usually combined with a conventional heating method to reduce processing time and the total processing cost.

REFERENCES

Bengtsson, N.E. and T. Ohlsson. 1974. Microwave heating in the food industry. *Proceedings of the IEEE* 62: 44–55.

Bouraoui, M., P. Richard, and T. Durance. 1994. Microwave and convective drying of potato slices. *Journal of Food Process Engineering* 17: 353–363.

Brachet, A., P. Christen, and J.L. Veuthey. 2002. Focused microwave-assisted extraction of cocaine and benzoylecgonine from *coca* leaves. *Phytochemical Analysis* 13: 162–169.

Campanone, L.A. and N.E. Zaritzky. 2005. Mathematical analysis of microwave heating process. *Journal of Food Engineering* 60: 359–368.

Clary, C.D., S.J. Wang, and V.E. Petrucci. 2005. Fixed and incremental levels of microwave power application on drying grapes under vacuum. *Journal of Food Science* 70: 344–349.

Csiktusnadi Kiss, G.A., E.F. Forgacs, T. Cserhati, T. Mota, H. Morais, and A. Ramos. 2000. Optimisation of the microwave-assisted extraction of pigments from *paprika* (*Capsicum annuum L.*) powders. *Journal of Chromatography A* 889: 41–49.

Cui, Z.W., S.Y. Xu, and D.W. Sun. 2004. Microwave-vacuum drying kinetics of carrot slices. *Journal of Food Engineering* 65: 157–164.

Durance, T.D. and J.H. Wang. 2002. Energy consumption, density, and rehydration rate of vacuum microwave- hot-air convection-dehydrated tomatoes. *Journal of Food Science* 67: 2212–2216.

Eskilsson, C.S., E. Bjorklund, L. Mathiasson, L. Karlsson, and A. Torstensson. 1999. Microwave-assisted extraction of felodipine tablets. *Journal of Chromatography A* 840: 59–70.

Eskilsson, C.S., and E. Bjorklund. 2000. Analytical-scale microwave-assisted extraction. *Journal of Chromatography A* 902: 227–250.

Font, N., F. Hernandez, E.A. Hogendoorn, R.A. Baumann, and P. van Zoonen. 1998. Microwave-assisted solvent extraction and reversed-phase liquid chromatography-UV detection for screening soils for sulfonylurea herbicides. *Journal of Chromatography A* 798: 179–186.

Funebo, T. and T. Ohlsson. 1998. Microwave-assisted air dehydration of apple and mushroom. *Journal of Food Engineering* 38: 353–367.

Gowen, A.A., N. Abu-Ghannam, J. Frias, and J. Oliveira. 2008. Modeling dehydration and rehydration of cooked soybeans subjected to combined microwave-hot-air drying. *Innovative Food Science & Emerging Technologies* 9: 129–137.

Gunasekaran, S. and H.W. Yang. 2007. Optimization of pulsed microwave heating. *Journal of Food Engineering* 78: 1457–1462.

Gunnasekaran, N., P. Mallikarjunan, J. Eifert, and S. Sumner. 2005. Effect of fat content and temperature on dielectric properties of ground beef. *Transactions of the ASAE* 48: 673–680.

Guo, Z., Q. Jin, G. Fan, Y. Duan, C. Qin, and M. Wen. 2001. Microwave-assisted extraction of effective constituents from a Chinese herbal medicine *Radix puerariae. Analytica Chimica Acta* 436: 41–47.

Hao, J.Y., W. Han, S.D. Huang, B.Y. Xue, and X. Deng. 2002. Microwave-assisted extraction of artemisinin from *Artemisia annua L. Separation and Purification Technology* 28: 191–196.

Hill, J.M. and T.R. Marchant. 1994. Modeling microwave heating. *Applied Mathematic Modeling* 20: 3–15.

Huang, L. and J. Sites. 2007. Automatic control of a microwave heating process for in-package pasteurization of beef frankfurters. *Journal of Food Engineering* 80: 226–233.

Kaufmann, B., P. Christen, and J.L. Veuthey. 2001. Parameters affecting microwave-assisted extraction of *withanolides. Phytochemical Analysis* 12: 327–331.

Keskin, S.O., G. Sumnu, and S. Sahin. 2004. Bread baking in halogen lamp-microwave combination oven. *Food Research International* 37: 489–495.

Kratchanova, M., E. Pavlova, and I. Panchev. 2004. The effect of microwave heating of fresh orange peels on the fruit tissue and quality of extracted pectin. *Carbohydrate Polymers* 56: 181–186.

Kudra, T., G.S.V. Raghavan, C. Akyel, R. Bosisio, and F.R. van de Voort. 1992. Electromagnetic properties of milk and its constituents at 2.45 MHz. *International Microwave Power Institute Journal* 27: 199–204.

Kwon, J.H., J.M.R. Belanger, J.R. Jocelyn Pare, and V.A. Yaylayan. 2003. Application of microwave-assisted process (MAP TM) to the fast extraction of ginseng saponins. *Food Research International* 36: 491–498.

Lakshmi, S., A. Chakkaravarthi, R. Subramanian, and V. Singh. 2007. Energy consumption in microwave cooking of rice and its comparison with other domestic appliances. *Journal of Food Engineering* 78: 715–722.

LeBail, A., T. Koutchma, and H.S. Ramaswamy. 2000. Modeling of temperature profiles under continuous tube-flow microwave and steam heating conditions. *Journal of Food Process Engineering* 23: 1–24.

Lee, M.Z.C. and T.R. Marchant. 2004. Microwave thawing of cylinders. *Applied Mathematical Modeling* 28: 711–733.

Molins, C., E.A. Hogendoorn, H.A.G. Heusinkveld, P. van Zoonen, and R.A. Baumann. 1997. Microwave assisted solvent extraction (MASE) of organochlorine pesticides from soil samples. *International Journal of Environmental Analytical Chemistry* 68: 155–169.

Ni, H., A.K. Datta, and R. Parmeswar. 1999. Moisture loss as related to heating uniformity in microwave processing of solid foods. *Journal of Food Process Engineering* 22: 367–382.

Oliveira, M.E.C. and A.S. Franca. 2002. Microwave heating of foodstuffs. *Journal of Food Engineering* 53: 347–359.

Orsat, V., W. Yang, V. Changrue, and G.S.V. Raghavan. 2007. Microwave-assisted drying of biomaterials. *Food and Bioproducts Processing* 85: 255–263.

Ozbek, B. and G. Dadali. 2007. Thin-layer drying characteristics and modeling of mint leaves undergoing microwave treatment. *Journal of Food Engineering* 83: 541–549.

Pan, X., H. Liu, G. Jia, and Y.Y. Shu. 2000. Microwave-assisted extraction of glycyrrhizic acid from *licorice* root. *Biochemical Engineering Journal* 5: 173–177.

Pan, X., G. Niu, and H. Liu. 2002. Comparison of microwave-assisted extraction and conventional extraction techniques for the extraction of tanshinones from *Salvia miltiorrhiza* Bunge. *Biochemical Engineering Journal* 12: 71–77.

Pan, X., G. Niu, and H. Liu. 2003. Microwave-assisted extraction of tea polyphenols and tea caffeine from green tea leaves. *Chemical Engineering Process* 42: 129–133.

Pedreno-Molina, J.L., J. Monzo-Cabrera, and M. Pinzolas. 2007. A new procedure for power efficiency optimization in microwave ovens based on thermographic measurements and load location search. *International Communications in Heat and Mass Transfer* 34: 564–569.

Piyasena, P., H.S. Ramaswamy, G.B. Awuah, and C. Defelice. 2003. Dielectric properties of starch solutions as influenced by temperature, concentration, frequency and salt. *Journal of Food Process Engineering* 26: 93–119.

Prabhanjan, D.G., H.S. Ramaswamy, and G.S.V. Raghavan. 1995. Microwave-assisted convective air drying of thin layer carrots. *Journal of Food Engineering* 25: 283–293.

Regier, M., E. Mayer-Miebach, D. Behsnilian, E. Neff, and A. Schuchmann. 2005. Influences of drying and storage of lycopene-rich carrots on the carotenoid content. *Drying Technology* 23: 989–998.

Riva, M., A. Schiraldi, and L. Di Cesare. 1991. Drying of *Agaricus biosporus* mushroom by microwave-hot air combination. *Lebensmittel Wissenschaft und Technologie* 24: 479–482.

Romano, V.R., F. Marra, and U. Tammaro. 2005. Modeling of microwave heating of foodstuff: Study on influence of sample dimensions with a FEM approach. *Journal of Food Engineering* 71: 233–241.

Ryynanen, S. 1995. The electromagnetic properties of food materials: A review of the basic principles. *Journal of Food Engineering* 26: 409–429.

Sakai, N., W. Mao, Y. Koshima, and M. Watanabe. 2005. A method for developing model food system in microwave heating studies. *Journal of Food Engineering* 66: 525–531.

Sharma, G.P. and S. Prasad. 2006. Specific energy consumption in microwave drying of garlic cloves. *Energy* 31: 1921–1926.

Shu, Y.Y., M.Y. Ko, and Y.S. Chang. 2003. Microwave-assisted extraction of ginsenosides from *ginseng* root. *Microchemical Journal* 74: 131–139.

Sumnu, G., S. Sahin, and M. Sevimli. 2005. Microwave, infrared and infrared-microwave combination baking of cakes. *Journal of Food Engineering* 71: 150–155.

Sun, H., H. Zhu, H. Feng, and L. Xu. 2007. Thermoelectromagnetic coupling in microwave freeze-drying. *Journal of Food Process Engineering* 30: 131–149.

Tang, J., H. Feng, and M. Lau. 2002. Microwave heating in food processing, In *Advances in Bioprocess Engineering*, Yangand X.H. and J. Tang (Eds.), Hackensack, New Jersey: World Scientific.

Venkatesh, M.S. and G.S.V. Raghavan. 2004. An overview of microwave processing and dielectric properties of agri-food materials. *Biosystems Engineering* 88: 1–18.

Vilayannur, R.S., V.M. Puri, and R.C. Anantheswaran. 1998. Size and shape effect on nonuniformity of temperature and moisture distributions in microwave heated food materials: Part II experimental validation. *Journal of Food Process Engineering* 21: 235–248.

Wang, L.J. and C.L. Weller. 2006. Recent advances in extraction of natural products from plants. *Trends in Food Science and Technology* 17: 300–312.

Williams, O.J., G.S.V. Raghavan, V. Orsat, and J. Dai. 2004. Microwave-assisted extraction of Capsaicinoids from Capsicum fruit. *Journal of Food Biochemistry* 28: 113–122.

Xu, Y.Y., Z. Min, and A.S. Mujumdar. 2004. Studies on hot air and microwave vacuum drying of wild cabbage. *Drying Technology* 22: 2201–2209.

Yang, H.W. and S. Gunasekaran. 2004. Comparison of temperature distribution in model food cylinders based on Maxwell's equation and Lambert's law during pulsed microwave heating. *Journal of Food Engineering* 64: 445–453.

Zhang, J. and A.K. Datta. 2006. Mathematical modeling of bread baking process. *Journal of Food Engineering* 75: 78–89.

Zhang, M., J. Tang, A.S. Mujumdar, and S. Wang. 2006. Trends in microwave-related drying of fruits and vegetables. *Trends in Food Science & Technology* 17: 524–534.

22 Energy Efficiency and Conservation in Supercritical Fluid Processing

22.1 INTRODUCTION

A supercritical state is achieved when the temperature and the pressure of a substance are raised over its critical values. A supercritical fluid has characteristics of both gases and liquids. Compared with liquid solvents, supercritical fluids have several major advantages: (1) the dissolving power of supercritical fluids depends on their density, which is highly adjustable by changing the pressure or temperature; (2) supercritical fluids have a higher diffusion coefficient and lower viscosity and surface tension than liquid solvents, leading to more favorable mass transfer (Wang and Weller, 2006).

Supercritical fluids, particularly supercritical carbon dioxide, have been used in extraction, drying, and particle formulation in the food industry (Sihvonen et al., 1999; Marr and Gamse, 2000). A typical supercritical fluid processing cycle requires a pressurizing step to increase the pressure of fluids to be higher than their critical pressure and a heating step to increase the temperature to be higher than the critical temperature of the fluids, a use step of supercritical fluids in processing of foods such as extraction, drying, and particle formulation, a separation step to remove solvent from products, and a solvent regeneration step (Sievers and Eggers, 1996).

Supercritical fluids have been used to extract nutraceuticals from plant materials (Wang and Weller, 2006), remove moisture from foods (Brown et al., 2008), and produce fine particles (Sihvonen et al., 1999; Marr and Gamse, 2000) in the food industry. Particularly, supercritical fluid technology can be used in processing of functional foods and nutraceuticals. In this chapter, the common supercritical fluids, supercritical fluid process, and applications of supercritical fluids in the food industry are reviewed. Finally, the energy consumption and the energy saving opportunities of a supercritical fluid process are discussed.

22.2 OVERVIEW OF SUPERCRITICAL FLUID PROCESSES

22.2.1 Supercritical Fluids

A fluid becomes supercritical after it passes its vapor–liquid critical point with the increase in temperature and pressure. After a fluid becomes supercritical, there is no phase change between gas and liquid with further changes in temperature and pressure

TABLE 22.1
Critical Points of Selected Substances

Solvent	Temperature (°C)	Pressure (MPa)
CO_2	31	7.3
Ethanol	241	6.1
Methanol	239	8.09
Water	374	21.8
Ammonia	133	11.1
Propane	97	4.2

above its critical point. A supercritical fluid has both gaseous properties such as high diffusivity, low viscosity, and high compressibility and liquid properties such as high density. The unique properties of supercritical fluids can enhance heat and mass transfer, reaction kinetics, and equilibrium between solid biomass and supercritical fluids. Furthermore, due to the high compressibility of supercritical fluids, their solvent properties can be easily adjusted by changing the pressure and temperature (Wang and Weller, 2006). Thus, the products can be easily recovered from the fluids by the reduction of the dissolving power or density (Sihvonen et al., 1999; Wang and Weller, 2006). The critical points of selected substances are given in Table 22.1.

22.2.2 Supercritical Fluid Process

A supercritical fluid processing system is shown in Figure 22.1. During supercritical fluid processing, a raw plant material is loaded into a high-pressure processing vessel, which is equipped with temperature controllers and pressure valves at both the inlet and outlet of the vessel to keep desired processing conditions. A pump or compressor is used to increase the pressure of the fluid to be higher than its critical pressure. A heat exchanger is used to increase the temperature of the fluid to be higher than its

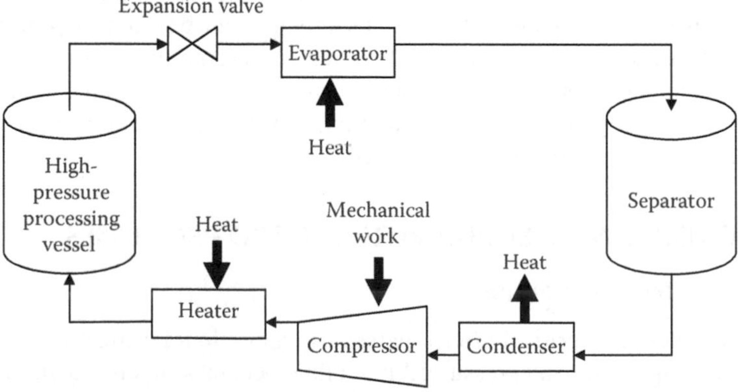

FIGURE 22.1 A typical single-stage supercritical fluid process. (Reprinted from Sievers, U. and Eggers, R., *Chem. Eng. Process.*, 35, 239, 1996. With permission.)

critical temperature. The supercritical fluid is then delivered into the high-processing vessel to extract, dry, or assist in formulating particle products. After the process, the supercritical fluid is usually changed into its vapor phase by reducing its pressure, increasing its temperature, or a combination to decrease the salvation power of the fluid. The pressure of the supercritical fluid can be decreased below its critical pressure (subcritical stage) through an expansion valve. During the fluid expansion, the temperature is decreased sharply. An evaporator is usually used to increase the fluid temperature after expansion. The fluid and the dissolved compounds are transported to the separator, where the fluid vapor flashes away from the product. The product is then collected via a valve located in the lower part of the separators. The fluid vapor is condensed in a condenser so that the liquid fluid can be easily compressed by a compressor to complete the cycle (Sihvonen et al., 1999).

22.2.3 Practical Issues of a Supercritical Fluid Process

A suitable solvent should be selected as a working fluid in a supercritical fluid process. With a reduction in the price of carbon dioxide and restrictions in the use of other organic solvents, carbon dioxide has begun to move from some marginal applications to being the major solvent for supercritical fluid processing, particularly extraction (Hurren, 1999). The critical state of carbon dioxide fluid is at a temperature of only 304 K and pressure of 7.3 MPa. In addition, carbon dioxide is non-flammable and nontoxic. Supercritical CO_2 is a good solvent for dissolving non-polar compounds such as hydrocarbons (Vilegas et al., 1997). Although supercritical water and superheated water have certain advantages such as higher dissolving ability for polar compounds, they are not suitable for thermally labile compounds (Lang and Wai, 2001). Addition of polar co-solvents (modifiers) to the supercritical CO_2 is known to significantly increase the solubility of polar compounds. Among all the modifiers including methanol, ethanol, acetonitrile, acetone, water, ethyl ether, and dichloromethane, methanol is the most commonly used because it is an effective polar modifier and is up to 20% miscible with CO_2. However, ethanol may be a better choice in supercritical fluid processing of foods because of its lower toxicity (Lang and Wai, 2001; Hamburger et al., 2004). Furthermore, the use of methanol as a modifier requires a slightly higher temperature to reach the supercritical state and this could be disadvantageous for thermolabile compounds in foods. A mixture of modifiers can also be used in a supercritical fluid process. The best modifier usually can be determined based on preliminary experiments. One disadvantage of using a modifier is that it can cause poor selectivity.

Preparation of food materials is another critical factor for a supercritical fluid process. The high moisture content can cause mechanical difficulties such as restrictor clogging due to ice formation. Although water is only about 0.3% soluble in supercritical CO_2, highly water-soluble solutes would prefer to partition into the aqueous phase, resulting in low efficiency of supercritical fluid processes. Some chemicals such as Na_2SO_4 and silica gel are thus mixed with the plant materials to retain the moisture for the supercritical fluid processing of fresh materials (Lang and Wai, 2001). If solid materials are used, the particle size of the materials is also important for a good supercritical fluid process. Large particles may result in a long process because the process may be controlled by internal diffusion. Fine powder can speed up the process but may also cause difficulty in maintaining a proper flow rate (Coelho et al., 2003).

The solubility of a target compound such as oil and water in a supercritical fluid is a major factor in determining its processing efficiency. The temperature and density of the fluid control the solubility. The choice of a proper density of a supercritical fluid such as CO_2 is the crucial point influencing solvent power and selectivity and the main factor determining the extract composition (Cherchi et al., 2001). By controlling the fluid density and temperature, a supercritical fluid could selectively extract target substances such as water during supercritical fluid drying in the food materials by controlling the conditions such as temperature and pressure (density).

The extraction time has been proven to be another parameter that determines extract composition. Lower molecular weight and less polar compounds are more readily extracted during supercritical CO_2 extraction since the extraction mechanism is usually controlled by internal diffusion (Poiana et al., 1999; Cherchi et al., 2001). Therefore, the extract composition varies with the extraction time. However, Coelho et al. (2003) reported that the increase in CO_2 flow rate did not seem to influence the composition for the supercritical CO_2 extraction of *Foeniculum vulgare* volatile oil, although it increased the extraction rate.

22.3 APPLICATIONS OF SUPERCRITICAL FLUID PROCESSES IN THE FOOD INDUSTRY

22.3.1 Supercritical Fluid Extraction

Supercritical fluid extraction is a potential alternative to conventional extraction methods using organic solvents for extracting biologically active components from plants (Modey et al., 1996; Dean et al., 1998; Dean and Liu, 2000; Szentmihalyi et al., 2002; Ellington et al., 2003; Hamburger et al., 2004; Andras et al., 2005). It has been used to extract plant materials, especially lipids (Bernardo-Gil et al., 2002), essential oils (Berna et al., 2000; Coelho et al., 2003; Marongiu et al., 2003), and flavors (Sass-Kiss et al., 1998; Giannuzzo et al., 2003). Overviews of fundamentals and applications of supercritical fluids in different processes have been prepared by Sihvonen et al. (1999), Marr and Gamse (2000), and Hauthal (2001). Turner et al. (2001) have reviewed supercritical fluids in the extraction and chromatographic separation of fat-soluble vitamins. An overview of published data for the supercritical fluid extraction of different materials was given by Marr and Gamse (2000), Lang and Wai (2001), and Meireles and Angela (2003).

Supercritical fluid extraction can prevent the oxidation of lipids. Bernardo-Gil et al. (2002) found that the contents of free fatty acids, sterols, triacylglycerols, and tocopherols in the hazelnut oil extracted by supercritical CO_2 were comparable with those obtained with n-hexane extraction. However, the supercritical CO_2-extracted oil was more protected against oxidation of unstable polyunsaturated fatty acids than the n-hexane-extracted oil. Oil extracted with supercritical CO_2 was clearer than the one extracted by n-hexane.

Supercritical CO_2 extraction can achieve higher yield and quality of essential oils, flavors, and natural aromas than conventional steam distillation. The mean percentage yields of cedar wood oil for supercritical CO_2 extraction and steam distillation were 4.4% and 1.3%, respectively (Eller and King, 2000). The yield of supercritical CO_2 extraction of essential oil from *juniper* wood at 50°C and 10 MPa was 14.7% (w/w), while hydrodistillation gave a yield of 11% (w/w) (Marongiu et al., 2003). Coelho et al. (2003) found that moderate supercritical CO_2 conditions (9 MPa and 40°C) could achieve an efficient extraction of *Foeniculum vulgare* volatile oil, enabling about 94% of the oil to be extracted within 150 min. Compared with hydrodistillation, supercritical CO_2 extraction at 20 MPa and 50°C led to higher concentrations of light oxygenated compounds in the oil extracted from *Egyptian marjoram* leaves, which gave the oil a superior aroma. The antioxidant property of the supercritical CO_2 extract was also markedly higher than that of hydrodistillation extract (El-Ghorab et al., 2004). However, it should be noted that the composition of the essential oil was determined by two important parameters: CO_2 density (pressure) and extraction time (Roy et al., 1996; Cherchi et al., 2001).

A comparison of different extraction methods for the extraction of oleoresin from dried onion showed that the yield after supercritical CO_2 extraction was 22 times higher than that after steam distillation, but it was only 7% of the yield achieved by the extraction using a polar solvent of alcohol at 25°C (Sass-Kiss et al., 1998). However, the flavor and biological activity of onion are attributed mainly to its sulfur-containing compounds. The concentration of sulfur was the highest in steam distilled onion oil while it was the lowest in the extract of alcohol at 25°C. The oleoresin produced by supercritical CO_2 extraction had the best sensory quality. Many active substances in plants such as phenolics, alkaloids, and glycosidic compounds are poorly soluble in CO_2 and hence not extractable (Hamburger et al., 2004).

The number of industrial-scale applications of supercritical fluids in plant extraction has remained very small due to the lipophilic nature of supercritical CO_2. Modifiers such as methanol and ethanol are therefore used in the supercritical CO_2 extraction of polar substances. Supercritical CO_2 modified with 15% ethanol gave higher yields than pure supercritical CO_2 to extract naringin (a glycosylated falvonoid) from *citrus paradise* at 9.5 MPa and 58.6°C (Giannuzzo et al., 2003). The use of a 10% ethanol co-solvent resulted in a much higher yield of *epicatechin* (13 mg/100 g seed coat) than that with pure CO_2 (22 μg/100 g seed coat) from sweet Thai *tamarind* seed coat (Luengthanaphol et al., 2004). Supercritical CO_2 extraction with methanol in the range of 3%–7% as modifier was proven to be a very efficient and fast method to recover higher than 98% of colchicines and colchicoside, and 97% of 3-demethylcolchicine from seeds of *Colchicum autumnale* (Ellington et al., 2003).

22.3.2 Supercritical Fluid Drying

Supercritical fluids have also been used to remove moisture from foods. Conventional air-drying is the most commonly used drying process in the food industry. However, the high air temperature, which is typical 65°C–85°C, may cause damage to the microstructure of food materials and have a negative effect on the color, texture, taste, aroma, and nutritional value of the dried products. The supercritical fluid can retain the microstructure of dried foods much better than that of air-dried foods (Brown et al., 2008). However, it should be addressed that supercritical CO_2 is a non-polar substance. The solubility of water in supercritical CO_2 at 20 MPa and 50°C can only be 4 mg/g. If supercritical CO_2 is used in the drying process, polar co-solvents such as ethanol, acetone, and ethyl acetate should be used to increase the solubility of water in the supercritical CO_2.

22.3.3 Supercritical Fluid for Particle Formulation

Supercritical fluid can be used to produce particles with micron and submicron size especially with a narrow size distribution (Sihvonen et al., 1999; Marr and Gamse, 2000). These particles can be obtained by milling but they have a wide size distribution. The dissolving power of supercritical fluids depends on their density. As the compressibility of the fluids near their critical points is remarkable, the dissolving power of the fluids can be adjusted to match either that of gas or a liquid solvent by modifying the pressure (Sihvonen et al., 1999). A rapid expansion of the supercritical solution process for production of fine powders or particles is shown in Figure 22.2. During particle formulation, the substance that will be powdered is first solved in a supercritical fluid. The mixture is then expanded into an atmospheric pressure or even vacuum spray chamber through a nozzle. By this sudden depressurization, the solvent power is lost immediately, causing the dissolved substance to precipitate in fine dispersed form. If CO_2 is used as the solvent, the end particle products are free of solvent residues. The rapid expansion of supercritical CO_2 is suitable for production of non-polar compound particles (Sihvonen et al., 1999; Marr and Gamse, 2000). The limitation of rapid expansion of supercritical CO_2 is that the solubility of polar or high molecular weight compounds is rather low in supercritical CO_2. For production of polar compound particles, other processes such as gas antisolvent can be used. In the gas antisolvent process, the polar compound is dissolved in an organic solvent. The mixture is blended into the supercritical CO_2 by bubbling the solution with the fluid so that the solution expands and the compound crystallizes. The role of supercritical CO_2 is to decrease the solvent power of the organic solvent (Sihvonen et al., 1999; Marr and Gamse, 2000).

22.4 ENERGY USE AND EFFICIENCY

During supercritical CO_2 processing, the solvent CO_2 is usually separated from the product at a subcritical pressure. A large amount of low-temperature heat is exchanged in supercritical fluid processing. The thermodynamic properties of CO_2, which are functions of pressure and temperature, can be used to calculate the energy consumption at each step of the supercritical fluid cycle (Smith et al., 1999). Prior to processing with

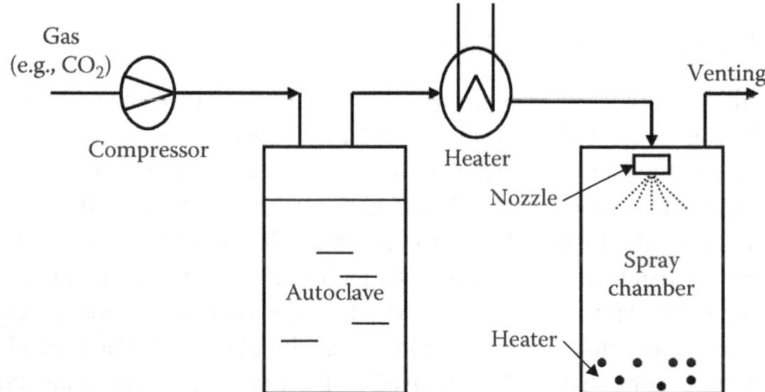

FIGURE 22.2 Rapid expansion of a supercritical solution process for production of fine powers.

supercritical CO_2, power is needed by the compressor to increase the pressure of CO_2 and thermal energy is needed by the heater to increase the temperature of CO_2. After the processing, the supercritical CO_2 must be changed to vapor and flash away from the product. For the liquid-to-vapor phase change of CO_2, the pressure of the fluid is decreased by the expansion. Thermal energy is needed to increase its temperature due to the cooling effect generated by the expansion. Before compression, the vapor should be condensed in a condenser with a cooling medium to increase the efficiency of the compressor.

Typical operating conditions for a single-stage supercritical CO_2 extraction of hops are the mass flow rate of the solvent was 10 kg CO_2/(kg hops·h); the extraction time was 4 h; the extraction pressure and temperature were 30 MPa and 60°C, respectively, while the separation pressure and temperature were 6 MPa and 40°C, respectively. In this process, about 40 kg of CO_2 was used to process 1 kg of hops at a mass ratio of CO_2 to food materials of 40. The amount of supercritical fluid required to process a kg of foods can be determined by the solubility of dissolved compounds in the supercritical fluid. The solubility is affected by the density of the fluid and thus the operating conditions. The solubility of selected compounds in supercritical CO_2 can be found in the literature (Gupta and Shim, 2006).

Sievers and Eggers (1996) analyzed the energy consumption for a single-stage supercritical CO_2 extraction process, which was used to process 15 ton of hops per 24 h. The energy consumption to cycle 1 kg of CO_2 in the supercritical CO_2 process and to process 1 kg of hops is given in Table 22.2 (Sievers and Eggers, 1996). The energy consumption by the supercritical CO_2 extraction of hops was compared with that of a conventional solid–liquid extraction with methylene chloride in an energetically optimized plant with a solid throughput of 50 ton of hops per 24 h, 500 ton of solvent per 24 h, and an extraction time of 1.5 h. It can be seen from Table 22.2 that the supercritical fluid process requires more mechanical and thermal energy than traditional solid–solvent extraction.

TABLE 22.2
Energy Flow in a Single-Stage Supercritical Fluid Extraction Process with 1 kg CO_2

Energy Source	Mechanical Energy	Heating			Cooling
Unit	Compressor	Heater	Evaporator	Total	Condenser
Energy consumed by supercritical CO_2 (kJ/kg CO_2)	30.7	33.7	142.0	175.7	−206.4
Energy consumed by supercritical CO_2 (kJ/kg hops)	1228	1348	5680	7028	−8256
Energy consumed by solid-solvent extraction (kJ/kg hops)	180	—	—	3300	−3400

Source: Data from Sievers, U. and Eggers, R., *Chem. Eng. Process.*, 35, 239, 1996. With permission.

22.5 POTENTIAL ENERGY CONSERVATION MEASURES

Production costs for supercritical fluid processes are significantly affected by the energy requirement. The production costs or the energy consumption can be reduced by

- Optimization of the processing condition (Sievers, 1998)
- Application of a heat recovery system (Sievers and Eggers, 1996)

Operating conditions such as the processing pressure and temperature, the separation pressure and temperature, and the mass ratio of fluid to materials to be processed directly determine the energy consumption of a supercritical fluid process. Process optimization can reduce the energy consumption and thus improve the production economics of a supercritical fluid process (Sievers, 1998). Sievers (1998) recommended that supercritical CO_2 extraction of soybean oil at 50 MPa and 100°C consumed the smallest amount of energy. Lower extraction temperature may be required to preserve the quality of the extracted oil. Separation at a supercritical pressure (above 7.4 MPa) was recommended (Sievers, 1998) but the separation pressure should not exceed 15 MPa.

Alternatively, a heat recovery unit can be added to a supercritical fluid cycle to optimize the energy flow in the cycle. In the supercritical fluid cycle, heat is added to the heater and evaporator while it is removed from the condenser. Therefore, the condenser can be a heat source while the heater and evaporator are heat sinks of a heat pump as shown in Figure 22.3. Heat is carried off in the condenser of the supercritical fluid cycle by the refrigerant in the heat pump cycle. The liquid refrigerant in the heat pump cycle becomes vapor in the condenser of the supercritical fluid cycle. The compressor of the heat pump increases the pressure of the refrigerant vapor. The high-pressure refrigerant vapor is condensed at a high temperature in the heater and evaporator of the supercritical fluid cycle, releasing the latent heat of condensation and supplying the heat to the heater and evaporator. The heat pump can significantly reduce the heat input to the heater and evaporator and cooling input to the condenser at the expense of increasing the mechanical energy for the compressor

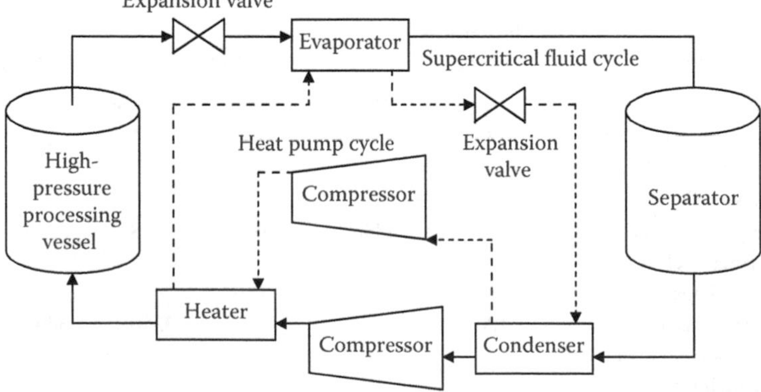

FIGURE 22.3 Single-stage supercritical fluid process with heat recovery by a heat pump using condenser as a heat source, and evaporator and heater as heat sink. (Reprinted from Sievers, U. and Eggers, R., *Chem. Eng. Process.*, 35, 239, 1996. With permission.)

of the heat pump. Since a supercritical fluid consumes a much larger amount of thermal energy than mechanical energy, the total energy cost of supercritical CO_2 processing with heat recovery using a heat pump should be comparable with that of traditional solid-solvent extraction.

22.6 SUMMARY

Supercritical fluids are alternative to toxic organic solvents for the extraction of food additives and nutraceuticals. Similar to supercritical fluid extraction, supercritical fluids have also been used to remove moisture from foods so that the foods are dried with supercritical fluids. Supercritical particle formation techniques could be a useful tool in the processing of functional food ingredients and nutraceuticals in the food industry. A supercritical fluid process requires mechanical energy in a compressor to increase the fluid pressure, a large amount of heat in a heater and evaporator to increase the fluid temperature, and cooling energy in a condenser to condense the fluid vapor. A heat pump can be used to facilitate the internal energy exchange between condenser, and heater and evaporator at the expense of mechanical energy for the compressor of the heat pump.

REFERENCES

Andras, C.D., B. Simandi, F. Orsi, C. Lambrou, D. Missopolinou-Tatala, C. Panayiotou, et al. 2005. Supercritical carbon dioxide extraction of okra (*Hibiscus esculentus L*) seeds. *Journal of the Science of Food and Agriculture* 85: 1415–1419.

Berna, A., A. Tarrega, M. Blasco, and S. Subirats. 2000. Supercritical CO_2 extraction of essential oil from orange peel: Effect of the height of the bed. *Journal of Supercritical Fluids* 18: 227–237.

Bernardo-Gil, M.G., J. Grenha, J. Santos, and P. Cardoso. 2002. Supercritical fluid extraction and characterization of oil from hazelnut. *European Journal of Lipid Science and Technology* 104: 402–409.

Brown, Z.K., P.J. Fryer, I.T. Norton, S. Bakalis, and R.H. Bridson. 2008. Drying of foods using supercritical carbon dioxide—investigations with carrot. *Innovative Food and Emerging Technologies* 9: 280–289.

Cherchi, G., D. Deidda, B. De Gioannis, B. Marongiu, R. Pompei, and S. Porcedda. 2001. Extraction of *santolina insularis* essential oil by supercritical carbon dioxide: Influence of some process parameters and biological activity. *Flavour and Fragrance Journal* 16: 35–43.

Coelho, J.A.P., A.P. Pereira, R.L. Mendes, and A.M.F. Palavra. 2003. Supercritical carbon dioxide extraction of *Foeniculum vulgare* volatile oil. *Flavour and Fragrance Journal* 18: 316–319.

Dean, J.R., B. Liu, and R. Price. 1998. Extraction of magnolol from *Magnolia officinalis* using supercritical fluid extraction and phytosol solvent extraction. *Phytochemical Analysis* 9: 248–252.

Dean, J.R. and B. Liu. 2000. Supercritical fluid extraction of Chinese herbal medicines: Investigation of extraction kinetics. *Phytochemical Analysis* 11: 1–6.

Eller, F.J. and J.W. King. 2000. Supercritical carbon dioxide extraction of cedarwood oil: A study of extraction parameters and oil characteristics. *Phytochemical Analysis* 11: 226–231.

El-Ghorab, A., A.F. Mansour, and K.F. El-massry. 2004. Effect of extraction methods on the chemical composition and antioxidant activity of Egyptian marjoram (*Majorana hortensis* Moench). *Flavour and Fragrance Journal* 19: 54–61.

Ellington, E., J. Bastida, F. Viladomat, and C. Codina. 2003. Supercritical carbon dioxide extraction of colchicines and related alkaloids from seeds of *Colchicum autumnale L*. *Phytochemical Analysis* 14: 164–169.

Giannuzzo, A.N., H.J. Boggetti, M.A. Nazareno, and H.T. Mishima. 2003. Supercritical fluid extraction of naringin from the peel of citrus paradise. *Phytochemical Analysis* 14: 221–223.

Gupta, R.B. and J.J. Shim. 2006. *Solubility in Supercritical Carbon Dioxide*. Boca Raton, Florida: CRC Press.

Hamburger, M., D. Baumann, and S. Adler. 2004. Supercritical carbon dioxide extraction of selected medicinal plants-effects of high pressure and added ethanol on yield of extracted substances. *Phytochemical Analysis* 15: 46–54.

Hauthal, W.H. 2001. Advances with supercritical fluids [review]. *Chemosphere* 43: 123–135.

Hurren, D. 1999. Supercritical fluid extraction with CO_2. *Filtration & Separation* 36: 25–27.

Lang, Q. and C.M. Wai. 2001. Supercritical fluid extraction in herbal and natural product studies- a practical review. *Talanta* 53: 771–782.

Luengthanaphol, S., D. Mongkholkhajornsilp, S. Douglas, P.L. Douglas, L. Pengsopa, and S. Pongamphai. 2004. Extraction of antioxidants from sweet Thai tamarind seed coat-Preliminary experiments. *Journal of Food Engineering* 63: 247–252.

Marr, R. and T. Gamse. 2000. Use of supercritical fluids for different processes including new developments—a review. *Chemical Engineering and Processing* 39: 19–28.

Marongiu, B., S. Porcedda, A. Caredda, B. De Gioannis, L. Vargiu, and P. La Colla. 2003. Extraction of *Juniperus oxycedrus* ssp. Oxycedrus essential oil by supercritical carbon dioxide: Influence of some process parameters and biological activity. *Flavour and Fragrance Journal* 18: 390–397.

Meireles, A. and M. Angela. 2003. Supercritical extraction from solid: Process design data (2001–2003). *Current Opinion in Solid State and Materials Science* 7: 321–330.

Modey, W.K., D.A. Mulholland, and M.W. Raynor. 1996. Analytical supercritical fluid extraction of natural products. *Phytochemical Analysis* 7: 1–15.

Poiana, M., R. Fresa, and B. Mincione. 1999. Supercritical carbon dioxide extraction of bergamot peels. Extraction kinetics of oil and its components. *Flavour and Fragrance Journal* 14: 358–366.

Roy, B.C., M. Goto, A. Kodama, and T. Hirose. 1996. Supercritical CO_2 extraction of essential oils and cuticular waxes from peppermint leaves. *Journal of Chemical Technology and Biotechnology* 67: 21–26.

Sass-Kiss, A., B. Simandi, Y. Gao, F. Boross, and Z. Vamos-Falusi. 1998. Study on the pilot-scale extraction of onion oleoresin using supercritical CO_2. *Journal of the Science of Food and Agriculture* 76: 320–326.

Sievers, U. 1998. Energy optimization of supercritical fluid extraction processes with separation at supercritical pressure. *Chemical Engineering and Processing* 37: 451–460.

Sievers, U. and R. Eggers. 1996. Heat recovery in supercritical fluid extraction process with separation at subcritical pressure. *Chemical Engineering and Processing* 35: 239–246.

Sihvonen, M., E. Jarvenpaa, V. Hietaniemi, and R. Huopalahti. 1999. Advances in supercritical carbon dioxide technologies. *Trends in Food Science & Technology* 10: 217–222.

Smith, R.L., H. Inomata, M. Kanno, and K. Arai. 1999. Energy analysis of supercritical carbon dioxide extraction presses. *Journal of Supercritical Fluids* 15: 145–156.

Szentmihalyi, K., P. Vinkler, B. Lakatos, V. Illes, and M. Then. 2002. Rose hip (*Rosa canina L.*) oil obtained from waste hip seeds by different extraction methods. *Bioresource Technology* 82: 195–201.

Turner, C., J.W. King, and L. Mathiasson. 2001. Supercritical fluid extraction and chromatography for fat-soluble vitamin analysis. *Journal of Chromatography A* 936: 215–237.

Vilegas, J.H.Y., E. de Marchi, and F.M. Lancas. 1997. Extraction of low-polarity compounds (with emphasis on coumarin and kaurenoic acid) from *Milania glomerata* ("Guaco") leaves. *Phytochemical Analysis* 8: 266–270.

Wang, L.J. and C.L. Weller. 2006. Recent advances in extraction of natural products from plants. *Trends in Food Science and Technology* 17: 300–312.

Part V

Conversion of Food Processing Wastes into Energy

23 Food Processing Wastes and Utilizations

23.1 INTRODUCTION

Food processing facilities consume large amounts of energy to convert raw materials to final food products. The food industry is one of seven energy-intensive industries. Large quantities of wastes in both liquid and solid forms are produced annually by the food processing industry. Solid organic waste is normally understood as organic biodegradable waste with moisture content below 85%–90%. Food wastes comprise approximately 40% of the total municipal solid wastes (Mata-Alvarez et al., 2000). The food processing waste materials contain principally biodegradable organic mater and their disposal may create serious environmental problems (Hang, 2004).

Various methods including landfill, incineration, and complete decomposition have been used for the treatment of food wastes (Park et al., 2002). Waste characteristics, governmental regulations, and disposal costs are the primary considerations for the methods of waste treatment in the food manufacturing industry. Traditionally, large amounts of solid food processing wastes are buried in landfills while liquid food processing wastes are released into rivers, lakes and oceans, and disposed in public sewer systems without treatment. Public concern and government legislation on environment protection have reduced the number of wastes that are considered safe for disposal. Use of incineration is restricted because it is expensive to operate. The increasing energy prices and waste disposal costs have dramatically increased food production costs in recent years. A survey of the United States Department of Agriculture showed that $50 million annually could be saved in solid waste disposal costs alone for landfills if 5% of the total amount of 43.54×10^9 kg food loss from processing, retail, food service, and consumer foods in 1995 were recovered (Kantor et al., 1997).

Conversion of food processing wastes into useful energy products such as bioethanol, biodiesel, bio-oil, biogas, syngas, steam, and electricity in a food processing facility could result in significant savings for the food manufacturing industry in terms of reducing the amount of energy purchased and waste disposal costs. The energy utilization of food processing wastes is dependent on a basic understanding of (1) operating conditions of a food processing facility, (2) waste types, availability and energy potential, and (3) process and equipment for handling and conversion of the waste. Comprehensive technical and cost analyses are essential to determine the feasibility and economics of an energy conversion process for utilization of food process wastes.

The types of raw food materials and the operations of a food processing facility greatly influence the kinds and amounts of wastes produced. In this chapter, wastes from fruit, vegetable, grain, oilseed, and meat processing facilities are reviewed in terms of their availability, quality, and current utilizations. Energy utilization of different food processing wastes and available energy conversion technologies are briefly discussed.

23.2 FRUIT AND VEGETABLE PROCESSING WASTES

23.2.1 AVAILABILITY AND QUALITY

Fruit and vegetable processing facilities generate large amounts of liquid and solid wastes. For processing each 1,000kg of raw apples, 20,861 L of wastewater is generated, which contains 2.5kg of suspended solids and 300kg of solid residues. For processing each 1,000kg of raw citrus, 12,517L of wastewater is produced, which contains 2.5kg of suspended solids and 220kg of solid residues (Woodruff and Luh, 1975). The United States fruit and vegetable processing industry annually generates approximately 11 million tons of by-product wastes along with over 430 billion L of effluent wastewater (Blondin et al., 1983). Table 23.1 gives the waste quantities in selected countries (Laufenberg et al., 2003).

TABLE 23.1
Fruit and Vegetable Processing Wastes in Selected Countries

Country	Year	Quantity and Waste Type
German	1997	380,000 t/a organic waste from potato, vegetable, and fruit processing
		1,954,000 t/a spent malt wand hops in breweries
		1,800,000 t/a grape pomace in viniculture
		3,000,000 t/a crude fiber residues in sugar production
		100,000 t of wet apple pomace (~25,000 t dry apple pomace) if 400,000 t apples are processed into apple juice
Belgium	1992	105,000 t/a biowaste including vegetable, garden, and fruit waste
		280,000 t/a estimations due to legislation of separate household collection
Thailand	1993	386,930 t/a empty fruit bunches
		165,830 t/a palm press fiber
		110,550 t/a palm kernel shells
		1,000,000 t/a cassava pulp
Spain	1997	> 250,000 t/a olive pomace
EEC	1996	14,000,000 t/a dry sugar beet pulp
Portugal	1994	14,000,000 t/a tomato pomace
Jordan	1999	36,000 t/a olive pomace
Malaysia	1996	2,520,000 t/a palm mesocarp fiber
		1,440,000 t/a oil palm shells
		4,140,000 t/a empty fruit bunches
Australia	1995	400,000 t/a pineapple peel
United States	—	300,000 t/a grape pomace in California only in 1994
		9,525 t/a cranberry pomace in 1998
		200,000 t/a almond shells in 1997
		3,300,000 t/a orange peel in Florida in 1994

Source: Reprinted from Laufenberg, G., Kunz, B., and Nystroem, M., *Bioresour. Technol.*, 87, 167, 2003. With permission.

TABLE 23.2
By-Product Wastes of Selected Vegetables

Vegetables	Amount of By-Product Waste (% of input)
Sweet corn	72.4
Potato	55
Carrot	47.5
Beet	45
Snap bean	28.2
Green pea	22

Source: Reprinted from Blondin, G.A., Comiskey, S.J., and Harkin, J.M., *Energy Agri.*, 2, 21, 1983. With permission.

Table 23.2 gives the mass percentage of by-product wastes generated from selected vegetables during processing (Blondin et al., 1983). On average, the waste is as high as 72.4% of the original mass for sweet corn. Approximately half of the potato, carrot, and beet enter their waste streams during processing. Snap bean and green pea generate less waste, which is 28.2% and 22% of their original weights, respectively.

During processing of fruit and vegetable, solid residues are produced from peelings, trimmings, cores, stems, pits, and culls of undesirable fruits or vegetables, nutshells, kernel fragments, and grain hulls. Liquid wastes are produced from operations such as hydro-handling, product cleaning, and blanching. Water is used in washers, blanchers, French pumps, graders, peeler flume, and expressates during processing of vegetables and fruits (Sargent and Steffe, 1986).

Solid processing wastes from fruits and vegetables have high moisture and carbohydrate contents. The moisture content of fresh fruits and vegetables is as high as 95% on a wet basis. Pomace or presscakes from juice production may have a moisture content below 50% depending on the method and efficiency of the press. Most of the carbohydrates in fruit and vegetable wastes are composed of soluble sugars and other easily hydrolysable polysaccharides (Cooper, 1976). A major portion of the carbohydrates is dissolved or suspended in the processing wastewaters. Appropriate disposal of these wastewaters is a costly burden to the fruit and vegetable processing industry (Blondin et al., 1983). Excessively dilute wastewaters (e.g., less than 0.2% sugar) would be too costly to recover the dissolved sugars.

Since a major portion of carbohydrates is dissolved or suspended in the processing wastewaters, a separation process such as reverse osmosis (RO) is needed to recover the fermentable sugars from the wastewaters. Wastewater volumes, soluble hexose concentrations in native and amylase treated wastewaters, and soluble hexose yields for six major U.S. process vegetables are given in Table 23.3 (Blondin et al., 1983).

As shown in Table 23.3, corn and potato wastewaters are the most significant sources of recoverable fermentable sugars; snap bean wastewaters are the poorest sources and wastewaters from pea, carrot, and beet are intermediate. The soluble

TABLE 23.3
Summary of Recoverable Fermentable Sugars from Vegetable Wastewater in the United States

Process Vegetable	Metric Tons of Vegetables	Wastewater Volume (L/ton)	Hexose Concentration (%)		kg Hexose per Ton Crop		Total Metric Tons Recoverable Hexose	
			Native	Enzyme treated	Native	Enzyme treated	Native	Enzyme treated
Corn	1,864,000	2027	2.67	3.50	54.2	70.9	101,000	132,200
Pea	519,600	6771	0.52	0.61	35.0	41.3	18,190	21,460
Snap bean	674,200	1690	0.62	0.77	10.6	13.0	7,147	8,765
Carrot	424,600	2515	0.83	0.92	20.8	23.1	8,832	9,808
Beet	218,300	2803	0.81	1.02	22.6	28.6	4,934	6,243
Potato	6,196,000	2403	2.98	3.77	71.6	90.6	443,600	561,400
Total	9,896,700						583,703	739,876

Source: Reprinted from Blondin, G.A., Comiskey, S.J., and Harkin, J.M., *Energy Agri.*, 2, 21, 1983. With permission.

hexose concentration in wastewater varies from 0.52% to 2.98% for native wastewater and from 0.61% to 3.77% for α-amylase treated wastewater. As shown in Table 23.3, approximately 584,000 metric tons of native fermentable sugars can be recovered. If α-amylase was used to treat the wastewaters, approximately 740,000 tons of fermentable sugars can be recovered industry-wide from these six selected crops. These fermentable sugars are contained in approximately 2.5 billion liters of wastewater.

23.2.2 Current Utilization

Most of the fruit and vegetable processing residuals are presently used for animal feeds. Cannery wastes such as potato and corn by-products and pomace could be used as a fiber substitute in animal feeds. The fruit and vegetable wastes can be fed to animals directly in the same form as they are produced. The wastes can also be used to make prepared animal feeds after drying and treatment with nutrient additives, which have longer storage periods. However, more research should be conducted to investigate any negative effects of the feeds made from fruit and vegetable wastes on the health of animals. Solid food processing wastes may need some physical or chemical pretreatment to increase their digestibility as animal feeds (Lahmar et al., 1994).

Part of the solid residuals from fruits and vegetables is disposed of as wastes in landfills, on the land, or burned on-site. The costs of disposal of solid wastes on land vary considerably and are dependent on the moisture content of the waste materials, the distance the wastes must be transported, and the methods of delivery and application. One of the most promising approaches to minimizing the disposal costs and environmental problems is to remove excess moisture from the waste materials using dewatering methods such as filtration and centrifugation (Hang, 2004). Some fruit and vegetable processing wastes are treated aerobically and converted into a hygienic, stable, and odor-free fertilizer product (Mohaibes and Heinonen-Tanski, 2004).

Most of the liquid wastes from fruits and vegetables are released into rivers, lakes, and oceans, and disposed in public sewer systems without any treatment. A small amount of liquid wastes are utilized in on-site treatment or irrigation. If the liquid wastes contain a high amount of suspended solids, some treatments are required before they are released to the environment. If the liquid wastes contain toxic constituents such as salt brines and lye, they must be detoxified before disposal (Sargent and Steffe, 1986).

Transport costs and sales problems due to the low quality of residual matter from fruit and vegetable processing facilities have led to alternative utilization of these wastes (Laufenberg et al., 2003).

23.2.3 Energy Utilization

Fruit and vegetable processing facilities generate solid organic wastes and processing wastewaters. Since fruit and vegetable processing wastes are rich in carbohydrates, they can be used as a source for the production of fermentable sugars. Both solid organic wastes and processing wastewaters can be used as energy feedstocks.

The solid organic residue can also be converted to liquid or gaseous fuel by either a biological conversion process or a thermochemical process. Ethanol can be produced directly from fruit pomace with a high sugar content (Hang et al., 1986; Ngadi and Correia, 1992; Nigam, 2000). Starchy and cellulosic waste materials from fruit and vegetable processes, however, must be hydrolyzed first to fermentable sugars before a yeast culture can ferment them to ethanol. Fisher and Bipp (2005) used acidic hydrolysis of organic acids and monosaccharides from several food processing residues including sugar beet molasses, whey powder, wine yeast, potato peel sludge, spent hops, malt dust, and apple marc. The yields of ethanol depend on the initial carbohydrate content in the wastes. Ethanol fermentation from food processing wastes is discussed in Chapter 25. Microorganisms can convert fruit and vegetable processing wastes into biogas under anaerobic conditions, which is discussed in Chapter 24. The biogas, which mainly consists of methane and carbon dioxide, can be used to generate steam required in food processing facilities (Hang, 2004). Dry, solid fruit and vegetable processing wastes can also be burned directly to generate steam in the food processing facilities or converted to gaseous and liquid fuels using a thermochemical conversion process, which is discussed in Chapter 27.

Since a major portion of carbohydrates is dissolved or suspended in the processing wastewaters during processing of fruits and vegetables, fruit and vegetable processing wastewater is also a potential feedstock for alcohol fermentation (Blondin et al., 1983). Reverse osmosis (RO) technology as discussed in Chapter 17 has been used to recover and concentrate soluble sugars from dilute effluent and juice wastewaters. The recovered sugar concentrates are sterile, easily storable or transportable, and excellent feedstocks for bioconversion processes. Moreover, the reverse osmosis process can remove most of the BOD-contributing solids from effluents in a fruit and vegetable processing facility and significantly reduce wastewater disposal costs. Using the reverse osmosis technology in the United States, approximately 1.42 million tons of fermentable sugars could be recovered as a 20% sugar concentrate, which is suitable for bioconversion to useful liquid fuels. The annual fuel alcohol potential of these sugars is between 750 and 900 million liters. Economic analysis indicates that overall alcohol production costs, on average, should be only 40% of current prices for bulk fuel alcohol. The coincident major reduction in wastewater biological oxygen demand (BOD) and associated disposal costs and the recovery of over 430 billion L of reusable RO permeate water afford added incentives for industrial participation in the production of sugar concentrates from fruit and vegetable by-product wastewaters (Blondin et al., 1983).

23.3 OILSEED PROCESSING WASTES

23.3.1 Availability and Quality

There are many oil plants such as soybean, rapeseed, sunflower, peanut, palm, and linseed. Vegetable oils and fats are important constituents of human foods and animal feeds. Vegetable oils and animal fats are also used to produce industrial products such as biodiesel (Ma and Hanna, 1999). Approximately 125 million tons of oils and fats are produced worldwide in 2003. In recent years, the amounts of oils and fats produced have increased continuously by approximately 3% per year. Soybean, palm, rapeseed, and sunflower are the main oil sources as shown in Table 23.4. In 2003, about 101 million tons of these oils and fats were used in human foods. Another 6.3 million tons of oils and fats were used as animal feeds. About 17.4 million tons of oils and fats were used for production of industrial products (Hill, 2006). In the United States, the average production capacity of vegetable oils was 10.7 million tons/year between 1995 and 2000. Soybean was the dominant oil source in the United States, which was 8.3 million tons/year or about 78% of the total oil production in the United States and 27% of the total soybean production in the world (Canakci, 2007).

The oil contents of oil seeds are related to their varieties, genotype, and environmental influences. Sunflower seed typically has 40%–50% of oil by weight (Gercel,

TABLE 23.4
World Production of Oils and Fats in 2003

Oil and Fat	Amount (million tons)
Soybean	31.3
Palm oil	27.8
Rapeseed oil	12.5
Sunflower oil	8.9
Palm kernel oil/coconut oil	6.5
Other plant oils	14.9
Tallow	8.0
Butter	6.3
Other animal fats	8.2
Total	125

Source: Reproduced from Hill, K., in *Biorefineries—Industrial Processes and Products*, Copyright Wiley-VCH Verlag GmbH & Co., Weinheim, 2006, 291–314. With permission.

TABLE 23.5
Composition of Oil Cakes

Oil Cake	Dry Matter (%)	Crude Protein (%)	Crude Fiber (%)	Ash (%)	Calcium (%)	Phosphorous (%)
Canola	90	33.9	9.7	6.2	0.79	1.06
Coconut	88.8	25.2	10.8	6.0	0.08	0.67
Cottonseed	94.3	40.3	15.7	6.8	0.31	0.11
Groundnut	95.6	49.5	5.3	4.5	0.11	0.74
Mustard	89.8	38.5	3.5	9.9	0.05	1.11
Olive	85.2	6.3	40.0	4.2	-	-
Palm kernel	90.8	18.6	37	4.5	0.31	0.85
Sesame	83.2	35.6	7.6	11.8	2.45	1.11
Soybean	84.8	47.5	5.1	6.4	0.13	0.69
Sunflower	91	34.1	13.2	6.6	0.30	1.30

Source: Reprinted from Ramachandran, S., Singh, S.K., Larroche, C., Soccol, C.R., and Pandey, A., *Bioresour. Technol.*, 98, 2000, 2007. With permission.

2002). Rapeseed has 40% of oil (Onay et al., 2001). Soybean has only 15%–25% of oil but it is a rich source of protein for foods and feeds. In oilseed milling facilities, oil cakes are the solid residue generated after oil extraction, which mainly include sunflower oil cake, soybean cake, coconut oil cake, palm kernel cake, cottonseed cake, canola oil cake, olive oil cake, rapeseed cake, mustard oil cake, groundnut oil cake, and sesame oil cake. There are two types of oil cakes: edible and nonedible. (Ramachandran et al., 2007). Oil cakes are rich in protein, fiber, and energy contents. The composition of oil cakes depends on their variety, growing condition, and extraction methods. The amino acid profiles significantly vary from one variety to another. The chemical compositions of selected oil cakes are given in Table 23.5.

23.3.2 Current Utilization

Edible oil cakes such as soybean cake are usually rich in protein. Edible oil cakes have been used for feed in the poultry, fish, and swine industry. Some oil cakes can also be considered for food supplementation. Nonedible cakes are used as organic nitrogenous fertilizers due to their high N, P, and K contents. Oil cakes are also an ideal source as a support matrix for various biotechnological processes for production of enzymes, antibiotics, vitamins, antioxidants, and mushrooms (Ramachandran et al., 2007).

23.3.3 Energy Utilization

An increasing amount of oil extracted from oilseeds has been used for biodiesel production. Oil cakes can also be used as energy feedstocks for production of biogas and bio-oil by either a biological conversion or thermochemical conversion process. Nonedible oil cakes have been used to adjust the carbon to nitrogen ratio of feedstocks to anaerobic digestion processes for maximum microbiological activity and fuel output as discussed in Chapter 24 (Lingaiah and Rajasekaran, 1986). Oil cakes such as olive oil cake, cottonseed oil cake, sunflower oil cake, and rapeseed cake are potential renewable energy sources for cleaner energy production through thermochemical conversion processes as discussed in Chapter 27 (Ozbay et al., 2001; Yorgun et al., 2001; Gercel, 2002; Atimtay and Topal, 2004; Ozcimen and Karaosmanoglu, 2004).

23.4 GRAIN PROCESSING WASTES

23.4.1 Availability and Quality

Corn, wheat, and rice account for 87% of all grain production worldwide. Grains have been used as both food and industrial commodities because of their high values in carbohydrates, proteins, fibers, and oils. The compositions of main grains are given in Table 23.6.

Grain processing facilities extract carbohydrates, proteins, fats, and other materials from grains and convert them into multiple products including foods, fuels, and high-value chemicals and materials. Grains such as corn and wheat are processed through either a wet milling or dry milling process (Johnson, 2006).

Wet milling has been used for many years in the starch industry. It has also been adapted and modified for ethanol production. In the corn wet milling process, corn is first soaked in water containing sulfur dioxide for 24–48 h. It is then ground and the components are physically separated using a series of centrifuges, screens, and washes. Some of the separated fractions then undergo additional refining steps. The corn wet milling process converts corn into starch, oil, gluten meal, gluten feed, and germ meal as shown in Figure 23.1. (Kraus, 2006). For wheat wet milling, the bran and germ are generally removed by dry processing in a flour mill before steeping in water.

The use of grain for ethanol production opens up a large new market for grains. A dry milling process is much simpler than a wet milling process (Rendleman

TABLE 23.6
Composition of Selected Grains

	Corn	Wheat	Rice	Potato	Tapioca
Moisture (%)	16	13	14	75	70
Starch (%)	62	60	77	19	24
Protein (%)	8.2	13	7	2	1.5
Fat (%)	4.2	3	0.4	0.1	0.5
Fiber (%)	2.2	1.3	0.3	1.6	0.7
Minerals/ash (%)	1.2	1.7	0.5	1.2	2
Sugars (%)	2.2	8	0.3	1.1	0.5

Source: Reprinted from Grull, D.R., Jetzinger, F., Kozich, M., and Wastyn, M.M., in *Biorefineries—Industrial Processes and Products*, Copyright Wiley-VCH Verlag GmbH & Co., Weinheim, 2006, 61–95. With permission.

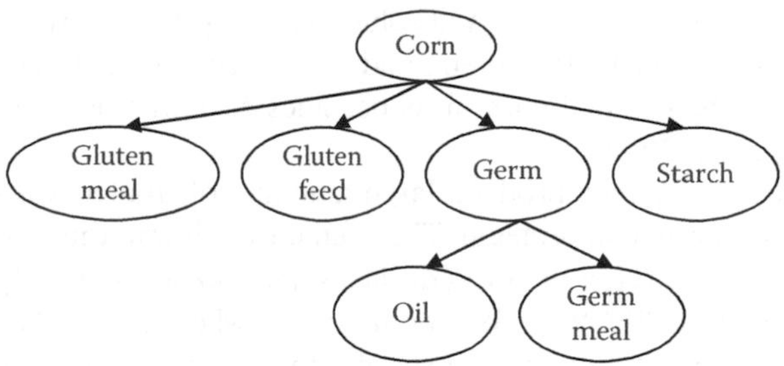

FIGURE 23.1 Corn wet milling. (Reproduced from Kraus, G.A., *Biorefineries—Industrial Processes and Products*, Copyright Wiley-VCH Verlag GmbH & Co., Weinheim, 2006. With permission.)

and Hohmann, 1993). The dry milling process has been widely used in the ethanol industry. The dry milling process does not fractionate the different components of the grains. However, since the germ contains high contents of proteins and lipids, the germ is removed by sieving and aspiration or by gravity methods for foods and feeds (Koseolu et al., 1991). A dry milling process involves grinding of the grain, followed by addition of water, and treatment with heat. Enzymes are added to the slurry and the sugar that results from starch hydrolysis is fermented to ethanol by the addition of yeast. Fuel ethanol is recovered by distillation and evaporation. The processing wastes after ethanol recovery are distillers' grains and solubles and traditionally used as animal feeds.

A large amount of corn is used to produce ethanol in the United States and wheat is considered the primary raw material for ethanol production in Europe and Australia. Grains are an abundant renewable resource of starch for production of foods, feeds, and fuels. The type of coproducts that are produced in the grain processing industry depends on a number of factors, including conversion technology, feedstock, and milling processing (Turhollow and Heady, 1986).

Grain processing facilities may generate four main coproducts besides starch:

- Gluten with a high protein content
- Bran with a high fiber content
- Germ with high protein and lipid contents
- Distillers' grains and solubles in ethanol production

Protein is seen as the major coproduct in ethanol production. Corn has about 8% of protein by weight, and a concentrate containing 90% protein has been isolated from corn gluten meal (Satterlee, 1981). Wheat has an approximate protein content of 13% by weight, 80–90 of which is made up of gluten (Weegels et al., 1992). Cereal bran has a high fiber content (Dexter et al., 1994). Fiber is composed mainly of non-starch polysaccharides and lignin. Different grains have different proportions of different types of fibers. The insoluble dietary fiber includes cellulose, lignin, and hemicelluloses while the soluble fiber consists of pectins, β-glucans, gums, and mucilages. Germ is particularly attractive because it contains high concentrations of protein and minerals and a number of high-value lipid compounds. The dry milling process generates distillers' grains and solubles as by-products, which contain high concentrations of dietary fiber, protein, and fats (Rasco et al., 1987). In a corn dry mill, one-third of the original corn mass becomes corn distillers' dried grains and solubles (Wang et al., 2005).

Starch processing plants produce large amounts of diluted wastewater, which may cause an environmental problem. The need for fresh water in a corn wet milling plant is as high as 1.5 m^3/ton of corn (Kollacks and Rekers, 1988). Typically, wheat milling results in 78%–80% flour, 19% bran, and 1% wheat germ. The flour is washed with water to produce starch and the starch processing residue is a commercially accepted feedstock for ethanol production (Nguyen, 2003). The starch processing wastewater contains some solids, which can be recovered as a potential renewable source (Nguyen, 2003; Verma et al., 2007). Nguyen (2003) reported that the distillery effluent from a starch-based ethanol plant has 3.3% total solids.

23.4.2 Current Utilization

It has been estimated that coproduct revenue can account for up to 40% of the income of a grain-based ethanol plant (Chang et al., 1995). However, many coproducts from grains have not yet established their markets. Traditionally, the animal feed industry has provided a market for coproducts from ethanol production.

Wheat gluten with a high protein content is an important additive in bakery and breakfast foods, fish and meat products, and dairy products (Gras and Simmonds, 1980; Satterlee, 1981). Wheat gluten has also been used in the nonfood industries for production of adhesives and films (Krull and Inglett, 1971). Corn protein has not established a market to be used in foods. Zein is the only corn protein that has been developed for industrial production of packaging films, linoleum tiles, coatings, ink, and textile fibers.

Corn germ in wet milling is extracted to produce corn oil, which is the most valuable coproduct of the corn wet milling process. Corn germ makes up 10%–20% of the total product generated when corn is dry milled (Blessin et al., 1973). Both corn and wheat germs are particularly attractive for use in food products because they contain a high concentration of proteins and minerals and a number of high-value lipid components.

One of the promising areas for coproducts from grain processing facilities is that nutraceuticals such as wheat fiber, phytosterols, policosanols, and free fatty acids extracted from grains or their coproducts have been proved to have either health or medical advantages (Wrick, 1993; DeFelice, 1995; Wang et al., 2005).

The wastewater from the grain processing facilities is usually disposed of by spray irrigation. The problems caused by spray irrigation include strong odor, insect invasion, increase in soil acidity, salt leaching, buildup of sulfates, and putrescity (Nguyen, 2003).

23.4.3 Energy Utilization

An increasing amount of starch from grains has been used directly for ethanol production. Grain residuals are rich in cellulose and hemicellulose, which can be a renewable source for enzymatic production of soluble sugars as feedstocks for ethanol fermentation (Hang, 2004). Wheat bran consists of three main components of residual starch, cellulose, and hemicellulose. Choteborska et al. (2004) produced a sugar solution from wheat bran for ethanol fermentation. They first treated the wheat bran with starch degrading enzymes to remove the starch from the bran. The maximum yield of sugars was 52.1 g/100 g of starch-free wheat bran obtained using 1% of sulfuric acid at 130°C for 40 min. Furfural and 5-hydroxy-methyl-2-furaldehyd, which cause inhibition of fermentation, were as low as 0.28 g/L and 0.05 g/L, respectively.

A dry mill generates a large amount of distillers' grains. Tucker et al. (2004) used a dilute acid pretreatment process to hydrolyze sugars from the residual starch and fiber in the distillers' grains for additional ethanol production. The pretreatment of distillers' grain at 140°C with 3.27% H_2SO_4 for 20 min could hydrolyze 77% of available carbohydrate in the distillers' grains. The yeast of *Saccharomyces cerevisiae* D_5A was further used to ferment to sugars. The ethanol yield was 73% of theoretical value from available glucans.

As wastewater from starch processing facilities contains some residual starch, it can be converted to ethanol by fermentation. The solids in the wastewater can economically be recovered by ultrafiltration (Nguyen, 2003).

Crop residues such as bran, husk, and bagasse can be used as energy feedstocks for the production of liquid and gaseous fuels and supply of heat and power in the processing facilities. Thermal energy consumption is significant in the grain milling facilities. Supplying heat and power from grain processing residues is a promising way (Eggeman and Verser, 2006). The thermochemical conversion of crop residues is discussed in Chapter 27.

23.5 MEAT PROCESSING WASTES

23.5.1 Availability and Quality

The consumption of pork, beef, and poultry in developed countries has been on the increase. Figure 23.2 shows the capita consumption of pork, beef, and poultry in several developed countries in 1997 (Salminen and Rintala, 2002).

In the meat industry, slaughterhouses produce increasing amounts of organic solid by-products and wastes. The organic solid wastes generated in a poultry slaughterhouse include blood, feather, offal, feet, head, trimmings, and bones. A broiler is about 1.8–1.9 kg before it is slaughtered. During slaughtering and processing, about 40% of the broiler mass enters into the waste stream. Table 23.7 gives the quantities and characteristics of organic solid wastes produced in poultry slaughterhouses (Salminen and Rintala, 2002).

Solid organic wastes from slaughterhouses may contain several species of microorganisms including potential pathogens (Salminen and Rintala, 2002). In addition, animals may accumulate various metals, drugs, and other chemicals added in their feed for nutritional and pharmaceutical purposes (Haapapuro et al., 1997). The disposal and processing of animal wastes should destroy potential pathogens present in the wastes for prevention of animal and public health.

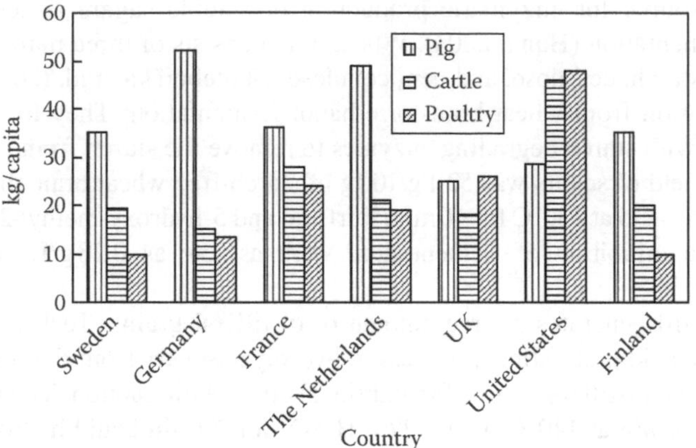

FIGURE 23.2 Capita consumption of pork, beef, and poultry in several developed countries. (Reprinted from Salminen, E. and Rintala, J., *Bioresour. Technol.*, 83, 13, 2002. With permission.)

23.5.2 Current Utilization

Slaughterhouse wastes have been used for animal feeds and organic fertilizers. Rendering is to separate fat from meat by heating the meat. Commission of the European Communities (1990) required that rendering at 133°C and 300 kPa for a minimum of 20 min is needed for high-risk materials such as dead and stillborn animals intended for animal feed or as an intermediate product for the manufacture of organic fertilizers and other products. The fat may be used for animal feeds or the production of chemicals and fuel products. Since slaughterhouse wastes are rich sources of proteins and vitamins, some low-risk wastes are preserved with formic acid and used as animal feed for fur animals and pets. However, caution should be exercised to reduce the risk of disease transmission via the feeds (Salminen and Rintala, 2002).

TABLE 23.7
Quantities and Characteristics of Organic Solid Wastes Produced in Poultry Slaughterhouses

	Live Weight (%)	Total Solid (%)	Volatile Solid (%)	Protein (% of Total Solid)	Fat (% of Total Solid)
Blood	2	22	91	48	2
Feather	10	24.3	96.7	91	1–10
Offal, feet, and heart	21	39	95	32	54
Trimmings and bone	7	22.4	68	51	22

Source: Adapted from Salminen, E. and Rintala, J., *Bioresour. Technol.*, 83, 13, 2002. With permission.

Incineration, landfilling, and composting have been used to treat slaughterhouse wastes. Incineration is a thermal degradation technology for effectively destroying potential infectious microorganisms in wastes (Ritter and Chinside, 1995). Caution should be exercised to control air emission, incineration conditions, and disposal of solid and liquid residues after incineration. Burial of slaughterhouse wastes should also be strictly controlled to avoid groundwater contamination. Composting, which is an aerobic biological process, has been used to reduce pathogens and decompose organic slaughterhouse and other food processing wastes (Mohaibes and Heinonen-Tanski, 2004). The composted materials can be used as soil conditioner or fertilizer. The emission to air, water, and soil associated with composting may cause a pollution problem and reduction of nitrogen in the compost fertilizer (Tritt and Schuchardt, 1992).

23.5.3 Energy Utilization

Meat wastes can be converted to liquid and gaseous fuels by a biological conversion, thermochemical conversion, or chemical conversion process. Anaerobic digestion, which is a biological process, has been widely used to treat slaughterhouse wastewaters. Anaerobic digestion of slaughterhouse wastes has been discussed in detail in Chapter 24. The fats generated during meat rendering can be used as a feedstock for biodiesel production, which is discussed in Chapter 26. Meat processing wastes can also be converted to bio-oil using a thermochemical liquefaction process, which is discussed in Chapter 27.

23.6 WASTE OILS AND FATS

23.6.1 Availability and Quality

Both hydrogenated and non-hydrogenated vegetable oils are used in commercial food frying operations. Recycled grease products are referred to as waste grease. Greases are generally classified in two categories, yellow grease and brown grease. Yellow grease is produced from vegetable oil or animal fat that has been heated and used for cooking a wide variety of meat, fish, or vegetable products. Yellow grease is required to have a free fatty acid (FFA) level of less than 15%. If the FFA level exceeds 15%, it is called brown grease, sometimes referred to as trap grease, and it may be sold at a discount. Brown grease is often cited as a potential feedstock for biodiesel production because it currently has very low value (Canakci, 2007). The main sources of animal fats are primarily meat animal processing facilities. Another source of animal fats is the collection and processing of animal mortalities by rendering companies. In general, all greases and oils are classified as lipids. Oils are generally considered to be liquids at room temperature, while greases are solid at room temperature. Many animal fats and hydrogenated vegetable oils tend to be solid at room temperature. Chemically, greases, oils, and fats are classified as triacylglycerides (Canakci, 2007).

The food processing industry generates a large amount of waste cooking oils and animal fats. Approximately 1.135 million tons of waste restaurant fats are collected annually from restaurants and fast-food establishments in the United States (Haumann, 1990). An annual average of 4.1 kg/person of yellow grease and 5.9 kg/

TABLE 23.8
Average Production Capacity of Nonedible Oils and Fats in the United States between 1995 and 2000 (Billion Pounds)

Animal Fats and Grease	Amount (Million Tons)
Tallow	2.489
Yellow grease	1.195
Lard and grease	0.593
Poultry fat	1.006
Total	5.284

Source: Adapted from Canakci, M., *Bioresour. Technol.*, 98, 183, 2007. With permission.

person of brown grease were produced in metropolitan areas in the United States in 1998 (Wiltsee, 1998). The production of inexpensive nonedible feedstocks including grease and animal fats was about 5.284 million tons/year, which represents one-third of the United States total oils and fats production as shown in Table 23.8 (Canakci, 2007). In France, fatty residues from both plants and animals represent an overall production of 0.55 million tons/year. The food industry generates 29% of the total fatty residues. The catering industry, wastewater treatment plants, and autonomous sanitation have shares about 32%, 23%, and 16%, respectively (Mouneimne et al., 2003).

Table 23.9 shows the fatty acid distribution of some common vegetable oils and animal fats (Ma and Hanna, 1999; Hill, 2006; Canakci, 2007). Saturated compounds such as lauric acid C12:0, myristic acid C14:0, palmitic acid C16:0, and stearic acid C18:0 have higher cetane numbers and are less prone to oxidation than unsaturated compounds such as oleic C18:1, linoleic C18:2, and linolenic C18:3 but they tend to crystallize at unacceptably high temperatures. Coconut oil contains a high proportion of fatty acids of short or medium chain length, mainly 12 and 14 carbon atoms, which are particularly suitable for further processing to surfactants, washing and cleansing agents, and cosmetics. Another characteristic of coconut is that its saturation level is as high as 91.0%.

Animal fats and greases contain a large amount of saturated fatty acids, which are almost 50% of the total fatty acids. Due to the higher content of saturated fatty acids, animal fats and greases have the unique properties of high melting point and high viscosity. They are solid at room temperature. Vegetable oils usually contain higher levels of unsaturated fatty acids. They are liquid at room temperature. However, they are more prone to oxidation due to the unsaturated bonds. Polyunsaturated fatty acids are very susceptible to polymerization and gum formation caused by oxidation during storage or by complex oxidative and thermal polymerization at the higher temperature and pressure of combustion. The gum does not combust completely, resulting in carbon deposits and lubricating oil thickening (Ma and Hanna, 1999).

The contents of free fatty acids and moisture of feedstocks are two main parameters used to evaluate the quality of feedstocks. Natural vegetable oils and

TABLE 23.9
Fatty Acid Distribution of Selected Vegetable Oils and Animal Fats

	Fatty Acid Distribution (% By Weight)							
Products	**Lauric (C12:0)**	**Myristic (C14:0)**	**Palmitic (16:0)**	**Stearic (18:0)**	**Oleic (18:1)**	**Linoleic (18:2)**	**Linolenic (18:3)**	**Saturation Level (%)**
Soybean oil	0.1	0.1	10.2	3.7	22.8	53.7	8.6	14.1
Cottonseed	0.1	0.7	20.1	2.6	19.2	55.2	0.6	23.5
Rapeseed oil	—	—	3.5	0.9	64.4	22.3	8.2	4.4
Sunflower oil	—	—	6.1	3.3	16.9	73.7	—	9.4
Palm oil	0.1	1.0	42.8	4.5	40.5	10.1	0.2	48.4
Coconut oil	48.8	20.0	7.8	3.1	4.4	0.8	—	91.0
Yellow grease	—	2.4	23.2	13.0	44.3	7.0	0.7	38.6
Brown grease	—	1.7	22.8	12.5	42.4	12.1	0.8	37.0
Lard	0.1	1.4	23.6	14.2	44.2	10.7	0.4	39.3
Tallow	0.1	2.8	23.3	19.4	42.4	2.9	0.9	45.6

Source: Modified from Ma, F. and Hanna, M.A., *Bioresour. Technol.*, 70, 1, 1999; Hill, K., in *Biorefineries—Industrial Processes and Products*, Wiley-VCH Verlag GmbH & Co., Weinheim, 2006, 291–314; Canakci, M., *Bioresour. Technol.*, 98, 183, 2007.

animal fats are extracted or pressed to obtain crude oil or fat. These usually contain free fatty acids, phospholipids, sterols, water, odorants, and other impurities. Even refined oils and fats contain small amounts of free fatty acids and water. Research conducted by Canakci (2007) showed that the content of free fatty acids in nonedible oils and fats varied from 0.7% to 41.8%, and moisture content from 0.01% to 55.38%.

During high-temperature cooking, various chemical reactions such as hydrolysis, polymerization, and oxidation can occur in vegetable oils. The physical and chemical properties of the oil change during cooking. The percentage of FFAs has been found to increase due to the hydrolysis of triglycerides in the presence of food moisture and oxidation. As an example, the FFA level of fresh soybean oil changed from 0.04% to 1.51% after 70h of frying at 190°C (Tyagi and Vasishtha, 1996). Increases in viscosity were also reported due to polymerization, which resulted in the formation of higher molecular weight compounds. Other observations were that the acid value, specific gravity and saponification value of the frying oil increased, but the iodine value decreased. The peroxide value increased to a maximum and then started to decrease (Mittelbach et al., 1992).

23.6.2 Current Utilization

Waste vegetable oils and fats are generally low in cost and are currently collected from large food processing and service facilities. They are then rendered and used almost exclusively in animal feed in developed countries (Canakci, 2007). Mouneimne et al. (2003) reported that 60% of solid fatty residues generated in France were stored in landfill sites. In most developing countries, a large proportion of the waste oils and fats are disposed of inappropriately (Al-Widyan and Al-Shyoukh, 2002). Fatty wastes are a major source of pollution and landfilling of fatty wastes is not acceptable in some developed countries.

23.6.3 Energy Utilization

Biodiesel is currently produced from food-grade vegetable oils such as soybean oil in the United States and rapeseed in Europe. Since food-grade vegetable oils are expensive, biodiesel produced from food-grade vegetable oil is not economically feasible. Animal fats, waste cooking oils, and restaurant grease are potential feedstocks for biodiesel production, which is discussed in Chapter 26. Waste vegetable oils from restaurants and rendered animal fats are inexpensive compared with food-grade vegetable oils. The price of yellow grease varied widely from $0.09 to $0.20/lb in 2000, compared to $0.35/lb for soybean oil. Brown grease is usually discounted $0.01–$0.03/lb below the price of yellow grease. One pound of most fats and oils can be converted to a pound of biodiesel. If all of the 5.284 million tons/year of greases and animal fats in the United States were converted to biodiesel, it would replace about 1.5 billion gal of diesel fuel (Canakci, 2007).

The free fatty acid and water contents have significant effects on the transesterification of glycerides with alcohols using alkaline in a traditional biodiesel production process. They also interfere with the separation of fatty acid esters and glycerol. Free fatty acids and moisture reduce the efficiency of transesterification reaction to convert a feedstock into biodiesel using traditional alkaline catalysts. Therefore, an efficient process for converting waste grease and animal fats must tolerate a wide range of feedstock properties.

23.7 ENERGY CONVERSION TECHNOLOGIES FOR FOOD PROCESSING WASTES

Several conversion processes, including biological conversion, thermochemical conversion, and chemical conversion, can transform different food processing wastes into different forms of energy products including heat and power, gaseous fuels including biogas and syngas, and liquid fuels including bioethanol, biodiesel, and bio-oil as shown in Figure 23.3.

Thermochemical conversion includes pyrolysis, gasification, combustion, and hydrothermal liquefaction. During thermochemical conversion, organic food processing wastes are broken down to small gaseous or liquid molecules at an elevated temperature and in the presence or absence of air or oxygen. If the combustion air (or oxygen) is insufficient or absent, the derived gaseous and liquid fuels are produced from the organic wastes. Gasification is an oxygen-insufficient process to

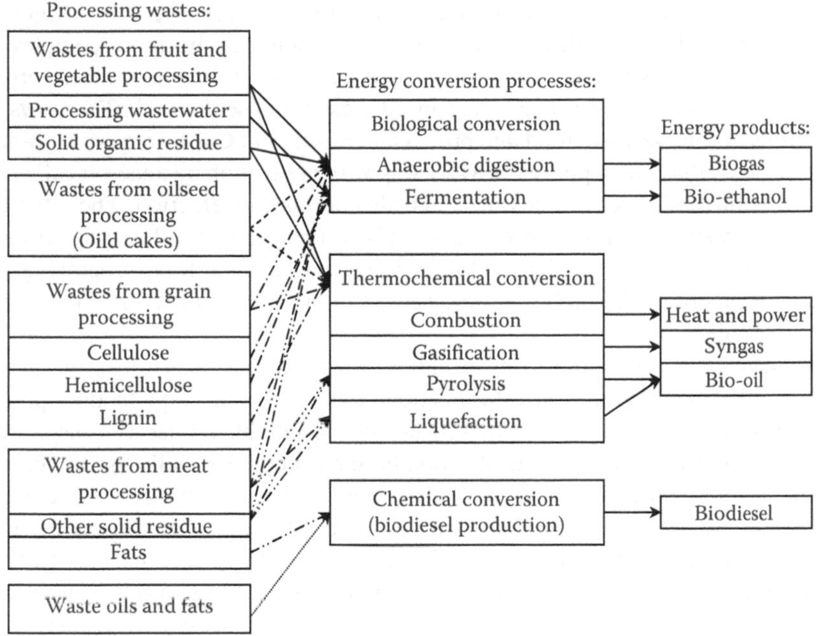

FIGURE 23.3 Energy conversion technologies for food processing wastes.

produce syngas, which is a gaseous mixture mainly including carbon monoxide, carbon dioxide, methane, and hydrogen. Pyrolysis is an oxygen-absent process to produce liquid tar and solid char. Thermochemical conversion can convert a variety of food processing wastes into usable energy such as process steam and electricity. Anaerobic digestion and fermentation are two primary biological conversion processes to convert processing wastes into energy. During anaerobic digestion, microorganisms breakdown organic waste materials to produce biogas, which is a gaseous mixture mainly consisting of methane and carbon dioxide. During fermentation, microorganisms such as yeasts ferment simple sugars into ethanol. Esterification is a popular chemical process to produce biodiesel from vegetable oil and animal fats by transesterifying one molecule of triacylglyceride in the oils and fats into three molecules of alkyl esters as biodiesel and one molecule of glycerol as by-product.

Selection of an appropriate process is dependent upon (1) the characteristics of wastes such as moisture content, heat value, and physical and chemical properties, (2) the quantity of wastes available, (3) the desired form of energy products, (4) the efficiency of the energy conversion process, (5) energy demands for the processing facilities and in the reachable market, and (6) economical feasibility. High moisture content food wastes (e.g., >50%) are usually better suited to biological processes such as anaerobic digestion while low moisture content wastes are better for thermochemical conversion processes such as combustion, gasification, and pyrolysis. Thermochemical liquefaction can also be used to convert wet food processing wastes at high pressure and moderate temperature into a bio-oil of partly oxygenated hydrocarbons. However, thermochemical liquefaction may be more complex and expensive than a pyrolysis process, which is another process to convert organic matter to bio-oil. A survey conducted by Matteson and Jenkins (2007) shows that two-thirds of the available food and processing residues in California are high-moisture resources that can support 134 MWe of power generation by anaerobic digestion and other conversion techniques such as hydrothermal liquefaction. The other one-third of the residues is low-moisture materials, which can be used as a direct fuel in combustion power plants. The food processing may include food processing solid wastes, meat processing solid wastes, vegetable crop residues, and food waste in municipal solid waste.

In many cases, it is the form of energy required that determines the process selection. Fermentation, transesterification, pyrolysis, and liquefaction produce liquid fuels including ethanol, biodiesel, and bio-oil suitable for use as transportation fuels. Fermentation and transesterification have been commercially used to produce bioethanol and biodiesel to replace gasoline and fossil diesel, respectively. Pyrolysis is a rapidly developing technology with a great potential to produce a liquid oil from dry biomass, which is suitable for use in diesel engines and gas turbines. A thermochemical liquefaction process can convert wet biomass to a liquid fuel but it is more complex and expensive than a pyrolysis process (McKendry, 2002). Combustion, gasification, and anaerobic digestion produce gaseous energy products such as hot gas for steam generation, syngas and biogas, which are suitable to be used at the production location. Gasification is likely to be commercially viable because it has the highest overall conversion efficiency for gaseous fuel production. Anaerobic digestion is considered as a conversion process to produce a gaseous fuel and treat high moisture content organic processing wastes for environmental protection (McKendry, 2002).

It is worth noting that high-value phytochemicals can be identified and extracted from the waste residuals in the food processing facilities before they are used for energy production for an increase in economic profitability (Wang et al., 2005).

23.8 SUMMARY

Food processing facilities generate large amounts of organic wastes in a solid or liquid form. Traditionally, part of those wastes is processed as animal feeds. Large amounts of solid food processing wastes are buried in landfills at a cost while liquid food processing wastes are released into rivers, lakes and oceans, and disposed in public sewer systems without treatment. The utilization of the energy in food process wastes can reduce both fossil fuel costs and waste disposal costs in a food processing facility. Integration of energy conversion processes into food processing facilities will achieve not only economic profitability but also environmental benefits. Conversion of food processing wastes into energy is a growing practice in the food manufacturing industry. Food processing wastes can be converted to heat and power, and liquid and gaseous fuels using a biological, thermochemical, or chemical conversion process depending on the characteristics of wastes, the quantity of wastes available, the desired form of energy products, the efficiency of the energy conversion process, energy demands in the market, and economical feasibility. Wet food processing wastes are usually better suited for biological processes such as anaerobic digestion while dry wastes are better for thermochemical conversion processes such as combustion, gasification, and pyrolysis. Thermochemical liquefaction can also be used to convert wet food processing wastes at high pressure and moderate temperature into a bio-oil of partly oxygenated hydrocarbons. Fermentation, transesterification, pyrolysis, and liquefaction produce liquid fuels for use as transportation fuels. Combustion, gasification, and anaerobic digestion produce gaseous energy products, which are suitable to be used at the production location.

REFERENCES

Al-Widyan, M.I. and A.O. Al-Shyoukh. 2002. Experimental evaluation of the transesterification of waste palm oil into biodiesel. *Bioresource Technology* 85: 253–256.

Atimtay, A.T. and H. Topal. 2004. Co-combustion of olive cake with lignite coal in a circulating fluidized bed. *Fuel* 83: 859–867.

Blessin, C.W., W.J. Garcia, W.L. Deatherage, J.F. Cavins, and G.E. Inglett. 1973. Composition of three food products containing defatted corn germ flour. *Journal of Food Science* 38: 602–606.

Blondin, G.A., S.J. Comiskey, and J.M. Harkin. 1983. Recovery of fermentable sugars from process vegetable wastewaters. *Energy in Agriculture* 2: 21–36.

Canakci, M. 2007. The potential of restaurant waste lipids as biodiesel feedstocks. *Bioresource Technology* 98: 183–190.

Chang, D., M.P. Hojilla-Evangelista, L.A. Johnson, and D.J. Myers. 1995. Economic-engineering assessment of sequential extraction processing of corn. *Transactions of the ASAE* 38: 1129–1138.

Choteborska, P., B. Palmarola-Adrados, M. Galbe, G. Zacchi, K. Melzoch, and M. Rychtera. 2004. Processing of wheat bran to sugar solution. *Journal of Food Engineering* 61: 561–565.

Commission of the European Communities. 1990. Council Directive 90/667/EEC. Official Journal, No. L 363: 51.

Cooper, J.L. 1976. The potential of food processing solid wastes as a source of cellulose for enzymatic conversion. *Biotechnology and Bioengineering* 6: 251–271.

DeFelice, S.L. 1995. The time has come for nutraceutical cereals. *Cereal Foods World* 40: 51–52.

Dexter, J.E., D.G. Martin, G.T. Sadaranganey, J. Michaelides, N. Mathieson, J.J. Tkac, and B.A. Marchylo. 1994. Preprocessing: Effect on durum wheat milling and spaghetti-making quality. *Cereal Chemistry* 71: 10–16.

Eggeman, T. and D. Verser. 2006. The importance of utility systems in today's biorefineries and a vision for tomorrow. *Applied Biochemistry and Biotechnology* 129–132: 361–381.

Fischer, K. and H.P. Bipp. 2005. Generation of organic acids and monosaccharides by hydrolytic and oxidative transformation of food processing residues. *Bioresource Technology* 96: 831–842.

Gercel, H.F. 2002. The production and evaluation of bio-oils from the pyrolysis of sunflower-oil cake. *Biomass and Bioenergy* 23: 307–314.

Gras, P.W. and D.H. Simmonds. 1980. The utilization of protein-rich products from wheat carbohydrate separation processes. *Food Technology Australia* 32: 470–472.

Grull, D.R., F. Jetzinger, M. Kozich, and M.M. Wastyn. 2006. Industrial starch platform-status quo of production, modification and application. In *Biorefineries—Industrial Processes and Products*, Kamm, B. Gruber, P.R., and M. Kamm (Eds.), pp. 61–95. Weinheim: Wiley-VCH Verlag GmbH & Co.

Haapapuro, E.R., N.D. Barnard, and M. Simon 1997. Review-animal waste used as livestock feed: danger to human health. *Preventive Medicine* 26: 599–602.

Hang, Y.D. 2004. Management and utilization of food processing wastes. *Journal of Food Science* 69: 104–107.

Hang, Y.D., Lee, C.Y., and Woodams, E.E. 1986. Solid state fermentation of grape pomace for ethanol production. *Biotechnology Letters* 8: 53–56.

Haumann, B.F. 1990. Renderers give new life to waste restaurant fats. *Inform* 1: 722–725.

Hill, K. 2006. Industrial development and application of biobased oleochemicals. Volume 2, In *Biorefineries—Industrial Processes and Products*, Kamm, B., P.R. Gruber, and M. Kamm (Eds.), pp. 291–314. Weinheim: Wiley-VCH Verlag GmbH & Co.

Johnson, D.L. 2006. The corn wet milling and corn dry milling industry-a base for biorefinery technology developments. In *Biorefineries—Industrial Processes and Products*, Kamm, B., P.R. Gruber, and M. Kamm (Eds.), pp. 344–352. Weinheim: Wiley-VCH Verlag GmbH & Co.

Kantor, L.S., K. Lipton, A. Manchester, and V. Oliveira. 1997. Estimating and addressing America's food losses. Economic Research Service, United States Department of Agriculture, prepublished on the web. URL: http://151.121.66.126:80/whatsnew/feature/ARCHIVES/JULAUG97/INDEX.HTM

Kollacks, W.A. and C.J.N. Rekers. 1988. Five years of experience with the application of reverse osmosis on light middlings in a corn wet milling plant. *Starch* 40: 88–94.

Koseolu, S.S., K.C. Rhee, and E.W. Lusas. 1991. Membrane separations and applications in cereal processing. *Cereal Foods World* 36: 376–383.

Kraus, G.A. 2006. Phytochemicals, dyes, and pigments in the biorefinery context. In *Biorefineries—Industrial Processes and Products*, Kamm, B., P.R. Gruber, and M. Kamm, (Eds.), pp. 315–323. Weinheim: Wiley-VCH Verlag GmbH & Co.

Krull, L.H. and G.E. Inglett. 1971. Industrial uses of gluten. *Cereal Science Today* 16: 232–236, 261.

Lahmar, M., V. Fellner, R.L. Belyea, and J.E. Williams. 1994. Increasing the solubility and degradability of food processing biosolids. *Bioresource Technology* 50: 221–226.

Laufenberg, G., B. Kunz, and M. Nystroem. 2003. Transformation of vegetable waste into value added products: (A) the upgrading concept; (B) practical implementations. *Bioresource Technology* 87: 167–198.

Lingaiah, V. and P. Rajasekaran. 1986. Biodigestion of cow dung and organic wastes mixed with oil cake in relation to energy. *Agricultural Wastes* 17: 161–173.

Ma, F. and M.A. Hanna. 1999. Biodiesel production: A review. *Bioresource Technology* 70: 1–15.

Mata-Alvarez, J., S. Mace, and P. Llabres. 2000. Anaerobic digestion of organic solid wastes. An overview of research achievements and perspectives. *Bioresource Technology* 74: 3–16.

Matteson, G.C. and B.M. Jenkins. 2007. Food and processing residues in California: Resource assessment and potential for power generation. *Bioresource Technology* 98: 3098–3105.

McKendry, P. 2002. Energy production from biomass (part 2): conversion technologies. *Bioresource Technology* 83: 47–54.

Mittelbach, M., B. Pokits, and A. Silberholz. 1992. Production and fuel properties of fatty acid methyl esters from used frying oil. In *Liquid Fuels from Renewable Resources. Proceedings of an Alternative Energy Conference*, pp. 74–78. Nashville: ASAE Publication.

Mohaibes, M. and H. Heinonen-Tanski. 2004. Aerobic thermophilic treatment of farm slurry and food wastes. *Bioresource Technology* 95: 245–254.

Mouneimne, A.H., H. Carrere, N. Bernet, and J.P. Delgenes. 2003. Effect of saponification on the anaerobic digestion of solid fatty residues. *Bioresource Technology* 90: 89–94.

Ngadi, M.O. and L.R. Correia. 1992. Kinetics of solid-state ethanol fermentation from apple pomace. *Journal of Food Engineering* 17: 97–116.

Nguyen, M.H. 2003. Alternatives to spray irrigation of starch waste based distillery effluent. *Journal of Food Engineering* 60: 367–374.

Nigam, J.N. 2000. Continuous ethanol production from pineapple cannery waste using immobilized yeast cells. *Journal of Biotechnology* 80: 189–193.

Onay, O., S.H. Beis, and O.M. Kochar. 2001. Fast pyrolysis of rape seed in a well-swept fixed-bed reactor. *Journal of Analytical and Applied Pyrolysis* 58–59: 995–1007.

Ozbay, N., A.E. Putun, and E. Putun. 2001. Biocrude from biomass: Pyrolysis and steam pyrolysis of cottonseed cake. *Journal of Analytical and Applied Pyrolysis* 60: 89–101.

Ozcimen, D. and F. Karaosmanoglu. 2004. Production and characterization of bio-oil and biochar from rapeseed cake. *Renewable Energy* 29: 779–787.

Park, J.I., Y.S. Yun, and J.M. Park. 2002. Long-term operation of slurry bioreactor for decomposition of food wastes. *Bioresource Technology* 84: 101–104.

Ramachandran, S., S.K. Singh, C. Larroche, C.R. Soccol, and A. Pandey. 2007. Oil cakes and their biotechnological applications-a review. *Bioresource Technology* 98: 2000–2009.

Rasco, B.A., F.M. Dong, A.E. Hashisaka, S.S. Gazzaz, S.E. Downey, and M.L. San Buenaventura. 1987. Chemical composition of distillers' dried grains with solubles (DDGS) from solt white wheat, hard red wheat and corn. *Journal of Food Science* 52: 235–237.

Rendleman, C.M. and N. Hohmann. 1993. The impact of production innovations in the fuel ethanol industry. *Agribusiness New York* 9: 217–231.

Ritter, W.F. and A.E.M. Chinside. 1995. Impact of dead bird disposal pits on groundwater quality on the Delmarva Peninsula. *Bioresource Technology* 53: 105–111.

Salminen, E. and J. Rintala. 2002. Anaerobic digestion of organic solid poultry slaughterhouse waste—a review. *Bioresource Technology* 83: 13–26.

Sargent, S.A. and J.F. Steffe. 1986. Energy generation from direct combustion of solid food processing wastes. In *Energy in Food Processing*, Singh, R.P. (Ed.), pp. 247–266. New York: Elsevier Science Publishing Company Inc.

Satterlee, L.D. 1981. Proteins for use in foods. *Food Technology* 35: 53–70.

Tritt, W.P. and F. Schuchardt. 1992. Materials flow and possibilities of treating liquid and solid wastes from slaughterhouses in Germany. A review. *Bioresource Technology* 41: 235–245.

Tucker, M.P., N.J. Nagle, E.W. Jennings, K.N. Ibsen, A. Aden, Q.A. Nguyen, K.H. Kim, and S.L. Noll. 2004. Conversion of distiller's grain into fuel alcohol and a higher-value animal feed by dilute-acid pretreatment. *Applied Biochemistry and Biotechnology* 113–116: 1139–59.

Turhollow, A.F. and E.O. Heady 1986. Large-scale ethanol production from corn and grain sorghum improving conversion technology. *Energy in Agriculture* 5: 309–316.

Tyagi, V.K. and A.K. Vasishtha. 1996. Changes in the characteristics and composition of oils during deep-fat frying. *Journal of the American Oil Chemists' Society* 73: 499–506.

Verma, M., S.K. Brar, R.D. Tyagi, R.Y. Surampalli, and J.R. Valero. 2007. Starch industry wastewater as a substrate for antagonist *Trichoderma viride* production. *Bioresource Technology* 98: 2154–2162.

Wang, L.J., C.L. Weller, and K.T. Hwang. 2005. Extraction of lipids from grain sorghum DDG. *Transactions of the ASABE* 48: 1883–1888.

Weegels, P.L., J.P. Marseille, and R.J. Hamer. 1992. Enzymes as a processing aid in the separation of wheat flour into starch and gluten. *Starch* 44: 44–48.

Wiltsee, G. 1998. Urban waste grease resource treatment. Final Report to the National Renewable Energy Laboratory, NREL/SR-50-26141.

Woodruff, J.G. and B.S. Luh. 1975. *Commercial Fruit Processing*, AVI Publisher, Westport, CT.

Wrick, K.L. 1993. Functional foods: cereal products at the food-drug interface. *Cereal Foods World* 38: 205–214.

Yorgun, S., S. Sensoz, and O.M. Kockar. 2001. Flash pyrolysis of sunflower oil cakes for production of liquid fuels. *Journal of Analytical and Applied Pyrolysis* 60: 1–12.

24 Anaerobic Digestion of Food Processing Wastes

24.1 INTRODUCTION

Food wastes such as fruit and vegetable processing wastes, meat processing wastes, and slaughterhouse wastes can be treated either aerobically or anaerobically to reduce pollutant and pathogen risk and recover materials from the wastes. An aerobic process is to convert the wastes into carbon dioxide and water by aerobic respiration. Aerobically treated organic wastes could be used as a fertilizer (Van Heerden et al., 2002). Ideal composting conditions are the carbon to nitrogen ratio of a composting material between 20 and 40, the moisture content between 50% and 60%, adequate oxygen supply, small particle sizes, and enough void space for air to flow through (Chang et al., 2006). Anaerobic digestion is a biological process, in which an organic matter is degraded to a gaseous mixture of biogas in the absence of oxygen. Biogas, which mainly consists of methane and carbon dioxide, can be used as an energy source to replace fossil natural gas. The methane in the biogas could be as high as 50%–60% by volume (Gebauer, 2004). If biogas produced by anaerobic digestion of biomass is used for electricity generation, the overall conversion efficiency from biomass to electricity is about 10%–16% (McKendry, 2002).

High moisture food processing wastes are difficult to be converted in a thermochemical conversion process without auxiliary fuels. Thermochemical conversion is discussed in Chapter 27. One great advantage of anaerobic digestion is that it can be used to treat very wet and pasty organic wastes or liquid wastes (Shih, 1993; Braber, 1995). Anaerobic digestion is a commercially proven technology and is widely used for treating organic wastes with a high moisture content (i.e., >80%–90% moisture) (McKendry, 2002). For the treatment of food processing wastes, anaerobic digestion can not only produce methane for energy but also destroy pathogenic bacteria presenting in the wastes and reduce pollutant emission from the wastes. Recent advances in anaerobic digestion technology have made it possible to compete well with other methods for the treatment of diverse food processing wastes. Several reviews on anaerobic digestion of organic wastes have been given by Zinder (1984); Walker et al. (1985); Shih (1993); Mata-Alvarez et al. (2000); and Salminen and Rintala (2002). Anaerobic digestion has been used to treat solid fish wastes (Gebauer, 2004), slaughterhouse wastes (Salminen and Rintala, 2002), and food processing wastewater (Berardino et al., 2000). In this chapter, the anaerobic digestion pathways and process are reviewed. The effects of waste characteristics and operating conditions for anaerobic digestion are discussed. Finally, applications of anaerobic digestion for the treatment of food processing wastes are addressed.

24.2 OVERVIEW OF ANAEROBIC DIGESTION TECHNOLOGIES

24.2.1 Anaerobic Digestion Pathways

Organic food wastes may contain carbohydrates, proteins, and lipids. Various microorganisms are involved in many steps of anaerobic digestion of those compounds in wastes. Each of those steps could be a rate-limiting step. There are three main categories of bacteria: aerobic bacteria, which must have oxygen to grow, facultatively anaerobic bacteria, which can metabolize and grow with or without oxygen, and anaerobic bacteria, which can grow only in the absence of oxygen. Although all digesters may contain aerobic and facultatively anaerobic bacteria, the bacteria involved in the metabolic pathways for conversion of wastes to biogas are anaerobic bacteria (Hobson et al., 1981). Figure 24.1 shows the anaerobic digestion pathways of different compounds that are present in food wastes (Salminen and Rintala, 2002).

Anaerobic digestion starts with the hydrolysis of different compounds in the organic wastes. Proteins are hydrolyzed to amino acids. Lipids are hydrolyzed to long-chain fatty acids and glycerol via β-oxidation. Carbohydrates are hydrolyzed to sugars. After the hydrolysis, the intermediates of long-chain fatty acids, amino acids, and sugars are converted to volatile fatty acids with three or more carbon, hydrogen, and carbon dioxide by fermentative bacteria. Ammonia and sulfurides are the by-products of amino acid fermentation. Part of long-chain fatty acids, volatile fatty acids, and neutral compounds such as sugars are metabolized to acetate, hydrogen,

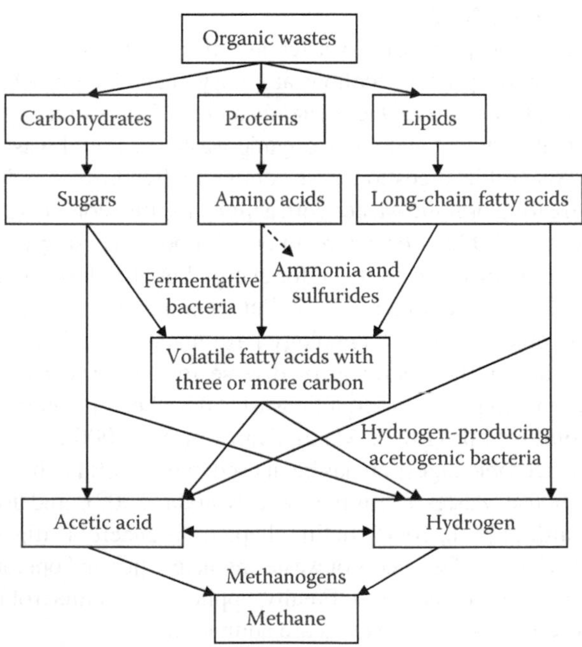

FIGURE 24.1 Anaerobic digestion pathways of different compounds in organic wastes. (Adapted from Salminen, E. and Rintala, J., *Bioresour. Technol.*, 83, 13, 2002. With permission.)

and carbon dioxide by hydrogen-producing acetogenic bacteria. The acetate, hydrogen, and carbon dioxide are ultimately converted to the biogas of methane and carbon dioxide by methanogens (Zinder, 1984; Salminen and Rintala, 2002). The hydraulic retention times (HRT) of wastes during anaerobic digestion could be several days to a month (Boubaker and Ridha, 2007).

The biogas can be used to generate electricity and meanwhile the by-products of heat can be recovered and used to heat the anaerobic digestion system and supply heat for the processing facilities (Pellerin et al., 1988).

24.2.2 Effects of Feedstocks on Anaerobic Digestion

The chemical composition of organic wastes and the reaction intermediates are important for an anaerobic digestion process. The anaerobic digestion of organic wastes may be inhibited by different compounds such as long-chain fatty acids, ammonia, and salt. The anaerobic digestion process may also be restricted by the ratio of carbon and nitrogen.

The growth of bacteria for the degradation of long-chain fatty acids is slow and the degradation of long-chain fatty acids requires low partial pressure of hydrogen. The accumulating long-chain fatty acids are toxic to anaerobic microorganisms such as acetogens and methanogens (Angelidaki and Ahring, 1995). The inhibition of anaerobic digestion by long-chain fatty acids depends on the type of bacteria present, the specific surface area of sludge, the carbon chain length, and the saturation of the long-chain fatty acids (Salminen and Rintala, 2002). Gram-positive microorganisms and methanogens are more vulnerably inhibited by long-chain fatty acids than gram-negative microorganisms (Roy et al., 1985). Suspended sludge is more vulnerable to cause inhabitation than granular sludge (Hwu et al., 1996). Saturated long-chain fatty acids with 12–14 carbons and unsaturated long-chain fatty acids with 18 carbons are more inhibitory (Komatsu et al., 1991). Some substances including albumin, starch, bile acids, cholesterol, bentonite, and calcium may reduce the toxicity and prevent the inhibition of long chain fatty acids during anaerobic digestion (Salminen and Rintala, 2002). The biological degradation of fats consists of two main steps according to Figure 24.1: (1) hydrolysis of triacyglycerides to long-chain fatty acids and glycerol by extracellular lipolytic enzymes and (2) transport of long-chain fatty acids into cells for successive breaks in the carbon chain via β-oxidation. The insolubility of fats in water limits their biological degradation. Mouneimne et al. (2003) used saponification treatment to enhance the emulsification and bioavailability of fatty residues in a liquid medium during anaerobic digestion. They found that the degradation of fat residues was enhanced by 40% at a pH of 8.5.

Ammonia produced from protein degradation may also inhibit anaerobic microorganisms such as methanogens because unionized ammonia is toxic and it can diffuse across the cell membrane (Kadam and Boone, 1996). Fujishima et al. (1999) investigated the effect of moisture content on the anaerobic digestion of dewatered sewage sludge under mesophilic conditions. They found that methane production decreased when the moisture content of sludge was lower than 91% due to the non-acclimatization to high ammonia concentration of hydrogenotrophic methanogenic bacteria. Hansen et al. (1999) found that a small amount of sulfuride (i.e., 23 mg S^{-2}/L) may increase ammonia inhibition while activated carbon (2.5%

by weight) or $FeCl_2$ (4.4 mM) could relieve inhibition by reduction of sulfuride concentration. Several approaches including air stripping to remove ammonia from feedstock, ammonia recovery by adding magnesium and orthophosphate, and addition of phosphorite ore have been attempted to prevent ammonia inhibition (Salminen and Rintala, 2002). Other methods have also been tested successfully: (1) dilution of digester content with water, and (2) adjustment of feedstock carbon and nitrogen ratio (Kayhanian, 1999).

Salts such as sodium present in feedstock may also inhibit anaerobic digestion (Gebauer, 2004). Gebauer (2004) found that for mesophilic anaerobic treatment of sludge from saline fish farm effluents with sodium of 10–10.5 g/L, the process was strongly inhibited. If diluting the sludge 1:1 with tap water to reduce the sodium concentration to 5.3 g/L), the inhibition could be overcome.

During anaerobic digestion, microorganisms are needed to convert complex macromolecules into low molecular weight compounds such as methane, carbon dioxide, water, and ammonia. The composition of the feedstock is important because there is a need for a suitable ratio between carbon and nitrogen (Mshandete et al., 2004). Co-digestion of different wastes has been proved an effective way to supply suitable nutrients and dilute toxicants during anaerobic digestion (Mata-Alvarez et al., 2000).

24.2.3 Effects of Operating Conditions on Anaerobic Digestion

Main intermediates in the conversion are volatile fatty acids. If a high concentration of volatile fatty acids is formed, pH will be reduced and that can reach levels where the methanogenic bacteria are first severely inhibited and may even die. A digester usually runs naturally at a pH of 7 or just over, about 7.2, which is optimum for methanogenic bacteria (Hobson et al., 1981). Therefore, it is important to have a buffering capacity in the system, i.e., products that will counteract the effects of the volatile fatty acids need also to be formed (Mshandete et al., 2004). However, at a high pH value, unionized ammonia dominates, which is more inhibitory than ionized ammonia (Kadam and Boone, 1996).

Solid concentration is an important parameter. The addition of large amounts of water requires a large digester volume and high posttreatment costs for the digester residue. In a continuously stirred tank digester, an influent substrate concentration of 3%–8% total solids is added daily and an equal amount of effluent is withdrawn. Dry anaerobic fermentation, which is also called high solids anaerobic digestion, can use the total solid concentration of more than 25% (Gunaseelan, 1997). However, Kalia et al. (2000) reported that banana steam slurry with 16% of total solids concentration inhibited anaerobic digestion and resulted in 50%–60% loss in biomethanation process efficiency.

Anaerobic digestion of organic wastes is usually operated at mesophilic and thermophilic temperatures (Gebauer, 2004; Boubaker and Ridha, 2007). Different bacteria grow at these two temperature ranges, which are from about 5°C to 45°C or 50°C and from about 55°C to about 70°C. In either case, there is optimum temperature for the fastest bacteria growth (Hobson et al., 1981). Relatively few studies have been conducted in the digestion of organic wastes at psychrophilic temperatures (i.e., below 20°C) (Lo and Liao, 1986). The benefits of psychrophilic digestion is less energy input for heating influents and maintaining the temperature of digesters. The main disadvantage of psychrophilic digestion is a low production rate of biogas (Lo and Liao, 1986). Kalia et al. (2000) found that thermophilic digestion rate at a temperature of 50°C–55°C was 2.4 times faster than mesophilic digestion rate at 37°C–40°C. Furthermore, Shih (1987) found that thermophilic digestion was usually more effective than mesophilic digestion for destroying pathogens.

The anaerobic digestion of organic wastes at a high temperature may be economically sustainable if inexpensive external heat such as heat from an electricity generation system is available for warming of the waste sludge to process temperature (Gebauer, 2004). Digesters should be designed to ensure more uniform digester temperatures. Obtaining this uniform temperature profile will require a better understanding of the modes of heat transfer that occur within and around the digester, and the fluid dynamics in the digester. An influent heat exchanger is usually used to heat the incoming feedstock to the digestion temperature (Walker et al., 1985).

24.2.4 Anaerobic Digester Design

There are five main types of digesters, which are completely mixed digesters (continuously stirred tank digesters, CSTD), fixed film digesters, plug flow digester, anaerobic filter, and fluidized bed digester (Converti et al., 1990; Gunaseelan, 1997; Berardina et al., 2000; Nishio and Nakashimada, 2007). The continuously stirred tank digester is the most popular design. In the CSTD, the feeding should be continuous for maximum efficiency. In a plug-flow digester, a volume of the medium with a suitable inoculum enters at one end of the tube and the digestion is completed when the medium reaches the other end. Unlike completely mixed digesters where the substrate concentration and temperature distribution are forced to uniform values with mechanical mixing, the performance of the plug flow digester is strongly dependent on how well the fluid dynamics and fermentation kinetics are integrated. The anaerobic filter is primarily used for the digestion of wastewaters produced in large quantities. The bacteria are attached to a solid support such as stones packed inside a tank. Wastewater flows upward through the tank and the retention time is only a few hours. The fluidized bed digester is a modified form of an anaerobic filter. In the fluidized bed digester, the bacteria are attached to small glass spheres, which are freely suspended in the up-flowing fed (Gunaseelan, 1997).

24.3 ANAEROBIC DIGESTION OF FOOD PROCESSING WASTES

Anaerobic digestion has been used in the treatment of fish processing wastes, slaughtering wastes, fruit and vegetable processing wastes, and food processing wastewater, as shown in Table 24.1.

TABLE 24.1
Applications of Anaerobic Digestion of Food Processing Wastes

Food Wastes	Digester Type	Temperature (°C)	Hydraulic Retention Time (Days)	Loading Rate	Yield	References
Sludge of saline fish	Continuously stirred tank digester	35	24–60	1.0–3.12 g COD/l d	0.114– 0.184 L/g COD	Gebauer, 2004
Cattle and lamb paunch contents, blood, and process wastewaters	Continuously stirred tank digester	—	43	0.36 kg COD/m^3d	0.18 m^3/kg COD	Bank, 1994
Cattle blood, rumen, and paunch contents	Continuously stirred tank digester	35	2–10	3.6 kg TS/m^3 d	0.27 m^3/kg TS	Banks and Wang, 1999
Solid poultry slaughterhouse waste	Continuously stirred tank digester	35	50	0.8 kg VS/m^3 d	0.52–0.55 m^3/kg VS	Salminen et al., 2000
Spinach waste	Continuously stirred tank digester	33	32	0.83–1.18 kg VS/m^3 d	0.316 m^3/kg VS	Knol et al., 1978
Apple slurry	Continuously stirred tank digester	33	32	1.02–1.60 kg VS/m^3 d	0.308 m^3/kg VS	Knol et al., 1978
Corn cobs	Continuously stirred tank digester	35–37	—	3.90 kg VS/m^3 d	0.267 m^3/kg VS	Lane, 1984
Apple cake	Continuously stirred tank digester	35–37	—	3.88 kg VS/m^3 d	0.252 m^3/kg VS	Lane, 1984
Tomato processing wastes	Continuously stirred tank digester	35	24	4.3 kg VS/m^3 d	0.420 m^3/kg VS	Sarada and Joseph, 1994

24.3.1 Anaerobic Digestion of Fish Wastes

Gebauer (2004) used a continuously stirred tank digester at 35°C for the mesophilic anaerobic treatment of sludge from saline fish farm effluents with total solids of 8.2%–10.2% by weight, chemical oxygen demand (COD) of 60–74 g/L, and sodium (Na) of 10–10.5 g/L. It was found that the COD content of the sludge was reduced by 36% to 60% and methane yields were between 0.114 and 0.184 L/g COD added. However, the process was strongly inhibited, presumably by sodium, and unstable, with propionic acid being the main compound of the volatile fatty acids (VFA). When diluting the sludge 1:1 with tap water (Na: 5.3 g/L), the inhibition could be overcome and a stable process with low VFA concentrations was achieved.

24.3.2 Anaerobic Digestion of Slaughterhouse Wastes

Anaerobic digestion has been used to treat solid slaughterhouse wastes such as cattle paunch wastes, blood, and settlement tank solids (Banks, 1994; Banks and Wang, 1999). Banks (1994) treated cattle and lamb paunch contents, blood, and process wastewaters with an organic loading of 0.36 kg COD/m^3 d in a 105 m^3 continuously stirred tank digester. About 0.18 m^3/kg COD was observed at a hydraulic retention time of 43 days. Bank and Wang (1999) further investigated the treatment of cattle blood and rumen paunch contents in a two-phase process with uncoupled solids and liquid retention times. The two-phase process allowed a higher overall loading at 3.6 kg total solids (TS)/m^3 d. Salminen et al. (2000) treated solid poultry slaughterhouse waste at a loading rate of 0.8 kg (volatile solid) VS/m^3 d and hydraulic retention time of 50 days in a 2 L continuously stirred tank digester. A methane yield of 0.55 m^3/kg VS was observed.

24.3.3 Anaerobic Digestion of Fruit and Vegetable Solid Waste

Fruit and vegetable processing wastes are characterized by high percentages of moisture (e.g., >80%) and volatile solid (e.g., >95%), and a very high biodegradability. Anaerobic digestion can be used to reduce the organic solids in the fruit and vegetable processing wastes and recover energy from the wastes (Kalia et al., 2000).

The methane yield of the anaerobic digestion of fruit and vegetable processing wastes is very high. The CH_4 yield in the anaerobic digestion of fruit and vegetable solid wastes was reported to be as high as 0.53 m^3/kg VS with 100% VS conversion for the digestion of damaged banana (Gunaseelan, 1997). Linke (2006) found that both biogas yield and CH_4 in the biogas for the anaerobic digestion of solid wastes from potato processing waste in a completely stirred tank reactor decreased with the increase in the organic loading rate. For the organic loading rate in the range from 0.8 to 3.4 g/L day, the biogas yield and CH_4 obtained were reduced from 0.85 to 0.65 L/g, and from 58% to 50%, respectively.

Knol et al. (1978) found that the maximum loading rate for the stable digestion of a variety of fruit and vegetable solid wastes ranging from 0.8 to 1.6 kg VS/m^3 day. Gollakota and Mether (1988) reported that the deoiled cake of nonedible oil seeds could be anaerobically digested at a loading rate of 8 kg TS/m^3 day, 15 days HRT, and 37°C with intermittent mixing.

24.3.4 Anaerobic Digestion of Food Processing Wastewater

Effluents from the food processing industry containing lipids, proteins, carbohydrates, and other organic materials might cause environmental damage if they are discharged in rivers and lakes without any treatment. Anaerobic digestion can be used to efficiently and effectively treat the wastewater from food processing facilities. Since food processing facilities produce diluted wastewater, the anaerobic digestion of food processing wastewater is more to reduce environmental pollution than produce energy.

Chavez et al. (2005) used an up-flow anaerobic sludge blanket reactor to remove organic materials from poultry slaughter wastewater with mean biochemical oxygen demand (BOD_5) of 5500 mg/L and COD of 7333 mg/L. They found that 95% of BOD_5 was removed from poultry slaughter wastewater at organic loading rates up to 31 kg BOD_5/m^3d, temperature between 25°C and 39°C, and hydraulic residence time between 3.5 and 4.5 h. Berardina et al. (2000) treated processing wastewater with an average polluting power between 630 and 2620 mg COD/L from a confectionery factory in a semi-continuous anaerobic digestion filter.

Starch processing plants usually produce diluted wastewater. Colin et al. (2007) used an anaerobic horizontal flow filter with bamboo as support to treat cassava starch extraction wastewater from a starch sedimentation basin. The anaerobic filter is based on immobilizing microorganisms on a support. At the maximum organic loading rate of 11.8 g COD/L d without the dilution of the wastewater, 87% of the COD was removed and produced 0.36 L biogas per gram COD removed at a hydraulic retention time of approximately 9.5 h. Methane content in the biogas was in the range of 69%–81%.

24.3.5 Co-Digestion

Co-digestion of wastes with different characteristics is an approach to dilute toxicants and supply the required nutrients (Mata-Alvarez et al., 2000). Mshandete et al. (2004) used a batchwise digester to co-digest sisal pulp and fish wastes for improving the digestibility of materials and biogas yield through adjusting the ratio between carbon and nitrogen. The methane yields from sisal pulp and fish waste alone were 0.32 and 0.39 m^3 CH_4/kg VS, respectively, at TS of 5%, co-digestion with 33% of fish waste, and 67% of sisal pulp representing 16.6% of TS gave a methane yield of 0.62 m^3 CH_4/kg VS added. Salminen and Rintala (1999) found that ammonia generated by the degradation of proteins and long-chain fatty acids caused the failure of the anaerobic digestion of mesophilic and thermophilic granular sewage sludge. However, co-digestion of poultry slaughterhouse wastes and wastes from a food packing plant with mesophilic digested sewage sludge produced up to 0.33 m^3/kg VS at a loading rate of 4.6 g VS/l d and HRT of 18 days. Rosenwinkel and Meyer (1999) successfully produced 0.23 m^3/kg TS from an anaerobic process for the co-digestion of slaughterhouse waste, hog, and cow stomach contents with sewage sludge in a pilot scale mesophilic digester at a loading rate of 2.9 kg TS/m^3 d and a HRT of 17 days.

Biswas et al. (2007) investigated and found that the effects of the concentrations of slurry, carbohydrate, protein, and fat on methane generation during the co-anaerobic digestion of food/vegetable residues mainly comprised various vegetable residues, oil cakes from mustard oil mills, and effluent acid whey of local sweet-meat shops. They observed that methane generation decreased with an increase in the initial slurry concentration (72–700 kg/m^3), carbohydrate concentration (ratios of carbohydrate, protein, and fat: 6.9:4.3:1–12.1:4.3:1), and protein concentration (ratios of carbohydrates, protein, and fat: 5.6:7.0:1.0–5.6:13.0:1). The methane generation increased with an increase in the fat concentration by varying the ratios of carbohydrates, protein, and fat from 7.2:10:1.6 to 7.2:10:2 and decreased with further increase in fat concentration.

Fruit and vegetable wastes and chicken manure are another promising combination to be used as co-digestion feedstock. Callaghan et al. (2002) investigated the co-digestion of fruit and vegetable wastes and chicken manure in a completely stirred tank reactor at 35°C, HRT of 21 days, and loading rate in the range of 3.19–5.01 kg VS/m^3 day. They found that by increasing the fruit and vegetable wastes from 20% to 50%, the methane yield increased from 0.23 to 0.45 m^3/kg VS added. The increase in fruit and vegetable wastes in the feedstock can decrease the ammonia inhibition from chicken manure.

24.4 ECONOMIC AND ENVIRONMENTAL IMPACTS OF ANAEROBIC DIGESTION OF FOOD PROCESSING WASTES

Anaerobic digestion effluents are generally not suitable for direct disposal on the land because they are too wet, contain some phytotoxic volatile fatty acids, and not hygenized if the digestion does not occur at a thermophilic temperature. Therefore, aerobic posttreatment after anaerobic digestion is needed (Mata-Alvarez et al., 2000). Compared to direct aerobic composting, anaerobic digestion technology is complex and requires a huge investment but it can recover an amount of energy and reduce pollutant emissions from the wastes. The economics of anaerobic digestion technology depends on energy costs. Furthermore, Mata-Alvarez et al. (2000) thought that the future of anaerobic digestion should be sought in the context of an overall sustainable waste management perspective. De Baere (1999) reported that the total emissions of volatile organic compounds, NH_3, and H_2S during combined anaerobic digestion and aerobic posttreatment were only 236 and 44 g/ton wastes, compared with 742 g/ton wastes for aerobic composting. The anaerobic digestion process can recover 99% of volatile organic compounds as biogas.

In 1985, Walker et al. analyzed the economics of a plug flow anaerobic digester for a dairy farm. Capital costs for a plug flow digester with dimensions of 22.6 m length × 6.1 m width × 2.4 m height for a biogas production capacity of 370–450 m^3/day in a dairy farm are summarized in Table 24.2. The total capital cost of $68,500 reflected the total investment for the digester system. The cost per volume of the digester was $204/m^3. The major operating costs for the digester system were propane for the digester start-up, the cost of seeding the digester when performance is poor or during start-up, the cost for maintaining the instrumentation and testing chemicals, and the labor cost. Annual propane usage varied depending on the type of problems encountered with the digester over a given year. On average, the annual operating costs were found to be between $300 and $1000.

TABLE 24.2
Capital Cost of Digester

Components	$
Digester and hoppers construction	44,300
Digester cover	2,400
Gas transport	400
Digester heat exchangers	6,800
Hopper heat exchanger	1,700
Backup heating system	3,200
Digester instrumentation	2,200
Utility building	7,000
Total	68,500

Source: Adapted from Walker, L.P., Pellerin, R.A., Heisler, M.G., Farmer, G.S., and Hills, L.A., *Energy Agr.*, 4, 347, 1985. With permission.

24.5 SUMMARY

Anaerobic digestion is a biological process to convert organic matter to a gaseous mixture of methane and carbon dioxide, which can be used as an energy source to replace fossil natural gas. Food wastes such as fruit and vegetable processing wastes, meat processing wastes, and slaughterhouse wastes have been treated anaerobically. The chemical composition of the organic wastes, the reaction intermediates, and the operating conditions are important for the performance of an anaerobic digestion process. Co-digestion of various processing wastes with different characteristics is an approach to dilute toxicants and supply required nutrients. Anaerobic digestion can recover an amount of energy and reduce pollutant emissions from the wastes. Meanwhile, the digested residues can be considered a stable organic matter for soil conditioning.

REFERENCES

Angelidaki, I. and B.K. Ahring. 1995. Establishment and characterization of an anaerobic thermophilic (55°C) enrichment culture degrading long-chain fatty acids. *Applied Environmental Microbiology* 61: 2442–2445.

Banks, C.J. 1994. Anaerobic digestion of solid and high nitrogen content fractions of slaughterhouse wastes. Environmentally responsible food processing. *AIChE Symp. Ser.* 90: 48–55.

Banks, C.J. and Z. Wang. 1999. Development of a two phase anaerobic digester for the treatment of mixed abattoir wastes. *Water Science and Technology* 40: 67–76.

Berardina, S.D., S. Costa, and A. Converti. 2000. Semi-continuous anaerobic digestion of a food industry wastewater in an anaerobic filter. *Bioresource Technology* 71: 261–266.

Biswas, J., R. Chowdhury, and P. Bhattacharya. 2007. Mathematical modeling for the prediction of biogas generation characteristics of an anaerobic digester based on food/vegetable residue. *Biomass and Bioenergy* 31: 80–86.

Boubaker, F. and B.C. Ridha. 2007. Anaerobic co-digestion of olive mill wastewater with olive mill solid waste in a tubular digester at mesophilic temperature. *Bioresource Technology* 98: 769–774.

Braber, K. 1995. Anaerobic digestion of municipal solid waste: A modern waste disposal option on the verge of breakthrough. *Biomass Bioenergy* 9: 365–376.

Callaghan, F.J., D.A.J. Wase, K. Thayanithy, and C.F. Forster. 2002. Continuous co-digestion of cattle slurry with fruit and vegetable wastes and chicken manure. *Biomass and Bioenergy* 27: 71–77.

Chang, J.I., J.J. Tsai, and K.H. Wu. 2006. Thermophilic composting of food waste. *Bioresource Technology* 97: 116–122.

Chavez, C.P., R.L. Castillo, L. Dendooven, and E.M. Escamilla-Silva. 2005. Poultry slaughter wastewater treatment with an up-flow anaerobic sludge blanket (UASB) reactor. *Bioresource Technology* 96: 1730–1736.

Colin, X., J.L. Farinet, O. Rojas, and D. Alazard. 2007. Anaerobic treatment of cassava starch extraction wastewater using a horizontal flow filter with bamboo as support. *Bioresource Technology* 98: 1602–1607.

Converti, A., M. Zilli, M. Del Borghi, and G. Ferraiolo. 1990. Fluidized bed reactor in the anaerobic treatment of wine wastewater. *Bioprocess Engineering* 5: 49–55.

De Baere, L. 1999. Anaerobic digestion of solid waste: State-of-the art. In *Proceedings of the Second International Symposium on Anaerobic Digestion of Solid Wastes* (Vol. 1), Mata-Alvarez, J., A. Tilche, and F. Cecchi, (Eds.), pp. 290–299.

Fujishima, S., T. Miyahara, and T. Noike. 1999. Effect of moisture content on anaerobic digestion of dewatered sludge: Ammonia inhibition to carbohydrate removal and methane production. In *Proceedings of the Second International Symposium on Anaerobic Digestion of Solid Wastes*, Mata-Alvarez, J., A. Tilche, and F. Cecchi. (Eds.), pp. 348–355.

Gebauer, R. 2004. Mesophilic anaerobic treatment of sludge from saline fish farm effluents with biogas production. *Bioresource Technology* 93: 155–167.

Gollakota, K.G. and K.K. Meher. 1988. Effect of particle size, temperature, loading rate and stirring on biogas production from castor cake (oil expelled). *Biological Wastes* 24: 243–249.

Gunaseelan, V.N. 1997. Anaerobic digestion of biomass for methane production: A review. *Biomass and Bioenergy* 13: 83–114.

Hansen, K.H., I. Angelidaki, and B.K. Ahring. 1999. Improving thermophilic anaerobic digestion of swine manure. *Water Resource* 33: 1805–1810.

Hobson, P.N., S. Bousfield, and R. Summers. 1981. *Methane Production from Agricultural and Domestic Wastes*. New York: John Wiley & Sons, Inc.

Hwu, C.S., B. Donlon, and G. Lettinga. 1996. Comparative toxicity of long-chain fatty acid to anaerobic sludges from various origins. *Water Science and Technology* 34: 351–358.

Kadam, P.C. and D.R. Boone. 1996. Influence of pH on ammonia accumulation and toxicity in halophilic, methylotrophic methanogens. *Applied Environmental Microbiology* 62: 4486–4492.

Kalia, V.C., V. Sonakya, and N. Raizada. 2000. Anaerobic digestion of banana stem waste. *Bioresource Technology* 73: 191–193.

Kayhanian, M. 1999. Ammonia inhibition in high-solids biogasification-an overview and practical solution. *Environmental Technology* 20: 355–365.

Knol, W., M.M. van der Most, and J. de Waart. 1978. Biogas production by anaerobic digestion of fruit and vegetable waste. A preliminary study. *Journal of Science of Food and Agriculture* 29: 822–830.

Komatsu, T., K. Hanaki, and T. Matsuo. 1991. Prevention of lipid inhibition in anaerobic processes by introducing a two-phase system. *Water Science and Technology* 23: 1189–1200.

Lane, A.G. 1984. Laboratory scale anaerobic digestion of fruit and vegetable solid waste. *Biomass* 5: 245–259.

Linke, B. 2006. Kinetic study of thermophilic anaerobic digestion of solid wastes from potato processing. *Biomass and Bioenergy* 30: 892–896.

Lo, K.V. and P.H. Liao. 1986. Psychrophilic anaerobic digestion of screened dairy manure. *Energy in Agriculture* 5: 339–345.

Mata-Alvarez, J., S. Mace, and P. Llabres. 2000. Anaerobic digestion of organic solid wastes. An overview of research achievements and perspectives. *Bioresource Technology* 74: 3–16.

McKendry, P. 2002. Energy production from biomass (part 2): Conversion technologies. *Bioresource Technology* 83: 47–54.

Mouneimne, A.H., H. Carrere, N. Bernet, and J.P. Delgenes. 2003. Effect of saponification on the anaerobic digestion of solid fatty residues. *Bioresource Technology* 90: 89–94.

Mshandete, A., A. Kivaisi, M. Rubindamayugi, and B. Mattiasson. 2004. Anaerobic batch co-digestion of sisal pulp and fish wastes. *Bioresource Technology* 95: 19–24.

Nishio, N. and Y. Nakashimada. 2007. Recent development of anaerobic digestion processes for energy recovery from wastes. *Journal of Bioscience and Bioengineering* 103: 105–112.

Pellerin, R.A., L.P. Walker, M.G. Heisler, and G.S. Farmer. 1988. Operation and performance of biogas-fueled cogeneration systems. *Energy in Agriculture* 6: 295–310.

Rosenwinkel, K.-H. and H. Meyer. 1999. Anaerobic treatment of slaughterhouse residues in municipal digesters. *Water Science and Technology* 40: 101–111.

Roy, F., G. Albagnac, and E. Samain. 1985. Influence of calcium addition on growth of highly purified syntrophic culture degradating long-chain fatty acids. *Applied Environmental Microbiology* 49: 702–705.

Salminen, E.A. and J.A. Rintala. 1999. Anaerobic digestion of poultry slaughtering wastes. *Environmental Technology* 20: 21–28.

Salminen, E. and J. Rintala. 2002. Anaerobic digestion of organic solid poultry slaughterhouse waste—a review. *Bioresource Technology* 83: 13–26.

Salminen, E., J. Rintala, L.Y. Lokshina, and V.A. Vavilin. 2000. Anaerobic batch degradation of solid poultry slaughterhouse waste. *Water Science and Technology* 41: 33–41.

Sarada, R. and R. Joseph. 1994. Studies on factors influencing methane production from tomato processing wastes. *Bioresource Technology* 47: 55–57.

Shih, J.C.H. 1987. Ecological benefits of anaerobic digestion. *Poultry Science* 66: 946–950.

Shih, J.C.H. 1993. Recent development in poultry waste digestion and feather utilization-a review. *Poultry Science* 72: 1617–1620.

Van Heerden, I., C. Cronje, S.H. Swart, and J.M. Kotze. 2002. Microbial, chemical and physical aspects of citrus waste composting. *Bioresource Technology* 81: 71–76.

Walker, L.P., R.A. Pellerin, M.G. Heisler, G.S. Farmer, and L.A. Hills. 1985. Anaerobic digestion on a dairy farm: Overview. *Energy in Agriculture* 4: 347–363.

Zinder, S.H. 1984. Microbiology of anaerobic conversion of organic wastes to methane: Recent developments. *ASM News* 50: 294–298.

25 Fermentation of Food Processing Wastes into Transportation Alcohols

25.1 INTRODUCTION

Ethanol has been a key industrial chemical for many years. Fuel ethanol, in particular, is considered as being more environmentally friendly than fossil fuels. It has been seen as a replacement for gasoline. Fuel ethanol can be produced from three main types of raw materials: sugars, starches, and lignocellulosic materials. Ethanol is currently produced from sucrose and starch. Supplies of sucrose and starch will not be sufficient to meet the feedstock demand in the ethanol industry. The gasoline market in the United States alone is 150 billion gallons per year. Replacement of methyl tertiary butyl ether (MEBE) as a gasoline additive at 6% of total gasoline volume in a short term requires approximately 10 billion gallons of ethanol per year. Production of 10 billion gallons of ethanol requires about 30% of the United States farmland currently growing corn (Dean et al., 2006).

Food processing wastes can be explored as a potentially low-cost and abundant feedstock for the production of ethanol. Some food processing wastes such as fruit and vegetable processing wastes, grain processing wastes, and sugar beet molasses are carbohydrate-rich biomass residues. Those wastes can be used as feedstocks for the production of fermentable sugars, which can be fermented to alcohols. Integration of fruit and vegetable processing, grain milling, and sugar processing facilities with an ethanol fermentation facility will be significant not only to produce food products and transportation fuel of ethanol but also to diversify high-value products whose manufacture could be scaled up or down depending on circumstances, economics, and demands. Compared with sugar and starch, plant originated food processing wastes such as fruit and vegetable processing wastes are abundant in cellulose, which is resistant to breakdown. In this chapter, the feedstocks, fermentation microorganisms, and fermentation processes for ethanol production are reviewed. Use of food processing wastes as a feedstock for ethanol production is discussed.

25.2 FEEDSTOCK FOR FERMENTABLE SUGAR

Sugarcane/beet (sucrose), starch grains (starch), and lignocellulosics (cellulose), which are all polysaccharides, are three main feedstocks for production of fermentable sugars. Food processing wastes may contain large amounts of sucrose, starch, and cellulose, which can be recovered for production of fermentable sugars. Fermentable sugars, whether derived from sucrose, starch or cellulose, will become cost competitive with petroleum-derived carbon for the production of transportation fuels. Ethanol is a promising alternative transportation fuel for a secure and independent energy supply and reduced greenhouse gas emissions. In the updated Vision for Bioenergy and Bio-based Products in the United States, published by the Biomass Research and Development Initiative (BRDI), a goal was set for the United States to increase the market share of 1.2% of biofuel in 2004 to 20% by 2030 (BRDI, 2006).

25.2.1 Sucrose

The simplest and currently most cost-competitive system to produce ethanol uses sucrose from sugar beet and cane as the feedstock. Sugars are first extracted from the feedstock. Sucrose is second only to cellulose in availability and the current output far exceeds all other commercial carbohydrates combined. It is estimated that only 1.7% of the annual sucrose production goes to nonfood uses. More than 50% of ethanol produced in the world today has sucrose as the feedstock (Dean et al., 2006).

25.2.2 Starch

Several grains such as corn, wheat, sorghum, rice, oats, and cassava contain around 60%–75% starch on a dry basis. Starch-based ethanol production has become more cost competitive as a result of innovations in farming and grain milling technologies. The wet and dry milling processes have widely been used to process grains into different food and energy products. In grain wet milling, the grain kernel such as corn is processed through a series of steps into a starch or glucose stream and co-products such as oil, protein, fiber, and other nutrients. The glucose can be further converted to ethanol and organic acids such as lactic acid by fermentation. In grain dry milling, many of the initial steeping and extraction steps used in grain wet milling are removed to reduce the complexity and capital cost of the milling facilities. The starch in a grain dry mill is enzymatically converted to glucose and fermented to ethanol with yeast. Compared to a grain wet mill, the co-products of distillers' dried grains with solubles or DDGS from a grain dry mill have a relatively low value (Dean et al., 2006).

25.2.3 Lignocelluloses

Supplies of sucrose and starch feedstocks for ethanol production will not be sufficient to meet the transportation ethanol demand. Among the three main types of raw materials, lignocellulosic materials represent the most abundant global source of biomass and still have not been largely used. It has become possible to produce ethanol from nonfood lignocellulosic biomass resources such as food processing wastes and agricultural residues with the recent advances in biotechnology, particularly in the area of enzyme and fermentation technology.

Considering the large amount of cellulosic biomass available for saccharification to fermentable sugars, there is a clear opportunity to develop commercial processes for the production of ethanol at a very large volume and low price from cellulosic biomass. However, unlike sucrose and starch, which can be converted to fermentable sugar with relative ease, cellulosic biomass is a complicated structure of cellulose, hemicellulose, lignin, and numerous minor components as a support element of plants. The two structural carbohydrates of cellulose and hemicellulose, which together make up as much as 65%–80% of the dry matter of lignocellulosic biomass, can be converted into sugars such as glucose (40%–50% of carbohydrates) and xylose (15%–20% of carbohydrates) (Roberto et al., 2003).

The number of glucose residues of a cellulose molecule can exceed 15,000. Cellulose has extensive hydrogen bonding. Only agents that can attack the glycosidic linkages between the glucose residues or which can disrupt the hydrogen bonding can solubilize cellulose. The carbohydrate polymers are tightly bound to lignin by hydrogen and covalent bonds. A bioprocess for conversion of the lignocellulosic biomass to ethanol involves three main reaction steps with the assistance of chemical catalysts or biocatalysts:

- Delignification of biomass to liberate cellulose and hemicellulose from their complex matrix with lignin
- Depolymerization of cellulose and hemicellulose to produce free sugars
- Fermentation of mixed hexose and pentose sugars to produce ethanol

The process for conversion of cellulosic biomass to fermentable sugars faces major technical and engineering challenges that have so far prevented the large-scale commercial use of cellulose as a source of fermentable sugar (Dean et al., 2006). The improvement has come from more effective pretreatment, reducing the cost for the production of enzymes, reducing the enzyme requirements, simultaneous saccharification and fermentation, and effective utilization of C5 sugars derived from hemicellulose components.

25.2.4 Fermentable Carbon Sources from Food Processing Wastes

Fermentation of sugars into commercial ethanol with microorganisms has been successful for many years. However, these fermentation processes used a relatively clean sugar stream from starch or sucrose that contains few impurities. Food processing wastes, such as fruit and vegetable wastes, oil cakes from oil mills, by-products from cereal mills, and whole crop refineries, are rich in lignocellulose. Lignocellulosic biomass consists of 65% to 85% carbohydrates, mainly polysaccharides.

After pretreatment and hydrolysis of food processing wastes, a mixture of hexose and pentose sugars, several degradation by-products such as furfural, hydroxymethylfurfural, phenols, formic, acetic and other organic acids, and salts may be obtained. Several of these compounds are well-known inhibitors to fermentation processes. It is necessary to develop microorganisms capable of performing better at a low pH and high temperature, under a high concentration of carbohydrates, and resisting the by-products generated during pretreatment and hydrolysis (Gonzalez et al., 2003).

25.3 MICROORGANISMS FOR ETHANOL PRODUCTION

25.3.1 Microorganisms

Microbes can be essentially divided into two categories: the prokaryotes and the eukaryotes. The prokaryotes, which embrace the bacteria, are substantially simple in their structure. They comprise a protective cell wall, surrounding a plasma membrane, within which is a nuclear region immersed in cytoplasm. The nuclear materials, DNA, are the genetic blueprint of the cell. The cytoplasm contains the enzymes that catalyze the reactions necessary for growth, survival, and reproduction of the organisms. The membrane regulates the entry and exit of materials into and from the cell. A eukaryotic cell, such as yeast, is substantially more complex. It is divided into organelles, the intracellular equivalent of our human organs. Each has its own function. The DNA is located in the nucleus, which is bounded by a membrane like all the organelles. Other major organelles in eukaryotes are the mitochondria, wherein energy is generated, and the endoplasmic reticulum, which is an interconnected network of tubules, vesicles, and sacs with various functions including protein and sterols synthesis, sequestration of calcium, production of the storage polysaccharide glycogen, and insertion of proteins into membranes. All the membranes in the eukaryotes and the prokaryotes contain lipid and protein. Both prokaryotes and eukaryotes have polymeric storage materials located in their cytoplasm (Bamforth, 2005).

Three main categories of microorganisms are used in the ethanol industry for hydrolysis of starch, hydrolysis of cellulose, and ethanol fermentation. Hydrolysis of starch requires amylase and glucoamylase. Amylase can be synthesized by microorganisms such as *Aspergillus niger.* Food processing facilities produce large quantities of wastewaters with a high chemical oxygen demand (COD) and biochemical oxygen demand (BOD). For example, the untreated effluent of a brewery facility has a BOD of 500–2600 mg/L and COD of 780–3500 mg/L, and untreated wastes from a meat processing plant has a BOD of 600–3000 mg/L and COD of 800–4000 mg/L. These wastewaters supplemented with starch can be used to produce amylase by *Aspergillus niger* (Hernandez et al., 2006). The potential microorganisms for lignocellulosic biomass hydrolysis will include *Trichoderma reesei, Clostridium thermocellum*, and recombinant *Escherichia coli* (Lin and Tanaka, 2006).

The potential microorganisms for ethanol fermentation will include *Saccharomyces cerevisiae, Zymomonas mobilis, C. thermocellum*, and recombinant *E. coli. S. cerevisiae*, also called brewer's yeast or baker's yeast, is a facultative anaerobe. *S. cerevisiae* can produce ethanol to a concentration as high as 18% of the fermentation broth by volume. The yeast can use both a simple sugar of glucose and disaccharide of sucrose. *S. cerevisiae* is a generally recognized as safe (GRAS) microorganism. *Z. mobilis* is an ethanol-producing bacterium widely known for its high ethanol yield (up to 1.9 mol ethanol/mol glucose) and productivities. *Z. mobilis* is also a GRAS microorganism (Yanase et al., 2002; Lin and Tanaka, 2006)). Also *Z. mobilis* tolerates high sugar and ethanol concentrations (Rogers et al., 1982). However, *S. cerevisiae* is still preferred by the industry because of its hardiness. Recombinant *E. coli* is another microorganism that can be used to produce ethanol. It can ferment a wide spectrum of sugars including both hexoses and pentoses. The major disadvantages with *E. coli* are a narrow and neutral pH growth range (6–8), less hardy cultures than yeast, and public perceptions regarding the safety of *E. coli* (Lin and Tanaka, 2006).

25.3.2 Nutritional Needs

The four elements required by microorganisms in the largest quantity are carbon, hydrogen, oxygen, and nitrogen. These elements are the main constituents of the key cellular components of carbohydrates, lipids, protein, and nucleic acids. Phosphorus, sulfur, calcium, magnesium, potassium, sodium, iron, and other minerals and vitamins are also needed for their well-being.

Before the microorganisms are used as biocatalysts for hydrolysis and fermentation, they are cultivated to increase the cell number using a well-characterized culture media. Popular culturing media include yeast extract and glucose solution (Neves et al., 2007). Micronutrients such as $(NH_4)_2SO_4$, KH_2PO_4, $FeSO_4{\cdot}7H_2O$, $MgSO_4{\cdot}7H_2O$, $MnCl_2{\cdot}2H_2O$, and mycological peptone should be added into the culturing media. Yeast extract could be a good source for these micronutrients.

25.3.3 Environmental Needs

A range of physical, chemical, and physicochemical parameters have impacts on the growth of microorganisms. These parameters may include temperature, pH, water activity, oxygen, radiation, pressure, and static agents. Cellular macromolecules, especially enzymes, are prone to denaturation by heat, which limits the temperatures that the enzyme can tolerate. On the other hand, at a very low temperature, lipids in the membranes of cells cannot sufficiently flow. The microorganisms that can thrive at a relatively high temperature (i.e., >40°C) are called thermophiles while those that are capable of growth at a very low temperature (i.e., <10°C) are called psychrophiles. The microorganisms that are in between thermophiles and psychrophiles are referred to as mesophiles. Most microorganisms have a relatively narrow range of pH for the best growth. Most microbes contain between 70% and 80% water. Maintaining this level is a challenge when the organisms are exposed to environments that contain too little water or excess water. Microbes differ substantially in their requirement for oxygen. Obligate aerobes must have oxygen as the terminal electron acceptor for their aerobic growth while obligate anaerobes are killed by any oxygen. Facultative anaerobes can use oxygen as a terminal electron acceptor but they can also function in the absence of oxygen (Bamforth, 2005).

25.4 ETHANOL FERMENTATION PROCESS

25.4.1 Overview of Ethanol Production Process

Fuel ethanol is currently produced from starch and sugar cane. There are two existing technologies in the starch-based ethanol industry: dry mills and wet mills. Most existing dry corn mills are small with capacities ranging from 5–30 million gallons per year. Dry mills produce one composite by-product: distillers' dried grains containing the residual protein, oil, and fiber after the carbohydrate in the grain kernel

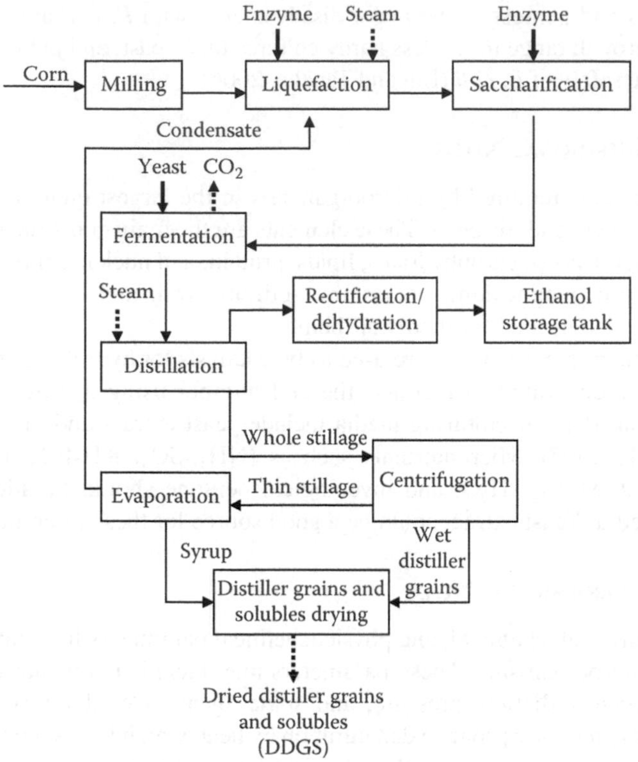

FIGURE 25.1 Typical dry-grind ethanol process.

is removed for ethanol production. The crude protein content of dried distillers' grains and dried distillers' grains and solubles ranges from 28.7% to 31.6%.

The wet mills are usually very large with capacities ranging from 50 to 330 million gallons per year. Wet mills are more complex than dry mills. By-products are separated into corn gluten feed with about 20% protein, corn gluten meal with about 60% protein, and edible oil. Many wet mills also produce sweeteners. Average capital costs are higher for wet mills (Gallagher et al., 2005). A typical dry-grind ethanol process is given in Figure 25.1 and a wet milling process is given in Chapter 11 (Figure 11.1).

25.4.2 Starch Hydrolysis

Starch is usually liquefied with α-amylase at a high temperature (e.g., 95°C) for a couple of hours and then further hydrolyzed into glucose with glucoamylase at a lower temperature (e.g., 55°C) for several hours (Montesinos and Navarro, 2000). Starch is first hydrolyzed by adding α-amylase to avoid gelatinization and then cooked at a high temperature. The liquefied starch is hydrolyzed to glucose with the glucoamylase at a lower temperature (Lin and Tanaka, 2006).

25.4.3 Pretreatment and Hydrolysis of Lignocellulosic Biomass

Plants are the main source of biomass. The lignocellulosic biomass, which is an intertwined network of cellulose, hemicellulose, and lignin, is resistant to breakdown. Enzymatic methods have been increasingly used for the hydrolysis of cellulose into glucose. Although enzymatic hydrolysis works successfully to produce glucose when applied to pure cellulose, without any pretreatment, only approximately 20% of the cellulose can be hydrolyzed to glucose with high doses of enzymes (Gharpuray et al., 1983). Hemicellulose is more easily depolymerized than the cellulose fraction in lignocellulosic biomass. To convert biomass into fermentable sugars, the purpose of the pretreatment is to disorder or remove cellulose, hemicellulose, and lignin interactions under the action of stress, temperature, pressure, and pH and improve the accessibility of hydrolytic enzymes to the sugar polymers of cellulose (Teter et al., 2006).

The physical pretreatment includes a reduction in the size of biomass material by milling, crushing and chopping. Lignin in the biomass not only acts as a block to enzyme action by coating cellulose microfibrils but also interferes with enzymatic hydrolysis by directly absorbing some cellulose active enzymes. After physical pretreatment, some chemical treatments may be used to generate cellulose with both improved solvent accessibility and reduced lignin interference with enzyme action. The chemical treatments include dilute acid pretreatment, ammonia fiber explosion, hot-water/steam pretreatment, and wet oxidation (Mosier et al., 2005; Karimi et al., 2006; Teter et al., 2006).

Most of the existing pretreatment technologies use extreme conditions (chemicals such as strong acids, elevated temperature, and pressures) to maximize the release of sugars, mainly glucose, from biomass for ethanol fermentation. However, the extreme conditions also result in sugar degradation, inhibitor formation, and the degradation of other non-fermentable components in the lignocellulosic biomass. The existing pretreatment technologies were designated mainly to increase the enzymatic hydrolysis of the cellulose component for glucose production, while neglecting and decreasing the economic values of other components such as hemicellulose-derived sugars (mainly xylose) and lignin. The chemical and thermochemical pretreatment process usually causes partial degradation of the hemicellulose-derived sugars to furfural and hydroxymethylfurfural (HMF) and significant solubilization and transformation of lignin to phenolics. These derived chemicals in pretreatment have been shown to inhibit fermentation of the biomass-derived sugars to ethanol. These inhibitors must be removed (detoxified) if the biomass-derived sugars are used in a biological process. Reducing the chemicals' uses and lowering the temperature of the pretreatment process can significantly reduce the degradation of hemicellulose and lignin and thus the generation of inhibitors. Research has shown that when the conditions of steam-SO_2 explosion pretreatment were increased from low severity (175°C, 4.5% SO_2, 7.5 min) to high severity (215°C, 2.38% SO_2, 2.38 min), the concentration of glucose hydrolyzed from cellulose increased, whereas the recoverable xylose hydrolyzed from hemicellulose decreased from 87% to 36%, and furfural and HMF derived from xylose increased from 0.5% to 2.0% and 1.8%, respectively (Boussiad et al., 1999). Therefore, novel pretreatment methods such as mechanical cavitation should be developed (Li et al., 2004).

After pretreatment, cellulase is used to hydrolyze sugars from cellulose. Cellulase is a general term encompassing a diverse set of enzymes that participate in the hydrolysis of cellulose into glucose. An effective industrial cellulase should include enzymes that can perform multi-tasks and the enzyme function should be collaborative and synergistic. Production of fermentable sugars from cellulose is currently more expensive than their production from starch. This is because amylase hydrolysis of starch is intrinsically faster and the kinetics of cellulase action requires relatively higher loading of enzymes.

25.4.4 Ethanol Fermentation

Ethanol is produced by fermentation with yeasts or bacteria. The well-known microorganisms for ethanol fermentation include *S. cerevisiae* and *Z. mobilis* (Delgenes et al., 1996). Yeasts (*S. cerevisiae*) have been used to produce ethanol from sucrose and glucose for thousands of years. Yeasts are grown in propagation tanks and are added to the broth at appropriate temperatures. New microorganisms are searched by genetic engineering to enlarge their substrate spectrum for ethanol production. The genetically modified *Z. mobilis* has the ability to use pentose such as xylose in addition to hexose such as glucose and fructose (Zhang et al., 1995). *E. coli* KO11 is another recombinant bacterial strain developed to ferment arabinose, xylose, and hexoses to ethanol (Ohta et al., 1990; Beall et al., 1991). Other microorganisms for xylose fermentation include *Pichia stipitis* and *Candida shehatae* (Delgenes et al., 1996). The fermented broth or beer is distilled to recover ethanol. The distillery waste, effluent, or spillage is centrifuged to recover the yeast cell fragments and the water is usually spray irrigated on forage crops. The solids in the spillage can be recovered with ultrafiltration, reverse osmosis, and nanofiltration while the treated water can be recycled.

25.5 ADVANCED CONCEPTS IN ETHANOL FERMENTATION

25.5.1 Simultaneous Saccharification and Fermentation

The performance of yeast and bacteria for ethanol fermentation may be inhibited by the product of ethanol. During simultaneous saccharification and fermentation, the sugars released during hydrolysis are promptly converted into ethanol. Therefore, the sugar level remains relatively stable. In simultaneous saccharification and fermentation, hydrolysis and fermentation occur simultaneously in the same vessel, and the end-product inhibition of the enzymes is relieved because the fermenting organisms immediately consume the released sugars. Process integration via simultaneous saccharification and fermentation can also reduce the capital cost (Hahn-Hagerdal et al., 2006; Kroumov et al., 2006; Lin and Tanaka, 2006). The challenges of simultaneous saccharification and fermentation are to develop genetically modified microorganisms that can perform both hydrolysis and fermentation and to find the best operating condition for both hydrolysis and fermentation, particularly in the case of two sets of microorganisms being used for hydrolysis and fermentation, respectively.

25.5.2 Solid-State Fermentation

Solid-state fermentation of food processing wastes into value-added products such as ethanol, methane, lactic acid, and citric acid has been of major interest to many researchers in recent years. Solid-state fermentation can deal with the utilization of water-insoluble materials for microbial growth and metabolism. It is usually carried out in solid or semisolid systems in the near absence of free water or reduced water content compared with traditional fermentation. Solid fermentation has already been investigated to produce ethanol from food processing wastes (Zheng and Shetty, 1998; Laufenberg et al., 2003). For the same ethanol yield, the investment cost of a solid-state fermentation reactor is only one-third of the cost of a traditional fermentation reactor (Laufenberg et al., 2003).

25.6 EXAMPLES OF ETHANOL FERMENTATION FROM FOOD PROCESSING WASTES

25.6.1 Ethanol Fermentation from Grain Milling By-Products

Grain milling may generate a large amount of by-products such as bran, which contain low-grade flours. The bran consists of three main components: residual starch, hemicellulose, and cellulose. Wheat bran usually accounts for 15%–20% of the grain. Hydrolysis of starch is a well-understood process. It is still a problem to hydrolyze hemicellulose and cellulose. A yield of 52.1 g sugar /100 g of starch-free bran residue can be achieved from a bran slurry at a solid concentration of 5% using 1% of sulfuric acid at 130°C for 40 min. The furfural and 5-hydroxy-methyl-2-furaldehyd, which inhibit the ethanol fermentation, are only 0.28 g/L and 0.05 g/L, respectively (Choteborska et al., 2004).

Wheat milling by-products at the breaking rolls and size reduction system, which were generally used as a supplement in animal feed (Neves, 2006), have been used as a substrate for bioethanol production. The major components of the by-products are 15.6% starch (w/w), 15% protein, 14% moisture, 2.7% ash, and 0.8% fiber. Slurries containing low-grade wheat flour at 100, 200, and 300 g/L were used for simultaneous saccharification and fermentation with *Z. mobilis*. Mashes containing 200 g low-grade wheat flour/liter produced about 52 g ethanol/liter with a productivity of 2.17 g/L h. Meanwhile, it was found that using *Z. mobilis* for the fermentation of wheat milling by-products, the ethanol concentration was about 30% higher than that obtained with *S. cerevisiae* (Neves et al., 2007).

Effluent from starch processing facilities can be used as a feedstock for ethanol production. Starch processing residue is a commercially accepted feedstock for industrial ethanol. Typically, wheat milling results in 78%–80% flour, 19% bran, and 1% wheat germ. The flour is washed with water to produce starch. As the effluent still contains some residual starch, it can be converted to ethanol by fermentation (Nguyen, 2003).

Corn is used to produce ethanol in the United States. The current ethanol production capacity is more than 5 billion gallons per year in the United States. The ethanol production facilities simultaneously co-produce 7 million tons of distillers' grains per year, which are mainly used as a cattle feed supplement. Distillers' grains with a higher protein and lower fiber content would allow to penetrate the swine and poultry feed markets. One way to increase the protein content of the distillers' grain is to convert the residual starch and fiber into additional ethanol. Pretreatment of corn distillers' grains with 3.27% H_2SO_4 at 140°C for 20 min can convert 77% of available carbohydrate in the distillers' grain into soluble sugars. Using *S. cerevisiae* D5A, the yield of ethanol is 73% of theoretical value from available glucans (Tucker et al., 2004).

25.6.2 Ethanol Fermentation from Vegetable and Fruit Processing Wastes

Most of the carbohydrates present in processing fruit and vegetable solid and liquid wastes are composed of soluble sugars and easily hydrolysable polysaccharides. There are two waste streams of solid wastes and effluent wastewater from fruit and vegetable processing facilities, which can be used to produce fermentable sugars. In the United States, the fruit and vegetable processing industry generates approximately 11 million tons of by-product wastes along with over 4.3×10^{11} L of effluent wastewaters. Discrete solids include culls, leaves, trimmings, stems, peels, pods, husks, cobs, silk, and defective parts of processed vegetables and fruits. Fruit and vegetable processing wastes have not been used as a feedstock for ethanol production though they have favorable features to be used as an ethanol feedstock. The discrete solids are usually dried, palletized, and sold as low-value animal feeds. The production of animal feeds from discrete solid wastes consumes large amounts of energy. The prices of animal feeds often are not high enough to cover the operating costs (Wilkins et al., 2007). Research has already been carried out to develop high-value energy products such as fuel ethanol from fruit and vegetable processing wastes.

Apple pomace is the main by-product of apple cider and juice processing industries and accounts for about 25% of the original fruit mass. Apple pomace typically contains between 66.4%–78.2% moisture and 9.5%–22.0% carbohydrates. Fermentable sugars in apple pomace such as glucose, fructose, and sucrose can be converted to ethanol (Hang et al., 1981; Hang, 1987; Ngadi and Correia, 1992). Ethanol yields on glucose and yeast cells are from 0.33 to 0.37 g/g and from 2.6×10^{-8} g/colony forming unit (CFU) to 3.14×10^{-8} g/CFU, respectively. As the ethanol concentration increases from 1% to 18% g/g (dry weight), the specific growth rate decreases to almost zero and ethanol inhibits the performance of the yeast (Ngadi and Correia, 1992). During grapefruit processing, about half of the fruit is expressed as juice with the rest being peel waste consisting of peels, seeds, and segment membranes (Braddock, 1999). Cellulose, pectin, and hemicellulose in grapefruit peel waste can be hydrolyzed by pectinase and cellulase enzymes to monomer sugars, which can be used by microorganisms to produce ethanol and other fermentation products. Wilkins et al. (2007) found that 5 mg pectinase/g peel dry matter and 2 mg cellulase/g peel dry matter supplemented with 2.1 mg β-glucosidase/g peel dry matter at an optimum pH value of 4.8 and temperature of 45°C were the lowest loadings to yield the most glucose from the grapefruit peel waste.

It was reported that the reverse osmosis method could recover 1.42 million tons of fermentable sugars at a 20% sugar concentrate from the 4.3×10^{11} L of effluent wastewaters in the fruit and vegetable processing facilities in the United States. The potential for the production of ethanol from the recovered sugars was between 750 and 900 million liters per year. Furthermore, the recovery of sugars from wastewater could reduce the biological oxygen demand of the wastewater and thus reduce the disposal cost (Blondin et al., 1983). Because a reverse osmosis process is relatively expensive, excessively dilute wastewater that contains less than 0.2% sugars is too costly to be processed (Blondin et al., 1983).

25.6.3 Ethanol Fermentation from Sugar Processing Wastes

Molasses is a by-product of the sugar industry. Beet and cane molasses have been used to produce ethanol (Moriya et al., 1989; Patil et al., 1989; Doelle and Doelle, 1990; Cachot and Pons, 1991; Doelle et al., 1991; Roukas, 1996). Roukas (1996) investigated the production of ethanol from non-sterilized beet molasses by free and immobilized *S. cerevisiae* cells in batch and fed-batch culture. In the fed-batch culture, both free and immobilized *S. cerevisiae* cells gave the same maximum ethanol concentration at 53 g/L at the initial sugar concentration of 250 g/L and feeding rate of 250 mL/h. The sugar utilization rates were 81% and 73.3% for free cells and immobilized cells, respectively.

25.6.4 Ethanol Fermentation from Cheese Whey

Among the food processing wastes, cheese whey may be the largest source of biomass suitable for the production of alcohol. Whey is one of the troublesome by-products in the dairy industry. Fluid whey contains about 6%–6.5% solids. Approximately 13 million tons of fluid whey is produced annually in the United States. The main components of dried whey powder include 42%–45% lactose (disaccharides), 20%–23% of protein, 10% of Na, K, and Ca, and 25% of ash (Fischer and Bipp, 2005). The yeast of *S. cerevisiae,* which is the microorganism used in the existing ethanol industry, lacks the ability to assimilate lactose. Co-immobilized β-galactosidase and yeast of *S. cerevisiae* were used to produce ethanol from whey (Staniszewski et al., 2007).

25.7 SUMMARY

Fuel ethanol has been used as a promising alternative to petroleum-based transportation fuels. The most common renewable sources for production of fuel ethanol are sugarcane/sugar beet and grains. However, these resources are also used to produce foods. By-products generated in food processing facilities represent one of the important biomass sources that have a potential to be converted into ethanol. Examples of these by-products for ethanol production may include low-grade starch from grain milling facilities, waste streams from fruit and vegetable processing facilities, wastes from sugar, and whey from dairy processing sectors. Conversion of the waste streams in the food processing facilities not only adds revenue to the facilities but also reduces the disposal costs of these wastes.

REFERENCES

Bamforth, C.W. 2005. *Food, Fermentation and Micro-organisms.* Ames: Blackwell Science Ltd.

Beall, D.S., K. Ohta, and L.O. Ingram. 1991. Parametric studies of ethanol production from xylose and other sugars by recombinant *Escherichia coli. Biotechnology and Bioengineering* 38: 296–303.

Biomass Research and Development Initiative (BRDI). 2006. *Vision for Bioenergy and Biobased Products in the United States.*

Blondin, G.A., S.J. Comiskey, and J.M. Harkin. 1983. Recovery of fermentable sugars from process vegetable wastewaters. *Energy in Agriculture* 2: 21–36.

Boussaid, A., J. Robinson, Y.J. Cai, D.J. Gregg, and J.N. Saddler. 1999. Fermentability of the hemicellulose-derived sugars from steam-exploded softwood (Douglas-fir). *Biotechnology and Bioengineering* 64: 284–289.

Braddock, R.J. 1999. *Handbook of Citrus By-Products Processing Technology.* New York: John Wiley and Sons, Inc.

Cachot, T. and M. Pons. 1991. Improvement of alcoholic fermentation on cane and beet molasses by supplementation. *Journal of Fermentation Bioengineering* 71: 24–27.

Choteborska, P., B. Palmarola-Adrados, M. Galbe, G. Zacchi, K. Melzoch, and M. Rychtera. 2004. Processing of wheat bran to sugar solution. *Journal of Food Engineering* 61: 561–565.

Dean, B., T. Dodge, F. Valle, and G. Chotani. 2006. Development of biorefineries-technical and economic considerations. In: *Biorefineries—Industrial Processes and Products, Status Quo & Future Directions*, Vol. I, Kamm, B., P. Gruber, and M. Kamm (Eds.), pp. 67–83. Weinheim: Wiley-YCH Verlag GmbH & Co. kGaA.

Delgenes, J.P., R. Moletta, and J.M. Navarro. 1996. Effects of lignocellulose degradation products on ethanol fermentations of glucose and xylose by *Saccharomyces cerevisiae, Zymomonas mobilis, Pichia stipitis* and *Candida shehatae. Enzyme and Microbial Technology* 19: 220–225.

Doelle, H.W., L.D. Kennedy, and M.B. Doelle. 1991. Scale-up of ethanol production from sugar cane using *Zymomonas mobilis. Biotechnology Letter* 13: 131–136.

Doelle, M.B. and H.W. Doelle. 1990. Sugar-cane molasses fermentation by *Zymomonas mobilis. Applied Microbiology and Biotechnology* 33: 31–35.

Fischer, K. and H.P. Bipp. 2005. Generation of organic acids and monosaccharides by hydrolytic and oxidative transformation of food processing residues. *Bioresource Technology* 96: 831–842.

Gallagher, P.W., H. Brubaker, and H. Shapouri. 2005. Plant size: Capital cost relationships in the dry mill ethanol industry. *Biomass and Bioenergy* 28: 565–571.

Gharpuray, M.M., Y.H. Lee, and L.T. Fan. 1983. Structural modification of lignocellulosics by pretreatments to enhance enzymatic hydrolysis. *Biotechnology and Bioengineering* 25: 157–172.

Gonzalez, R., H. Tao, J. Purvis, S. York, K. Shanmugam, and L. Ingram. 2003. Gene array-based identification of changes that contribute to ethanol tolerance in ethnaologenic *Escherichia coli*: Comparison of KO11 (parent) to LY01 (resistant mutant). *Biotechnol. Prog.* 19: 612–623.

Hahn-Hagerdal, B., M. Galbe, M.F. Gorwa-Grauslund, G. Liden, and G. Zacchi. 2006. Bio-ethanol-the fuel of tomorrow from the residues of today. *Trends in Biotechnology* 24: 549–556.

Hang, Y.D. 1987. Production of fuels and chemicals from apple pomace. *Food Technology* 41: 115–117.

Hang, Y.D., C.Y. Lee, E.E. Woodams, and H.J. Cooley. 1981. Production of alcohol from apple pomace. *Applied Environmental Microbiology* 42: 1128–1129.

Hernandez, M.S., M.R. Rodriguez, N.P. Guerra, and R.P. Roses. 2006. Amylase production by *Aspergillus niger* in submerged cultivation on two wastes from food industries. *Journal of Food Engineering* 73: 93–100.

Karimi, K., S. Kheradmandinia, and M.J. Taherzadeh. 2006. Conversion of rice straw to sugars by dilute-acid hydrolysis. *Biomass and Biotechnology* 30: 247–253.

Kroumov, A.D., A.N. Modenes, and M.C. de Araujo Tait. 2006. Development of new unstructured model for simultaneous saccharification and fermentation of starch to ethanol by recombinant strain. *Biochemical Engineering Journal* 28: 243–255.

Laufenberg, G., B. Kunz, and M. Nystroem. 2003. Transformation of vegetable waste into value added products: (A) the upgrading concept; (B) practical implementations. *Bioresource Technology* 87: 167–198.

Li, Y., R.R. Ruan, P.L. Chen, X.J. Pan, and Z. Liu. 2004. Enzymatic saccharification of corn stover pretreated by combined diluted alkaline and homogenization. *Transactions of the ASAE* 47: 821–825.

Lin, Y. and S. Tanaka. 2006. Ethanol fermentation from biomass resources: Current state and prospects. *Applied Microbiology and Biotechnology* 69: 627–642.

Montesinos, T. and J.M. Navarro. 2000. Production of alcohol from raw wheat flour by amyloglucosidase and *S. cerevisiae. Enzyme Microb. Technol.* 27: 362–370.

Moriya, K., H. Shimoi, S. Sato, K. Saito, and M. Tadenuma. 1989. Ethanol fermentation of beet molasses by a yeast resistant to distillery waste water and 2-deoxy-glucose. *Journal of Fermentation Bioengineering* 67: 321–323.

Mosier, N., R. Hendrickson, N. Ho, M. Sedlak, and M.R. Ladisch. 2005. Optimization of pH controlled liquid hot water pretreatment of corn stover. *Bioresource Technology* 96: 1986–1993.

Neves, M.A., T. Kimura, N. Shimizu, and K. Shiiba. 2006. Production of alcohol by simultaneous saccharification and fermentation of low-grade wheat flour. *Brazilian Arch. Biol. Technol.* 49: 481–490.

Neves, M.A.D., N. Shimizu, T. Kimura, and K. Shiiba. 2007. Kinetics of bioethanol production from wheat milling by-products. *Journal of Food Process Engineering* 30: 338–356.

Ngadi, M.O. and L.R. Correia. 1992. Kinetics of solid-state ethanol fermentation from apple pomace. *Journal of Food Engineering* 17: 97–116.

Nguyen, M.H. 2003. Alternatives to spray irrigation of starch waste based distillery effluent. *Journal of Food Engineering* 60: 367–374.

Ohta, K., F. Alterthum, and L.O. Ingram. 1990. Effect of environmental conditions on xylose fermentation by recombinant *Escherichia coli. Applied Environmental Microbiology* 56: 463–465.

Patil, S.G., D.V. Gonhale, and B.G. Patil. 1989. Novel supplements enhance the ethanol production in cane molasses fermentation by recycling yeast cell. *Biotechnology letter* 11: 213–216.

Roberto, I.C., S.I. Mussatto, and R.C.L.B. Rodrigues. 2003. Dilute-acid hydrolysis for optimization of xylose recovery from rice straw in a semi-pilot reactor. *Industrial Crops and Products* 7: 171–176.

Rogers, P.L., K.J. Lee, M.L. Skotnichi, and D.E. Tribe. 1982. Ethanol production by Z. mobilis. In: *Advances in Biochemical Engineering*, Fiechter, A. (Ed.), pp. 37–84. Berlin, Germany: Springer.

Roukas, T. 1996. Ethanol production from non-sterilized beet molasses by free and immobilized *Saccharomyces cerevisiae* cells using fed-batch culture. *Journal of Food Engineering* 27: 87–96.

Staniszewski, M., W. Kujawski, and M. Lewandowska. 2007. Ethanol production from whey in bioreactor with co-immobilized enzyme and yeast cells followed by pervaporative recovery of product-kinetic model predictions. *Journal of Food Engineering* 82: 618–625.

Teter, S.A., F. Xu, G.E. Nedwin, and J.R. Cherry. 2006. Enzymes for biorefineries. In: *Biorefineries-Industrial Processes and Products, Status Quo & Future Directions*, Vol. I, Kamm, B., P. Gruber, and M. Kamm (Eds.), pp. 357–83. Weinheim: Wiley-YCH Verlag GmbH & Co. kGaA.

Tucker, M.P., N.J. Nagle, E.W. Jennings, K.N. Ibsen, A. Aden, Q.A. Nguyen, K.H. Kim, and S.L. Noll. 2004. Conversion of distiller's grain into fuel alcohol and a higher-value animal feed by dilute-acid pretreatment. *Applied Biochemistry and Biotechnology* 1139: 113–116.

Wilkins, M.R., W.W. Widmer, K. Grohmann, and R.G. Cameron. 2007. Hydrolysis of grapefruit peel waste with cellulase and pectinase enzymes. *Bioresource Technology* 98: 1596–1601.

Yanase, H., M. Maeda, E. Hagiwara, H. Yagi, K. Taniguchi, and K. Okamoto. 2002. Identification of functionally important amino acid residues in *Zymomonas mobilis* levansucrase. *Journal of Biochemistry* 132: 565–572.

Zhang, M., C. Eddy, K. Deanda, M. Finkelstein, and S. Picataggio. 1995. Metabolic engineering of a pentose metabolism pathway in *Zymomonas mobilis*. *Science* 267: 240–243.

Zheng, Z. and K. Shetty. 1998. Cranberry processing waste for solid state fungal inoculant production. *Process Biochemistry* 33: 323–329.

26 Biodiesel Production from Waste Oils and Fats

26.1 INTRODUCTION

Biodiesel has become more attractive recently because it has low pollutant emission and it is made from renewable resources such as vegetable oils and animal fats. Crude vegetable oils and animal fats mainly consist of triacylglycerides, which consist of three long-chain fatty acids esterified to a single glycerol molecule. They usually have high viscosity and low volatility, which lead to severe engine deposits, injector coking, and piston ring sticking (Canakci, 2007). There are four primary ways to make biodiesel, direct use and blending, microemulsions, thermal cracking or pyrolysis, and transesterification. The most commonly used method to produce biodiesel from crude vegetable oils and animal fats is to transesterify vegetable oils and animal fats with light alcohols such as methanol and ethanol to an alkyl ester for the decrease of their viscosity and increase of their volatility (Ma and Hanna, 1999). Alkali catalysts such as sodium hydroxide, potassium hydroxide, and sodium methoxide are the most commonly used in traditional biodiesel production. For alkali-catalyzed transesterification of triglycerides, the starting materials including the oil or fat and light alcohol must meet certain specifications including low free fatty acid and moisture contents to prevent the saponification reaction (Ma and Hanna, 1999).

The main challenges for traditional biodiesel production from edible oils and fats are its high cost and limited availability of fat and oil resources. The cost of raw materials of edible oils and fats accounts for 60% to 75% of the total cost of biodiesel fuel (Krawczyk, 1996). Waste cooking oils, restaurant grease, and animal fats are potential feedstocks for biodiesel production. The use of waste oils and fats can significantly lower the cost of biodiesel (Zhang et al., 2003a). Production of biodiesel from waste oils and fats can also help to solve their disposal problem. However, waste oils and fats cannot be used to produce biodiesel using a traditional alkaline-catalyzed biodiesel production process due to its high free fatty acid and moisture contents (Murayama, 1994). If an alkaline catalyst is used to produce biodiesel from waste oils and fats, an acid pretreatment is required to reduce the free fatty acids in the feedstock. Several novel methods have been investigated to produce biodiesel from waste oils and fats. The acid-catalyzed process using waste oils and fats has been proved to be technically feasible. However, the large alcohol requirement for an acid catalyzed biodiesel production process results in larger transesterification reactors and excess alcohol distillation column (Zhang et al., 2003a). Several technologies have been attempted to reduce the production cost of biodiesel. A continuous transesterification process is one choice to lower the production cost. The recovery of high-quality glycerol is another way to lower production cost. Another choice is to produce biodiesel from waste oils and fats in supercritical alcohols.

Knothe et al. (1997), Ma and Hanna (1999), and Zhang et al. (2003a) have reviewed biodiesel production. In this chapter, the biodiesel fuel and its production process are briefly reviewed. Technical and economical assessment of the production of biodiesel from waste oils and fats is then discussed.

26.2 BIODIESEL FUELS

Biodiesel fuel can be used in existing diesel engines with little or no modification. Biodiesel is an alternative, nontoxic, biodegradable, and renewable diesel fuel. Biodiesel, primarily rapeseed methyl ester, has been in commercial use as an alternative fuel in many European countries since 1988. German biodiesel standard DIN V 51606 requires a rapeseed ester to have density of 0.875–0.900 g/mL at 15°C, viscosity of 3.5–5.0 mm^2/s at 15°C, acid number below 0.5 mg KOH/g, and iodine value below 115 g iodine/100 g among many specifications (Lang et al., 2001). The recent European Standard of EN 14214 was approved by the European Committee for Standardization in 2003. In the United States, biodiesel, primarily soybean methyl ester, must meet American Society of Testing and Materials (ASTM) specifications designated in ASTM D-6751. The ASTM D-6751 and EN 14214 specifications for biodiesel Fuel (B100) blend stock for distillate fuels are given in Table 26.1.

The alkyl monoesters of fatty acids from vegetable oils and animal fats, which have similar properties as diesel fuel, are known as biodiesel. The various fatty acid profiles of the different oil and fat feedstocks with different chain length, degree of unsaturation, and branching of the chain influence the properties of the biodiesel fuel, such as cetane number, exhaust emission, heat of combustion, cold flow, oxidative stability, viscosity, and lubricity (Knothe, 2005). One drawback of biodiesel is that there is an inverse relationship between the oxidative stability of biodiesel and its cold flow properties. Saturated compounds are less prone to oxidation than unsaturated compounds but they raise the cloud point of the fuel. Biodiesel from vegetable oils is highly unsaturated. It is very prone to oxidation and its cloud point is about 0°C. The impact of fuel oxidation on an engine's performance and emissions is not currently understood. However, the use of waste oils may provide some improvement in oxidative stability and cetane value over that seen with first-use vegetable oil or a temperature reduction in the cold flow properties of the fuel compared with animal fats, which have a higher proportion of saturated fatty acids.

It is worth noting that most standards for biodiesel fuel usually specify the biodiesel to meet the performance requirements of engines without specifying the actual composition of the fuel. Therefore, biodiesel can be made from any feedstock as long as the standard can be met. The feedstocks in current commercial biodiesel production are exclusively vegetable oils including soybean oil, rapeseed oil, palm oil, sunflower oil, coconut oil, and tung oil (Ma and Hanna, 1999). Oils from algae, bacteria, and fungi also have been investigated as feedstock in biodiesel production (Shay, 1993). The saturated fatty acid in animal fats such as beef tallow is almost 50% of the total fatty acids, giving the animal fats the unique properties of high melting point and high viscosity. Not much research has been conducted on production of biodiesel from animal fats (Ma and Hanna, 1999). Table 26.2 shows selected properties of biodiesels produced from various bio-oils (Lang et al., 2001). The viscosities of biodiesels in

TABLE 26.1
Standard Specification for Biodiesel Fuel (B100) Blend Stock for Distillate Fuels (ASTM D 6751-02 and EN 14214)

Property	ASTM Method	Limits	Units	EN Method	Limits	Units
Flash point (closed up)	D 93	130 min	°C	ISO CD 3679e	101 min	°C
Cetane number	D 613	47 min	—	EN ISO 5165	51 min	—
Kinematic viscosity, 40°C	D 445	1.9–6.0	mm^2/s	EN ISO 3104	3.5–5.0	mm^2/s
Sulfated ash	D 874	0.02 max	Mass%	ISO 3987	0.02 max	Mass%
Sulfur	D 5453	0.05 max	Mass%	—	10 max	mg/kg
Phosphorous content	D 4951	0.001 max	Mass %	pr EN 14107p	10 max	mg/kg
Copper strip corrosion	D 130	No. 3 max	—	EN ISO 2160	Class 1	—
Water and sediment	D 2709	0.050 max	Vol %	EN ISO 12937	500 max	mg/kg
Cloud point	D 2500	Report	°C	—	—	—
Carbon residue, 100% sample	D 4530	0.050 max	Mass %	—	—	—
Acid number	D 664	0.80 max	mg KOH/g	pr EN 14104	0.5 max	mg KOH/g
Iodine value	—	—	—	pr EN 14111	120 max	—
Distillation temperature, atmospheric equivalent temperature, 90% recovered	D 1160	360 max	°C	—	—	—
Free glycerin	D 6584	0.02 max	Mass %	Pr EN 14105m/pr EN 14106	0.02 max	Mass %
Total glycerin	D 6584	0.24 max	Mass %	pr EN 14105m	0.25 max	Mass %

(continued)

TABLE 26.1 (continued)

Standard Specification for Biodiesel Fuel (B100) Blend Stock for Distillate Fuels (ASTM D 6751-02 and EN 14214)

Property	ASTM Method	Limits	Units	EN Method	Limits	Units
Monoglyceride content	—	—	—	pr EN 14105m	0.8 max	Mass %
Diglyceride content	—	—	—	pr EN 14105m	0.2 max	Mass %
Triglyceride content	—	—	—	pr EN 14105m	0.2 max	Mass %
Linolenic acid methylester	—	—	—	pr EN 14103d	12	Mass %
Polyunsaturated (> = 4 double bonds) methylester	—	—	—	—	1	Mass %
Methanol content	—	—	—	pr EN 141101	0.2 max	Mass %
Alkali metals (Na and K)	—	—	—	pr EN 14108/pr EN 14109	5 max	mg/kg
Ester content	—	—	—	pr EN 14103d	96.5 min	Mass %
Tar remnant at 10% distillation remnant	—	—	—	EN ISO 10370	0.3 max	Mass %
Total contamination	—	—	—	EN 12662	24 max	mg/kg
Density at 15°C	—	—	—	EN ISO 3675/N ISO 12185	860–900	Kg/m^3
Oxidation stability, 110°C	—	—	—	pr EN 14112k	6 min	Hours

TABLE 26.2
Selected Properties of Biodiesels from Various Bio-Oils

Property	Product	Linseed	Canola	Sunflower	Rapeseed
Density at 25°C (g/mL)	Pure oil	0.925	0.912	0.914	0.908
	Methyl ester	0.887	0.875	0.882	0.877
	Ethyl ester	0.884	0.869	0.876	0.873
	2-Propyl ester	0.888	0.874	—	—
	Butyl ester	0.877	0.861	—	—
Kinetic viscosity (cP)	Pure oil	22.4	33.4	28.9	45.1
	Methyl ester	3.32	3.79	4.24	5.18
	Ethyl ester	3.64	3.91	4.4	7.6
	2-Propyl ester	4.88	6.24	—	—
	Butyl ester	4.06	4.39	—	—
Heating value (MJ/kg)	Pure oil	39.51	39.78	39.46	40.27
	Methyl ester	40.00	40.07	39.71	40.43
	Ethyl ester	39.65	40.41	39.80	40.97
	2-Propyl ester	39.56	40.04	—	—
	Butyl ester	40.38	40.24	—	—
Acid value (mg KOH/g)	Methyl ester	0.335	0.163	0.179	0.430
	Ethyl ester	0.324	0.265	0.610	0.152
	2-Propyl ester	0.586	0.469	—	—
	Butyl ester	0.254	0.280	—	—
Cloud point (°C)	Methyl ester	0	1	1	0
	Ethyl ester	−2	−1	−1	−2
	2-Propyl ester	3	7	—	—
	Butyl ester	−10	−6	—	—
Pour point (°C)	Methyl ester	−9	−9	−8	−15
	Ethyl ester	−6	−6	−5	−15
	2-Propyl ester	−12	−12	—	—
	Butyl ester	−13	−16	—	—

Source: Adapted from Lang, X., Dalai, A.K., Bakhshi, N.N., Reaney, M.J., and Hertz, P.B., *Biores. Technol.*, 80, 53, 2001. With permission.

the range of 3.3–7.6 cP at 40°C are much lower than those of pure oils at 22.4–45.1 cP and twice those of summer and winter diesel fuels at 1.72–3.50 cP. The heating value of biodiesels is approximately 40 MJ/kg, which is about 11% lower than that of diesel fuels at 45 MJ/kg.

26.3 TRADITIONAL BIODIESEL PRODUCTION

26.3.1 Traditional Biodiesel Production Process

A flow diagram for producing biodiesel via transesterification of oils and fats using an alkaline catalyst is shown in Figure 26.1. A typical alkali-catalyzed biodiesel

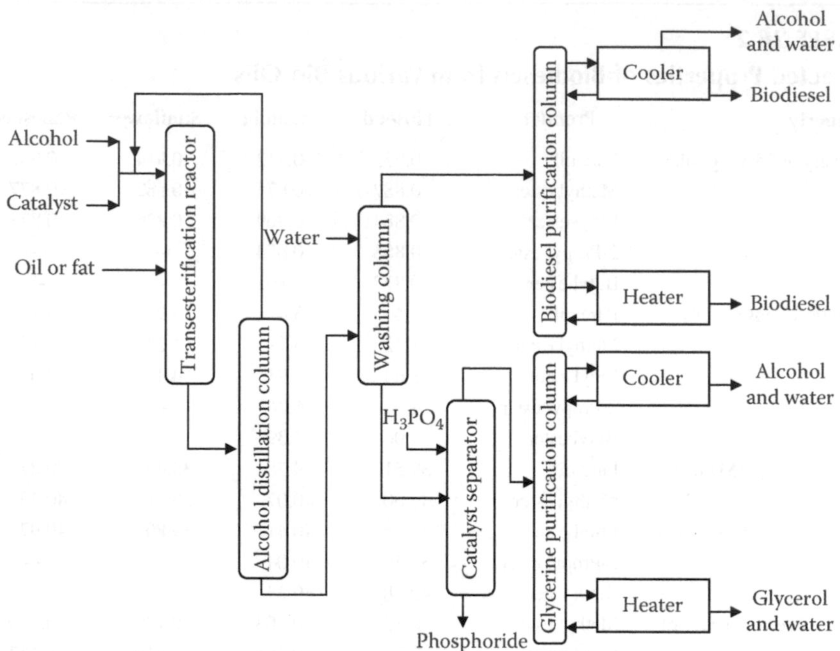

FIGURE 26.1 Flow diagram of a typical alkali-catalyzed biodiesel production process.

production process mainly consists of unit operations of transesterification reaction, distillation for excess alcohol recovery, water washing for separation of biodiesel from glycerol, catalyst, and alcohol, distillation for crude biodiesel purification, catalyst removal, and glycerol purification. An acid-catalyzed process is similar to the alkali-catalyzed process. However, because excess alcohol is used in an acid-catalyzed process, the transesterification reactor and alcohol distillation column of an acid-catalyzed process should be larger than that of an alkaline-catalyzed process for the same biodiesel production capacity. Zhang et al. (2003a) gave an overall review of four different continuous alkali- and acid-catalyzed processes for biodiesel production from edible vegetable oil and waste cooking oil.

26.3.2 Transesterification Reaction

Oils and fats are primarily water-insoluble, hydrophobic substances of triglycerides, which are made up of one mole of glycerol esterified with three moles of long-chain fatty acids in plants and animals, respectively. The most commonly used method to produce biodiesel from crude vegetable oils and animal fats is to transesterify vegetable oils and animal fats to an alkyl ester for a decrease in their viscosity and an increase in their volatility (Ma and Hanna, 1999). During transesterification, the glycerin is removed from the triglycerides and replaced with light alcohol molecules in the presence of a catalyst. Methanol and ethanol are two main light alcohols used in transesterification. The transesterification of triglyceride using methanol is shown in Figure 26.2 (Canakci, 2007).

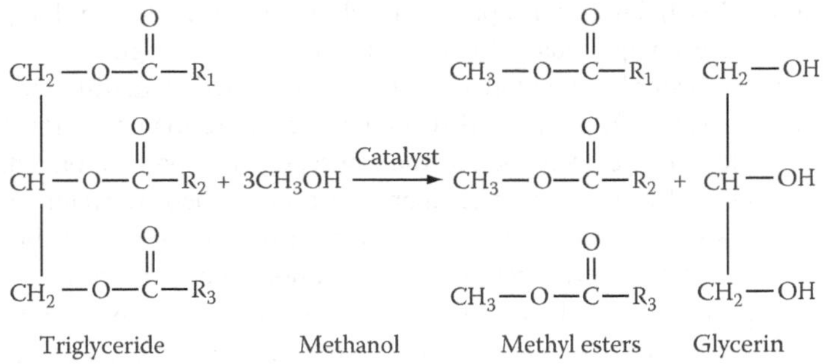

$$
\begin{array}{l}
CH_2-O-\overset{\displaystyle O}{\overset{\|}{C}}-R_1 \\
| \\
CH-O-\overset{\displaystyle O}{\overset{\|}{C}}-R_2 \\
| \\
CH_2-O-\overset{\displaystyle O}{\overset{\|}{C}}-R_3
\end{array}
+ 3CH_3OH \xrightarrow{\text{Catalyst}}
\begin{array}{l}
CH_3-O-\overset{\displaystyle O}{\overset{\|}{C}}-R_1 \\
CH_3-O-\overset{\displaystyle O}{\overset{\|}{C}}-R_2 \\
CH_3-O-\overset{\displaystyle O}{\overset{\|}{C}}-R_3
\end{array}
+
\begin{array}{l}
CH_2-OH \\
| \\
CH-OH \\
| \\
CH_2-OH
\end{array}
$$

Triglyceride Methanol Methyl esters Glycerin

FIGURE 26.2 Transesterification of triglyceride using methanol and catalyst.

Because the reaction is reversible, excess alcohol is used to shift the equilibrium to the products side. To complete a transesterification, a 3:1 molar ratio of alcohol to triglycerides is needed according to the stoichiometrical formula given in Figure 26.2. In practice, the ratio needs to be higher to drive the equilibrium to a maximum ester yield. The practical molar ratio is usually as high as 6:1 alcohols to vegetable oils depending on the quality of the oils (Freedman et al., 1984; Noureddini and Zhu, 1997). However, more excess alcohols present in the final product mixture tend to prevent the gravity separation of the glycerol from biodiesel, thus adding more cost to the process.

Transesterification can occur at different temperatures depending on the oil used. Higher reaction temperatures can increase the reaction rate and shorten the reaction time. The reaction temperature is usually lower than the boiling points of the alcohol used (e.g., 60°C for methanol and 78°C for ethanol). However, Vicente et al. (2007) reported that the biodiesel yield decreased with the increase in temperature due to the increase in triglyceride saponification and the subsequent dissolution of alkyl ester into glycerol.

The transesterification reaction for biodiesel production can be catalyzed by alkalis, acids, or enzymes. Alkali catalysts such as sodium hydroxide, potassium hydroxide, and sodium methoxide are the most commonly used in the traditional biodiesel industry because the process with an alkali catalyst has been proven to be fast and the reaction conditions are moderate. The addition of an alkali catalyst is usually 0.1%–1% of oils or fats by weight. Vicente et al. (2007) found that an increase in the amount of alkali catalyst increased the amount of soaps produced through triglyceride saponification and reduced the biodiesel yield.

The conversion rate increases with reaction time. Base catalyzed transesterifications are usually completed within 1 h. The transesterification of soybean, sunflower, peanut, and cotton seed oils with a 6:1 molar ratio of methanol to oil and 0.5% of sodium methoxide catalyst by weight at 60°C achieved a conversion efficiency of 93%–98% after 1 h. An approximate yield of 80% was observed after 1 min for soybean and sunflower oils (Freedman et al., 1984).

26.3.3 Recovery and Purification of Biodiesel

After transesterification of triglycerides with an alkali catalyst, the products are a mixture of esters, glycerol, alcohol, catalyst, soap, and tri-, di-, and monoglycerides

(Ma and Hanna, 1999). Different separation techniques have been used to purify the biodiesel from other compounds in the mixture. As shown in Figure 26.1, unreacted alcohol in the alkyl ester layer and glycerol layer is removed by distillation or evaporation. Vacuum is usually to keep the distillation or evaporation under 150°C (Zhang et al., 2003a). Alcohol is cycled back to the transesterification reactor. The reset of the mixture is cooled near to room temperature and washed by water in a simple gravity settler or a water-washing column. Water washing is used to further separate the biodiesel from the glycerol, catalyst, and alcohol. The glycerol with a specific gravity of 1.26 g/mL goes to the bottom of the washing column while the biodiesel with a specific gravity of 0.8–0.9 g/mL goes to the top. The catalyst and part of the residual alcohol are dissolved into the water and goes to the bottom. Zhang et al. (2003a) reported that the total amount of unconverted oil, alcohol, and water in the biodiesel from the top of a water-washing column with four theoretical stages was less than 6% by weight. The stream from the bottom of the column contained 81% glycerol, 8% water, 3% alcohol, and 9% catalyst. If more water is present in both biodiesel and glycerol streams after washing, the load for further purification of biodiesel and glycerol will increase.

After water washing, the biodiesel and glycerol are further purified in distillation columns as shown in Figure 26.1. For purification of biodiesel, a partial condenser is used at the top of the distillation column to separate biodiesel from a small amount of water and alcohol by condensing biodiesel as a liquid distillate and releasing water and alcohol as vent gases since the boiling point of biodiesel is higher than that of water and alcohol. The biodiesel can further be fractionally distilled at atmospheric pressure or under reduced pressure by distillation, crystallization, or other processes. A small amount of unconverted oil or fat remains at the bottom of the column, which can be recycled back to the reactor. The catalyst is dissolved in the glycerol stream after water washing. Phosphoric acid is usually used to neutralize and remove alkali catalysts by a gravity separator. For each liter of biodiesel produced, approximately 0.08 kg of crude glycerol is produced. The glycerol stream from the top of the separator can be distilled from further removal of water and alcohol. The crude glycerol is usually contaminated by residual catalysts. Purification of glycerol is very costly and generally out of the range of economic feasibility for small- to medium-sized plants. Alternative uses for the crude glycerol should be explored (Thompson and He, 2006).

To avoid the formation of emulsion during water washing, some organic solvents such as hexane or petroleum ether were also used to separate the biodiesel from other components via a liquid–liquid extraction process after the transesterification reaction (Nye et al., 1983).

26.4 CATALYSTS FOR BIODIESEL PRODUCTION

26.4.1 Alkaline Catalysts

Sodium hydroxide, potassium hydroxide, and sodium methoxide are the most commonly used alkali catalysts. Singh et al. (2006) found that methoxide catalysts could achieve higher biodiesel yields than hydroxide catalysts, and potassium-based catalysts gave better yields than the sodium-based catalysts ($KOCH_3 > NaOCH_3 > KOH > NaOH$). However, methoxide catalysts are more expensive than the hydroxides.

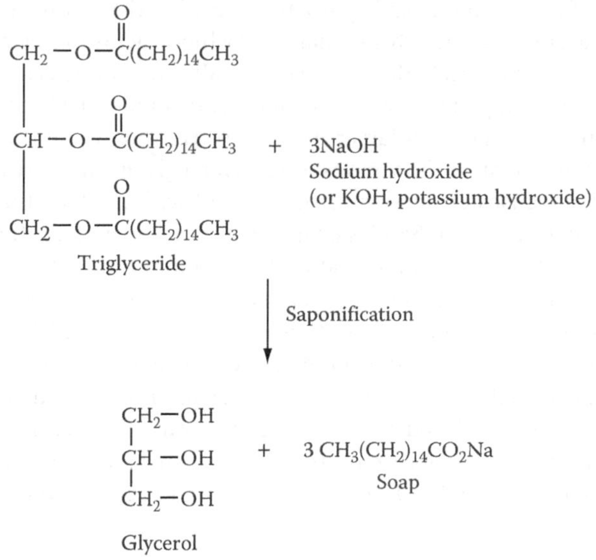

$$\begin{array}{l} CH_2-O-\overset{O}{\overset{\|}{C}}(CH_2)_{14}CH_3 \\ | \\ CH-O-\overset{O}{\overset{\|}{C}}(CH_2)_{14}CH_3 \\ | \\ CH_2-O-\overset{O}{\overset{\|}{C}}(CH_2)_{14}CH_3 \end{array} + \begin{array}{l} 3NaOH \\ \text{Sodium hydroxide} \\ \text{(or KOH, potassium hydroxide)} \end{array}$$

Triglyceride

$$\Big\downarrow \text{Saponification}$$

$$\begin{array}{l} CH_2-OH \\ | \\ CH-OH \\ | \\ CH_2-OH \end{array} + 3\ CH_3(CH_2)_{14}CO_2Na$$

Glycerol Soap

FIGURE 26.3 Saponification of triglyceride.

They are also more difficult to manipulate because they are very hygroscopic. Methoxides are preferred catalysts for a large continuous-flow biodiesel production process while hydroxides are preferred by small biodiesel producers (Singh et al., 2006). Potassium hydroxide has an advantage over sodium hydroxide because it can be neutralized with phosphoric acid after the reaction to produce potassium phosphate, which can be used as a fertilizer.

The alkali-catalyzed transesterification reaction is affected by free fatty acids and the water content of oils or fats. For an alkali-catalyzed transesterification, the triglycerides and alcohol must be substantially anhydrous because water makes the reaction partially change to the saponification path, as shown in Figure 26.3 (Liu, 1994). Low free fatty acid content in triglycerides is also required for alkali-catalyzed transesterification. Free fatty acids can react with an alkali catalyst to produce soaps and water. The formation of soaps not only lowers the yield of alkyl esters of biodiesel but also consumes the catalyst and reduces catalyst efficiency. Soap also increases the difficulty in the separation of biodiesel and glycerol and the water washing because soap causes the formation of emulsions (Zhang et al., 2003a). Therefore, the crude oil is usually first refined to remove a certain amount of water, free fatty acids, mucilaginous matter, protein, coloring matter, and sugars to meet the requirements of transesterification with an alkali catalyst.

26.4.2 Acid Catalysts

If more water and free fatty acids are in the triglycerides, acid-catalyzed transesterification can be used (Ma and Hanna, 1999). The most commonly preferred acid catalysts are sulfuric, sulfonic, and hydrochloric acids. Al-Widyan and Al-Shyoukh (2002) found that sulfuric acid is better than hydrochloric acids as a catalyst to convert waste palm oil into biodiesel. However, acid-catalyzed transesterification is much slower than alkali-catalyzed transesterification (Freedman et al., 1984). An acid-catalyzed reaction also needs a larger alcohol to triglyceride ratio than an alkali-catalyzed reaction to achieve the same ester yield for a given reaction time (Freedman et al., 1986). The transesterification of soybean oil with methanol in the presence of 1% sulfuric acid catalyst at 65°C is unsatisfactory if the molar ratios of methanol to oil are 6:1 and 20:1. At the molar ratio of 30:1, more than 90% oil is converted to methyl esters after 69 h reaction (Freedman et al., 1984). For acid-catalyzed transesterification, a high conversion efficiency can be obtained by increasing the molar ratio of alcohol to oil, reaction temperature, concentration of acid catalyst, and the reaction time (Canakci and Gerpen, 1999). Furthermore, Freedman et al. (1986) pointed out that acid-catalyzed transesterification was more corrosive to process equipment than the alkali-catalyzed process. If sulfuric acid is used as the catalyst, calcium oxide is usually used to neutralize and remove the catalyst in the downstream process.

26.4.3 Biocatalysts

Immobilized lipases can also be used as biocatalysts to catalyze the transesterification reaction of oils in supercritical carbon dioxide with an ester conversion of >98%. This process combines the extraction and transesterification of oil (Jackson and King, 1996). A continuous process may be possible for the simultaneous extraction and transesterification of oil (Ooi et al., 1996). However, the enzyme-catalyzed reaction usually requires a much longer reaction time than the alkali- or acid-catalyzed reactions. There is no commercial enzyme-catalyzed biodiesel production process available up to now (Zhang et al., 2003a).

26.4.4 Heterogeneous Catalysts

During traditional biodiesel production process, alkaline catalysts such as sodium hydroxide and potassium hydroxide or acid catalysts such as sulfuric acid can be dissolved into the alcohol, which increases the complexity and cost of downstream separation after transesterification reaction. Heterogeneous catalysts such as calcium oxide and calcium methoxide are insoluble in an organic solvent. Use of a heterogeneous catalyst in biodiesel production can significantly simplify the separation process and thus reduce the production cost (Gryglewicz, 1999). Peterson and Scarrach (1984) attempted to use heterogeneous catalysts in biodiesel production. However, the reaction occurred at a relatively low rate compared with an alkaline-catalyzed reaction because the three-phase mixture of oil–alcohol–heterogeneous catalyst has poor mass transport. A co-solvent such as tetrahydrofuran was used to transform the oil/alcohol two-phase reactants into a one-phase reactant. As a result of enhanced mass transfer between the reactants, the transesterification reaction was rapid and was completed in a few minutes (Boocock et al., 1996). The basicity of earth metal-based heterogeneous catalysts such as calcium oxide and calcium methoxide is also weaker than sodium hydroxide and potassium hydroxide. Gryglewicz (1999) reported that for the same conversion efficiency, the reaction time for the calcium oxide-catalyzed process increased 2–4 times compared to that for sodium hydroxide-catalyzed process.

26.5 BIODIESEL PRODUCTION FROM WASTE OILS AND FATS

26.5.1 Technical Challenges for Biodiesel Production from Waste Oils and Fats

Waste oils and rendered animal fats have been used as feedstocks to produce biodiesel. The problem with waste oils and fats is that they usually contain large amounts of free fatty acids that cannot be converted to biodiesel using an alkaline catalyst because of the formation of fatty acid salts (soap). The level of free fatty acids in waste cooking oil is usually greater than 2% by weight (Watanabe et al., 2001). The moisture levels and the free fatty acid levels of the collected waste restaurant oils and animal fats vary widely and were as high as 18% and 41.8%, respectively.

Any water and free fatty acids in feedstock will slow the alkali-catalyzed transesterification reaction. The soaps formed also prevent the separation of the biodiesel from the glycerin fraction. Ma et al. (1998a) investigated the effects of free fatty acids and water on transesterification of beef tallow with methanol. Their results showed that the water content of beef tallow should be kept below 0.06% w/w and free fatty acid content of beef tallow should be kept below 0.5%, w/w (acid value less than 1) to achieve the best conversion. Water content was a more critical variable in the transesterification process than free fatty acids. The requirements for oils and fats with low free fatty acids and water contents limits the use of waste oils and fats as a low-cost feedstock in traditional biodiesel production with an alkali catalyst.

26.5.2 Alkaline Catalytic Process for Biodiesel Production from Waste Oils and Animal Fats

If the waste oils and fats have low free fatty acid and water contents, they can be used as feedstocks in traditional alkaline catalytic process for biodiesel production. Zhang (1994) transesterified edible beef tallow with a free fatty acid content of 0.27%. The tallow was heated to remove moisture under vacuum at 60°C. Transesterification was conducted using 6:1 molar ratio of methanol/tallow, 1% (by the weight of tallow) sodium hydroxide dissolved in the methanol, and 60°C for about 30 min. After separation of glycerol, the ester layer was transesterified again using 0.2% sodium hydroxide and 20% methanol at 60°C for about 1 h. The mixture was washed with distilled water until the wash water was clear. The purified ester was heated again to 70°C under vacuum to remove residual moisture. The two steps of alkaline-catalyzed transesterification process yielded 400 g of tallow ester from 500 g of beef tallow at a conversion efficiency of 80%.

Ma et al. (1998b and 1999) studied the transesterification process of beef tallow with methanol. Because the solubility of methanol in beef tallow was only 19% w/w at 100°C, mixing was essential to disperse the methanol in beef tallow in order to start the reaction. They also pointed out that once the two phases were mixed and the reaction was started, stirring was no longer needed. After the reaction was finished, there was 60% w/w of unreacted methanol in the beef tallow ester phase and 40% w/w in the glycerol phase. The optimum operation sequence was to recover the unreacted methanol using vacuum distillation after transesterification, separate ester and glycerol phases, and then purify beef tallow methyl esters.

26.5.3 Acid Catalytic Process for Biodiesel Production from Waste Oils and Fats

26.5.3.1 Acid Pretreatment of Waste Oils and Fats

Since the free fatty acid content of waste oils and fats is usually too high, it is impossible to convert those oils and fats to biodiesel using a single alkaline-catalyzed process. Published results suggest that acid catalysis must decrease the acid value of the feedstock to less than 2 mg KOH/g before alkaline catalysis will give satisfactory results (Freedman and Pryde, 1982; Liu, 1994). Lepper and Friesenhagen (1986) recommended a pretreatment step to reduce the free fatty acid content via an esterification reaction with methanol in the presence of sulfuric acid catalyst. After the pretreatment, the oil phase, which had a free fatty acid content less than 0.5% by weight, was further used in an alkali-catalyzed process for biodiesel production. Canakci and Van Gerpen (2003) developed a two-step process for the production of biodiesel from brown grease. After the acid value of the brown grease that had 40% free fatty acids was reduced to less than 2 mg KOH/g with the pretreatment process, the reaction was continued with alkaline-catalyzed transesterification.

26.5.3.2 Acid Posttreatment of Soap in Crude Biodiesel

For high free fatty acid value oils, alkali-, and then acid-catalyzed transesterifications were also used (Sprules and Price, 1950). The free fatty acids were neutralized with alkali to form soap during the reaction. After the triglycerides were converted to esters, 5% by oil weight of sulfuric acid was added to neutralize the alkali catalyst, release the free fatty acids from the soap formed, and acidify the system. The esters were then made from the free fatty acids for 3–4 h. The mixture was neutralized with an alkali salt such as calcium carbonate.

26.5.3.3 Acid-Catalyzed Transesterification Reaction

Instead of using an acid for pretreatment of free fatty acids in the waste oils and fats or posttreatment of soap formed from free fatty acids, an acid-catalyzed process can be used to produce biodiesel from waste oils and fats. Acid catalysts are able to esterify free fatty acids in waste oils and fats (Freedman and Pryde, 1982; Aksoy et al., 1988; Liu, 1994). Zhang et al. (2003a) reported that acid-catalyzed process from waste cooking oil was potentially a competitive alternative to the commonly used alkali-catalyzed process. The acid-catalyzed process using waste cooking oil was more economically feasible by providing a lower total manufacturing cost than the alkali-catalyzed process using edible vegetable oils (Zhang et al., 2003b). The further sensitivity analyses conducted by Zhang et al. (2003b) showed that the production scale, the price of raw oil or fat feedstocks, and the price of biodiesel were the major factors affecting the economic feasibility of biodiesel production.

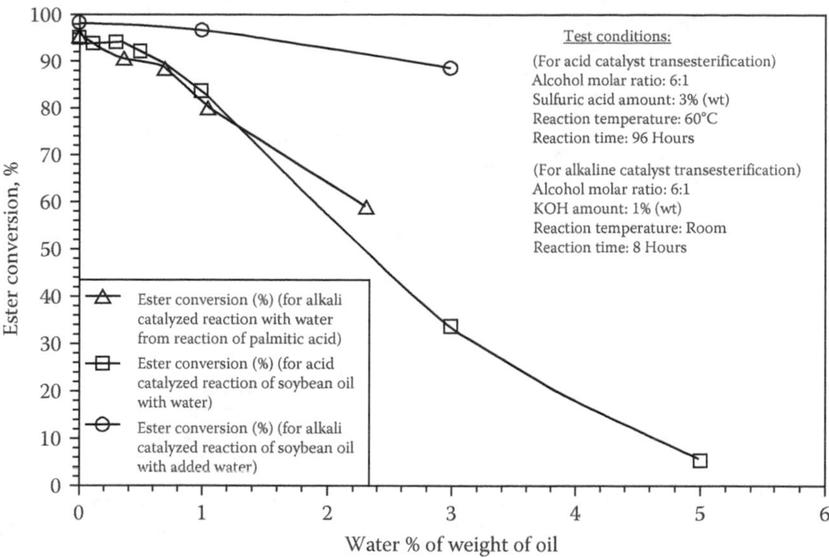

FIGURE 26.4 Effects of water content on acid-catalyzed ester production. (Reprinted from Canakci, M. and van Gerpen, J., *Trans ASAE* 42, 1203, 1999. With permission.)

Canakci and Van Gerpen (1999) investigated the relationship between free fatty acid level and triglyceride transesterification during acid-catalyzed biodiesel production. Sample mixtures were prepared by adding palmitic acid to soybean oil to obtain free fatty acid levels between 5% and 33%. The conversion rate of soybean oil to methyl ester dropped from 90.54% to 58.77% as the free fatty acid level increased from 5% to 33%. This is due to the effect of the water produced when the free fatty acids react with the alcohol to form alkyl esters. Figure 26.4 shows a comparison between the degree of ester conversion when water is produced by free fatty acids esterification and when water is deliberately added to soybean oil during transesterification. The coincidence of the lines indicates that water formation is the primary mechanism limiting the completion of the acid-catalyzed esterification reaction with free fatty acids.

26.5.4 Biodiesel Production from Waste Oils and Fats in Supercritical Alcohols

Alkaline- or acid-catalyzed transesterification of oils and fats for biodiesel production needs a complicated separation and purification process after transesterification reaction, which increases the energy consumption and production costs. One effective way to reduce the production cost and energy consumption is to transesterify the oils and fats into biodiesel without a catalyst. Supercritical alcohols have been used to transesterify the oils and fats into biodiesel without the need for catalysts (Saka and Kusdiana, 2001; Kusdiana and Saka, 2001a; Bunyakiat et al., 2006). Free fatty acids and water present in oils and fats have significant effects on the alkaline- or acid-catalyzed transesterification reaction for biodiesel production. Kusdiana and

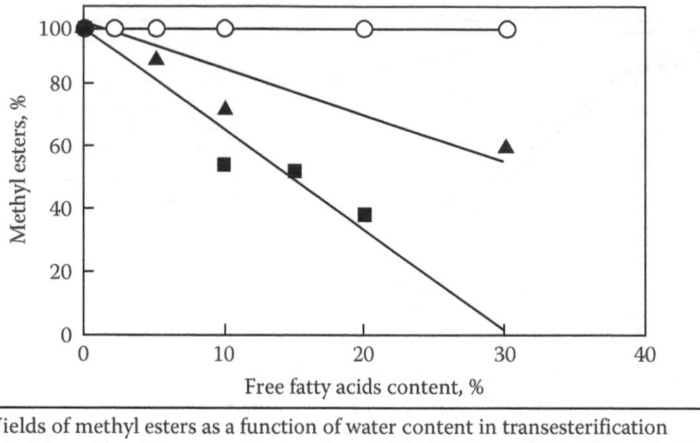

FIGURE 26.5 Effects of fatty acids content on the yields of methyl esters in biodiesel production. (Reprinted from Kusdiana, D. and Saka, S., *Biores. Technol.*, 91, 289, 2004. With permission.)

Saka (2001b and 2004) investigated the effects of free fatty acids and water on the transesterification reaction by supercritical methanol treatment.

In order to compare the effect of free fatty acid content of oil feedstock on transesterification conversion efficiency, oleic acid was added to rapeseed oil to adjust its free fatty acid content. The rapeseed oil with different free fatty acid contents was treated by (1) supercritical methanol at 350°C, 43 MPa, and 42:1 molar ratio of methanol to oil for 4 min; (2) methanol at 65°C, a 6:1 molar ratio of methanol to oil, and 1.5 wt% sodium hydroxide as a catalyst for 1 h; and (3) methanol at 65°C, a 6:1 molar ratio of methanol to oil, and 3 wt% sulfuric acid as a catalyst for 48 h, respectively (Kusdiana and Saka, 2004). Figure 26.5 shows the comparison of the yields of methyl esters from triglycerides with various free fatty acid contents using different transesterification methods. As shown in Figure 26.5, Kusdiana and Saka (2004) found that free fatty acids present in the oils can be simultaneously esterified in supercritical alcohol. A high and constant conversion efficiency was obtained by the supercritical methanol method for rapeseed oil with various free fatty acid contents. However, the conversion efficiency decreased to around 50% for the acid-catalyzed process and 35% for the alkaline-catalyzed process if 20% of free fatty acids presented in the rapeseed oil.

In order to compare the effect of water content of oil feedstock on transesterification conversion efficiency, rapeseed oil with different water contents was treated by (1) supercritical methanol at 350°C, 43 MPa, and 42:1 molar ratio of methanol to oil for 4 min; (2) methanol at 65°C, a 6:1 molar ratio of methanol to oil, and 1.5 wt% sodium hydroxide as a catalyst for 1 h; and (3) methanol at 65°C, a 6:1 molar ratio of methanol to oil, and 3 wt% sulfuric acid as a catalyst for 48 h(Kusdiana and Saka, 2004). Figure 26.6 shows the comparison of the yields of methyl esters from triglycerides with various water contents using different transesterification methods. For the alkaline-catalyzed process, the conversion efficiency was slightly decreased

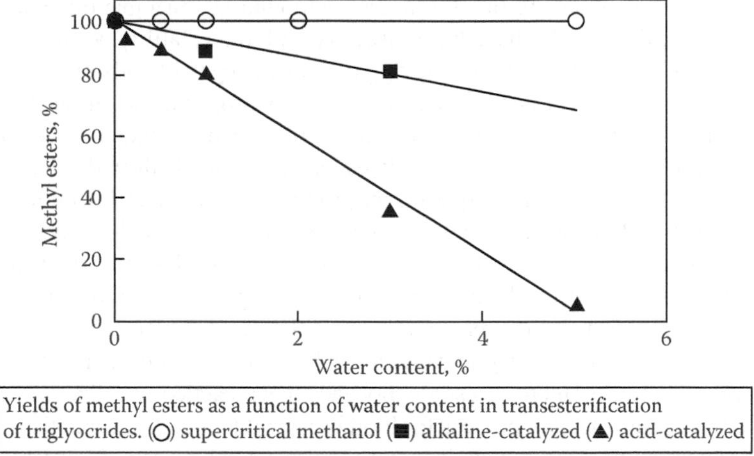

FIGURE 26.6 Effects of water content on the yields of methyl esters in biodiesel production. (Reprinted from Kusdiana, D. and Saka, S., *Biores. Technol.*, 91, 289, 2004. With permission.)

with the increase in water in the oil feedstock. For the acid-catalyzed process, the conversion efficiency was decreased to 6% if 5% of water presented in the oil feedstock. For supercritical methanol method, complete conversion of oil was observed at a water content as high as 50% (Kusdiana and Saka, 2004). During supercritical alcohol esterification of oil for biodiesel production, a certain amount of water could even enhance the methyl esters formation by hydrolyzing fatty acids from triglycerides. Therefore, for the vegetable oils containing water, transesterification of triglycerides, hydrolysis of triglycerides, and methyl esterification of fatty acids occur simultaneously during supercritical methanol treatment (Kusdiana and Saka, 2001b). Kusdiana and Saka (2004) found that the reaction time was a critical parameter if a large amount of water presented in the reaction system. If the reaction time was too long, part of alkyl esters were hydrolyzed back to fatty acids, reducing the conversion efficiency.

The transesterification of rapeseed oil by supercritical methanol was found to be completed within 240 s at 350°C, 43 MPa, and 42:1 molar ratio of methanol to oil (Saka and Kusdiana, 2001; Kusdiana and Saka, 2001a). Demirbas (2002) found that 95% of hazelnut kernels and cotton seed oil were converted to methyl esters within 300 s at 250°C and 24:1 molar ratio of methanol to oil. The properties of the biodiesel produced in supercritical methanol were found to be similar to those of No. 2 diesel fuel but were slightly more viscous. Bunyakiat et al. (2006) found that 95% and 96% of coconut oil and palm kernel oil were converted into methyl esters within 400 s at 350°C, 19 MPa, and a molar ratio of methanol to oil of 42.

26.6 SUMMARY

Vegetable oils and animal fats have been used as feedstocks to produce biodiesel. Triacylglycerides, which are formed by esterifying three long-chain fatty acids to a single glycerol molecule, are the dominant compound in vegetable oils and animal fats. Since triacylglycerides have high viscosity and low volatility, a transesterification process is needed to convert triacylglycerides in vegetable oils and animal fats to the biodiesel of alkyl esters, which have similar properties as diesel fuels. The transesterification process can be catalyzed by alkaline, acid, and enzyme catalysts. Alkaline catalysts such as sodium hydroxide, potassium hydroxide, and sodium methoxide are the most commonly used for production of biodiesel from food-grade vegetable oils with low free fatty acid and moisture contents. An acid-catalyzed process can tolerate the free fatty acids in the oil and fat feedstock. However, the transesterification efficiency of an acid-catalyzed process is significantly reduced if a large amount of water presents in the reaction system either from the feedstock or generated by the esterification of free fatty acids with alcohol. Furthermore, an acid transesterification process requires more excess alcohol and longer reaction time than an alkaline-catalyzed transesterification process for biodiesel production. A supercritical alcohol transesterification process, which does not require a catalyst, can efficiently convert triacylglycerides with high free fatty acid and moisture content to alkyl esters. Since waste oils and fats usually contain large amounts of water and free fatty acids, supercritical alcohols provide a promising approach for the production of biodiesel from waste oils and fats.

REFERENCES

Aksoy, H.A., I. Kahraman, F. Karaosmanoglu, and H. Civelekoglu. 1988. Evaluation of Turkish sulphur olive oil as an alternative diesel fuel. *Journal of the American Oil Chemists' Society* 65: 936–938.

Al-Widyan, M.I. and A.O. Al-Shyoukh. 2002. Experimental evaluation of the transesterification of waste palm oil into biodiesel. *Bioresource Technology* 85: 253–256.

Boocock, D.G.B., S.K. Konar, V. Mao, and H. Sidi. 1996. Fast one-phase oil-rich processes for the preparation of vegetable oil methyl esters. *Biomass and Bioenergy* 11: 43–50.

Bunyakiat, K., S. Makmee, R. Sawangkeaw, and S. Ngamprasertsith. 2006. Continuous production of biodiesel via transesterification from vegetable oils in supercritical methanol. *Energy & Fuels* 20: 812–817.

Canakci, M. 2007. The potential of restaurant waste lipids as biodiesel feedstocks. *Bioresource Technology* 98: 183–190.

Canakci, M. and J. van Gerpen. 1999. Biodiesel production via acid catalysis. *Transactions of the ASAE* 42: 1203–1210.

Canakci, M. and J. van Gerpen. 2003. A pilot plant to produce biodiesel from high free fatty acid feedstocks. *Transactions of the ASAE* 46: 945–954.

Demirbas, A. 2002. Biodiesel from vegetable oils via transesterification in supercritical methanol. *Energy Conversion and Management* 43: 2349–2356.

Freedman, B. and E.H. Pryde. 1982. Fatty esters from vegetable oils for use as a diesel fuel. In *Vegetable Oils Fuels-Proceedings of the International Conference on Plant and Vegetable Oils as Fuels*, 117–122. Fargo: ASAE Publication 4–82.

Freedman, B., R.O. Butterfield, and E.H. Pryde. 1986. Transesterification kinetics of soybean oil. *Journal of the American Oil Chemists' Society* 63: 1375–1380.

Freedman, B., E.H. Pryde, and T.L. Mounts. 1984. Variables affecting the yields of fatty esters from transesterified vegetable oils. *Journal of the American Oil Chemists' Society* 61: 1638–1643.

Gryglewicz, S. 1999. Rapeseed oil methyl esters preparation using heterogeneous catalysts. *Bioresource Technology* 70: 249–253.

Knothe, G. 2005. Dependence of biodiesel fuel properties on the structure of fatty acid alkyl esters. *Fuel Processing Technology* 86: 1059–1070.

Knothe, G., R.O. Dunn, and M.O. Bagby. 1997. Biodiesel: The use of vegetable oils and their derivatives as alternative diesel fuels. In *Fuels and Chemicals from Biomass*, Saha, B.C. and J. Woodward (Eds.), pp. 172–208. Washington DC: American Chemical Society symposium series 666.

Krawczyk, T. 1996. Biodiesel. *INFORM* 7: 801–822.

Kusdiana, D. and S. Saka. 2001a. Kinetics of transesterification in rapeseed oil to biodiesel fuels as treated in supercritical methanol. *Fuel* 80: 693–698.

Kusdiana, D. and S. Saka. 2001b. Methyl esterification of free fatty acids of rapeseed oil as treated in supercritical methanol. *Journal of Chemical Engineering of Japan* 34: 383–387.

Kusdiana, D. and S. Saka. 2004. Effects of water on biodiesel fuel production by supercritical methanol treatment. *Bioresource Technology* 91: 289–295.

Lang, X., A.K. Dalai, N.N. Bakhshi, M.J. Reaney, and P.B. Hertz. 2001. Preparation and characterization of bio-diesels from various bio-oils. *Bioresource Technology* 80: 53–62.

Lepper, H. and L. Friesenhagen. 1986. Process for production of fatty acid esters of short-chain aliphatic alcohols from fats and/or oils containing free fatty acids. U.S. Patent 4608202.

Liu, K. 1994. Preparation of fatty acid methyl esters for gas-chromatographic analysis of lipids in biological materials. *Journal of the American Oil Chemists' Society* 71: 1179–1187.

Ma, F. and M.A. Hanna. 1999. Biodiesel production: A review. *Bioresource Technology* 70: 1–15.

Ma, F., L. D. Clements, and M.A. Hanna. 1998a. The effects of catalyst, free fatty acids and water on transesterification of beef tallow. *Transactions of the ASAE* 41: 1261–1264.

Ma, F., L.D. Clements, and M.A. Hanna. 1998b. Biodiesel fuel from animal fat. Ancillary studies on transesterification of beef tallow. *Industrial and Engineering Chemistry Research* 37: 3768–3771.

Ma, F., L.D. Clements, and M.A. Hanna. 1999. The effect of mixing on transesterification of beef tallow. *Bioresource Technology* 69: 289–293.

Murayama, T. 1994. Evaluating vegetable oils as a diesel fuel. *Inform* 5: 1138–1145.

Noureddini, H. and D. Zhu. 1997. Kinetics of transesterification of soybean oil. *Journal of the American Oil Chemists' Society* 74: 1457–1463.

Nye, M.J., T.W. Williamson, S. Deshpande, J.H. Schrader, and W.H. Snively. 1983. Conversion of used frying oil to diesel fuel by transesterification: Preliminary test. *Journal of the American Oil Chemists' Society* 60: 1598–1601.

Ooi, C.K., A. Bhaskar, M.S. Yener, D.Q. Tuan, J. Hsu, and S.S.H. Rizvi. 1996. Continuous supercritical carbon dioxide processing of palm oil. *Journal of the American Oil Chemists' Society* 73: 233–237.

Peterson, G.R. and W.P. Scarrach. 1984. Rapeseed oil transesterification by heterogenous catalysis. *Journal of the American Oil Chemists' Society* 61: 1593–1597.

Saka, S. and D. Kusdiana. 2001. Biodiesel fuel from rapeseed oil as prepared in supercritical. methanol. *Fuel* 80: 225–231.

Shay, E.G. 1993. Diesel fuel from vegetable oils: Status and opportunities. *Biomass and Bioenergy* 4: 227–242.

Singh, A., B. He, J. Thompson, and J. Van Gerpen. 2006. Process optimization of biodiesel production using alkaline catalysts. *Applied Engineering in Agriculture* 22: 597–600.

Sprules, F.J. and D. Price. 1950. Production of fatty esters. U.S. Patent 2, 366–494.

Thompson, J.C. and B.B. He. 2006. Characterization of crude glycerol from biodiesel production from multiple feedstocks. *Applied Engineering in Agriculture* 22: 261–265.

Vicente, G., M. Martinez, and J. Aracil. 2007. Optimization of integrated biodiesel production. Part II: A study of the material balance. *Bioresource Technology* 98: 1754–1761.

Watanabe, Y., Y. Shimada, A. Sugihara, and Y. Tominaga. 2001. Enzymatic conversion of waste edible oil to biodiesel fuel in a fixed-bed bioreactor. *Journal of the American Oil Chemists' Society* 78: 703–707.

Zhang, D. 1994. Crystallization characteristics and fuel properties of tallow methyl esters. Master thesis, University of Nebraska–Lincoln.

Zhang, Y., M.A. Dube, D.D. McLean, and M. Kates. 2003a. Biodiesel production from waste cooking oil: 1. process design and technological assessment. *Bioresource Technology* 89: 1–16.

Zhang, Y., M.A. Dube, D.D. McLean, and M. Kates. 2003b. Biodiesel production from waste cooking oil: 2. Economic assessment and sensitivity analysis. *Bioresource Technology* 90: 229–240.

27 Thermochemical Conversion of Food Processing Wastes for Energy Utilization

27.1 INTRODUCTION

Biomass such as forest residues, agricultural residues, and organic food processing wastes can be converted to chemical and energy products via either biological (Lin and Tanaka, 2006) or thermochemical processes (Caputo et al., 2005; Yoshioka et al., 2005). Biological conversion of low-value lignocellulosic biomass to commercial chemical and energy products, particularly ethanol, still faces challenges in low economy and efficiency (Lin and Tanaka, 2006). Thermochemical conversion provides a competitive way to produce chemical and energy products from low-value and highly distributed biomass resources with large variations in properties (Caputo et al., 2005). Combustion, pyrolysis, gasification, and thermochemical liquefaction are four main thermochemical conversion methods (Knoef and Stasse, 1995).

Combustion is the conversion of chemical energy stored in an organic matter into heat, generating carbon dioxide and water as final products. Combustion usually produces hot gas at temperatures around 800°C–1000°C. Pyrolysis is the conversion of biomass to liquid, solid, and gaseous fractions by heating the biomass in the absence of air or oxygen to around 500°C. The gasification process falls between complete combustion and pyrolysis. Gasification is the partial oxidation of organic matter at a high temperature to convert the organic matter into a combustible gas mixture called syngas, which mainly consists of carbon monoxide, hydrogen, methane, and carbon dioxide. Gasifiers are operated at approximately 800°C–900°C although a non-catalytic entrained flow gasifier could be operated at a temperature as high as 1300°C. Solvents such as water and alcohols at an elevated temperature and pressure can convert solid biomass into liquid fuels via a thermochemical liquefaction process.

Thermochemical conversion technologies have been used for the reduction of environmental impact of food processing wastes and the recovery of energy from the wastes (Shinogi and Kanri, 2003). It is possible to convert any type of organic food wastes in a thermochemical process. However, it could be practical and economic to thermochemically convert a feedstock with a low moisture content (e.g., <50%–60%) if the biomass is not dried before the conversion.

27.2 COMBUSTION

27.2.1 Combustion Process

Biomass combustion is to directly burn the biomass in the presence of sufficient air to convert the chemical energy stored in the biomass into heat, which can be further used to generate mechanical power and electricity. During combustion, the organic wastes are oxidized with oxygen above their ignition temperature. The minimum oxygen requirement for the combustion is the theoretical (stoichiometric) amount of oxygen necessary to completely oxidize the carbon and hydrogen and other trace elements in the wastes to produce primarily carbon dioxide and water, and generate heat. The oxygen is usually supplied in excess of the stoichiometric requirements to assure complete combustion.

Dry organic food wastes (i.e., the moisture content <50% w.b.) are suitable for combustion. In most of cases, the wastes require some pretreatments such as drying, chopping, and grinding before combustion. Fluidized-bed combustion is known for its fuel flexibility and favorable emission control characteristics. It was originally developed for coal conversion but it has also been used as a converter for other solid fuels such as biomass. A fluidized bed is suitable for the combustion of various fuels. However, for optimization with respect to minimum emissions, the properties of the various fuels have to be considered (Leckner and Lyngfelt, 2002).

Co-firing of biomass in existing combustors has been used for the generation of heat and electricity from biomass. Biomass, such as bagasse, is traditionally combusted to supply heat and power in the process industry. The net efficiency for electricity generation from biomass combustion is usually very low, ranging from 20% to 40% (Caputo et al., 2005). Larger power generation systems (i.e., over 100 MWe) or power plants with co-combustion of biomass and coal have higher efficiencies (McKendry, 2002). Due to the concern about plugging of a coal feeding system, biomass feed is usually limited to 5%–10% of the total feedstock (Yoshioka et al., 2005). From an economic point of view, co-firing biomass in an existing boiler is an investment of low risk, as it requires a small capital.

27.2.2 Combustion of Food Processing Wastes

Solid food processing wastes can either directly be combusted or first converted into gaseous and liquid fuels for further combustion. Rice husk was used as a fuel in a 30 kWh bubbling fluidized-bed combustor. Combustion efficiency itself was higher than 97% (Armesto et al., 2003). Liquid processing wastes such as waste vegetable oils and fats can be converted to alcohol ester by transesterification to be used as a combustion fuel (Tashtoush et al., 2003). For the combustion of ethyl ester from waste palm oil at an air to fuel ratio ranging from 10:1 to 20:1 in a water-cooled furnace, the combustion efficiency and exhaust temperature are 66% and 600°C for the ratio of 10:1, and 56% and 560°C for the ratio of 20:1, respectively.

27.3 PYROLYSIS

27.3.1 Pyrolysis Process

Pyrolysis is the conversion of solid biomass to liquid, solid, and gaseous fractions by heating the biomass in the absence of air or oxygen to around 500°C or above. Pyrolysis is different from incineration in terms of supply of oxygen. Pyrolysis is suited for the conversion of dry biomass. Pyrolytic chemistry is difficult to characterize because of the large variety of reaction pathways and the large variety of reaction products that may be obtained from the complex reactions.

Pyrolysis can be further divided into conventional pyrolysis at a relatively low heating rate of less than 10°C/s and flash pyrolysis at a high heating rate of 100°C/s–1000°C/s. The percentages of each of solid, liquid, and gaseous fractions generated depend on the heating rate and the pyrolysis temperature (Islam et al., 1999). Biomass pyrolyzed at a low heating rate and a low temperature results in less liquid and gaseous products and more solid char. Flash and high-temperature pyrolysis can produce more liquid and gaseous fractions than slow pyrolysis (Goyal et al., 2008). The yield of pyrolytic oil could be in the range of 50%–70% depending on feedstocks and operating conditions.

Pyrolysis reactor designs mainly include fixed bed, moving bed, and fluidized bed. Pyrolysis in a fixed-bed reactor is usually slow. Fluidized-bed reactors including bubbling fluidized-bed and circulating fluidized-bed reactors can achieve fast pyrolysis (Meier and Faix, 1999).

27.3.2 Pyrolysis Products and Utilization

Pyrolysis of biomass produces three main products: solid char, liquid bio-oil, and gas. The pyrolysis gas, which contains mainly carbon dioxide and methane, can be used as a fuel in combustors for heat and power generation. The solid is generally considered to be a porous material that can improve soil physical properties such as soil texture, permeability, and water holding capacity. In addition, high ash content products such as cattle manure can also provide not only nitrogen but also inorganic salts, such as phosphorus, potassium, calcium, and magnesium to the crops (Shinogi and Kanri, 2003). Char can also be directly used as a solid fuel or further gasified to a gaseous fuel. It can also be processed to more value-added chemicals such as activated carbons (Bridgwater and Bridge, 1991).

Biomass pyrolysis is attractive because pyrolytic oil can be easily stored and transported compared to bulky solid biomass and gaseous fuels. However, pyrolytic oil normally contains a high proportion of oxygenates and their physical and chemical characteristics change rapidly during condensation and under storage conditions (Gercel, 2002). Therefore, pyrolytic oil is required to be upgraded to improve storage stability and calorific value if it is used as fuel or as a source of chemic feedstock (Samolada et al., 1998). Limited use and difficulty in downstream processing of bio-oil currently restrict the wide application of biomass pyrolysis technology (Faaij, 2006).

The liquid stream of bio-oil, which is the main product from a pyrolysis process, can be upgraded to higher quality liquid transportation fuel or used as a feedstock for production of various chemicals. Specifically, pyrolytic oil can be used to (Goyal et al., 2008)

- Generate heat and power in combustors
- Produce chemicals and resins
- Produce adhesives and binders
- Produce transportation fuels by upgrading

27.3.3 Pyrolysis of Food Processing Wastes

Pyrolysis has been used to convert solid food processing wastes such as oil seed cakes into liquid fuel. However, different feedstocks and operating conditions lead to large variations in bio-oil yield and compositions.

Gercel (2002) used a fixed-bed tubular reactor to pyrolyze sunflower oil cake. The maximum bio-oil yield was 48.89% by weight obtained at a pyrolysis temperature of 550°C and a heating rate of 5°C/s. Under these conditions, the yields of char and gas were 30.99% and 20.1% by weight, respectively. At a lower temperature, the gas and char were the major products while at a higher temperature, the bio-oil and char yields decreased and gas yield increased. The calorific value of bio-oil produced from sunflower cake was measured at 32.15 MJ/kg, which is very close to that of petroleum fractions.

Onay et al. (2001) conducted fast pyrolysis of rapeseed with a particle size range of 0.6–0.85 mm in a well-swept fixed-bed reactor at different heating rates. They observed that the maximum oil yield of 68% was obtained at a heating rate of 300°C/min and the pyrolysis temperature of 550°C, compared to 43% obtained at a heating rate of 100°C/min and the same temperature. If the heating rate was further increased above 300°C/min, the increase in the oil yield was negligible.

Zabaniotou et al. (2000) used a rapid pyrolysis process at a heating rate of 200°C/s to convert olive residues. The maximum bio-oil yield was 30% of dry biomass obtained at 450°C–550°C. The major gaseous products were found to be CO and CO_2.

27.4 GASIFICATION

27.4.1 Gasification Process

Both anaerobic digestion and gasification can convert solid biomass into a gaseous fuel. Anaerobic digestion, which is discussed in Chapter 24, involves microorganisms at a relatively low temperature (e.g., 35°C) under anaerobic conditions to convert biomass into methane rich biogas. Gasification is the partial oxidation of organic matter at a high temperature (e.g., 800°C) to convert the organic matter into a combustible gas mixture called syngas, which mainly consists of carbon monoxide, hydrogen, methane, and carbon dioxide (Higman and van der Burgt, 2003; Knoef, 2005). Comprehensive information on biomass gasification research, development, demonstration, and commercialization has been provided by the International Energy Agency (IEA), Bioenergy (http://www.gastechnology.org/iea), and European Gasification Network (Gasnet at http://www.gasnet.uk.net). Guides to gasification design and operation have been provided by Knoef (2005) and Higman and van der Burgt (2003). Wang et al. (2008) have given a critical review of the contemporary issues in thermal gasification of biomass and its application to electricity and fuel production.

Gasification is a partial oxidation process. The oxidant or gasifying agents can be air, pure oxygen, steam, carbon dioxide, or their mixtures. Different gasifying agents affect product gas yield, composition, and quality. Since the main components of cellulose, hemicellulose, and lignin in biomass have different gasification kinetics, the yield and composition of the syngas is also dependent on the component fractions of biomass feedstock (Hanaoka et al., 2005). Air is a cheap and widely used gasifying agent. The equivalence ratio (ER), which is the ratio of oxygen required for gasification to oxygen required for full combustion of a given amount of biomass, is usually 0.2–0.4 (Narvaez et al., 1996; Gabra et al., 2001b; Zainal et al., 2002). The large amount of nitrogen from air dilutes the syngas and thus lowers its heating value. Narvaez et al. (1996) reported a gas composition of 10% H_2, 14% CO, 15% CO_2, and about 50% of N_2 (by volume) from gasification of biomass at a temperature of 800°C and ER of 0.35. If pure oxygen is used as the gasifying agent, the heating value of syngas will increase but the operating costs will also increase due to the requirement for oxygen production. Heating value and hydrogen content of syngas can be increased if steam is used as the gasifying agent. Heating value of the product gas with steam as the gasifying agent is about 10–15 MJ Nm^{-3} (Gil et al., 1999; Rapagna et al., 2000), compared to 3–6 MJ Nm^{-3} for air gasification of biomass (Gabra et al., 2001b; Zainal et al., 2002) and 38 MJ Nm^{-3} for natural gas. The use of CO_2 as the gasifying agent is promising because of its presence in the syngas. Carbon dioxide with a catalyst such as Ni/Al can transform char, tar, and CH_4 into H_2 and/or CO, thus increasing H_2 and CO contents and decreasing CO_2 in the product gas (Garcia et al., 2001).

Biomass gasification requires heat supply. Partial combustion of biomass with air or oxygen is usually used to provide heat for drying the biomass, raising the biomass temperature, driving the endothermic gasification reactions, and generating H_2O and CO_2 for further reduction reactions (Basu, 2006). If pure steam, water, or carbon dioxide is used as the gasifying agent, an indirect or external heat supply is required for the endothermic gasification reactions (Hofbauer et al., 1997; Pletka et al., 2001a,b; Cummera and Brown, 2005). Indirectly heated biomass gasification systems are still under development. Alternatively, a mixture of steam, water, or carbon dioxide and air or oxygen can be used as the gasifying agent, and the partial combustion of biomass with air/oxygen provides the heat required for the steam or carbon dioxide gasification (Gil et al., 1999; Lv et al., 2004; Lucas et al., 2004).

A gasifier is designed to promote the interaction between biomass and gasifying agent to enhance the endothermic and heterogeneous reactions during gasification. There are three main types of gasifiers: fixed-bed, moving-bed, and fluidized-bed gasifiers (Higman and van der Burgt, 2003; Knoef, 2005; Basu, 2006). The fixed-bed gasifier has three sub-types of designs: downdraft, updraft, and crossdraft according to the method with which the gasifying agent is introduced into the gasifier (Basu, 2006). Both fixed-bed and moving-bed gasifiers produce syngas with large quantities of either tar or char due to the low- and nonuniform heat and mass transfer between solid biomass and gasifying agent. However, they are simple and reliable designs and can be used to gasify very wet biomass economically on a small scale (Basu, 2006). Fluidized-bed gasifiers, which consist of a large percentage of hot, inert bed materials such as sand and a small percentage of biomass (e.g., 1%–3% by mass), have been used widely in biomass gasification (Basu, 2006). Specific advantages of fluidized-bed gasification include high yield and heating value of syngas due to low tar and char formation at a high heating rate and uniform heating, low ash aggregation due to a low and uniform temperature, compact design and high productivity due to a short residence time, and flexibility in gasifying a range of solid biomass with large variations in properties (van der Drift et al., 2001). Due to the fluidization of biomass particles in the gasifiers, some fine particles could be entrained above the fluidizing bed and large bubbles may result in gas bypass through the bed, lowering the biomass conversion efficiency. Pan et al. (1999) reported that 20% of secondary air injection to the primary air injection above the biomass feeding point in a fluidized-bed gasifier reduced 88.7% (wt) of the total tar for the gasification in the temperature range from 840°C to 880°C. Alternatively, the circulating fluidized-bed gasifiers can recycle solids to increase the residence time and thus increase biomass conversion efficiency, compared to bubbling fluidized-bed gasifiers (Yin et al., 2002; Li et al., 2004). The productivity, gas heating value, and gasifier efficiency for gasification of rice husk in an industrial scale circulated fluidized-bed gasifier (ϕ 1.8 m) were 960 kg $m^{-2}h^{-1}$, 4.6–6.3 MJ Nm^{-3}, and 65%, compared to 127 kg $m^{-2}h^{-1}$, 3.8–4.6 MJ Nm^{-3}, and 47% for gasification of rice husks in a downdraft fixed-bed gasifier (ϕ 2 m) (Yin et al., 2002).

27.4.2 Gasification Products and Utilization

During gasification, a part of biomass is converted to char and tar instead of syngas. Gabra et al. (2001a) reported an average char yield of 17.5% of the input fuel during bagasse gasification. Herguido et al. (1992) found that only 80% of the carbon in the feedstock was converted to syngas during steam fluidized-bed gasification of wood sawdust at atmospheric pressure and 775°C. Four percent of the carbon was in liquid tar and the remaining carbon was in solid char. Use of syngas as a fuel for internal combustion engines, gas turbines, and fuel cells for heat and power generation, and as a feedstock for the synthesis of liquid fuels and chemicals depends mainly on the cleaning technologies used to remove particulate dust and condensable tar in the syngas. Reduction and conversion of char and tar can also increase syngas gas yield and overall conversion efficiency (Wang et al., 2008).

The syngas can be used to generate heat and power like natural gas (Kinoshita et al., 1997; Rodrigues et al., 2003). Low heat value syngas from air gasification of biomass can be used in a combustor (Raskin et al., 2001). Compared to direct biomass combustion, syngas from biomass gasification can increase the bio-based fuel percentage used in existing pulverized coal combustors without any concern about plugging of the coal feeding system during co-firing of biomass—coal. Biomass gasification can reduce the potential of ash slugging or other ash-related problems because the gasification temperature is lower than combustion and clean syngas is supplied to the combustor. A gasification process can use a variety of biomasses with large variations in their properties such as moisture content and particle size. However, if syngas is combusted directly to generate steam for power generation via a steam turbine, the electricity efficiency is limited by the theoretical limit of a steam turbine (Raskin et al., 2001). High-quality syngas almost free of tar and dust, and at a high heating value can be directly fed to gas engines (Sridhar et al., 2001; Wander et al., 2004), gas turbines (Miccio, 1999) or fuel cells (Donolo et al., 2006) for combined heat and power generation.

High-quality syngas can also be used to synthesize other chemicals and liquid fuels such as Fischer–Tropsch liquids, butanol, ethanol, and methanol (Vega et al., 1989; Worden et al., 1991; Grethlein and Jain, 1993; Maness and Weaver, 1994; Ng et al., 1999; Tao et al., 2001; Tijmensen et al., 2002; Brown, 2003; Datar et al., 2004; Brown, 2006), or to produce hydrogen (Rapagna et al., 2000; Watanabe et al., 2002; Rapagna et al., 2002; Hanaoka et al., 2005). The application of biomass gasification technology strongly depends on syngas quality control technologies (Gerhard et al., 1994; Aznar et al., 1998; Devi et al., 2003; Abu El-Rub et al., 2004; Ma et al., 2005; Srinakruang et al., 2005).

27.4.3 Gasification of Food Processing Wastes

Food wastes with a low moisture content are suitable for gasification. Rice husks are the coating for rice seeds. During the milling process, the husks are removed from the grain to create white rice. The total rice production worldwide is about 500 million tons. Since rice husk is about 20% of the mass of rice grains, the annual rice husk production is about 100 million tons (Mansaray et al., 1999). Rice husks can be converted into liquid and gaseous fuel through a thermochemical conversion process (Mansaray and Ghaly, 1999; Mansaray et al., 1999).

Mansaray et al. (1999) used a fluidized-bed gasifier for air gasification of rice husk at a fluidization velocity from 0.22 to 0.33 m/s, air equivalence ratio from 0.25 to 0.35, and temperature from 665°C to 830°C. The gas yield and carbon conversion were in the range of 1.30 to 1.98 Nm^3/kg and 55% to 81%, respectively. The higher heating value of the syngas was 3.09–5.03 MJ/Nm^3. Yin et al. (2002) investigated a circulating fluidized-bed biomass gasification and gas engine power generation system as shown in Figure 27.1 to provide 800 kW power from rice husk for a rice mill. The system was operated at a temperature from 700°C to 850°C and biomass loading capacity of 1500 kg/h. The gasification efficiency was 65% and rice consumption was 1.7–1.9 kg/kWh at an overall electricity generation efficiency of 17%. The unit investment was 370 $/kW.

A sugar factory produces nearly 30% of bagasse out of its total crushing. Many research efforts have been attempted to use the bagasse as a renewable feedstock for

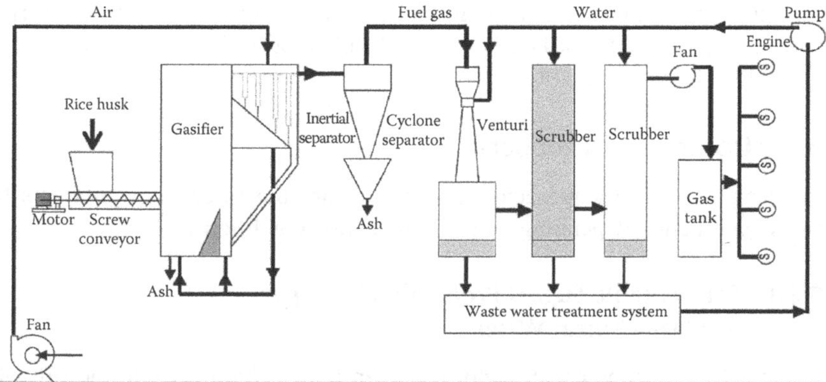

FIGURE 27.1 A gasification and power generation system. (Reprinted from Yin et al., *Biomass and Bioenergy* 23, 181, 2002. With permission.)

power generation and for the production of bio-based materials. A cyclone gasifier was used to gasify bagasse powder at 39–53 kg/h and 820°C–850°C. The heating values of the syngas at oxygen equivalence ratio from 0.18 to 0.25 were in the range of 3.5–4.5 MJ/Nm^3 dry gas (Gabra et al., 2001a,b).

27.5 THERMOCHEMICAL LIQUEFACTION

27.5.1 Liquefaction Process

Solvents at an elevated pressure and temperature can hydrolyze and liquefy biomass into liquid fuels by attacking the glycosidic linkages and cleavage of the molecules into smaller soluble fragments. Liquefaction is the conversion of biomass into a stable liquid hydrocarbon using moderate temperatures typically in the range from 300°C to 400°C and high pressures. During the liquefaction, biomass feedstocks are converted to bio-oil through a complex sequence of physical structure and chemical changes including solvolysis, depolymerization, decarboxylation, hydrogenolysis, and hydrogenation.

The popular solvents used for biomass hydrolysis and liquefaction are water and alcohols, which are environmentally friendly. A fluid becomes supercritical after it passes its vapor-liquid critical point with the increase in temperature and pressure. After a fluid becomes supercritical, there is no phase change between gas and liquid with further changes in temperature and pressure above the critical point. A supercritical fluid has both gaseous properties such as high diffusivity, low viscosity, and high compressibility, and liquid properties such as high density. The unique properties of supercritical fluids can enhance heat and mass transfer, reaction kinetics, and equilibrium between solid biomass and supercritical fluids. Furthermore, due to the high compressibility of supercritical fluids, their solvent properties can be easily adjusted by changing the pressure and temperature. Thus, the products can be easily recovered from the fluids by the reduction of the dissolving power (Sihvonen et al., 1999; Wang and Weller, 2006). The critical points of selected substances are given in Chapter 22 (Table 22.1).

Many novel processes and products have been developed using the inherent physical and chemical properties of supercritical fluids such as supercritical CO_2 extraction of plant materials, dyeing, and production of fine powders (Marr and Gamse, 2000; Hauthal, 2001). Supercritical fluids have also been used to liquefy biomass (Ehara and Saka, 2005).

27.5.2 Liquefaction Products

Fermentable sugars, organic acids, and bio-oil are the three main products from biomass liquefaction depending on the liquefaction conditions.

27.5.2.1 Fermentable Sugars Production by Super- or Near-Critical Water

Water, at near- or supercritical conditions, can liquefy biomass into bio-oil, which can be further refined to sugars, organic acids, and other valuable products (Saka,

2006). Near- or supercritical water liquefaction of biomass is a relatively novel process. Several researchers found that at temperatures above 190°C, a part of lignin and hemicelluloses macromolecules undergo liquefaction after a few minutes of exposure to hot liquid water. At higher temperatures, the remaining lignocelluloses continued to be hydrothermally liquefied (Mok and Antal, 1992; Antal and Varhegyi, 1995; Antal et al., 1998; Gronli et al., 1999). Supercritical or near-critical water treatment of cellulose has been studied to obtain fermentable sugars for alcohol production (Sasaki et al., 1998).

Research has been conducted to investigate the reaction mechanism of biomass in near- or supercritical water. If biomass is immersed in hot water with the increase in temperature, the cellulose in the biomass is hydrolyzed slightly into sugars over 200°C. At around 250°C, the cellulose is decomposed sharply to form water-soluble sugars and non-sugars, gas, oil, and char. Over 300°C, the sugars and oil are decomposed and more char is produced (Minowa et al., 1998). Sodium carbonate can inhibit char formation from oil, resulting in a high oil yield even at a high reaction temperature of 350°C for a 1 h reaction (Sjostrom, 1981; Minowa et al., 1995).

Due to the high critical temperature of water (T_c=374°C), hydrolysis products such as glucose were found to decompose rapidly in supercritical water (Minami and Saka, 2005). The decomposition of saccharides is accelerated by increasing the processing temperature, thus reducing the yield of fermentable saccharides. Ehara and Saka (2005) found that a combined process of short supercritical water treatment followed by sub-critical water treatment could effectively inhibit the decomposition of fermentable saccharides and thus increase the yield of fermentable saccharides.

27.5.2.2 Organic Acids Production by Super- or Near-Critical Water

Schutt et al. (2002) used sub-critical water to partially oxidize cellulose and other biomass-derived compounds with a palladium catalyst for the selective production of organic acids such as malic and acetic acids, which are value-added chemical precursors. They observed that at 150°C, nearly 100% conversion of the cellulose was achieved in approximately 5h. Furthermore, the measurable catalyst decomposition and leached metal in the aqueous phase were small. Saka (2006) investigated the yield and profiles of organic acids produced by supercritical water processing of biomass under various processing conditions. He found that the maximum yield of the total organic acids was about 25% of the biomass by mass obtained by supercritical water treatment at 380°C and 25MPa for 4min. The main acids included formic acid, acetic acid, glycolic acid, and lactic acid.

27.5.2.3 Bio-Oil Production by Thermochemical Liquefaction

Direct solvent liquefaction provides a method for production of liquid fuels, chemical feedstocks, and carbon materials from biomass. Biomass liquefaction is a depolymerization process of the constitutive compounds in biomass to form smaller fragments. Several kinds of reactions such as decarboxylation, dehydration, dehydrogenation, and deoxygenation can occur during biomass liquefaction with solvents. The unstable fragments could rearrange through condensation, cyclization, and polymerization to form new compounds such as aromatic compounds. Bio-oils obtained by high-pressure liquefication of biomass result in a complex mixture of volatile organic acids, alcohols, aldehydes, ethers, ketones, furans, phenols, hydrocarbons, and other non-volatile components. The oils could be upgraded catalytically to yield an organic distillate product, which is rich in hydrocarbons and useful chemicals such as toluene, xylenes, and phenols (Demirbas, 2004).

Kucuk (2001) used supercritical methanol, ethanol, and acetone to liquefy verbascum and sunflower stalks. The liquid yields varied from 40% to 60.5% of the feedstock mass at temperatures of 260°C and 300°C and with a catalyst of 10% NaOH or without catalyst. Minami and Saka (2005) reported that more than 95% of the wood was decomposed and liquefied using supercritical methanol (T_c = 239°C and P_c = 8.09 MPa) at 350°C for 30 min. If 10% water (by mass) was added into the supercritical methanol at 350°C, the liquefaction time was shortened to 5 min (Minami and Saka, 2005). Saka (2006) found that the molecular weight distribution of the product oil could be regulated by changing the kind of alcohol or reaction conditions.

When high-temperature and high-pressure solvents are used to liquefy solid mass, a gaseous product containing H_2, CO, CH_4, and CO_2 will be generated simultaneously (Matsumura et al., 2005). The yield and composition of the syngas depend on the operating conditions, reactor design, and biomass characteristics (Yu et al., 2006).

27.6 SUMMARY

Food processing wastes are abundant renewable energy sources in food processing facilities. Thermochemical methods including combustion, pyrolysis, gasification, and liquefaction can convert solid biomass such as food processing wastes into heat, power, and liquid and gaseous fuels. Pyrolysis converts solid biomass into char, liquid oil, and gas. The pyrolytic oil can be upgraded to transportation fuels or used as a feedstock to produce chemicals and materials and generate heat and power. Pyrolytic oil can be easily transported and stored. Gasification provides a competitive way to convert diverse, highly distributed, and low-value lignocellulosic biomass to syngas for combined heat generation, synthesis of liquid fuels, and production of hydrogen. Co-firing of syngas in existing pulverized coal and natural gas combustors has been successfully commercialized. However, more research is needed to improve syngas quality for its commercial uses in a high energy-efficient heat and power generator such as gas turbines or fuel cells and the production of liquid fuels and hydrogen. Solvents such as water and alcohols at an elevated temperature and pressure can extract high-value chemicals such as sugars from biomass and liquefy solid biomass into a liquid fuel of bio-oil via a thermochemical liquefaction process.

REFERENCES

Abu El-Rub, Z., E.A. Bramer, and G. Brem. 2004. Review of catalysts for tar elimination in biomass gasification processes. *Industrial and Engineering Chemistry Research* 43: 6911–6919.

Antal, M.J. and G. Varhegyi. 1995. Cellulose pyrolysis kinetics: The current state of knowledge. *Industrial and Engineering Chemistry Research* 34: 703–717.

Antal, M.J., G. Varhegyi, and E. Jakab. 1998. Cellulose pyrolysis kinetics: Revisited. *Industrial and Engineering Chemistry Research* 37: 1267–1275.

Armesto, L., A. Bahillo, K. Veijonen, A. Cabanillas, and J. Otero. 2003. Combustion behavior of rice husk in a bubbling fluidized bed. *Biomass and Bioenergy* 23: 171–179.

Aznar, M.P., M.A. Caballero, J. Gil, J.A. Martin, and J. Corella. 1998. Commercial steam reforming catalysts to improve biomass gasification with steam-oxygen mixtures. 2. Catalytic tar removal. *Industrial and Engineering Chemistry Research* 37: 2668–2680.

Basu, P. 2006. *Combustion and Gasification in Fluidized Beds.* pp. 59–101, Boca Raton, FL: CRC Press.

Bridgwater, A.V. and S.A. Bridge. 1991. A review of biomass pyrolysis and pyrolysis technologies, Chapter 2. In *Biomass Pyrolysis Liquids Upgrading and Utilization*, Bridgwater, A.V. and G. Grassi (Eds.), London: Elsevier Applied Science.

Brown, R.C. 2003. *Biorenewable Resources: Engineering New Products from Agricultural.* Ames IA: Blackwell Publishing.

Brown, R.C. 2006. Biorefineries based on hybrid thermochemical-biological processing-an overview. In *Biorefineries—Industrial Processes and Products. Status Quo and Future Directions* (vol.1), Kamm, B., P.R. Gruber, and M. Kamm (Eds.), pp. 227–251. Weinheim: Wiley-VCH Verlag GmbH & Co.

Caputo, A.C., M. Palumbo, P.M. Pelagagge, and F. Scacchia. 2005. Economics of biomass energy utilization in combustion and gasification plants: Effects of logistic variables. *Biomass and Bioenergy* 28: 35–51.

Cummera, K. and R.C. Brown. 2005. Indirectly heated biomass gasification using a latent-heat ballast-part 3: Refinement of the heat transfer model. *Biomass and Bioenergy* 28: 321–330.

Datar, R.P., R.M. Shenkman, B.G. Cateni, R.L. Huhnke, and R.S. Lewis. 2004. Fermentation of biomass-generated producer gas to ethanol. *Biotechnology and Bioengineering* 86: 587–594.

Demirbas, A. 2004. Current technologies for the thermo-conversion of biomass into fuels and chemicals. *Energy Source* 26: 715–730.

Devi, L., K.J. Ptasinski, and F.J.J.G. Janssen. 2003. A review of the primary measures for tar elimination in biomass gasification processes. *Biomass and Bioenergy* 24: 125–140.

Donolo, G., G. de Simon, and M. Fermeglia. 2006. Steady state simulation of energy production from biomass by molten carbonate fuel cells. *Journal of Power Sources* 158: 1282–1289.

Ehara, K. and S. Saka. 2005. Decomposition behavior of cellulose by supercritical water, subcritical water and their combined treatments. *Journal of Wood Science* 51: 148–153.

Faaij, A.P.C. 2006. Bio-energy in Europe: Changing technology choices. *Energy Policy* 34: 322–342.

Gabra, M., E. Pettersson, R. Backman, and B. Kjellstrom. 2001a. Evaluation of cyclone gasifier performance for gasification of sugar cane residue-Part 1: Gasification of bagasse. *Biomass and Bioenergy* 21: 351–369.

Gabra, M., E. Pettersson, R. Backman, and B. Kjellstrom. 2001b. Evaluation of cyclone gasifier performance for gasification of sugar cane residue-Part 2: Gasification of cane trash. *Biomass and Bioenergy* 21: 371–380.

Garcia, L., M.L. Salvador, J. Arauzo, and R. Bilbao. 2001. CO_2 as a gasifying agent for gas production from pine sawdust at low temperature using Ni/Al coprecipitated catalyst. *Fuel Process Technology* 69: 157–174.

Gercel, H.F. 2002. The production and evaluation of bio-oils from the pyrolysis of sunflower-oil cake. *Biomass and Bioenergy* 23: 307–314.

Gerhard, S.C., D.N. Wang, R.P. Overend, and M.A. Paisley. 1994. Catalytic conditioning of synthesis gas produced by biomass gasification. *Biomass and Bioenergy* 7: 307–313.

Gil, J., J. Corella, M.P. Aznar, and M.A. Caballero. 1999. Biomass gasification in atmospheric and bubbling fluidized bed: Effect of the type of gasifying agent on the product distribution. *Biomass and Bioenergy* 17: 389–403.

Goyal, H.B., D. Seal, and R.C. Saxena. 2008. Bio-fuels from thermochemical conversion of renewable resources: A review. *Renewable and Sustainable Energy Reviews* 12: 504–517.

Grethlein, A.J. and M.K. Jain. 1993. Bioprocessing of coal-derived synthesis gases by anaerobic bacteria. *Trends in Biotechnology* 10: 418–423.

Hanaoka, T., S. Inoue, S. Uno, T. Ogi, and T. Minowa. 2005. Effect of woody biomass components on air-steam gasification. *Biomass and Bioenergy* 28: 69–76.

Hanaoka, T., T. Yoshida, S. Fujimoto, K. Kamei, M. Harada, Y. Suzuki, H. Hatano, S. Yokoyama, and T. Minowa. 2005. Hydrogen production from woody biomass by steam gasification using a CO_2 sorbent. *Biomass and Bioenergy* 28: 63–68.

Hauthal, W.H. 2001. Advances with supercritical fluids. *Chemosphere* 43: 123–125.

Herguido, J., J. Corella, and J. Gonzalez-Saiz. 1992. Steam gasification of lignocellulosic residues in a fluidized bed at a small pilot scale-effect of the type of feedstock. *Industrial and Engineering Chemistry Research* 31: 1274–1282.

Higman, C. and M. van der Burgt. 2003. *Gasification*. Burlington, MA: Elsevier.

Hofbauer, H., T. Fleck, G. Veronik, R. Rauch, H. Mackinger, and E. Fercher. 1997. The FICFB-gasification process. In *Developments in Thermochemical Biomass Conversion*, Bridgwater, A.V. and D.G.B. Boocock (Eds.), pp. 1016–1025. London: Blackie.

Islam, M.N., R. Zailani, and F.N. Ani. 1999. Pyrolytic oil from fluidized bed pyrolysis of oil palm shell and its characterization. *Renewable Energy* 17: 73–84.

Kinoshita, C.M., S.Q. Turn, R.P. Overend, and R.L. Bain. 1997. Power generation potential of biomass gasification systems. *Journal of Energy Engineering* 23: 88–99.

Knoef, H.A.M. 2005. *Handbook Biomass Gasification*. Netherlands: Biomass Technology Group.

Knoef, H.E.M. and H.E.M. Stassen. 1995. Energy generation from biomass and waste in the Netherlands: A brief overview and perspective. *Renewable Energy* 6: 329–334.

Kucuk, M.M. 2001. Liquefaction of biomass by supercritical gas extraction. *Energy Sources* 23: 363–368.

Leckner, B. and A. Lyngfelt. 2002. Optimization of emissions from fluidized bed combustion of coal, biofuel and waste. *International Journal of Energy Research* 26: 1191–1202.

Li, X.T., J.R. Grace, C.J. Lim, A.P. Watkinson, H.P. Chen, and J.R. Kim. 2004. Biomass gasification in a circulating fluidized bed. *Biomass and Bioenergy* 26: 171–193.

Lin, Y. and S. Tanaka. 2006. Ethanol fermentation from biomass resources: Current state and prospects. *Applied Microbiol Biotechnol* 69: 627–642.

Lucas, C., D. Szewczyk, W. Blasiak, and S. Mochida. 2004. High-temperature air and steam gasification of densified biofuels. *Biomass and Bioenergy* 27: 563–575.

Lv, P.M., Z.H. Xiong, J. Chang, C.Z. Wu, Y. Chen, and J.X. Zhu. 2004. An experimental study on biomass air-steam gasification in a fluidized bed. *Bioresource Technology* 95: 95–101.

Ma, L., H. Verelst, and G.V. Baron. 2005. Integrated high temperature gas cleaning: Tar removal in biomass gasification with a catalytic filter. *Catal Today* 105: 729–734.

Maness, P.C. and P.F. Weaver. 1994. Production of poly-3-hydroxyalkanoates from CO and H_2 by a novel photosynthetic bacterium. *Applied Biochemistry Biotechnology* 45&46: 395–406.

Mansaray, K.G. and A.E. Ghaly. 1999. Determination of kinetic parameters of rice husks in oxygen using thermogravimetric analysis. *Biomass and Bioenergy* 17: 19–31.

Mansaray, K.G., A.E. Ghaly, A.M. Al-Taweel, F. Hamdullahpur, and V.I. Ugursal. 1999. Air gasification of rice husk in a dual distributor type fluidized bed gasifier. *Biomass and Bioenergy* 17: 315–332.

Marr, R. and T. Gamse. 2000. Use of supercritical fluids for different processes including new developments—a review. *Chemical Engineering and Processing* 39: 19–28.

Matsumura, Y., T. Minowa, B. Potic, S.R.A. Kersten, W. Prins, W.P.M. van Swaaij, B. van de Beld, D.C. Elliott, G.G. Neuenschwander, A. Kruse, and M.J. Antal Jr. 2005. Biomass gasification in near- and super-critical water: Status and prospects. *Biomass and Bioenergy* 29: 269–292.

McKendry, P. 2002. Energy production from biomass (part 2): Conversion technologies. *Bioresource Technology* 83: 47–54.

McKendry, P. 2002. Energy production from biomass (part 3): Gasification technologies. *Bioresource Technology* 83: 55–63.

Meier, D. and O. Faix. 1999. State of the art of applied fast pyrolysis of lignocellulosic materials-a review. *Bioresource Technology* 68: 71–77.

Miccio, F. 1999. Gasification of two biomass fuels in bubbling fluidized bed. In *Processing of the 15th International Conference on Fluidized Bed Combustion*. Savannah, GA.

Minami, E. and S. Saka. 2005. Decomposition behavior of woody biomass in water-added supercritical methanol. *Journal of Wood Science* 51: 395–400.

Minowa, T., Z. Fang, T. Ogi, and G. Varhegyi. 1998. Decomposition of cellulose and glucose in hot-compressed water under catalyst-free conditions. *Journal of Chemical Engineering of Japan* 31: 131–134.

Minowa, T., M. Murakami, Y. Dote, T. Ogi, and S. Yokoyama. 1995. Oil production from garbage by thermochemical liquefaction. *Biomass and Bioenergy* 8: 117–120.

Mok, W.S.L. and M.J. Antal. 1992. Uncatalyzed solvolysis of whole biomass hemicellulose by hot compressed liquid water. *Industrial and Engineering Chemistry Research* 31: 1157–1161.

Narvaez, I., A. Orio, M.P. Aznar, and J. Corella. 1996. Biomass gasification with air in an atmospheric bubbling fluidized bed-effect of six operational variables on the quality of produced raw gas. *Industrial and Engineering Chemistry Research* 35: 2110–2120.

Ng, K.L., D. Chadwick, and B.A. Toseland. 1999. Kinetics and modeling of dimethyl ether synthesis from synthesis gas. *Chemical Engineering Science* 54: 3587–3592.

Onay, O., S.H. Beis, and O.M. Kochar. 2001. Fast pyrolysis of rape seed in a well-swept fixed-bed reactor. *Journal of Analytical and Applied Pyrolysis* 58–59: 995–1007.

Pan, Y.G., X. Roca, E. Velo, and L. Puigjaner. 1999. Removal of tar by secondary air injection in fluidized bed gasification of residual biomass and coal. *Fuel* 78: 1703–1709.

Pletka, R., R.C. Brown, and J. Smeenk. 2001. Indirectly heated biomass gasification using a latent heat ballast Part 1: Experimental evaluations. *Biomass and Bioenergy* 20: 297–305.

Rapagna, S., N. Jand, A. Kiennemann, and P.U. Foscolo. 2000. Steam-gasification of biomass in a fluidised-bed of olivine particles. *Biomass and Bioenergy* 19: 187–197.

Rapagna, S., H. Provendier, C. Petit, A. Kiennemann, and P.U. Foscolo. 2002. Development of catalysts suitable for hydrogen or syn-gas production from biomass gasification. *Biomass and Bioenergy* 22: 377–388.

Raskin, N., J. Palonen, and J. Nieminen. 2001. Power boiler fuel augmentation with a biomass fired atmospheric circulating fluid-bed gasifier. *Biomass and Bioenergy* 20: 471–481.

Rodrigues, M., A. Walter, and A. Faaij. 2003. Co-firing of natural gas and Biomass gas in biomass integrated gasification/combined cycle systems. *Energy* 28: 1115–1131.

Saka, S. 2006. Recent progress in supercritical fluid science for biofuel production from woody biomass. *Forestry Studies in China* 8: 9–15.

Samolada, M.C., W. Baldauf, and I.A. Vasalos. 1998. Production of a bio-gasoline by upgrading biomass flash pyrolysis liquids via hydrogen processing and catalytic cracking. *Fuel* 77: 1667–1675.

Sasaki, M., B. Kabyemela, R. Malaluan, S. Hirose, N. Takeda, T. Adschiri, and K. Arai. 1998. Cellulose hydrolysis in subcritical and supercritical water. *Journal of Supercritical Fluid* 13: 261–268.

Schutt, B.D., B. Serrano, R.L. Cerro, and M.A. Abraham. 2002. Production of chemicals from cellulose and biomass-derived compounds through catalytic sub-critical water oxidation in a monolith reactor. *Biomass and Bioenergy* 22: 365–375.

Shinogi, Y. and Y. Kanri. 2003. Pyrolysis of plant, animal and human waste: Physical and chemical characterization of the pyrolytic products. *Bioresource Technology* 90: 241–247.

Sihvonen, M., E. Jarvenpaa, V. Hietaniemi, and R. Huopalahti. 1999. Advances in supercritical carbon dioxide technologies. *Trends in Food Science & Technology* 10: 217–222.

Sjostrom, E. 1981. *Wood Chemistry Fundamentals and Applications.* Orland, FL: Academic Press.

Sridhar, G., P.J. Paul, and H.S. Mukunda. 2001. Biomass derived producer gas as a reciprocating engine fuel-an experimental analysis. *Biomass and Bioenergy* 21: 61–72.

Srinakruang, J., K. Sato, T. Vitidsant, and K. Fujimoto. 2005. A highly efficient catalyst for tar gasification with steam. *Catal Commun* 6: 437–440.

Tao, J.L., K.W. Jun, and K.W. Lee. 2001. Co-production of dimethyl ether and methanol from CO_2 hydrogenation: Development of a stable hybrid catalyst. *Applied Organometallic Chemistry* 15: 105–108.

Tashtoush, G., M.I. Al-Widyan, and A.O. Al-Shyoukh. 2003. Combustion performance and emissions of ethyl ester of a waste vegetable oil in a water-cooler furnace. *Applied Thermal Engineering* 23: 285–293.

Tijmensen, M.J.A., A.P.C. Faaij, C.N. Hamelinck, and M.R.M. van Hardeveld. 2002. Exploration of the possibilities for production of Fischer Tropsch liquids and power via biomass gasification. *Biomass and Bioenergy* 23: 129–52.

van der Drift, A., J. van Doorn, and J.W. Vermeulen. 2001. Ten residual biomass fuels for circulating fluidized-bed gasification. *Biomass and Bioenergy* 20: 45–56.

Vega, J.L., S. Prieto, B.B. Elmore, E.C. Clausen, and J.L. Gaddy. 1989. The biological production of ethanol from synthesis gas. *Applied Biochemistry and Biotechnology* 20&21: 781–797.

Wander, P.R., C.R. Altafini, and R.M. Barreto. 2004. Assessment of a small sawdust gasification unit. *Biomass and Bioenergy* 27: 467–476.

Wang, L.J. and C.L. Weller. 2006. Recent advances in extraction of natural products from plants. *Trends in Food Science and Technology* 17: 300–312.

Wang, L.J., C.L. Weller, M.A. Hanna, and D.D. Jones. 2008. Contemporary issues in thermal gasification of biomass and its application to electricity and fuel production. *Biomass and Bioenergy* 32: 573–581.

Watanabe, M., H. Inomata, and K. Arai. 2002. Catalytic hydrogen generation from biomass (glucose and cellulose) with ZrO_2 in supercritical water. *Biomass and Bioenergy* 22: 405–410.

Worden, R.M., A.J. Grethlein, M.K. Jain, and R. Datta. 1991. Production of butanol and ethanol from synthesis gas via fermentation. *Fuel* 70: 615–619.

Yin, X.L., C.Z. Wu, S.P. Zheng, and Y. Chen. 2002. Design and operation of a CFB gasification and power generation system for rice husk. *Biomass and Bioenergy* 23: 181–187.

Yorgun, S., S. Sensoz, and O.M. Kockar. 2001. Characterization of the pyrolysis oil produced in the slow pyrolysis of sunflower-extracted bagasse. *Biomass and Bioenergy* 20: 141–148.

Yoshioka, T., S. Hirata, Y. Matsumura, and K. Sakanishi. 2005. Woody biomass resources and conversion in Japan: The current situation and projections to 2010 and 2050. *Biomass Bioenergy* 29: 336–346.

Yu, F., R. Ruan, P. Chen, S. Deng, Y. Liu, and X. Lin. 2006. Liquefaction of corn cobs with supercritical water treatment. *Transactions of the ASABE* 50: 175–180.

Zabaniotou, A.A., G. Kalogiannis, E. Kappas, and A.J. Karabelas. 2000. Olive residues (cuttings and kernels) rapid pyrolysis product yields and kinetics. *Biomass and Bioenergy* 18: 411–420.

Zainal, Z.A., A. Rifau, G.A. Quadir, and K.N. Seetharamu. 2002. Experimental investigation of a downdraft biomass gasifier. *Biomass Bioenergy* 23: 283–289.

Index

A

B

C

D

E

F

G

H

I

K

L

M

N

O

P

R

S

T

U

V

W

X

Y

Z

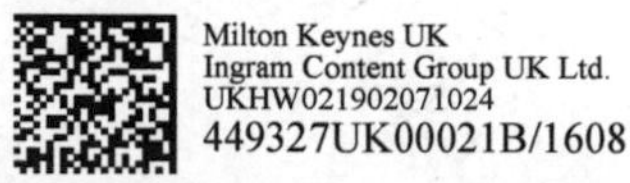

Milton Keynes UK
Ingram Content Group UK Ltd.
UKHW021902071024
449327UK00021B/1608

9 780367 386252